Coordinated Operation and Planning of Modern Heat and Electricity Incorporated Networks

Coordinated Operation and Planning of Modern Heat and Electricity Incorporated Networks

Edited by

Mohammadreza Daneshvar
Faculty of Electrical and Computer Engineering,
University of Tabriz, Tabriz, Iran

Behnam Mohammadi-Ivatloo
Faculty of Electrical and Computer Engineering,
University of Tabriz, Tabriz, Iran

Kazem Zare
Faculty of Electrical and Computer Engineering,
University of Tabriz, Tabriz, Iran

IEEE Press Series on Power and Energy Systems

Published by John Wiley & Sons, Inc., Hoboken, New Jersey.
Published simultaneously in Canada.

For general information on our other products and services or for technical support, please contact our Customer Care Department within the United States at (800) 762-2974, outside the United States at (317) 572-3993 or fax (317) 572-4002.

Wiley also publishes its books in a variety of electronic formats. Some content that appears in print may not be available in electronic formats. For more information about Wiley products, visit our website at www.wiley.com.

Library of Congress Cataloging-in-Publication Data Applied for
Hardback ISBN: 9781119862123

Cover Design: Wiley
Cover Image: © zf L/Getty Images

Set in 9.5/12.5pt STIXTwoText by Straive, Pondicherry, India

Contents

Editor Biographies

Mohammadreza Daneshvar does research in the field of energy and electrical engineering at the Smart Energy Systems Lab of the Department of Electrical and Computer Engineering at the University of Tabriz. He is the author of more than 40 top journal and conference papers in the field of multi-energy systems, grid modernization, transactive energy, and optimizing the multi-carrier energy grids. Moreover, he is the author and editor of more than four books in the mentioned fields of study that are published and are in the publication process in the Springer, Elsevier, and Wiley-IEEE Press publishers. He serves as an active reviewer with more than 40 top journals of the IEEE, Elsevier, Springer, Wiley, Taylor & Francis, and IOS Press and was ranked among the top 1% of reviewers in Engineering and Cross-Field based on Publons global reviewer database. His research interests include Smart Grids, Transactive Energy, Energy Management, Renewable Energy Sources, Multi-Carrier Energy Systems, Grid Modernization, Electrical Energy Storage Systems, Microgrids, Energy Hubs, Machine Learning and Deep Learning, Blockchain Technology, and Optimization Techniques.

Behnam Mohammadi-Ivatloo is a member of the Faculty of Engineering with the Department of Electrical and Electronics Engineering at Muğla Sıtkı Koçman University, Turkey. He was previously a Senior Research Fellow at Aalborg University, Denmark. He is also a Professor at the University of Tabriz, from where he is currently on leave. Before joining the University of Tabriz, he was a research associate at the Institute for Sustainable Energy, Environment and Economy at the University of Calgary. He obtained MSc and Ph.D. degrees in electrical engineering from the Sharif University of Technology. His main research interests are renewable energies, microgrid systems, and smart grids.

Kazem Zare received the B.Sc. and M.Sc. degrees in electrical engineering from the University of Tabriz, Tabriz, Iran, in 2000 and 2003, respectively, and a Ph.D. degree from Tarbiat Modares University, Tehran, Iran, in 2009. Currently, he is a Professor in the Faculty of Electrical and Computer Engineering, University of Tabriz, Tabriz, Iran. He was PI or CO-PI in 10 national and international-funded research projects. He is included in the 2018, 2019, and 2021 Thomson Reuters' list of the top 1% most cited researchers. He is the associate editor of the Sustainable Cities and Society journal and the editor of the e-prime journal. His research areas include power system economics, distribution networks, microgrid, energy management, smart building, demand response, and power system optimization.

List of Contributors

Mohammad Taghi Ameli
Department of Electrical Engineering
Shahid Beheshti University
Tehran, Iran
and
Electrical Networks Research Institute
Shahid Beheshti University
Tehran, Iran

Jaber Fallah Ardashir
Department of Electrical Engineering
Tabriz Branch, Islamic Azad
University
Tabriz, Iran

Arash Asrari
School of Electrical, Computer, and
Biomedical Engineering, Southern
Illinois University, Carbondale
IL, USA

Sasan Azad
Department of Electrical Engineering
Shahid Beheshti University
Tehran, Iran
and
Electrical Networks Research Institute
Shahid Beheshti University
Tehran, Iran

Valentina Cecchi
Department of Electrical and
Computer Engineering, University of
North Carolina, Charlotte
NC, USA

Yumin Chen
China Southern Grid Digital Power
Grid Co., Ltd.
Digital Power Grid Branch
Engineering and Technology Division
Building C, Yunsheng Science Park
Guangzhou, China

Mojtaba Dadashi
Faculty of Electrical and Computer
Engineering
University of Tabriz
Tabriz, Iran

Mohammadreza Daneshvar
Faculty of Electrical and Computer
Engineering, University of Tabriz
Tabriz, Iran

Mohammad Reza Dehbozorgi
Department of Power and Control
School of Electrical and Computer
Engineering, Shiraz University
Shiraz, Iran

Sobhan Dorahaki
Department of Electrical Engineering
Shahid Bahonar University of Kerman
Kerman, Iran

Poria Fajri
Department of Electrical and
Biomedical Engineering
University of Nevada, Reno
NV, USA

Xue Feng
Engineering Cluster
Singapore Institute of Technology
Dover, Singapore

Hadi Vatankhah Ghadim
Department of Electrical Engineering
Tabriz Branch, Islamic Azad
University
Tabriz, Iran

Ensieh Ghanbari
Department of Electrical Engineering
Iran University of Science and
Technology
Tehran, Iran

Reza Gharibi
Department of Electrical Engineering
Amirkabir University of Technology
(Tehran Polytechnic)
Tehran, Iran

Hessam Golmohamadi
Department of Computer Science
Aalborg University
Aalborg, Denmark

Aminabbas Golshanfard
Energy Modelling and Sustainable
Energy System (METSAP) Research
Lab., Faculty of New Sciences and
Technologies
University of Tehran
Tehran, Iran

Sara Haghifam
Faculty of Electrical and Computer
Engineering
University of Tabriz
Tabriz, Iran
and
School of Technology and Innovations
Flexible Energy Resources

University of Vaasa
Vaasa, Finland

Seyed Mehdi Hakimi
Department of Electrical Engineering
and Renewable Energy Research
Center, Damavand Branch
Islamic Azad University
Damavand, Iran

Arezoo Hasankhani
Department of Computer and
Electrical Engineering and Computer
Science, Florida Atlantic University
Boca Raton, FL, USA

Hamid HassanzadehFard
Department of Electrical Engineering
Miyaneh Branch, Islamic Azad
University
Miyaneh, Iran

Alireza Heidari
School of Electrical Engineering and
Telecommunication, University of
New South Wales, Sydney, Australia

Ali Jalilian
Deputy of Operation and Dispatch
Kermanshah Power Electrical
Distribution Company (KPEDC)
Kermanshah, Iran

S. Mahdi Kazemi-Razi
Department of Electrical Engineering
Amirkabir University of Technology
(Tehran Polytechnic), Tehran, Iran

Javad Khazaei
Department of Electrical and
Computer Engineering, Lehigh
University, Bethlehem
PA, USA

Hannu Laaksonen
School of Technology and Innovations
Flexible Energy Resources
University of Vaasa, Vaasa
Finland

Zhengmao Li
School of Electrical and Electronic
Engineering, Nanyang Technological
University, Nanyang, Singapore

Sahar Mobasheri
Department of Electrical Engineering
Shahid Bahonar University of Kerman
Kerman, Iran

Mohammad Mohammadi
Department of Power and Control
School of Electrical and Computer
Engineering, Shiraz University
Shiraz, Iran

Behnam Mohammadi-Ivatloo
Faculty of Electrical and Computer
Engineering
University of Tabriz
Tabriz, Iran

Milad Mohammadyari
Department of Electrical Engineering
University of Tehran
Tehran, Iran

Ehsan Naderi
School of Electrical, Computer, and
Biomedical Engineering
Southern Illinois University
Carbondale, IL
USA

Hamed Nafisi
Department of Electrical Engineering
Amirkabir University of Technology
(Tehran Polytechnic)
Tehran, Iran

Mehrdad Setayesh Nazar
Faculty of Electrical Engineering
Shahid Beheshti University
Tehran, Iran

Younes Noorollahi
Energy Modelling and Sustainable
Energy System (METSAP) Research
Lab., Faculty of New Sciences and
Technologies, University of Tehran
Tehran, Iran

Saba Norouzi
Faculty of Electrical and Computer
Engineering, University of Tabriz
Tabriz, Iran

Mohammad Hossein Pafeshordeh
Department of Power and Control
School of Electrical and Computer
Engineering, Shiraz University
Shiraz, Iran

Fereidoun H. Panahi
Department of Electrical Engineering
University of Kurdistan
Sanandaj, Iran

Farzad H. Panahi
Department of Electrical Engineering
University of Kurdistan
Sanandaj, Iran

Navid Talaei Pashiri
Department of Electrical Engineering
Shahid Beheshti University
Tehran, Iran

Masoud Rashidinejad
Department of Electrical Engineering
Shahid Bahonar University of Kerman
Kerman, Iran

Mohammad Rastegar
Department of Power and Control
School of Electrical and Computer
Engineering
Shiraz University
Shiraz, Iran

Mohammad Reza Salehizadeh
Department of Electrical Engineering
Marvdasht Branch
Islamic Azad University
Marvdasht, Iran

Miadreza Shafie-khah
School of Technology and Innovations
Flexible Energy Resources
University of Vaasa
Vaasa, Finland

Ali Sharifzadeh
Department of Electrical Engineering
Shahid Beheshti University
Tehran, Iran

Behrooz Vahidi
Department of Electrical Engineering
Amirkabir University of Technology
(Tehran Polytechnic)
Tehran, Iran

Yan Xu
School of Electrical and Electronic
Engineering, Nanyang Technological
University, Nanyang, Singapore

Seyedeh Soudabeh Zadsar
Department of Electrical Engineering
Shahid Bahonar University of Kerman
Kerman, Iran

Kazem Zare
Faculty of Electrical and Computer
Engineering, University of Tabriz
Tabriz, Iran

Preface

Nowadays, the significant development in energy production systems has created an evolutionary trend in coupling multi-carrier energy networks (MCENs) to each other. In addition to this progress in energy networks with different vectors, the day-by-day growth in energy consumption has also more highlighted the inevitable dependencies between MCENs. To this end, future modern energy grids not only are targeted to be developed as a couple of multi-energy systems, but also they will be structured with a high/full level of renewable energy resources (RERs). In this structure, hybrid energy systems such as combined heating, cooling, and power units play an undeniable role in reliable meeting the energy demand. However, how different types of energy networks can be integratively operated in the modern energy infrastructure is a key question that needs to be addressed in deep detail. As the reliable electrical and heating energy supply is critical for the energy network, a great need is felt for coordinated operation and planning of the heat and electricity networks (HENs) under the modern structure of MCENs. Therefore, as a pioneering book that presents the fundamental theories, technologies, and solutions for real-world problems in modern heat and electricity incorporated networks, this book is targeted to not only cover the coordinated operation of HENs but also to support the planning of HENs and more clarify the HENs presence in the future modern MCENs.

The current book consists of 18 chapters. Chapter 1 wants to clarify what are the characteristics of future modern energy networks and why grid modernization is essential for the interdependent structure of networks that are going to be engaged with a large number of stochastic clean energy production devices. The related challenges and opportunities are also covered in this chapter to give a clear overview for the future modern multi-carrier energy grid. Chapter 2 reviews the transition from conventional energy networks to multi-carrier energy grids, introduces the background of multi-carrier energy systems and definitions in the literature and investigates the benefits and challenges of these systems. Chapter 3 covers the definition and benefits of MCENs as well as proposes innovative solutions to overcome the uncertainties in optimal operation and planning of modern MCENs. Chapter 4 reviews the main topics of modern MCENs and the state of the arts by mostly focusing on implementing new financial concepts for providing broad flexibility to better coordination between operating and planning of modern heat and power incorporated networks. Chapter 5 provides an overview of the optimal operation of MCENs by concentrating on components, energy conversion technologies, advantages, challenges, and applications. Moreover, the economic and environmental benefits, as well as reliability and flexibility improvements of utilizing energy hubs systems, are discussed. Chapter 6 scrutinizes a cyberattack model, which can simultaneously target transmission and distribution sectors of a modern heat and electricity incorporated network leading to system congestions and possible power outages. Chapter 7 discusses an intelligent monitoring and maintenance system for multi-unmanned aerial vehicle teams applicable to modern heat and electricity incorporated networks. Chapter 8 deals with the stages of operation and planning of heat and electricity incorporated networks with the aim of proposing a model for the coordinated operation and planning of them. Chapter 9 defines optimal operation in the heat and electricity incorporated networks as an operation with maximized flexibility and sustainability along with minimized cost and CO_2 emission with the aim of optimal coordinated operation of them. Chapter 10 mainly presents the optimal energy management scheme of the combined-heat-and-electrical microgrid by focusing on the operation side of modern heat and electricity incorporated networks. Chapter 11 investigates the role of intelligent heat controllers in power and heat networks by suggesting an economic heat controller for residential heat pumps with water tanks. Chapter 12 aims to model an electricity-heat incorporated energy system in the presence of various sector coupling and storage technologies and investigate the impact of a hybrid energy storage system on the optimal performance of the considered network. Chapter 13 discusses the optimal operation of a microgrid including a hybrid energy system in the presence of a demand response program and also presents an intuitive perception of the microgrid through an optimization model. Chapter 14 provides a techno-economic analysis of hydrogen technologies, waste converters, and demand response in the coordinated operation of heat and electricity systems. Chapter 15 introduces an algorithm for optimal operational planning of heat and electricity systems integrating the commitment scenarios of smart buildings. Chapter 16 provides effective information about the meaning of optimal planning of heat and electricity incorporated networks, its basic structure, advantages, and possible challenges. Chapter 17 presents a general approach to the simultaneous optimization and planning of the electrical and thermal energy systems. Chapter 18 evaluates the energy production of various types of (distributed generations) DGs and hybrid storage systems in the heat and electricity incorporated network set up for minimizing the total costs of the applied DGs, fuel consumption, and pollutant emissions considering reliability analysis.

As any research achievement may not be free of gaps, the Editors kindly welcome any suggestions and comments from the respectful readers to improve the quality of this work. The interested readers can share their valuable comments with the Editors via m.r.daneshvar95@gmail.com.

Editors

Mohammadreza Daneshvar
Faculty of Electrical and Computer Engineering
University of Tabriz
Tabriz, Iran

Behnam Mohammadi-Ivatloo
Faculty of Electrical and Computer Engineering
University of Tabriz
Tabriz, Iran

Kazem Zare
Faculty of Electrical and Computer Engineering
University of Tabriz
Tabriz, Iran

1

Overview of Modern Energy Networks

Mohammadreza Daneshvar, Behnam Mohammadi-Ivatloo, and Kazem Zare

Faculty of Electrical and Computer Engineering, University of Tabriz, Tabriz, Iran

1.1 Introduction

Nowadays, energy grids witness accelerating changes in their different domains from the generation sector to the distribution area. This evolution mostly relies on rapid developments in the technology of smart devices that is occurring in line with variations in the well-being range of people. Another aspect of this transition backs to the need for delivering efficient, reliable, affordable, and sustainable energy to the consumer side [1]. The main motivation for this transformation was flourished when the drawbacks of traditional energy systems have appeared, especially in terms of technical, economical, and environmental aspects [2]. The smart grid was the first common term devoted for the pass that its main goal is built on overcoming the disadvantages of the conventional energy structure, especially focusing on improving power quality, resilience, efficiency, reliability, security, and environmental aspects [3]. In light of the main mission to answer the need for smartization of the power grid, the smart grid promoted widespread developments in the different sectors of energy grids. It has driven the focus of energy generation from the centralized mechanism to the decentralized one and has allowed end-users to contribute to the grid services and participate in a variety of energy management programs [4]. Another key criterion of smartization lies in extending the appropriate volume of RERs in the generation sector [5]. The applicability, availability, and environmentally friendly nature of RERs have made it a thrust area of research, and the smart grid structure has made their applicability vast and more promising [6]. In this respect, increasing demand for energy has created a fundamental need for reinforcement of the energy generation sector, making it reliable in energy supply. Modern energy grids (MEGs) are targeted to equip their

Coordinated Operation and Planning of Modern Heat and Electricity Incorporated Networks,
First Edition. Edited by Mohammadreza Daneshvar, Behnam Mohammadi-Ivatloo, and Kazem Zare.
© 2023 The Institute of Electrical and Electronics Engineers, Inc.
Published 2023 by John Wiley & Sons, Inc.

energy production part by a high/full level of RERs for maximum clean energy production [7]. The structure of MEGs is planned to be designed in a way to allow for involving advanced technologies for facilitating people's daily life as well as address the challenges of the previous conventional system. MEGs are famous for their capabilities in advancing innovative ways for covering the current shortages and expected upcoming challenges by adopting intelligent devices and coordinated platforms [8]. As properly integrating different energy grids in the coupled structure alongside producing a high/full renewable energy are recognized as the great requirements for the future energy infrastructure [9], one of the main focuses of MEGs belongs to enable future grids to benefit from the mentioned advantages. This chapter aims at discussing the overview of MEGs and clarifying some critical aspects of their presence in the near future that can be considered in designing future roadmaps or modifying current ones for future modern grids. Additionally, the related opportunities and challenges are specified to give a simplistic and complete overview of MEGs properties.

1.2 Reliability and Resilience of Modern Energy Grids

Reliability and resilience are two key characteristics of MEGs that refer to the ability of the system to operate in a sustainable way under diverse operation conditions [10]. By targeting a high range of RERs for presence in MEGs, the reliability of the system needs to be kept under the huge uncertain fluctuations in the energy production process. Given the reliability lies on the function of the system under normal/prescribed operation conditions [11], its satisfaction requires more promising solutions in the presence of high RERs of MEGs. How several promising strategies can be coordinately engaged to provide an adequate level of reliability in the highly smartized area will be challengeable for the future operation of MEGs. Several main factors affect the system's reliability that focusing on their effectiveness on the reliability index can properly highlight critical tips for filling the related system gaps against their bad effects. Some of the important ones are environmental barriers like vegetation and trees, autonomous monitoring of the system, effects of weather changes and their threats to the system's reliability, the strength of the grid in providing a sufficient level of reliability, especially in the presence of numerous stochastic producers, the accuracy rate of predictions for the critical factors like RERs' outputs and energy consumption, damages created by animals, humans, and their vehicles, etc.

On the other hand, the system's resilience refers to the ability of the system to efficiently and swiftly recover its performance from its previous deterioration condition and resume normal operation [12]. The rapid advancements in energy technologies and widespread adoption of them along with RERs in the body of the grid expose them in the randomization operation, making it more prone to failures and instabilities [13]. The system's situation can be even more acute in some cases when

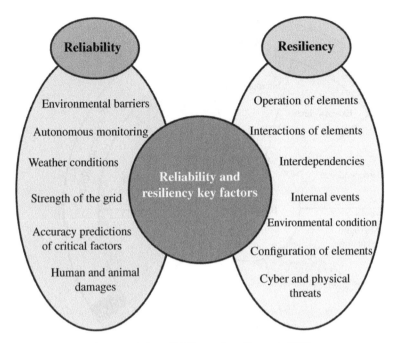

Figure 1.1 Key factors of the reliability and resiliency in MEGs.

inevitable disruption of services has occurred under extreme environmental conditions that require a recovering in an efficient and agile manner. The inability property of the grid in withstanding adverse events is recognized as its vulnerability [14]. The effective factors in the reliability and resiliency of MEGs are illustrated in Figure 1.1.

From the resilience perspective of the system, recovering ability, adaptability, and vulnerability are the properties that affect the system's resilience as the three major components. In addition, several other factors also influence the aforementioned domains that are depicted in Figure 1.2 [12]. Such factors need to be deeply scrutinized to develop reliable frameworks for MEGs with a full/high level of RERs to realize a sustainable energy structure.

1.3 Renewable Energy Availability in Modern Energy Grids

Eco-friendly energy production has become the first priority in the energy generation sector when fossil-fuel-based units highlighted their remarkable bad effects on environmental and economic issues [15]. In this regard, different technologies of RERs are advanced day by day to enable the energy grid to produce clean energy at a high level. Although RERs offer affordable and environmentally friendly energy

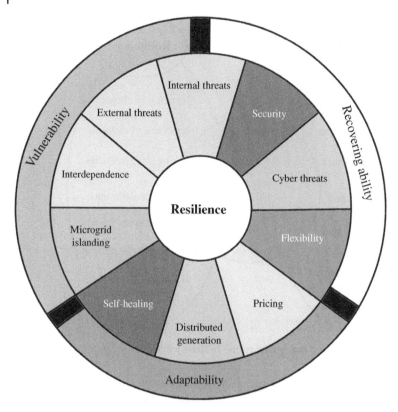

Figure 1.2 Components of MEGs resilience. *Source:* Adapted from [12].

generation, the main problem for them is their availability in energy production. Wind, solar, and hydro are three typical RERs that are widely planned for clean energy generation, and their technology has become enough matured for MEGs' goals. Unlike the traditional controllable energy production units, the main drawback of RERs is that they are not dispatchable and cannot provide regulated energy to the grid [16]. Limited predictability along with the daily and seasonal changes have given intermittent sprite for their energy production [17]. As MEGs are promised for the operation of full/high RERs, innovative mechanisms and frameworks are required for realizing this great goal. This issue not only needs advanced information and communication technologies but also relies on the essentiality usage of several confident ways for ensuring the flexibility of energy supply in the system [18]. How such ways should be effectively developed and implemented is a key challenge regarding the widespread usage of RERs. However, the research world has witnessed significant advancements in the practical solutions for increasing the penetration of RERs in the body of the power generation process. Some of the most prominent of these solutions are indicated in Figure 1.3.

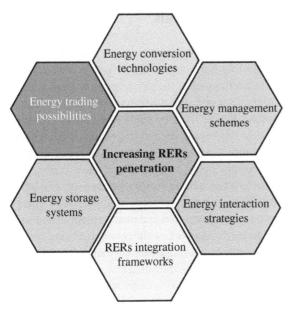

Figure 1.3 Some of the most prominent solutions for increasing RERs penetration in MEGs.

The denoted solutions in Figure 1.3 are the usual practical ways for facilitating the integration of RERs in the grid. RERs outputs directly depend on the climate conditions that provide uncertain situations for scheduling of the system. In the moments with desirable climate changes, the surplus clean produced power can be effectively used under the different processes. Energy conversion technologies like power-to-gas systems and hydrogen-based procedures enable the grid to convert the surplus power to other carriers of energy for supporting various energy grids such as natural gas, district heating network, etc. [19]. On the other hand, energy storage systems in their diverse types such as batteries, pumped storage, and compressed air energy storage can store excess energy and play a supportive role for the system when it suffers from lower-level energy production. Energy management schemes can effectively allow end-users to be actively involved in the customer-side services and provide a certain degree of flexibility by load shifting/shedding in return for specific rewards [20]. The common type of energy management scheme is demand response programs that are categorized into incentive-based and price-based programs. From another side, a variety of energy interaction strategies can be employed in the scheduling of the system that allows the grid to dynamically balance energy in the highly deregulated environment. In addition to the aforementioned approaches, optimal integration of RERs is known as one of the proper ways of using the benefits of RERs on a large scale. However, its feasibility requires coordinated operation of several intelligent devices that developing the advanced smart agents are expected as a primary condition for the implementation of related plans.

1.4 Modern Multi-Carrier Energy Grids

Nowadays, by increasing the deployment of hybrid devices as well as the importance of flexibility due to the becoming of RERs as the target units for energy production, a strong coupling between the grid industries such as the electric power system, district heating and cooling networks, natural gas grid, and the water distribution system is necessitated more than ever before. Figure 1.4 shows different sectors of multi-carrier energy networks (MCENs). In this regard, co- and tri-generation energy systems alongside the multi-carrier energy consumption devices have created undeniable dependences among multi-vector energy structures and have made their relationship tight [21]. MCENs are the integrated version of several energy grid infrastructures that enable the system for cooperative interactions, multi-vector energy sharing and conversion, and multi-energy generation and management in line with the objectives of MEGs. The MCENs mechanisms allow the renewable-based system to produce a huge volume of clean energy by offering sustainable ways for managing produced energy in all time periods. In the MCENs infrastructure, each energy sector can engage with other energy structures and develop energy interactions for maintaining the grid's sustainability and reliability of the energy supply while pursuing specific economic, technical, and environmental goals. The future MEGs are targeted to be developed under MCENs paradigms for using different energy layers aiming to make the overall infrastructure of the grid reliable in energy supply, resilient and self-healing, secure and stable, efficient in various energy processes, and flexible in the energy management in the presence and high penetration of uncontrollable and eco-friendly energy generation systems.

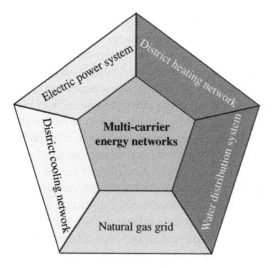

Figure 1.4 Different sectors of multi-carrier energy networks.

1.5 Challenges and Opportunities of Modern Energy Grids

As earlier described, MEGs are not only highly interdependent but also self-sufficient in their operation. In light of the main targets of MEGs in equipping to the full/high level of RERs as well as coupling different energy structures and creating MCENs infrastructure, future MEGs require advanced technologies, innovative solutions, and a strong roadmap that clearly state what are the requirements for the grid modernization, how they can be procured for implementing the relevant plans, and which actions need to be adopted to close to overcome the challenges ahead. Due to this, thinking about the general and basic challenges can be the primary step in realizing the goals of modern grids. Effectively detecting and analyzing these challenges not only can able the system for making up the practical solutions but also can facilitate reaching the confident scheduling. On the other hand, identifying critical challenges can give suitable overviews regarding the feasibility of MEGs frameworks procuring opportunities for the possible modifications. Some of the most important of these challenges are listed as:

- Controlling and managing this great transition.
- Optimally evolving planning, operations, and marketing.
- Optimally adopting informed, prudent, and future-looking decision-making.
- Handling the complexity of MEGs with large-scale and hybrid components.
- Handling the computational burden of numerous control systems and expert computing.
- Adopting diverse kinds of computational language and developing unique processes.
- Developing a complete and sufficient system with the capability of addressing the requirements of future MEGs.
- Efficiently recovering and rapidly fostering the large disruptions.
- Integrally and optimally monitoring MCENs.
- Developing a holistic platform to adopt emerging technologies.
- Structuring an environment conducive to all energy market players.
- Building market structure for different owners with various expectations and opinions.
- Making an active energy interactions system for consumers with diverse consumption patterns.
- Manipulating incorporated frameworks conducive to governments with diverse political policies and suitable for regions with various geographical forms.
- Developing a comprehensive platform with logical justifications for persuading stakeholders, producers, and consumers with multifarious requirements to pursue MEGs protocols.

- Advancing a fair energy area aiming to allow different players to participate in MEGs' interactions, smartization schemes, and energy management programs.
- Developing an innovative mechanism of interdependent MEGs.
- Entirely presenting the dynamic process of restoration and failure.
- Presenting a holistic system assessment mechanism for analyzing the availability/reliability of highly complex interdependent MEGs.
- Optimally reconstructing the dynamic topology in the integrated MEGs.
- Developing sustainable architecture of MEGs in line with grid modernization criteria.

Despite MEGs face some critical challenges in their implementation, realizing them can provide tremendous opportunities that can revolute all energy sectors. Investigating these opportunities can also provide a convenient overview of the modern structure benefits, especially in terms of economic, convenience, technical, and environmental. The most important of these opportunities is illustrated in Figure 1.5 [22].

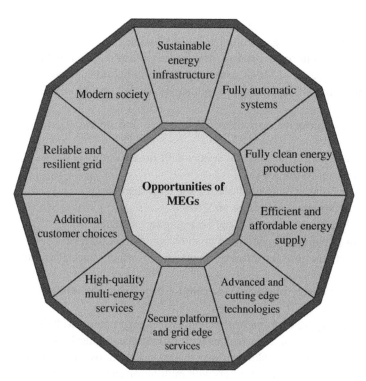

Figure 1.5 Opportunities of MEGs.

1.6 Summary

This chapter stands its state of the art on providing a whole overview for future MEGs. The importance of switching from traditional grids to coupled MEGs has made necessities for deeply exploring the key issues ahead of this transition. In light of the main mission of MEGs in fully clean energy production, the significance of the reliability and resiliency of MEGs is investigated by focusing on their key factors. The availability of RERs is discussed by evaluating the potential solutions for increasing the penetration of RERs while maintaining the system's flexibility in the acceptable range. In the next step, the chapter is concentrated on clarifying the necessity of operating energy grids interdependently due to the huge advancements in hybrid energy systems and energy conversion technologies, as well as the rapid emergence of them in the body of the grid. At the end of the chapter, the challenges and opportunities of MEGs are briefly described, and the overall lists for them are provided that can be used for planning the future schemes as well as modifying the current plans.

References

1 Choi, J., Lee, J.-I., Lee, I.-W., and Cha, S.-W. (2021). Robust PV-BESS scheduling for a grid with incentive for forecast accuracy. *IEEE Transactions on Sustainable Energy* 13 (1): 567–578.

2 Daneshvar, M., Mohammadi-Ivatloo, B., Asadi, S., and Galvani, S. (2020). Short term optimal hydro-thermal scheduling of the transmission system equipped with pumped storage in the competitive environment. *Majlesi Journal of Electrical Engineering* 14 (1): 77–84.

3 Shobole, A.A. and Wadi, M. (2021). Multiagent systems application for the smart grid protection. *Renewable and Sustainable Energy Reviews* 149: 111352.

4 Gao, H., Liu, J., Wang, L., and Wei, Z. (2017). Decentralized energy management for networked microgrids in future distribution systems. *IEEE Transactions on Power Systems* 33 (4): 3599–3610.

5 Daneshvar, M., Mohammadi-Ivatloo, B., Zare, K. et al. (2021). Chance-constrained scheduling of hybrid microgrids under transactive energy control. *International Journal of Energy Research* 45 (7): 10173–10190.

6 Hossain, M., Madlool, N., Rahim, N. et al. (2016). Role of smart grid in renewable energy: an overview. *Renewable and Sustainable Energy Reviews* 60: 1168–1184.

7 Daneshvar, M., Mohammadi-Ivatloo, B., Zare, K., and Asadi, S. (2020). Two-stage robust stochastic model scheduling for transactive energy based renewable microgrids. *IEEE Transactions on Industrial Informatics* 16 (11): 6857–6867.

8 Zhang, C., Dong, M., and Ota, K. (2020). Enabling computational intelligence for green internet of things: data-driven adaptation in LPWA networking. *IEEE Computational Intelligence Magazine* 15 (1): 32–43.

9 Daneshvar, M., Mohammadi-Ivatloo, B., Abapour, M. et al. (2020). Distributionally robust chance-constrained transactive energy framework for coupled electrical and gas microgrids. *IEEE Transactions on Industrial Electronics* 68 (1): 347–357.

10 Quint, R.D. (2021). Data-driven engineering: the reliability and resilience of the North American bulk power system [technology leaders]. *IEEE Electrification Magazine* 9 (1): 5–9.

11 Daneshvar, M. and Babaei, E. (2018). Exchange market algorithm for multiple DG placement and sizing in a radial distribution system. *Journal of Energy Management and Technology* 2 (1): 54–65.

12 Das, L., Munikoti, S., Natarajan, B., and Srinivasan, B. (2020). Measuring smart grid resilience: methods, challenges and opportunities. *Renewable and Sustainable Energy Reviews* 130: 109918.

13 Teimourzadeh, S., Aminifar, F., Davarpanah, M., and Guerrero, J.M. (2016). Macroprotections for microgrids: toward a new protection paradigm subsequent to distributed energy resource integration. *IEEE Industrial Electronics Magazine* 10 (3): 6–18.

14 Arghandeh, R., Von Meier, A., Mehrmanesh, L., and Mili, L. (2016). On the definition of cyber-physical resilience in power systems. *Renewable and Sustainable Energy Reviews* 58: 1060–1069.

15 Daneshvar, M., Eskandari, H., Sirous, A.B., and Esmaeilzadeh, R. (2021). A novel techno-economic risk-averse strategy for optimal scheduling of renewable-based industrial microgrid. *Sustainable Cities and Society* 70: 102879.

16 Qiu, H., Gu, W., Xu, Y. et al. (2018). Interval-partitioned uncertainty constrained robust dispatch for ac/dc hybrid microgrids with uncontrollable renewable generators. *IEEE Transactions on Smart Grid* 10 (4): 4603–4614.

17 Daneshvar, M., Mohammadi-Ivatloo, B., Zare, K. et al. (2020). A novel operational model for interconnected microgrids participation in transactive energy market: a hybrid IGDT/stochastic approach. *IEEE Transactions on Industrial Informatics* 17 (6): 4025–4035.

18 Rehmani, M.H., Reisslein, M., Rachedi, A. et al. (2018). Integrating renewable energy resources into the smart grid: recent developments in information and communication technologies. *IEEE Transactions on Industrial Informatics* 14 (7): 2814–2825.

19 Malinowski, M., Milczarek, A., Kot, R. et al. (2015). Optimized energy-conversion systems for small wind turbines: renewable energy sources in modern distributed power generation systems. *IEEE Power Electronics Magazine* 2 (3): 16–30.

20 Wang, K., Ouyang, Z., Krishnan, R. et al. (2015). A game theory-based energy management system using price elasticity for smart grids. *IEEE Transactions on Industrial Informatics* 11 (6): 1607–1616.

21 Bruno, S. and Nucci, C.A. (2017). Optimization of multi-energy carrier systems in urban areas. In: *From Smart Grids to Smart Cities: New Challenges in Optimizing Energy Grids* (ed. C.A. Nucci, M. La Scala, S. Lamonaca, et al.), 177–230. Wiley.

22 Daneshvar, M., Asadi, S., and Mohammadi-Ivatloo, B. (2021). *Grid Modernization— Future Energy Network Infrastructure*. Springer.

2

An Overview of the Transition from One-Dimensional Energy Networks to Multi-Carrier Energy Grids

Navid Talaei Pashiri[1], Mohammad Taghi Ameli[1,2], and Sasan Azad[1,2]

[1] Department of Electrical Engineering, Shahid Beheshti University, Tehran, Iran
[2] Electrical Networks Research Institute, Shahid Beheshti University, Tehran, Iran

Abbreviations

CCHP   Combined cooling and heat and power
CHP   Combined heat and power
DER   Distributed energy resource
DG   Distributed generation
DR   Demand response
DS   Distributed storage
GHG   Greenhouse gas
LHV   Low heat value
MDEN   Multi-dimensional energy network
RES   Renewable energy source

2.1 Introduction

Population growth, rising living standards, and economic growth lead to elevated energy demands. Taking environmental impacts into notice is essential to supply energy demand growth. The energy production processes pollute the environment by releasing carbon dioxide and greenhouse gases (GHGs) [1]. Energy systems emit 42% of the world's GHGs, so the energy industry has the largest share in global warming. Recent trends in GHG emissions show that global warming will exceed 1.5 °C and lead to irreversible destruction of ecosystems and consecutive

Coordinated Operation and Planning of Modern Heat and Electricity Incorporated Networks,
First Edition. Edited by Mohammadreza Daneshvar, Behnam Mohammadi-Ivatloo, and Kazem Zare.
© 2023 The Institute of Electrical and Electronics Engineers, Inc.
Published 2023 by John Wiley & Sons, Inc.

catastrophes if no immediate measures and policies are taken in the coming decades to reduce emissions. Countries worldwide ambitiously agreed to restrict global warming to less than 2°, or mere precisely 1.5 °C, in the 2015 Paris Agreement [2]. In the Cop26 summit in Glasgow in November 2021, it was concluded that if countries adhere to their restrictive commitments, the goals of the Paris Agreement will be within reach by 2030. These commitments are not using coal, reducing deforestation, accelerating the replacement of electric vehicles, and investing in RESs [3]. The decarbonization policies and measures must be implemented not only in the electricity sector but also in transportation, heating and cooling, industry, and the residential sector [2].

Due to the flexibility and reliability of fossil fuels, traditional energy sources are designed based on this type of energy carrier. Fossil fuels can be stored massively in liquid, gas, and solid forms and are consumed based on demand if the stored amounts are sufficient [4]. However, current traditional energy systems face limitations such as low fossil fuel-to-electricity conversion efficiency, transmission system losses, and high GHG emissions [5]. Energy systems are moving away from fossil fuels to renewable energy due to various reasons, such as reducing carbon emissions and fossil fuel consumption, increasing energy efficiency, creating local resources and jobs, and reducing energy production costs [4]. Figure 2.1 demonstrates the potential of RESs and their contribution to generating electricity and heat. The use of renewable energies (including wind and solar) has escalated in recent years due to the ambitious goals of the European Commission. From 1997 to 2018, the capacity of renewable energies, except hydropower, has increased from less than 1% to about 22% due to the increase in the capacity of photovoltaic energies. According to the European climate and energy framework, one of the goals for 2030 is reaching 32% of total energy production capacity for the share of renewable energy production capacity [6]. Such goals have been stated in other world regions to decarbonize electricity generation. Still, the problem is that increasing the share of renewable sources

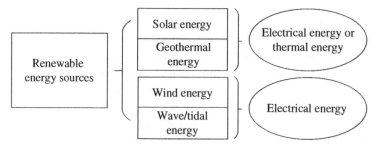

Figure 2.1 Renewable energy sources potentials in generating electrical and thermal energy.

will cause electricity generation to exceed the demand [7]. Unlike fossil fuels, one of the significant challenges of switching to new energy systems in which renewable energies such as photovoltaics and wind have a more significant share is that they are not capable of high volume storage [4]. Renewable energies are naturally bound with uncertainty, and their production rate is unsteady. Therefore, the increase in renewable energies worldwide elevates the risk of reducing the flexibility of power systems and sustainable energy supply [8].

Interactions of various energy carriers such as heat/cold, gas, and electricity provide great flexibility. These flexibilities can increase energy efficiency and lower carbon emissions in electricity generation [7]. Mancarella [9] suggested some interactions between energy carriers, such as electricity, heat, gas, hydrogen, transportation, etc., and opportunities to increase efficiency by the flexibility created by the interaction of electricity and other energy sectors. For example, in the interaction of the electricity and gas sectors, gas-fired power plants are a supporting factor in balancing production and consumption when RESs are fluctuating. Therefore, power system fluctuations are transmitted to the gas system through gas power plants, which increases the interdependency between the gas and electricity sectors [10, 11]. Electrifying the heating sector through heat pumps, increasing the use of cogeneration systems in different voltage levels, and controlling air conditioners by intelligent systems for managing the demand side and controlling the frequency are such interactions in energy sectors. Converting excess electricity into hydrogen or methane and injecting it into the gas grid is another interaction between the electricity and gas systems. Storing methane is a better option compared to electricity. It is also viable to convert gas to electricity in increased load times. Using the interactions between energy carriers in energy systems is mentioned in different articles titled Multi-energy Systems, Intelligent Systems, Integrated and Hybrid Systems, and it is said to provide more flexibility. In this chapter, this concept is called multi-carrier energy grids. Multi-carrier energy grids are systems that adapt the amount of production, which is based on the behavior of renewable energy, to the amount of consumption, by providing flexibility to increase efficiency and the share of renewable energy [7]. Combining electricity, heat, and transportation sectors creates new flexibilities that compensate for the lack of flexibility in renewable energies due to the lack of storage [4]. Multi-carrier energy grids have a slow development due to a lack of necessary infrastructure and utilization complexity [12].

This chapter presents an overview of the concept of multi-carrier energy systems and the necessity to switch from isolated (centralized) and traditional energy grids to integrated multi-carrier energy grids. Different parts of traditional energy systems are briefly introduced, followed by the backgrounds, definitions, benefits, and challenges of multi-carrier energy systems. Finally, the conclusions are presented in the last section. The main goals of this chapter are as follows:

- Understanding the need for decarbonization of energy sectors.
- Introducing traditional energy systems and energy sectors.
- Introducing the background of multi-carrier energy grids.
- Understanding the concept of multi-carrier energy grids.
- Investigating the benefits and challenges of transition to multi-carrier energy grids.

2.2 Traditional Energy Systems

In the early ages, houses were heated by burning wood. Because there were no transportation systems at the time, the wood was provided from nearby places. With the advancement of technology, mass production of energy, and its distribution by the power grid and gas network, the nature of decentralized energy production and consumption became extinct, and energy systems developed centrally. Today, in the age of communication, societies are more focused on production and consumption with the growth of societies and globalization. It is the end of centralized single-input/output energy systems. Future energy systems will be multiple input/output with integrated and decentralized energy sectors. Energy systems cannot be fully centralized or decentralized, and future energy systems will be a combination of centralized and decentralized systems that operate parallel with each other [13].

2.2.1 Electricity Grid

The first extensive use of electrical energy was for lighting. Shortly after Edison invented the incandescent light bulb, he made efforts to generate electricity on a larger scale. His efforts led to implementing the first central generation station on Pearl Street in lower Manhattan, New York, that provided lighting of several square blocks of the city. With this novel movement, the benefits and importance of electrification gradually became known [14]. The need to provide electricity to more consumers led to the formation of a traditional centralized power grid over time. The traditional power grid consists of three main parts of production, transmission, and distribution. In the beginning, the production of electricity by traditional electricity generation units was centralized and based on fossil fuels. The generated electricity was transmitted through distribution lines in large volumes to distribution systems. Large power plants in traditional power systems that use fossil fuels have an efficiency of about 37% and serve end-users through large and complex networks of transmission and distribution lines [15]. Usually, 26% of electricity is wasted in transmission and distribution lines. Wide and long transmission and distribution networks are exposed to natural catastrophes and destructive events. Therefore, the security of traditional and centralized power systems is weaker than distributed networks [16]. Due to the simple radial structure, the traditional and centralized method of generating electricity has several advantages, such as simplicity in planning, monitoring, creating energy balance, and troubleshooting. On the other hand, due to disadvantages, such as low efficiency, lack of fossil fuels, and environmental pollution, centralized power system needs revision in design and development.

2.2.2 Gas Grid

Gas Light and Coke Company, founded in 1812 by Frederik Winsor in London, provided natural gas to the public for the first time. The company provided street lighting with gas produced by heating coal. Exploring natural gas in the Northern Sea and its many benefits led to its substitution for coal gas in the gas industry. Natural gas is one of the most famous fossil fuels globally due to its advantages, such as having the highest low heat value (LHV = 50 000 Kj/kg) among fossil fuels, easy transfer even thousands of kilometers away, and minimal environmental footprints. According to [17], 1 295 563.6 km of natural gas pipelines have been installed in the world so far. Natural gas network is a complex network that distributes gas from the wellhead to the end-users through various equipment such as pipes, valves, compression stations, pressure regulation stations, etc. It ensures the proper condition of gas pressure and flow rate. There are three main types of pipelines in the natural gas network. The collecting pipelines extract the raw gas from the gas wells, transferring pipelines transfer the gas to the preprocessing plant, and the distribution system delivers the natural gas to the end-users. Gas demand is divided into residential, commercial/industrial, and electrical power generation categories. Residential users mainly use gas to heat their houses, and their consumption is related to weather and seasonal conditions. There is a wide range of industrial users, and they use the heat from burning gas or the gas itself as feed, depending on the type of the product. Power plants are also considered one of the major natural gas consumers and generate electricity and heat simultaneously with high efficiency [18].

2.2.3 Heating and Cooling Grid

District heating is an interconnected network of pipes that connects houses in a neighborhood, area, or town to the centralized heating plants or distributed heat generation units. The idea of supplying the heating demands of houses and commercial places through district heating was originated from different motivations. These motivations were creating an alternative solution for separate boilers in residential buildings, reducing energy losses that existed separately in each building and creating better safety. So far, three generations of heating and cooling networks have been introduced and installed. The first generation of the heating network used steam as a carrier to transfer heat through the pipes and was first implemented in the United States in the 1880s. Later on, this solution was discarded due to high losses and the dangers of high steam pressure. The second generation of the heating network used water above 100 °C as a heat transfer carrier instead of steam. The new method was more efficient, and many of the installed heating networks worldwide are still using the second-generation heating networks. The third generation, called Scandinavian district heating technology, was introduced in the 1970s, after the oil crisis and the growing attention to energy efficiency and replacing other fuels with oil. The third generation of the heating network still uses water as an energy carrier, but thanks to the prefabricated and insulated pipes and better technology, the water temperature can be kept below 100° [19]. The first cooling networks were implemented in Hartford, Hamburg, and Paris in the 1960s. Although the number of district cooling networks in the world is relatively small compared to district heating networks, the development of these networks has been simultaneous, and about 150 cooling systems are now in operation in European countries [20].

2.3 Background of Multi-Carrier Energy Systems

The evolution of distributed energy resources (DERs), cogeneration, and multi-generation led to the appearance of the multi-carrier energy grids concept. So in the following, we introduce the DERs, cogeneration, and multi-generation as the evolution path of multi-carrier energy grids.

2.3.1 Distributed Energy Resources Background

The concept of DERs is formed by gathering the three definitions of distributed generation (DG), demand response (DR), and distributed storage (DS). DG generates electricity locally using different sources; DR involves consumers to manage and reduce peak load; DS is locally storing energy by different storage technologies. The concept and framework of DERs formed about two decades ago. But, its implementation was not possible at that time due to the lack of necessary infrastructure and technologies. The employment of DERs in power systems has many advantages, so here are some of these advantages in the following [21]:

- Increasing energy efficiency.
- Increasing customer's participation in determining electricity prices based on demand.

- Increasing power system reliability.
- Reducing unsupplied energy in case of grid failures.
- Decreasing GHG emissions through efficient and clean energy production.

2.3.2 Cogeneration and Trigeneration Background

Cogeneration is the simultaneous production of two energy carriers, such as electricity or mechanical power and heat, which provides more significant energy savings and efficiency than the separate production of energy carriers [22]. Figure 2.2 compares the efficiency of traditional power plants with the cogeneration of electricity and heat for the same fuel consumption. Cogeneration systems consume less fuel than the separate generation of heat and electricity. Renewable energy-based cogeneration technologies such as biomass-based cogeneration, solar energy-based cogeneration, and fuel cell-based cogeneration are in the spotlight. Raj [22] investigates modeling and simulation, economics and environmental issues, and literature review of mentioned renewable energy-based cogeneration technologies. Simultaneous generation of energy carriers requires a constant and steady demand for both types of energy carriers over time. So, cogeneration of electricity and heat was applicable at the scale of thermal power plants in the past. Combined heat and power (CHP) as a cogeneration technology provides much higher energy efficiency by generating heat and electricity simultaneously.

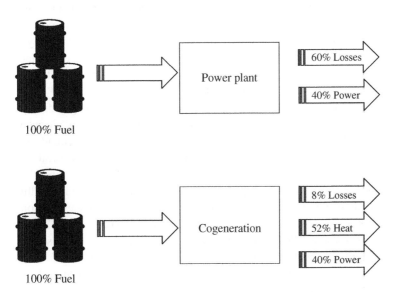

Figure 2.2 Comparing the efficiency of traditional power plants and the cogeneration of electricity and heat.

CHP technology prevents heat energy wastage and significantly reduces carbon emissions in the energy production process. Cogeneration is feasible in urban areas, commercial buildings, and residential buildings through DG technologies such as CHP [21].

The simultaneous demand for electricity, heating, and cooling led to the invention of power plants capable of responding to these three needs, called trigeneration. Combined cooling and heat and power (CCHP) technology provides three electricity, heat, and cold demands by burning natural gas. CCHP technology facilitates cooling production by coupling the prime mover of a CHP with an absorbent chiller. One challenge of cogeneration is the lack of heat demand in summer. In trigeneration technology, this problem is overcome by replacing the heating demand with cooling demands. Therefore, these systems are considered seasonal operation systems. Merging CCHP systems with renewable energy technologies effectively reduces environmental pollution and increases sustainable energy production [21]. Yang [23] investigates the optimal design of a CCHP solar hybrid system, considering the CCHP system, photovoltaic panels, and solar thermal collectors in three different types of buildings and climates.

2.3.3 Quad Generation

Li [16] presents the modeling of the solar-powered energy system for applying in residential buildings, which increases efficiency and reduces carbon dioxide emissions. This quad generation system responds to electricity, heating, cooling, and domestic hot water demands. Another product of this system is synthesis gas. Synthesis gas is a fuel gas mixture consisting primarily of hydrogen, carbon monoxide, and very often some carbon dioxide. Hospitals are among the major electricity consumers, leading to significant emissions of GHGs. With the global spread of the covid19 pandemic, most hospitals have an overloaded status, increasing their need for cooling, heating, electricity, and medical gas sources such as oxygen, nitrogen, and carbon dioxide. Chen et al. [24] investigate a concept called hospital quad generation, aiming at providing the heating, cooling, electricity, and medical gas needs by using RESs and energy conversion technologies. This study claims that this method has higher efficiency, higher capacity, better security, and less pollution than gas-based energy systems. With energy conversion technologies, the concept of trigeneration and quad generation can be evolved into a more comprehensive concept called multi-generation. In this case, the energy system provides different outputs of the required forms of energy with higher efficiency and fewer losses by receiving different inputs from energy carriers in a cheap way [21].

2.4 The Definition of Multi-Carrier Energy Grids

Fixed and unique titles are not used for the concept of multi-carrier energy grids, and titles such as Multi-energy Systems, Distributed Multi-generation, Integrated Energy Systems, Hybrid Energy Systems, Energy Hub, etc. are used in literature. Multi-energy systems provide customers' energy requirements with strong coordination between different sectors and energy carriers, with high reliability and minimal environmental impacts. They also have high efficiency and flexibility, but their management and control are complicated [25]. Multi-energy systems, in addition to fossil fuels and hydropower energy, can use energy sources such as wind energy, photovoltaic energy, solar collectors, biogas, fuel cells, geothermal energy, and other modern RESs [16]. Multi-energy systems use at least two infrastructures from electricity, gas, and heating/cooling infrastructures and are controlled centrally. These systems make optimal use of various forms of energy and use conversion capabilities to reduce the uncertainty of renewable energy through interactions between energy carriers [12]. Multiple energy systems are systems that operate networks, sectors, and different energy carriers seamlessly with the aim of decarbonization of energy sectors [26]. Multi-generation systems generate different energy carriers locally by combining cogeneration systems, heat pumps, electric/absorption chillers, and interaction with RESs. Developing these systems will increase efficiency, reduce carbon emissions, and grow the economy [21]. Energy systems in which the parts of electricity, heating, cooling, fuel, transportation, etc., interact optimally at different levels are called multi-energy systems [27]. A multiple carrier energy system is an interconnected network of energy carriers that, by coupling energy sectors, make optimal use of infrastructure to produce sustainable energy [28]. Multi-carrier energy systems, which are a combination of two or more energy conversion tools, are a way to meet the challenges of energy systems such as fossil fuel shortages, increasing energy consumption, significant climate change, and carbon emissions [29]. An energy hub is a system that integrates different energy infrastructures and meets a wide range of energy needs [30].

Therefore, the basis of the multi-carrier energy grids concept is the coordination and integration of different parts and carriers of energy, such as electricity, heat, cold, clean fuels, etc., to serve the end-users. Also, this concept emphasizes the prevention of environmental pollution and the transition from traditional energy systems based on single energy carriers to energy systems based on multiple energy carriers to produce sustainable energy. Inherent interaction of energy carriers by creating new solutions and flexibility increases efficiency and reduces the risk of passing through centralized energy systems. Multi-carrier energy grids are an excellent way to store the surplus electricity generated by RESs. For example, storage is possible by converting electricity into another energy carrier such as low

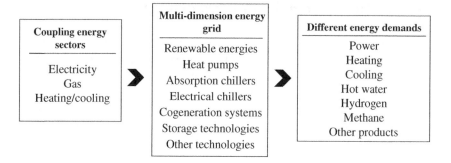

Figure 2.3 Sector coupling in multi-carrier energy grids.

carbon fuels (hydrogen and methane). Restrictions in the energy sector can also be resolved and managed in multi-carrier energy systems. For example, if there is a limitation in the electricity sector, the gas sector will be able to support the power grid. There are obstacles to implementing multi-carrier energy systems, such as high initial investment, communication, control infrastructure development, and complex design and operation. Multi-carrier energy grids are not limited to physical complexities; organizational complexities such as market mechanisms and regulatory and social regulation would be created due to the participation of end-users [25]. The development of modern decentralized energy systems, in addition to creating rules and incentives, requires investment in the development of high-efficiency energy conversion, clean energy generation technologies, and further research in optimization, operation, and design of multi-carrier energy systems [31]. Figure 2.3 shows the integration of different energy sectors and the use of high-efficiency technologies of energy production and conversion to meet different energy needs in multi-carrier energy grids.

2.5 Benefits of Multi-Carrier Energy Grids

1) Efficiency Increase
   Today, despite recent advances in energy generation and conversion, due to the evolution of relevant technologies, the efficiency of energy systems has increased significantly, and the final customers can meet their energy needs with high efficiency without putting pressure on upstream networks [32].
2) Power Loss Reduction
   Employing DERs and RESs near the end-users prevents multiple congestions and the loss of electrical energy into heat in transmission lines. Also, the amount of energy wasted in generating different forms of energy is significantly

reduced due to the advancement of technology and the use of cogeneration and multi-generation technologies such as CCHP [33].

3) Security and Resiliency Increase

In multi-carrier energy grids, due to the use of DERs, energy production units at different levels are decentralized, and the probability of their failure due to natural disasters and destructive attacks is reduced. So in case of failure, its impact on the whole system is small.

4) Environmental Pollution Reduction

In recent decades, concerns about the lack of fossil fuels, global warming, and GHG emissions due to energy sector activity have attracted much attention. One of the most appropriate approaches to reduce and mitigate the environmental impacts of GHG emissions is to expand multi-carrier energy grids based on RESs. Due to the use of clean and efficient technologies in generating renewable energies such as solar energy, photovoltaics, wind energy, geothermal energy, etc., these systems have a significant role in preventing increased carbon emissions. Also, these systems can utilize new technologies for hydrogen production, CO_2 trapping, and methane production, which are among the clean fuels [7, 34].

5) Reliability Increase

Reliability of power systems means "designing, operating, and maintaining an energy grid, aiming to provide safe, adequate, and stable energy flow" [35]. Flexibilities are created in integrated energy systems, such as storage systems, increased supply-demand reliability by mitigating the fluctuating nature of RESs, and peak load management [36]. By developing the energy trading and markets suited to the multi-carrier energy grids, the end customers contribute to the production and exchange of energy per hour of the day, resulting in the desired level of reliability in the energy system [37]. Gargari [38] evaluated a multi-carrier energy system, including several types of energy converters, such as CHP units, water desalination, electrical storage, and heat storage. It shows that the reliability of the loads increased significantly.

6) Sustainable Energy Development

Sustainability is subjected to creating a balance between the three fields of economics, the environment, and the community. Sustainable development of energy means a fair use of limited energy resources to respond to the energy requirements of communities concerning its environmental and economic consequences. Sustainable energy development depends on the community's energy demand over time. Considering the high percentage of RESs in the multi-carrier energy grids, these systems have much less dependence on fossil fuels than traditional energy systems to store fossil fuels for future generations. RESs improve the living standards of communities by meeting a range of diverse energy needs, increasing efficiency, and reducing emissions that affect the lives of future generations [13].

2.6 Challenges of Moving Toward Multi-Carrier Energy Grids

As mentioned earlier, one of the requirements for the transition from traditional energy systems to multi-carrier energy networks is reliance on RESs. The intermittent nature of RESs and their increasing participation in energy production pose severe power systems challenges, such as frequency and voltage regulation problems due to imbalances and inconsistencies in production and consumption [39]. The presence of generation resources in distribution systems leads to bidirectional power flow, which makes providing safety, reliability, and control of the distribution system challenging [40]. Demand response is one of the flexibilities of multi-carrier energy grids, which is helpful in reducing the effect of the intermittent nature of RESs. Demand response changes the behavior of end-users due to economic and non-economic incentives and dynamic prices [39]. Although infrastructures such as monitoring and communication technologies are now available, the widespread implementation of the demand response poses the power system with severe challenges in terms of control and optimization. Some of these challenges include the lack of efficient markets, invisible signal prices in most current markets and regulatory frameworks, conflict of interest with power plants participating in demand's peak, heavy investment costs for automation of existing infrastructure, lack of adequate incentives to persuade customers to participate in demand response program, and complexity of end-customer behavior [41].

RESs are often connected to the grid in remote places. Due to the network's weakness in remote areas, power generation fluctuation of RESs imposes unacceptable voltage changes on the network. Energy storage resources are another flexibility of multi-carrier energy grids that can compensate for the intermittent nature of renewable energy [42]. The development of energy storage technologies faces technological and economic obstacles. Technological challenges are achieving capacity increase, long life span, high security, high efficiency, increasing studies in optimization of storage systems, and development of standards to open the way for commercialization and industrialization of energy storage technologies. Also, establishing supportive policies, high investment costs, and lack of markets appropriate to energy storage systems are the economic challenges of developing energy storage technologies [43]. Modern multi-carrier energy grids will be confronted with the vast amount of information exchanged between intelligent components, generally related to the production, operation, and consumption of energy sectors. Modernization and integration of energy systems require advanced systems and platforms for communication between advanced intelligent systems. Modern energy grid optimization, coordination, and planning require infrastructures for storage, classification, and information analysis [44].

2.7 Conclusions

Achieving the limiting objectives of GHG emission requires substantial changes and redesign of traditional energy systems that rely on fossil fuels to supply the needs of different energy sectors. As mentioned earlier, the energy sector contributes to global warming with a 42% share of the global GHG emissions. Thus, reducing carbon emissions in electricity and heat generation and transportation will play a significant role in decarbonizing the energy supply process in the world. In order to reduce carbon emissions, energy sectors have a prospect to rely entirely on RESs, such as solar energy, wind energy, and hydropower. RESs have high efficiency and are cost-effective and eco-friendly, but the generated energy does not have large-scale storage capability. The fluctuating nature of RESs energy generation exposes the energy system to uncertainties. The interactions between sectors and different energy carriers generate the potential to increase the power system's efficiency and flexibility, which is known as a solution to relieve the fluctuating nature of renewable sources. Focusing on this issue led to the invention of a concept called multi-carrier energy grids. The main idea is sector coupling, which means linking the sectors of electricity, gas, cooling/heating, energy carriers, and end-users. This requires the reconsideration of the design, modeling, and operation of energy systems and the use of new energy conversion and generation technologies, which various researchers have addressed. Multi-carrier energy grids have advantages over traditional energy systems, such as high efficiency, low losses, high security and resiliency, high reliability, and low GHG emissions. There are some obstacles to moving toward multi-carrier energy grids, including frequency and voltage regulation problems due to the intermittent nature of RESs, control and safety problems due to the presence of generation resources in distribution systems, challenges of widespread implementation of the demand response, and challenges related to implementing modern communication systems and processing a large amount of information. Therefore, with the growing population of the world and the increasing energy demand, sustainable energy development requires a switch from traditional energy systems and the replacement of multi-carrier energy systems. Finally, this chapter explains the necessity of decarbonizing energy sectors and moving from traditional energy systems to multi-carrier energy grids by investigating the concept of multi-carrier energy grids and related challenges and benefits.

References

1 Farahmand-Zahed, A., Nojavan, S., Zare, K., and Mohammadi-Ivatloo, B. (2020). Economic and environmental benefits of renewable energy sources in multi-generation systems. In: *Integration of Clean and Sustainable Energy Resources and Storage in Multi-Generation Systems* (ed. F. Jabari, B. Mohammadi-Ivatloo and M. Mohammadpourfard), 1–14. Cham: Springer https://doi.org/10.1007/978-3-030-42420-6_1.

2 Sgobba, A. and Meskell, C. (2021). Impact of combined heat and power on the goal of decarbonizing energy use in Irish manufacturing. *J. Clean. Prod.* 278: 123325.

3 United Nations Climate Change. COP26 The Glasgow Climate Pact. https://ukcop26.org/wp-content/uploads/2021/11/COP26-Presidency-Outcomes-The-Climate-Pact.pdf (accessed 9 February 2022).

4 Connolly, D., Lund H, Mathiesen BV et al. (2013). Smart energy systems: holistic and integrated energy systems for the era of 100% renewable energy.

5 Ayele, G.T., Haurant, P., Laumert, B., and Lacarriere, B. (2018). An extended energy hub approach for load flow analysis of highly coupled district energy networks: illustration with electricity and heating. *Appl. Energy* 212: 850–867.

6 E. Commission (2014). A policy framework for climate and energy in the period from 2020 to 2030, *Tech. Rep. COM*, p. 15.

7 Ramsebner, J., Haas, R., Auer, H. et al. (2021). From single to multi-energy and hybrid grids: historic growth and future vision. *Renew. Sustain. Energy Rev.* 151: 111520.

8 Worighi, I., Maach, A., Hafid, A. et al. (2019). Integrating renewable energy in smart grid system: architecture, virtualization and analysis. *Sustain. Energy, Grids Networks* 18: 100226.

9 Mancarella, P. (2012). Smart multi-energy grids: concepts, benefits and challenges. *2012 IEEE Power and Energy Society General Meeting*, San Diego, CA, USA (22–26 July 2012), pp. 1–2. doi: 10.1109/PESGM.2012.6345120.

10 Ameli, H., Qadrdan, M., Strbac, G., and Ameli, M.T. (2020). Investing in flexibility in an integrated planning of natural gas and power systems. *IET Energy Syst. Integr.* 2 (2): 101–111.

11 Shabazbegian, V., Ameli, H., Ameli, M.T. et al. (2021). Co-optimization of resilient gas and electricity networks; a novel possibilistic chance-constrained programming approach. *Appl. Energy* 284: 116284.

12 Kleinschmidt, V., Hamacher, T., Perić, V., and Hesamzadeh, M.R. (2020). Unlocking flexibility in multi-energy systems: a literature review. *2020 17th*

International Conference on the European Energy Market (EEM), Stockholm, Sweden (16–18 September 2020), pp. 1–6. doi: 10.1109/EEM49802.2020.9221927.

13 Alanne, K. and Saari, A. (2006). Distributed energy generation and sustainable development. *Renew. Sustain. Energy Rev.* 10 (6): 539–558.

14 Ambrosio, R. (2016). Transactive energy systems. *IEEE Electrif. Mag.* 4 (4): 4–7.

15 Pehnt, M. (2008). Environmental impacts of distributed energy systems—the case of micro cogeneration. *Environ. Sci. Policy* 11 (1): 25–37.

16 Li, Z.X., Ehyaei, M.A., Kasmaei, H.K. et al. (2019). Thermodynamic modeling of a novel solar powered quad generation system to meet electrical and thermal loads of residential building and syngas production. *Energy Convers. Manag.* 199: 111982.

17 North America has the highest oil and gas pipeline length globally. https://www.offshore-technology.com/comment/north-america-has-the-highest-oil-and-gas-pipeline-length-globally/ (accessed 9 February 2022).

18 Qadrdan, M., Abeysekera, M., Wu, J. et al. (2020). The future of gas networks. In: *The Future of Gas Networks* (ed. M. Qadrdan, M. Abeysekera, J. Wu, et al.), 49–68. Cham: Springer.

19 Lund, H., Werner, S., Wiltshire, R. et al. (2014). 4th Generation District Heating (4GDH): integrating smart thermal grids into future sustainable energy systems. *Energy* 68: 1–11.

20 Werner, S. (2017). International review of district heating and cooling. *Energy* 137: 617–631.

21 Chicco, G. and Mancarella, P. (2009). Distributed multi-generation: a comprehensive view. *Renew. Sustain. Energy Rev.* 13 (3): 535–551.

22 Raj, N.T., Iniyan, S., and Goic, R. (2011). A review of renewable energy based cogeneration technologies. *Renew. Sustain. Energy Rev.* 15 (8): 3640–3648.

23 Yang, G. and Zhai, X.Q. (2019). Optimal design and performance analysis of solar hybrid CCHP system considering influence of building type and climate condition. *Energy* 174: 647–663.

24 Chen, X., Chen, Y., Zhang, M. et al. (2021). Hospital-oriented quad-generation (HOQG)—a combined cooling, heating, power and gas (CCHPG) system. *Appl. Energy* 300: 117382.

25 O'Malley, M.J., Anwar, M.B., Heinen, S. et al. (2020). Multicarrier energy systems: shaping our energy future. *Proc. IEEE* 108 (9): 1437–1456.

26 Corsetti, E., Riaz, S., Riello, M., and Mancarella, P. (2021). Modelling and deploying multi-energy flexibility: the energy lattice framework. *Adv. Appl. Energy* 2: 100030.

27 Mancarella, P. (2014). MES (multi-energy systems): an overview of concepts and evaluation models. *Energy* 65: 1–17.

28 Moghaddas-Tafreshi, S.M., Mohseni, S., Karami, M.E., and Kelly, S. (2019). Optimal energy management of a grid-connected multiple energy carrier micro-grid. *Appl. Therm. Eng.* 152: 796–806.

29 Nazari-heris, M., Jabari, F., Mohammadi-ivatloo, B. et al. (2020). An updated review on multi-carrier energy systems with electricity, gas, and water energy sources. *J. Clean. Prod.* 123136.

30 Hemmati, S., Ghaderi, S.F., and Ghazizadeh, M.S. (2018). Sustainable energy hub design under uncertainty using Benders decomposition method. *Energy* 143: 1029–1047.

31 Neyestani, N. (2021). Modeling of multienergy carriers dependencies in smart local networks with distributed energy resources. In: *Distributed Energy Resources in Local Integrated Energy Systems* (ed. G. Graditi and M. Di Somma), 63–87. Elsevier.

32 Daneshvar, M., Abapour, M., Mohammadi-ivatloo, B., and Asadi, S. (2019). Impact of optimal DG placement and sizing on power reliability and voltage profile of radial distribution networks. *Majlesi J. Electr. Eng.* 13 (2): 91–102.

33 Fu, L., Zhao, X.L., Zhang, S.G. et al. (2009). Laboratory research on combined cooling, heating and power (CCHP) systems. *Energy Convers. Manag.* 50 (4): 977–982.

34 Mehrpooya, M., Rahbari, C., and Moosavian, S.M.A. (2017). Introducing a hybrid multi-generation fuel cell system, hydrogen production and cryogenic CO_2 capturing process. *Chem. Eng. Process. Intensif.* 120: 134–147.

35 Kovalev, G.F. and Lebedeva, L.M. (2019). *Reliability of Power Systems*, vol. 1. Springer.

36 Ameli, M.T., Ameli, H., Strbac, G., and Shahbazbegian, V. (2021). Chapter 17 – Reliability and resiliency assessment in integrated gas and electricity systems in the presence of energy storage systems. In: *Energy Storage in Energy Markets* (ed. B. Mohammadi-Ivatloo, A.M. Shotorbani and A. Anvari-Moghaddam), 369–397. Academic Press.

37 Daneshvar, M., Mohammadi-Ivatloo, B., Asadi, S. et al. (2019). A transactive energy management framework for regional network of microgrids. *2019 International Conference on Smart Energy Systems and Technologies (SEST)*, Porto, Portugal (9–11 September 2019), pp. 1–6. doi: 10.1109/SEST.2019.8849075.

38 Gargari, M.Z. and Ghaffarpour, R. (2020). Reliability evaluation of multi-carrier energy system with different level of demands under various weather situation. *Energy* 196: 117091.

39 Cruz, M.R.M., Fitiwi, D.Z., Santos, S.F., and Catalão, J.P.S. (2018). A comprehensive survey of flexibility options for supporting the low-carbon energy future. *Renew. Sustain. Energy Rev.* 97: 338–353.

40 Trebolle, D., Gómez, T., Cossent, R., and Frías, P. (2010). Distribution planning with reliability options for distributed generation. *Electr. Power Syst. Res.* 80 (2): 222–229.

41 Pinson, P. and Madsen, H. (2014). Benefits and challenges of electrical demand response: a critical review. *Renew. Sustain. Energy Rev.* 39: 686–699.

42 Mohd, A., Ortjohann, E., Schmelter, A. et al. (2008). Challenges in integrating distributed energy storage systems into future smart grid. *2008 IEEE International Symposium on Industrial Electronics*, Cambridge, UK (30 June–2 July 2008), pp. 1627–1632. doi: 10.1109/ISIE.2008.4676896.

43 Yao, L., Yang, B., Cui, H. et al. (2016). Challenges and progresses of energy storage technology and its application in power systems. *J. Mod. Power Syst. Clean Energy* 4 (4): 519–528.

44 Daneshvar, M., Asadi, S., and Mohammadi-Ivatloo, B. (2021). Data management in modernizing the future multi-carrier energy networks. In: *Grid Modernization— Future Energy Network Infrastructure* (ed. M. Daneshvar, S. Asadi and B. Mohammadi-Ivatloo), 117–174. Cham: Springer International Publishing.

3

Overview of Modern Multi-Dimension Energy Networks

Saba Norouzi[1], Mojtaba Dadashi[1], Sara Haghifam[1,2], and Kazem Zare[1]

[1] Faculty of Electrical and Computer Engineering, University of Tabriz, Tabriz, Iran
[2] School of Technology and Innovations, Flexible Energy Resources, University of Vaasa, Vaasa, Finland

Nomenclature

Acronyms

CO_2	Carbon dioxide
CHP	Combined heat and power
EI	Energy internet
EVs	Electric vehicles
ESS	Energy storage system
GDH	Generation district heating
ICT	Information and communication technologies
IMO	International Maritime Organization
IoT	Internet of things
PEM	Proton exchange membrane
PtG	Power-to-gas
PtH	Power-to-heat
PtX	Power-to-X
P2P	Peer-to-peer
RERs	Renewable energy resources
PV	Solar photovoltaic
MDENs	Multi-dimension energy networks
VESS	Virtual energy storage system
WT	Wind turbine

Coordinated Operation and Planning of Modern Heat and Electricity Incorporated Networks,
First Edition. Edited by Mohammadreza Daneshvar, Behnam Mohammadi-Ivatloo, and Kazem Zare.
© 2023 The Institute of Electrical and Electronics Engineers, Inc.
Published 2023 by John Wiley & Sons, Inc.

3.1 Introduction

Under the Paris agreement in 2015, to cope with global warming and climate change effects, a large number of countries were committed to implementing policies to reduce greenhouse gas emissions; and as of April 2021, 44 countries and the European Union have pledged to reach carbon neutrality targets [1]. To this end, a transition in the global energy sector to fulfill the world's climate goals is inevitable. Thereby, the penetration levels of RERs in power systems rise dramatically in recent years. As a considerable proportion of emissions are related to electricity generation from gas and coal-fired power plants, a significant reduction in global emissions is expected by increasing the penetration levels of RERs in the energy systems [1]. Also, deploying the high-efficient multidisciplinary energy technologies and electrification of the end-users play a key role in emission reduction across all sectors to accelerate realizing CO_2 net-zero society [1].

On the other hand, the swift rise of RERs' penetration levels in power systems has highlighted the necessity of more flexibility in the power system to mitigate generation uncertainty and availability [2]. In this context, MDENs are widely concerned as a solution for improving energy system efficiency and flexibility [3]. The MDENs can be realized by physically and virtually connecting various energy sectors, including electricity, heat, gas, etc., across infrastructures and markets to widely exploit synergies among different types of energy carriers [4]. In the MDENs, interconnecting and interactions between different energy networks have brought advantages as well as disadvantages. It is derived substantial benefits from increasing the flexibility, efficiency, resiliency, and emission reduction of the MDENs; however, the emergence of new interactions and interconnecting has resulted in more complexity in the management and operation of the MDENs [5].

There has been a rapid acceleration in moving toward modern MDENs because of advances in sector coupling technologies, ICT, and the advent of new market opportunities. In the modern MDENs, due to the high penetration level of RERs, sector coupling technologies provide a promising way to exploit or store the surplus electricity generated by RERs. Not only do recent advances in sector coupling technologies enable the realization of modern MDENs, but they also provide high levels of flexibility for energy systems by increasing the possibility of energy conversions [6]. Moreover, advancements in ICT and the internet of things (IoT) are expected to provide sufficient interconnection between MDENs and proper management tools. As one of the features of modern MDENs is their ability to make automated and optimized decisions in real-time, the role of ICT is absolutely crucial in modern energy systems [7]. On the other hand, modern MDENs can establish new opportunities for separate energy businesses to enter into partnerships on the market level. Hence, new market environments help to accelerate moving toward modern MDENs [8]. Notably, ICT development in modern MDENs can result in new market opportunities and consequently change consumer behaviors. Consumers transform from passive users to influential parts can directly interact with the rest of the system by changing their usage patterns or participating in the energy markets as a prosumer [9]. Furthermore, it is worth noting that by advancements in communication technologies, further expectations are aroused by consumers. For example, they expect to have the ability to manage their energy surplus and shortage through peer-to-peer (P2P) online services [10].

Based on these descriptions, there is a great necessity to assess the features of modern MDENs from different aspects. Therefore, in this chapter, after scrutinizing the benefits of MDENs, the coordinated operation and planning of the modern MDENs are evaluated through innovative solutions, including new technologies and energy trading mechanisms. Then, different plans of various countries to increase the penetration of RERs and MDENs are presented. Finally, challenges related to modern MDENs are addressed to facilitate moving toward future energy systems.

3.2 Multi-Dimension Energy Networks

Scarcity of fossil fuels, pollutant emissions associated with them, low efficiency, and high loss in the conventional energy networks have led to the transition toward modern energy systems based on high penetration of RERs [11, 12]. According to [13], decarbonization, decentralization, and digitalization are major drivers of energy system transition. Decarbonization refers to reducing CO_2 emissions from energy production and consumption processes, which can be achieved on the supply-side by switching from fossil fuel-based conventional power plants to renewable-based energy generation. On the demand side, electrification of end-user services is a possible solution to move toward deep decarbonization as end-use electricity technologies are highly efficient with no CO_2 emissions. In this regard, electric vehicles (EVs) and heat pumps are highly efficient end-use electricity technologies [13]. The decentralization of energy systems is the other driver of the energy transition to make energy production closer to energy consumption sites. Increasing the penetration levels of RERs in power systems facilitates the decentralization of energy generation since these technologies can be employed in the consumption points on more minor scales. The decentralized energy systems can be realized by using more distributed energy resources, including renewable-based power generation units as well as energy storage technologies [14]. The other factor facilitating energy system transition is digitalization, known as exploiting advanced ICT in the energy systems' appliances and networks. Digitalization makes possible the adoption of smart controllability strategies on energy systems' management. In modern energy systems, digitalization enables the energy systems' operators to automatically control the fluctuations associated with RERs by adopting flexible options while allowing end-users to be involved in the demand side management [15]. It is expected that digitalization will become more prominent in future energy systems by the advent of new market opportunities for end-users.

The uncertainties associated with integrated energy systems can lead to an imbalance between supply and demand, which can be managed via flexibility provision options. In this regard, MDENs with the ability to provide high levels of flexibility can be considered a practical solution to realize energy transition targets [16]. According to [17], the integration of heating and electricity sectors not only improves the efficiency and flexibility of a holistic energy system but also fulfills decarbonization targets. In this respect, the integration of heating and electricity sectors comes into sharp focus. Incorporating of these two sectors in the form of an integrated energy system can be realized through co-generation plants such as combined heat and power (CHP) plants and sector coupling technologies like Power-to-X options (PtX) [18]. Notably, a CHP simultaneously produces electricity and heat from fuel sources like natural gas, waste, and biomass. However, apart from the high efficiency, the integration of CHP plants in the power systems with high penetration of RERs is likely to limit renewable energy integration in the power systems [19]. Obligations to meet a given heat demand not only can reduce the flexibility of the CHP plant but also may restrict high levels of RERs penetration in the energy system [20]. In this regard, thermal storage can be utilized along with a CHP plant as a practical solution to increase the energy system's flexibility [21]. Moreover, low-cost thermal energy storage in the integrated heat and electricity sectors can be used to maintain balance against the fluctuating generation of electricity from RERs [17, 22]. The other option that links the electricity and heating sectors is Power-to-Heat (PtH) technologies which are key elements of PtX options. It is worth noting that PtH technologies can categorize into two groups, including centralized and decentralized options [23]. According to the centralized approach, produced heat by electricity is located far from the location of heat demand. Consequently, there is a need for a district heating network to distribute the heat to the site of demand. In contrast, electricity is converted to heat right at/or near the heat demand location under the decentralized approach [23]. The most common PtH technologies are various kinds of heat pumps and electric boilers, mostly falling into two categories of centralized and decentralized options according to their scale [23]. A heat pump utilizes electricity to transport heat from a lower temperature region to a higher temperature region. Due to improving the efficiency of the heating sector as well as providing decarbonization of energy systems, heat pumps have recently attracted more attention in the building sector [17]. In [24], the integrated 2050 Danish energy system is analyzed, and the results reveal the importance of heat pumps and municipality solid waste CHP technologies on district heat networks to meet the emission reduction targets. Electric boilers are the other leading PtH technologies categorized into electrode boilers and electric resistance boilers [23]. The high efficiency of both kinds of electric boilers has made them attractive options that can contribute to the decarbonization of the energy system. According to [25, 26], the exploitation of electric boilers and CHP plants can relax CHP units' heat-electric constraints provide more integration of wind power into the power grid, and effectively address the accommodation of curtailed wind power issues.

It is worth noting that historically heating was the central focus; however, the new generation of district heating networks has to satisfy both heating and cooling demands. In this regard, 4th- and 5th generation district heating (GDH) systems that integrate into a smart MDEN are introduced to fulfill heating and cooling needs. These systems can operate at low temperatures with high efficiencies, so implementing of 4GDH and 5GDH systems is crucial to fulfilling decarbonization targets [27, 28]. Moreover, both 4GDH and 5GDH systems facilitate the integration of RERs into the power grid and the cooperation of CHP plants and thermal storage systems. Consequently, the development of the new generation of district heating can strengthen the interconnectedness between heat and electricity networks [29].

In addition to the capability of PtH options to integrate between heating and electricity sectors, PtX technologies play a crucial role in realizing the integration of heating, electricity, and gas sectors [6]. By exploiting PtX technologies, further processing of renewable energy generation takes place via electrolysis to produce hydrogen, synthetic liquid and gases fuels, and synthetic chemicals, which their storage and transportation over long distances are much cheaper than electricity. Therefore, PtX productions can be easily used in every place energy is needed, and consequently, exploiting PtX technologies in renewable-based energy systems can provide further levels of flexibility [30]. Among the PtX options, Power-to-Gas (PtG) technology is highly focused owing to its potential to offer long-term seasonal storage of renewable electrical energy [31]. Extra electricity produced by RERs can be converted to hydrogen and synthetic natural gas via PtG technology and stored in the gas networks [32]. In this respect, not only do PtG technologies increase the connectedness between power and gas networks, but they also provide additional flexibility options. On the other hand, in the last decade, the power system has witnessed a rise in the proportion of natural gas in electricity production due to natural gas's continuous price reduction and low-rate greenhouse emissions, leading to increasing the interconnectedness between power and gas sectors [33]. It should note that produced gas by the PtG options can be utilized in the heating sector to fulfill high-temperature demands with reduced carbon emissions [34]. Furthermore, with increasing the penetration levels of RERs and PtX technologies, the deployment of different types of electrolyzers is on the rise. Currently, several electrolysis methods are on the focus; among them, proton exchange membrane (PEM) electrolysis is acknowledged as a promising technology for PtG technology [35]. It is worth noting that PEM electrolysis while

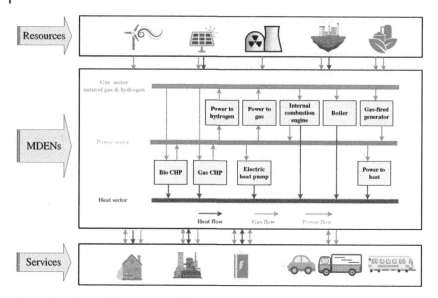

Figure 3.1 Interactions between MDENs technologies and networks.

producing hydrogen produces excess heat. The harnessing and integration of the produced excess heat in the district heating network can provide the other supply to meet heat demands. The method of capturing and exploiting of produced excess heat via a PEM electrolyzer is investigated in [36]. To sum up, it is concluded that MDENs enable the full exploitation of synergies between different energy carriers to accelerate moving toward decarbonized energy systems. Also, the high levels of uncertainty that have been a feature of energy transition can be fully addressed by MDENs, due to their high flexibility provision options. A simple overview of interactions between MDENs technologies and networks is depicted in Figure 3.1.

3.3 Benefits of MDENs

MDENs, as a potential solution for future energy systems, offer considerable advantages, shown in Figure 3.2, stemming from exploiting synergies between multi-energy carriers. In the following subsections, the main benefits of MDENs are provided.

3.3.1 Enhancing System Efficiency

The energy crisis problem has been challenging as energy consumption is rapidly increasing and traditional fossil fuel is quickly exhausted. In this context, the need for efficient usage of energy resources has been an utmost problem [37]. The

Figure 3.2 Benefits of MDENs.

efficiency of overall energy utilization can be improved via MDENs. In other words, higher integration of MDENs and a holistic approach toward the whole energy system may allow more efficient usage of resources. Furthermore, polygeneration systems with the ability to capture and reuse waste heat in the framework of MDENs are the other options to improve the energy system's efficiency [38].

3.3.2 Decarbonization

One of the most critical drivers of the energy transition is governments' policies to decrease greenhouse gas emissions from the energy sector, known as decarbonization [12]. To this end, the penetration level of RERs into the power system is increased to reduce the energy production from traditional systems with a high share of pollutant emissions. In this context, MDENs enable raising RERs' accommodation to foster fulfilling climate and energy targets. In addition, efficient usage of energy resources directly affects the decreasing pollutant emissions. Furthermore, electrification of the heat sector is vital in decarbonizing energy systems [39], which can be realized through MDENs.

3.3.3 Reducing System Operation Cost

First, MDENs enable efficient usage of energy resources which directly affects the reducing operation cost of the energy system. In addition, a holistic perspective toward the energy system and consequently optimizing the entire energy system operation cost rather than separately optimizing each energy sector is an effective way of simultaneously reducing the MDENs operation cost [3].

3.3.4 Improving System Flexibility and Reliability

Increasing the penetration level of intermittent RERs in the power system may result in demand and supply imbalance challenges. In this context, the need for flexible resources to provide the system with greater flexibility and reliability is of utmost importance. Employing diverse inputs and outputs in MDENs lets demand shift from one energy sector to another. As a result, one practical solution that offers more flexibility and reliability to the system is exploiting MDENs [3].

3.4 Moving Toward Modern Multi-Dimension Energy Networks

To successfully fulfill the energy transition, moving toward modern MDENs is inevitable. Nevertheless, future modern MDENs will confront a high level of complexity requiring revolutionary solutions in technological, policies, regulation, and societal aspects of meeting energy transition targets. Mostly, the design of the markets and regulatory structures are hardly keeping up with technological advancements [40]. Therefore, to accelerate moving toward future modern MDENs, both the technical side and regulatory and social side, must be concentrated. The following sections describe the pathways that can make progress toward modern MDENs.

3.4.1 Technology Advancements

Flexibility solutions in power systems are utilized to balance supply and demand, maintain the bus voltages, and increase transfer capacity [41]. Flexibility requirements and solutions can be categorized into three timescales: seconds to minutes, hourly, and seasonal/interannual fractions [42]. In the following, flexible solutions that have the potential to be used in the modern MDENs are described. At seconds to minutes timescales, synchronous condensers [43] and fast frequency response services from energy storage systems (ESS) [44] or other innovative solutions will be utilized to provide stability for the modern electricity sector equipped with 100% RERs. According to [45], coordination between networks or energy carriers in the modern MDENs can provide flexibility at longer timescales. Furthermore, exploiting different types of ESS and trade-based balancing solutions will be the other options to present flexibility in the hourly timescale for the modern MDENs [45].

Additionally, demand-side resources like EVs and various types of responsive loads will provide flexibility at an hourly timescale. It is expected in the modern MDENs, the concept of virtual energy storage system (VESS) will have attracted more attention. VESS aggregates various energy systems' controllable components to provide a considerable amount of ESS capacity with cheaper capital cost than the deployment of such real high capacity ESS [46]. Flexible loads, conventional ESS, multi-vector energy systems, microgrids, distributed generators all are examples of controllable components in power systems that can be incorporated into the framework of a VESS [46]. However, it will be crucial to adopt the advanced smart control and management tools and design novel market frameworks to fully exploit the potential of incorporated devices in the platform of the VESS in the modern MDENs. Finally, advanced long-term storage solutions like P2G and Power-to-Liquid will provide promising flexibility provision options for seasonal/interannual timescale [45]. Notably, financial issues and the economic feasibility of deploying novel long-term storage solutions create critical barriers to exploiting such technologies. According to [5], one of the main motivations for expanding electrolyzer technology is financial tax exemption. For this purpose, the electrolyzer technology does not need to be classified as an end-user, but it should be classified as a producer or energy conversion technology.

The decentralization trend will transform conventional centralized energy systems into distributed ones in future energy systems. In this context, to successfully meet energy transition targets, it is vital to create an environment in which such big data from distributed energy systems can be stored and processed. To this end, digitalization can provide an environment to support fulfilling decentralization targets. IoT is one of the recently emerged communication technologies that enable communication between devices employing different data types [47, 48]. In digitalized future power systems, exploiting advanced ICT as well as the IoT will enable making the optimal decision and fast processing by storage and process of extensive real-time data, improving control and monitoring, and also creating two-way energy and information flow between various parts of energy systems including generation, transmission, distribution, and end-users [49, 50]. In this regard, the concept of energy internet (EI) with the ability to provide digital interconnection between MDENs will be of absolutely central importance for modern MDENs [51]. It is worth noting that the concept of EI is different from the smart grid. The core of EI is smart grids; however, EI provides more interconnection between various energy networks through a cyber-network [52]. According to [53], future energy systems will be realized via EI with prominent features, including achieving flexibility of MDENs physical space and digitalization of cyber-space, which is based on real-time data and increasing the involvement of aware customers of social space. As the most recent emerging technology, IoT will be one of the promising options to create two-way energy and communication signal flow between different devices that can be employed in the framework of EI in future interconnected energy systems [47].

3.4.2 Policy-Regulatory-Societal Framework

Along with technological advancement, there is a pressing need to form a new policy- regulatory-societal framework to enable the shift from conventional separate energy networks to modern MDENs while satisfying energy transition targets. In modern MDENs, the implementation of clear policies will be paved the way to overcome the challenges related to the lack of hardware and technology standards, inappropriate energy markets, tariff design, high energy taxation, and data protection insecurity [40, 45, 54]. Furthermore, it will be essential to shape policies to provide a stable environment for increasing the investment in modern MDENs [54].

It is worth noting that the incorporation of an immense amount of renewable energy generation into existing liberalized power markets has resulted in very low or even negative electricity prices. Hence, this situation may hinder investors from increasing investments in renewable energy generation technologies and participating in power markets [55]. Therefore, to move toward renewable-based modern MDENs, a new market-clearing mechanism that stimulates investors to raise their involvement in the power market is of the utmost importance. In addition, new retail tariff designs should be focused on modern MDENs to incentivize consumers and energy-related established businesses providing products and services [56].

The other influential factor contributing to successful energy transition and facilitating progress toward modern MDENs is adopting new regulations. The outdated laws bring considerable benefits to network infrastructure rather than other cost-effective alternatives. They thus can limit the involvements of other market players such as consumers and prosumers in energy markets and demand response programs [40]. Moreover, outdated local regulations can impose restrictions on exploiting advanced ICT [40]. Notably, making the change in the regulatory environment is a highly complex matter, so the regulatory reform process can be particularly carried out locally during a relatively long period of time.

Realizing the digitalized future energy systems will bring out significant advantages as well as cybersecurity challenges. In the modern MDENs, cybersecurity challenges will become more complex as multiple digital systems will be connected in new ways, such as IoT. Furthermore, energy trading capabilities in modern MDENs seem to intensify cybersecurity challenges. In this context, the privacy of various types of stakeholders' data will need to be the main focus of attention in modern MDENs [57]. Private information leakage will have harmful consequences, including a degrading experience for market participants, cyber-attack threats, lack of confidence in services, economic losses, etc. [58]. In this regard, to ensure the privacy of market participants, especially customers, new policies must be shaped in the modern MDENs, aiming to trade-off between the benefits and risks received by market participants. To sum up, there is a pressing need for the adoption of an efficient policy-regulatory-societal framework to support profitable investment in RERs while facilitating the safe participation of prosumers and consumers in the energy markets and demand response programs by the advanced ICT.

3.5 Coordinated Operation of Modern MDENs

Traditionally, all energy distribution networks are operated independently based on the existing regulatory and market frameworks. The move toward MDENs has led to the widespread use of PtX technologies such as heat pumps, PtG, etc., resulting in increased synergies among different energy sectors [59]. Recently, considerable efforts have been expended to investigate the operation and scheduling of MDENs as an integrated energy system [60, 61]. From this perspective, the operation cost of MDENs is minimized in the day-ahead period rather than optimizing each energy distribution network's operation cost. According to [62, 63], coordinated operation of different sector coupling technologies, as well as the associated networks, can be beneficial in terms of facilitating the integration of RERs, improving the reliability and efficiency of the whole energy system, making balance between demand and supply, and providing congestion relief services in the various networks. However, the coordinated operation of different energy networks can be highly challenging due to the high rate of uncertainties and complexities. Increasing the interconnectedness between energy sectors will require advancements to address the issues related to high variability, unpredictability, stability, and complexity in modern MDENs' coordinated operation problem. In this regard, the following subsections are provided to investigate the requirements to be adopted.

3.5.1 Technologies

In different energy sectors, scheduling tools supplying demand at the lowest cost while considering technical constraints. From the holistic perspective toward modern MDENs' coordinated operation problem, exploiting advanced scheduling tools will be essential to propose a comprehensive solution while considering more detailed information about all energy sectors. Accordingly, in the following subsections, requirements of technological advancements are provided to effectively address the problem.

3.5.1.1 Enhanced Optimization Tools and Methods
The current status of the electricity, heat, and gas systems' infrastructures is reviewed in [64]. The results reveal that, despite many similarities, different

characteristics of electricity, gas, and heat networks lead to dissimilar technical challenges as various nonlinearities resulting from various dynamic phenomena are considered in the modeling. In this regard, different technical aspects are involved in the modeling approaches. To effectively address the coordinated operation problem of modern MDENs, multi-energy modeling considering multiple technical aspects should be focused on methodologies. In addition, in modern MDENs, optimization tools need to be developed to assess the stochastic nature of the resources. In this regard, probabilistic, risk-based, and innovative multi-stage approaches will be required to economically and efficiently address the coordinated operation problem of modern MDENs. There is a pressing need for enhanced calculation methods to manage the complexity of the problem. Compared to traditional offline deterministic optimization methods, online and real-time stochastic optimized decisions will be beneficial in maintaining the reliable operation of modern MDENs under increasing uncertainty.

3.5.1.2 Improved Forecasting Tools

Due to the significant advantages of modern MDENs, these networks will play a vital role in future energy systems. In this regard, it is of utmost importance to consider the complex coupling relationships between various energy demands in load forecasting [65]. The load forecasting methods are used to define patterns based on historical data to predict specific future data in different periods [66]. Considering the tight coupling of modern MDENs, accurately forecasting multiple loads will be a crucial prerequisite to providing a more efficient solution for scheduling future energy systems. Adopting appropriate forecasting tools will likely be essential in reducing overall errors and the impact of uncertainties in the operation of modern MDENs. To this end, in the modern MDENs, machine learning methods and high-performance deep learning methods deriving from them will be focused on accurately forecasting multiple loads [67].

3.5.2 Markets

In modern-MDENs, the high rate of uncertainties will highlight the necessity for valuable solutions to increase the reliability of the energy supply while ensuring a dynamic energy balance. In this regard, the multi-energy trading possibility can be one of the most effective solutions which will help ensure the energy supply's reliability while creating a dynamic energy balance [68]. It is crucial to redesign the end-user side to incentive them to participate in a large-scale multi-energy market. Strengthening demand-side integration in scheduling applications requires the introduction of new market mechanisms and effective regulations, which enable the involvement of customers and encourage the emergence of agents aggregating demand-side resources to provide a large number of consumers with participating in energy markets [45]. In this regard, different market mechanisms in which energy trading becomes possible are proposed in the following subsections.

3.5.2.1 Real-time Market Mechanisms

Advancements in communication technologies have led to a continuous transition from traditional energy grids to smart grids [69]. The use of smart grid-related technologies enables two-way communication of information flow, resulting in realizing real-time energy management [70]. In other words, the ability to exchange information provides distribution network operators with advanced management and monitoring systems to carry out real-time automated operations. In modern MDENs, the high share of RERs can create challenges in establishing real-time dynamic energy balance. Therefore, by exploiting advanced ICT, developing a real-time market mechanism will ensure a hedge against the high rate of uncertainties to achieve reliable energy supply and dynamic balance in modern MDENs.

With the proven necessity of real-time market mechanisms, different approaches can be taken into account for real-time energy management in the framework of a real-time market. Methods like multi-objective optimization and mixed-integer linear programming include a central operator that enables transaction management in real-time. In these methods, all participating units upload their information to the central operator, where then the energy management strategy will be determined and sent to each participant [71]. However, as the number of participants increases, real-time scheduling of resources is getting more difficult in these methods. Meanwhile, adopting these real-time energy management methods with a central operator makes it difficult for participants to adjust their bidding strategy based on received real-time information. Participants have to upload all their information to the central operator in these methods; therefore, energy management mechanisms based on the central part lack secure private communication. Thus, adopting real-time energy management methods with a central operator in the framework of real-time market mechanisms can reduce real-time transaction rates and participating units' overall profit while participants' privacy is invaded. As in the modern MDENs, the real-time energy market will be an effective solution for enhancing the system's reliability and stability; the aforementioned challenges must be addressed effectively. In this regard, the following subsection presents more flexible and efficient energy trading environments.

3.5.2.2 Peer-to-Peer Market Mechanisms

Prosumers who are both energy producers and consumers can play a crucial role in modern MDENs. Prosumers can use their production to satisfy their demands, while the excess energy can be sold to consumers or stored for later use. To encourage the prosumers to increase renewable energy generation, creating local energy markets in which the production of RERs can be traded between prosumers and consumers without any intermediary entity is essential. P2P transactions have turned out to stimulate prosumers to actively participate in the energy market [72]. In this market mechanism, a peer can be referred to one or a group of generators, consumers, and prosumers who are enabled to sell or buy energy directly with each other. P2P trading provides prosumers with the opportunity to sell their surplus renewable energy production and leads to energy cost reduction by giving end-users more flexibility in choosing suppliers [73]. Furthermore, a P2P-based market can be beneficial in terms of improving the local balance of energy production and consumption [74]. To sum up, P2P energy trading as a transparent market mechanism that enables real-time energy management will play a key role in mitigating the impacts of uncertainties on the coordinated operation of modern MDENs. In this regard, exploiting advanced digital technologies is essential for developing P2P energy transactions based on online services.

Recently emerging blockchain technology has received significant attention for successfully addressing security and transparency issues in the energy sector [75]. Blockchain technology can be applied to different areas of the energy system. In wholesale [76] or local level energy trading applications such as P2P energy trading [73], blockchain technology prove capable of providing a reliable and transparent energy trading environment without any central operator acting as a trusted intermediary. Blockchain technology can give prosumers more incentives to engage in P2P energy trading applications and provide local dynamic energy balance in the operation of distributed energy systems. A simple overview of the P2P market mechanism in the blockchain-based network is depicted in Figure 3.3. In terms of demand-side energy management, blockchain technology demonstrates the potential for realizing real-time energy management [77]. EVs, as one of the essential demand-side resources, can both play the role of energy suppliers and consumers in the distributed power system. Two-way communication of EVs with modern power systems can help alleviate the problem of imbalances in demand

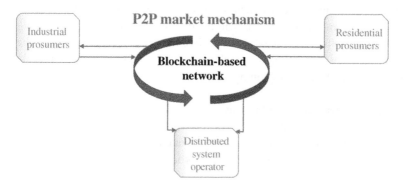

Figure 3.3 P2P market mechanism in blockchain-based network.

and supply. Blockchain technology can provide EVs with secure charging services and an easy payment mechanism to create a demand-supply energy balance in the operation of modern energy systems [78].

3.6 Coordinated Planning of Modern MDENs

Traditionally, the energy infrastructures such as natural gas, electricity, and district heating networks are designed and planned separately. However, to fully benefit from the advantages of MDENs and realize energy transition targets, the coordinated planning of various energy networks is essential [79]. The targets of coordinated planning of MDENs are to find optimal investment plans of all the assets associated with MDENs, including multi-energy generation, transmission, distribution, storage, and coupling components to meet growing demand, replace retiring infrastructures, and strengthen existing networks [80]. In this regard, the objective function of the coordinated planning of MDENs is minimizing the investment, operational, and maintenance costs while considering technical constraints and satisfying current and future demand at different time horizons. Furthermore, environmental performance, energy reliability, energy losses, and system quality can be considered in the objective function along with other terms [80, 81]. In the coordinated planning of modern MDENs, with a high penetration level of RERs, it is crucial to consider more ESS as well as conversion technologies to store and transform renewable energy generation.

Furthermore, as modern MDENs will be more digitalized and decentralized, energy and data security must be focused on the coordinated planning problem of modern MDENs [82]. Future scenarios with advanced integrated modeling of MDENs and associated sector-coupling technologies must be investigated to effectively address the coordinated planning problem. Notably, advanced analytical and optimization methods are required to handle the high complexity and long-term uncertainties related to the problem. In the future, the regulatory strategies, the emergence of energy markets, and fluctuations in energy prices can be influential factors in employing various investment strategies of companies. Therefore, the impacts of regulatory and market factors must be considered in the coordinated planning of modern MDENs to obtain more investment savings and efficient energy utilization [81].

3.7 Future Plans for Increasing RERs and MDENs

Over the last few decades, RERs have become an obvious alternative to realize decarbonization targets. Besides solar photovoltaic (PV) and onshore wind farms, offshore wind farms are attractive options for increasing the RERs penetration in the energy system. The significant advantage of offshore wind farms is the consistency in generation power [83]. Currently, offshore wind farms are highly focused in the EU, with 84% of global installations, due to technical and commercial maturity in this industry. The United Kingdom, Germany, Denmark, the Netherlands, and Belgium are large European offshore wind energy producers, respectively. In addition, the offshore wind industry in China has been rapidly expanding; as a result, China is placed third in global offshore wind energy producers after the United Kingdom and Germany [84].

Notably, hydrogen production from offshore wind farms can be exploited in the transport and heat sectors to fulfill decarbonization targets. Hydrogen can be stored and transported long distances with fewer costs than power transmission lines [85]. On the other hand, according to European Commission's hydrogen strategy released in 2020, developing a viable hydrogen market at a significant scale is vital to meet carbon neutrality targets. In this regard, exploiting high-potential offshore wind farms to produce hydrogen is essential to fulfilling decarbonization targets [86].

According to International Maritime Organization (IMO), 3–5% of the total Green House Gases emissions are related to maritime transportation; however, this portion will increase to 18% by 2050 if no actions are taken into account. Thereby, decarbonization of maritime operations is essential to realize the decarbonization targets of the energy transition. In this regard, the concept of seaport microgrid is proposed to increase RER's penetration in energy systems and consequently hit the decarbonization targets [87]. Therefore, renewable energy generation technologies such as PV systems, wind turbines (WTs), ocean energy, geothermal sources, biomass, and hydrogen contribute to power generation in seaports. As the world's largest solar power generation port, the Jurong port, Singapore, with more than 10 MW of PV modules, plans to become the first zero-carbon footprint port worldwide. To this end, it is vital to focus on the seaport microgrid concept to facilitate renewable energy integrations and meet decarbonization targets [87].

Improving efficiency and flexibility is essential to achieve ambitious goals in expanding the use of RESs and reducing greenhouse gas emissions. In this regard, MDENs are suitable solutions for the adaptation of future energy systems with efficiency and flexibility demands. The future MDEN-rich energy systems will face high levels of complexity that require fundamental changes in technological, policy, regulation, and social perspectives [45].

3.8 Challenges

As large-scale RERs will be the heart of future energy systems, a high share of uncertainty will be one of the most critical challenges of modern MDENs. According to [5], one of the most critical requirements moving toward future modern MDENs is substantial infrastructure investments to increase the electricity distribution grid's capacity and sufficiently provide renewable energy storage and transformation. Exploiting transformation technologies, such as P2G, in the modern MDENs will be essential but struggling due to their high capital cost and limited-hours full-load application in response to excess renewable energy generation [88]. Therefore, technological advancements in transformation technologies, such as P2G, are necessary to increase transformation efficiency and benefits of investors.

From the modeling perspective of modern MDENs, long-term and short-term uncertainties effects must be considered [89]. Notably, modern MDENs modeling is influenced by multi-energy systems' infrastructures and associated sector coupling technologies and is also affected by the regulatory strategies and various market stakeholders' attitudes [90]. Therefore, modern MDENs require increased research on holistic energy systems modeling while considering social and regulatory aspects. However, the effects of regulations, customers, and market stakeholders have rarely been included in the future energy system's modeling scenarios [91]. Accordingly, it is necessary to define new evaluation frameworks of modern MDENs' modeling to investigate their effectiveness from different aspects and various points of view, including end-users, utilities, and market stakeholders [92].

On the other hand, the digitalization trend has led to deploying advanced smart devices and producing a large amount of data from different parts of modern MDENs. Hence, it is notable that the risk of data leakage or destruction during transmission is potential. These produced data are valuable for energy producers and companies, grids' operators, prosumers, and consumers to spur participation in energy markets and also can be used to forecast future energy supply and demand [93]. As a result, modern MDENs will face the challenges of large amounts of data management, including information technology infrastructures, data collection and sharing, data processing, analysis, and data security and privacy issues [89]. Furthermore, the emergence of new market mechanisms in modern MDENs will intensify the importance of real-time data management and data security. For example, while P2P-based market mechanisms, as a new market mechanism, provide both prosumers and grid operators with significant benefits, the privacy of individual participants and the security of the grid are critical challenges that modern MDENs will face in the presence of new market mechanisms [94]. Notably, modern MDENs will face regulatory problems apart from technical and technological challenges. Notably, exploiting technological advancement can be done in the framework of the applicable legislative rules. For instance, the establishment of P2P energy markets and encouraging prosumers and retailers to participate in energy trading are all related to decisions of the regulatory board [95]. As a result, to successfully deal with challenges related to modern MDENs, both technological and regulatory aspects must be considered, as shown in Figure 3.4.

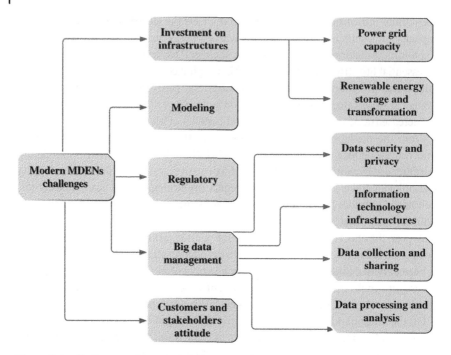

Figure 3.4 Challenges of modern MDEN.

3.9 Summary

The energy system transition has started in response to three drivers: decarbonization, decentralization, and digitalization. In this regard, the proven capabilities of MDENs make them a practical option for fulfilling energy transition targets. On the other hand, advances in ICTs and sector coupling technologies accelerate moving toward modern MDENs. According to these matters, after evaluating MDENs and their associated benefits in this chapter, technological and regulatory drivers accelerating moving toward modern MDENs are discussed. Furthermore, after investigating technological advancements and market-based solutions to address the modern MDENs' coordinated operation, significant issues related to the coordinated planning problem of modern MDENs are presented. Then, different plans of various countries to increase the penetration of RERs, and as a result, to realize energy transition targets are discussed. Finally, the primary challenges that modern MDENs will face are provided.

References

1 Bouckaert, S., Fernandez Pales, A., McGlade C. et al. (2021). *Net Zero by 2050: A Roadmap for the Global Energy Sector.* Paris, France: IEA.

2 Impram, S., Nese, S.V., and Oral, B. (2020). *Challenges of renewable energy penetration on power system flexibility: a survey. Energy Strategy Reviews* 31: 100539.

3 Hanna, R., Gazis, E., Rhodes, A., and Gross, R. (2018). *Unlocking the Potential of Energy Systems Integration.* London, UK: Imperial College London.

4 O'Malley, M., Kroposki, B., Hannegan, B. et al. (2016). *Energy Systems Integration. Defining and Describing the Value Proposition.* Golden, CO (United States): National Renewable Energy Lab.(NREL).

5 Arent, D.J., Bragg-Sitton, S.M., Miller, D.C. et al. (2020). Multi-input, multi-output hybrid energy systems. *Joule* 5 (1): 47–58. https://doi.org/10.1016/j.joule.2020.11.004.

6 Gea-Bermúdez, J., Jensen, I.G., Münster M. et al. (2021). The role of sector coupling in the green transition: A least-cost energy system development in Northern-central Europe towards 2050. *Applied Energy* 289: 116685.

7 Hossein Motlagh, N., Mohammadrezaei, M., Hunt, J., and Zakeri, B. (2020). Internet of Things (IoT) and the energy sector. *Energies* 13 (2): 494.

8 Abeysekera, M., Jenkins, N., and Wu, J. (2016). *Integrated Energy Systems: An Overview of Benefits, Analysis, Research Gaps and Opportunities.* UK: HubNet.

9 Abeysekera, A.M., Wu, J., and Jenkins, N. (2016). HubNet Position Paper Series: Integrated Energy Systems: An Overview of Benefits, Analysis Methods, Research Gaps and Opportunities. *Tech. Rep.* p. 1–49.

10 Zhang, C., Wu, J., Long, C., and Cheng, M. (2017). Review of existing peer-to-peer energy trading projects. *Energy Procedia* 105: 2563–2568.

11 Mohammadi, M., Noorollahi, Y., Mohammadi-Ivatloo, B., and Yousefi, H. (2017). Energy hub: from a model to a concept–a review. *Renewable and Sustainable Energy Reviews* 80: 1512–1527.

12 Capros, P., Kannavou, M., Evangelopoulou, S. et al. (2018). Outlook of the EU energy system up to 2050: the case of scenarios prepared for European commission's "clean energy for all Europeans" package using the PRIMES model. *Energy Strategy Reviews* 22: 255–263.

13 Mai, T., Steinberg, D., Logan, J. et al. (2018). An electrified future: initial scenarios and future research for US energy and electricity systems. *IEEE Power and Energy Magazine* 16 (4): 34–47.

14 Farid, A.M., Jiang, B., Muzhikyan, A., and Youcef-Toumi, K. (2016). The need for holistic enterprise control assessment methods for the future electricity grid. *Renewable and Sustainable Energy Reviews* 56: 669–685.

15 Heinen, S. and O'Malley, M.J. (2018). Complementarities of supply and demand sides in integrated energy systems. *IEEE Transactions on Smart Grid* 10 (1): 1156–1165.

16 Kondziella, H. and Bruckner, T. (2016). Flexibility requirements of renewable energy based electricity systems–a review of research results and methodologies. *Renewable and Sustainable Energy Reviews* 53: 10–22.

17 Ayele, G.T., Mabrouk, M.T., Haurant, P. et al. (2019). Optimal placement and sizing of heat pumps and heat only boilers in a coupled electricity and heating networks. *Energy* 182: 122–134.

18 Bernath, C., Deac, G., and Sensfuss, F. (2019). Influence of heat pumps on renewable electricity integration: Germany in a European context. *Energy Strategy Reviews* 26: 100389.

19 Rinne, S. and Syri, S. (2015). The possibilities of combined heat and power production balancing large amounts of wind power in Finland. *Energy* 82: 1034–1046.

20 Wang, H., Yin, W., Abdollahi, E. et al. (2015). Modelling and optimization of CHP based district heating system with renewable energy production and energy storage. *Applied Energy* 159: 401–421.

21 Navarro, J.P.J., Kavvadias, K.C., Quoilin, S., and Zucker, A. (2018). The joint effect of centralised cogeneration plants and thermal storage on the efficiency and cost of the power system. *Energy* 149: 535–549.

22 Lund, H., Duic, N., Østergaard, P.A., and Mathiesen, B.V. (2018). Future district heating systems and technologies: on the role of smart energy systems and 4th generation district heating. *Energy* 165: 614–619.

23 Bloess, A., Schill, W.-P., and Zerrahn, A. (2018). Power-to-heat for renewable energy integration: a review of technologies, modeling approaches, and flexibility potentials. *Applied Energy* 212: 1611–1626.

24 Lester, M.S., Bramstoft, R., and Münster, M. (2020). Analysis on electrofuels in future energy systems: a 2050 case study. *Energy* 199: 117408.

25 Ma, Y., Yu, Y., and Mi, Z. (2021). Accommodation of curtailed wind power by electric boilers equipped in different locations of heat-supply network for power system with CHPs. *Journal of Modern Power Systems and Clean Energy* 9 (4): 930–939. https://doi.org/10.35833/MPCE.2019.000151.

26 Liu, B., Li, J., Zhang, S. et al. (2020). Economic dispatch of combined heat and power energy systems using electric boiler to accommodate wind power. *IEEE Access* 8: 41288–41297.

27 Buffa, S., Cozzini, M., D'antoni, M. et al. (2019). 5th generation district heating and cooling systems: a review of existing cases in Europe. *Renewable and Sustainable Energy Reviews* 104: 504–522.

28 Lund, H., Østergaard, P.A., Chang, M. et al. (2018). The status of 4th generation district heating: research and results. *Energy* 164: 147–159.

29 Sorknæs, P., Østergaard, P.A., Thellufsen, J.Z. et al. (2020). The benefits of 4th generation district heating in a 100% renewable energy system. *Energy* 213: 119030.

30 Jannasch, A.-K., H. Pihl, M. Persson, E. Svensson, S. Harvey, and H. Wiertzema (2020). Opportunities and barriers for implementation of Power-to-X (P2X) technologies in the West Sweden Chemicals and Materials Clusters process industries. Lund.

31 Clegg, S. and Mancarella, P. (2016). Storing renewables in the gas network: modelling of power-to-gas seasonal storage flexibility in low-carbon power systems. *IET Generation, Transmission and Distribution* 10 (3): 566–575.

32 Nazari-Heris, M., Mirzaei, A., Mohammadi-Ivatloo, B. et al. (2020). Economic-environmental effect of power to gas technology in coupled electricity and gas systems with price-responsive shiftable loads. *Journal of Cleaner Production* 244: 118769.

33 Alabdulwahab, A., Abusorrah, A., Zhang, X., and Shahidehpour, M. (2015). Coordination of interdependent natural gas and electricity infrastructures for firming the variability of wind energy in stochastic day-ahead scheduling. *IEEE Transactions on Sustainable Energy* 6 (2): 606–615.

34 Nastasi, B., Basso, G.L., Garcia, D.A. et al. (2018). Power-to-gas leverage effect on power-to-heat application for urban renewable thermal energy systems. *International Journal of Hydrogen Energy* 43 (52): 23076–23090.

35 Schmidt, O., Gambhir, A., Staffell, I. et al. (2017). Future cost and performance of water electrolysis: an expert elicitation study. *International Journal of Hydrogen Energy* 42 (52): 30470–30492.

36 Burrin, D., Roy, S., Roskilly, A.P., and Smallbone, A. (2021). A combined heat and green hydrogen (CHH) generator integrated with a heat network. *Energy Conversion and Management* 246: 114686.

37 Chen, C. et al. (2020). Optimal coordinative operation strategy of the electric–thermal–gas integrated energy system considering CSP plant. *IET Energy Systems Integration* 2 (3): 187–195.

38 Gong, Y., Yang, X., Xu, J. et al. (2021). Optimal operation of integrated energy system considering virtual heating energy storage. *Energy Reports* 7: 419–425.

39 Madeddu, S. et al. (2020). The CO_2 reduction potential for the European industry via direct electrification of heat supply (power-to-heat). *Environmental Research Letters* 15 (12): 124004.

40 Dong, Z.Y. and Zhang, Y. (2021). Interdisciplinary vision of the digitalized future energy systems. *IEEE Open Access Journal of Power and Energy* 8: 557–569.

41 Hillberg, E., Zegers, A., Herndler, B. et al. (2019). *Flexibility needs in the future power system*. doi: 10.13140/RG.2.2.22580.71047

42 Lew, D., Bartlett, D., Groom, A. et al. (2019). Secrets of successful integration: operating experience with high levels of variable, inverter-based generation. *IEEE Power and Energy Magazine* 17 (6): 24–34.

43 Igbinovia, F.O., G. Fandi, Z. Muller, and J. Tlusty (2018). Reputation of the synchronous condenser technology in modern power grid. *2018 International Conference on Power System Technology (POWERCON)*, Guangzhou, China (6–8 November 2018). IEEE. doi: 10.1109/POWERCON.2018.8601540.

44 Meng, L., Zafar, J., Khadem, S.K. et al. (2019). Fast frequency response from energy storage systems—a review of grid standards, projects and technical issues. *IEEE Transactions on Smart Grid* 11 (2): 1566–1581.

45 O'Malley, M.J., Anwar, M.B., Heinen, S. et al. (2020). Multicarrier energy systems: shaping our energy future. *Proceedings of the IEEE* 108 (9): 1437–1456.

46 Cheng, M., Sami, S.S., and Wu, J. (2017). Benefits of using virtual energy storage system for power system frequency response. *Applied Energy* 194: 376–385.

47 Kabalci, Y., Kabalci, E., Padmanaban, S. et al. (2019). Internet of things applications as energy internet in smart grids and smart environments. *Electronics* 8 (9): 972.

48 Alrawais, A., Alhothaily, A., Hu, C., and Cheng, X. (2017). Fog computing for the internet of things: security and privacy issues. *IEEE Internet Computing* 21 (2): 34–42.

49 Reka, S.S. and Dragicevic, T. (2018). Future effectual role of energy delivery: a comprehensive review of Internet of Things and smart grid. *Renewable and Sustainable Energy Reviews* 91: 90–108.

50 Masera, M., Bompard, E.F., Profumo, F., and Hadjsaid, N. (2018). Smart (electricity) grids for smart cities: assessing roles and societal impacts. *Proceedings of the IEEE* 106 (4): 613–625.

51 Liu, T., Zhang, D., and Wu, T. (2020). Energy internet: concept, structure and its potential future development in China [J]. *International Journal of Smart Grid and Clean Energy* 9 (6): 1019–1026.

52 Hussain, H.M., Narayanan, A., Nardelli, P.H., and Yang, Y. (2020). What is energy internet? Concepts, technologies, and future directions. *IEEE Access* 8: 183127–183145.

53 Wu, Y., Wu, Y., Guerrero, J.M., and Vasquez, J.C. (2021). A comprehensive overview of framework for developing sustainable energy internet: from things-based energy network to services-based management system. *Renewable and Sustainable Energy Reviews* 150: 111409.

54 Paiho, S., Kiljander, J., Sarala, R. et al. (2021). Towards cross-commodity energy-sharing communities–a review of the market, regulatory, and technical situation. *Renewable and Sustainable Energy Reviews* 151: 111568.

55 Cieplinski, A., D'Alessandro, S., and Marghella, F. (2021). Assessing the renewable energy policy paradox: a scenario analysis for the Italian electricity market. *Renewable and Sustainable Energy Reviews* 142: 110838.

56 Blazquez, J., Fuentes-Bracamontes, R., Bollino, C.A., and Nezamuddin, N. (2018). The renewable energy policy Paradox. *Renewable and Sustainable Energy Reviews* 82: 1–5.

57 Jørgensen, P.-A., J. E. Simensen, C. Esnoul, X. Gao, S. A. Olsen, and B. A. Gran (2020). Addressing cybersecurity in energy Island. Paper accepted for ESREL. Singapore: Research Publishing. doi: 10.3850/978-981-14-8593-0_5428-cd

58 Vozikis, D., E. Darra, T. Kuusk, D. Kavallieros, A. Reintam, and X. Bellekens (2020). On the importance of cyber-security training for multi-vector energy distribution system operators. *Proceedings of the 15th International Conference on Availability, Reliability and Security*. New York, NY: Association for Computing Machinery.

59 Zhou, B., Xu, D., Li, C. et al. (2018). Optimal scheduling of biogas–solar–wind renewable portfolio for multicarrier energy supplies. *IEEE Transactions on Power Systems* 33 (6): 6229–6239.

60 Liu, J., Wang, A., Qu, Y., and Wang, W. (2018). Coordinated operation of multi-integrated energy system based on linear weighted sum and grasshopper optimization algorithm. *IEEE Access* 6: 42186–42195.

61 Mirzaei, M.A., Nazari-Heris, M., Zare, K. et al. (2020). Evaluating the impact of multi-carrier energy storage systems in optimal operation of integrated electricity, gas and district heating networks. *Applied Thermal Engineering* 176: 115413.

62 Qin, Y., Wu, L., Zheng, J. et al. (2019). Optimal operation of integrated energy systems subject to coupled demand constraints of electricity and natural gas. *CSEE Journal of Power and Energy Systems* 6 (2): 444–457.

63 Zhang, M., Wu, Q., Wen, J. et al. (2021). Optimal operation of integrated electricity and heat system: a review of modeling and solution methods. *Renewable and Sustainable Energy Reviews* 135: 110098.

64 Guelpa, E., Bischi, A., Verda, V. et al. (2019). Towards future infrastructures for sustainable multi-energy systems: a review. *Energy* 184: 2–21.

65 Zheng, J., Zhang, L., Chen, J. et al. (2021). Multiple-load forecasting for integrated energy system based on copula-DBiLSTM. *Energies* 14 (8): 2188.

66 Hwang, J.S., Fitri, I.R., Kim, J.-S., and Song, H. (2020). Optimal ESS scheduling for peak shaving of building energy using accuracy-enhanced load forecast. *Energies* 13 (21): 5633.

67 Zhou, D., Ma, S., Hao, J. et al. (2020). An electricity load forecasting model for integrated energy system based on BiGAN and transfer learning. *Energy Reports* 6: 3446–3461.

68 Daneshvar, M., Asadi, S., and Mohammadi-Ivatloo, B. (2021). *Energy Trading Possibilities in the Modern Multi-Carrier Energy Networks, in Grid Modernization—Future Energy Network Infrastructure*, 175–214. Springer.

69 Misra, S., Bera, S., Ojha, T. et al. (2016). *ENTRUST: Energy trading under uncertainty in smart grid systems. Computer Networks* 110: 232–242.

70 Bahrami, S. and Amini, M.H. (2017). *A decentralized framework for real-time energy trading in distribution networks with load and generation uncertainty. arXiv preprint arXiv:1705.02575* https://doi.org/10.48550/arXiv.1705.02575.

71 Wang, L., Liu, J., Yuan, R. et al. (2020). Adaptive bidding strategy for real-time energy management in multi-energy market enhanced by blockchain. *Applied Energy* 279: 115866.

72 Morstyn, T., Farrell, N., Darby, S.J., and McCulloch, M.D. (2018). Using peer-to-peer energy-trading platforms to incentivize prosumers to form federated power plants. *Nature Energy* 3 (2): 94–101.

73 Jing, Z., Pipattanasomporn, M., and Rahman, S. (2019). Blockchain-based Negawatt trading platform: Conceptual architecture and case studies. *2019 IEEE PES GTD Grand International Conference and Exposition Asia (GTD Asia),* Bangkok, Thailand (19–23 March 2019). IEEE.

74 Zhang, C., Wu, J., Zhou, Y. et al. (2018). Peer-to-peer energy trading in a microgrid. *Applied Energy* 220: 1–12.

75 Dong, Z., Luo, F., and Liang, G. (2018). Blockchain: a secure, decentralized, trusted cyber infrastructure solution for future energy systems. *Journal of Modern Power Systems and Clean Energy* 6 (5): 958–967.

76 Tesfamicael, A.D., Liu, V., Mckague, M. et al. (2020). A design for a secure energy market trading system in a national wholesale electricity market. *IEEE Access* 8: 132424–132445.

77 Kolahan, A., Maadi, S.R., Teymouri, Z., and Schenone, C. (2021). Blockchain-based solution for energy demand-side management of residential buildings. *Sustainable Cities and Society* 75: 103316.

78 Erenoğlu, A.K., Şengör, İ., Erdinç, O., and Catalão, J.P. (2020). *Blockchain and its Application Fields in Moth Power Economy and Demand Side Management, in Blockchain-based Smart Grids,* 103–130. Elsevier.

79 Yang, W., Liu, W., Chung, C.Y., and Wen, F. (2019). Coordinated planning strategy for integrated energy systems in a district energy sector. *IEEE Transactions on Sustainable Energy* 11 (3): 1807–1819.

80 Hosseini, S.H.R., Allahham, A., Walker, S.L., and Taylor, P. (2020). Optimal planning and operation of multi-vector energy networks: a systematic review. *Renewable and Sustainable Energy Reviews* 133: 110216.

81 Fan, H., Yu, Z., Xia, S., and Li, X. (2021). Review on coordinated planning of source-network-load-storage for integrated energy systems. *Frontiers in Energy Research* 9: 138.

82 Anvari-Moghaddam, A., Mohammadi-Ivatloo, B., Asadi, S. et al. (2019). *Sustainable Energy Systems Planning, Integration, and Management.* Multidisciplinary Digital Publishing Institute.

83 Ali, S.W., Sadiq, M., Terriche, Y. et al. (2021). Offshore wind farm-grid integration: a review on infrastructure, challenges, and grid solutions. *IEEE Access* 9: 102811–102827.

84 DeCastro, M., Salvador, S., Gómez-Gesteira, M. et al. (2019). Europe, China and the United States: three different approaches to the development of offshore wind energy. *Renewable and Sustainable Energy Reviews* 109: 55–70.

85 Leahy, P., McKeogh, E., Murphy, J., and Cummins, V. (2021). Development of a viability assessment model for hydrogen production from dedicated offshore wind farms. *International Journal of Hydrogen Energy* 46 (48): 24620–24631.

86 Espegren, K., Damman, S., Pisciella, P. et al. (2021). The role of hydrogen in the transition from a petroleum economy to a low-carbon society. *International Journal of Hydrogen Energy.*

87 Fang, S., Wang, Y., Gou, B., and Xu, Y. (2019). Toward future green maritime transportation: an overview of seaport microgrids and all-electric ships. *IEEE Transactions on Vehicular Technology* 69 (1): 207–219.

88 Chandrasekar, A., Flynn, D., and Syron, E. (2021). Operational challenges for low and high temperature electrolyzers exploiting curtailed wind energy for hydrogen production. *International Journal of Hydrogen Energy*.

89 Xu, Y., Yan, C., Liu, H. et al. (2020). *Smart energy systems: a critical review on design and operation optimization. Sustainable Cities and Society* 62: 102369. https://doi. org/10.1016/j.scs.2020.102369.

90 Fazendeiro, L.M. and Simões, S.G. (2021). Historical variation of IEA energy and CO_2 emission projections: implications for future energy modeling. *Sustainability* 13 (13): 7432.

91 Ramsebner, J., Haas, R., Auer, H. et al. (2021). From single to multi-energy and hybrid grids: historic growth and future vision. *Renewable and Sustainable Energy Reviews* 151: 111520.

92 Berjawi, A., Walker, S., Patsios, C., and Hosseini, S. (2021). An evaluation framework for future integrated energy systems: a whole energy systems approach. *Renewable and Sustainable Energy Reviews* 145: 111163.

93 Lopes, J.A.P., Madureira, A.G., Matos, M. et al. (2020). The future of power systems: challenges, trends, and upcoming paradigms. *Wiley Interdisciplinary Reviews: Energy and Environment* 9 (3): e368.

94 Khorasany, M., Dorri, A., Razzaghi, R., and Jurdak, R. (2021). Lightweight blockchain framework for location-aware peer-to-peer energy trading. *International Journal of Electrical Power & Energy Systems* 127: 106610.

95 Tushar, W., Yuen, C., Saha, T.K. et al. (2021). Peer-to-peer energy systems for connected communities: a review of recent advances and emerging challenges. *Applied Energy* 282: 116131.

4

Modern Smart Multi-Dimensional Infrastructure Energy Systems – State of the Arts

Ali Sharifzadeh[1], Sasan Azad[1,2], and Mohammad Taghi Ameli[1,2]

[1] *Department of Electrical Engineering, Shahid Beheshti University, Tehran, Iran*
[2] *Electrical Networks Research Institute, Shahid Beheshti University, Tehran, Iran*

Abbreviations

DG	Distributed generation
ES	Energy system
IEMS	Integrated energy management system
E.INFRA	Energy infrastructure
G2H	Gas-to-heat
G2P	Gas-to-power
H&E	Heat and electricity
MDES	Multi-dimensional energy systems
MeG	Modern energy generation
PS	Power system
P2G	Power-to-gas
P2H	Power-to-heat

4.1 Introduction

Climate protection dominates the social debate on sustainable development from the Paris Agreement and the designated National Partnerships to the European Green Agreement [1]. As the most significant carbon emission, the evolution of the energy sector is in the spotlight. However, the political discussion on the sustainable development of energy supply is primarily limited to climate protection

Coordinated Operation and Planning of Modern Heat and Electricity Incorporated Networks,
First Edition. Edited by Mohammadreza Daneshvar, Behnam Mohammadi-Ivatloo, and Kazem Zare.
© 2023 The Institute of Electrical and Electronics Engineers, Inc.
Published 2023 by John Wiley & Sons, Inc.

goals through the management of energy resources and increased productivity at affordable costs for citizens and industry. For this purpose, multi-dimensional energy systems are proposed. These integrated infrastructure management systems can alleviate energy crises, improve energy usage, and manage various types of energy in a global solution. Multiple subsystems, including power electricity, gas pipeline networks, cold/heat networks, transportation networks, and cyber and physical energy systems, provide connection characteristics such as multi-energy source modeling. Therefore, multi-dimensional energy systems and energy hubs in different networks will be an approach that will be addressed in the future industry [2, 3].

On the other hand, the interaction between systems and performance challenges determines their effectiveness. The main challenge for owners is management and operation issues of energy infrastructures. Energy exchange and financial settlement are also among the issues that need to be addressed. New tools in operation, such as intelligent methods and machine learning, are well used in energy systems. Some concepts such as blockchain [4] and modern financial models are also effective for increasing the capabilities of these systems. In general, any innovative methods in other engineering fields, especially computer sciences, will be instrumental in energy field problem-solving. Therefore, mastering the concepts of new energy systems and familiarity with modern tools will make the management and operation of infrastructures easier. Modeling and mathematical developments and free model approaches have also been widely used in this field. However, moving simultaneously with innovations using new structures and tools has created profound changes in current and future networks. It has made it easier to achieve different goals based on policies and fundamental concepts.

As mentioned above, past studies review will be more effective in future researches. These studies have been performed in subsystem interaction, global management systems, reliabilities and evaluation indicators. On the other hand, energy infrastructure and its challenges have also been considered by researchers in the past and future.

The purpose of this chapter is to examine the current state of energy infrastructure concerning electrical, heating, and gas systems and to examine multi-energy infrastructures. Also, the description of energy infrastructure from the point of view of technology and modeling was briefly reviewed with concerning to the gas pipelines, electricity networks and thermal. It also focused on different types of converters by studying energy conversions like gas-to-heat, gas-to-power, power-to-heat, and power-to-gas technologies.

This chapter briefly explains multi-dimensional energy systems and reviews the state of the art and future issues and challenges. Also, it examines recent researches in multi-dimensional energy systems, the interaction between subsystems, and management structures. In addition, modeling methods and evaluation indicators, including power gas systems, power heating, power network, and physical and cyber energy systems, are generally reviewed. It also looks at the future challenges associated with these systems and the prospects for this type.

The remainder of this chapter is organized as follows: Introduction presents a brief of the work; Section 4.2 presents a review of energy networks. An overview of the infrastructure of multi-dimensional systems and energy hubs with a discussion about applications and advantages addresses is provided in Section 4.3. Section 4.4 Aggregates the findings and identifies the modeling of energy systems. Section 4.5 reviews the advantages of integrated management for energy infrastructures. Sections 4.6 and 4.7 present a brief review of energy conversion and economic/environmental impacts. Finally, Sections 4.8 and 4.9 present a summary of the advantages in future networks and a conclusion of the chapter and the work.

4.2 Energy Networks

Conventional energy networks consist of different parts and are usually independent of different types of energy. Generation type and resources are the main differences between forms of energy. Some types of energy, such as electrical, are easily transmissible over long distances and have lower losses. Others are produced at a lower cost, and some, like natural gas, can be easily converted to other forms [5]. The advantages of each different energy type affect how it is used in an extensive energy network.

On the other hand, some problems that other types may have, such as difficult heat transfer, create challenges in the network. Converters also help reap the benefits of various energies and may interconnect several types of energy networks to each other. In the conventional situation, each network is managed independently, and the governing organizations generally plan the exchanges.

Conventional energy networks include the main following:

- Electrical energy networks
  An electrical grid includes all power sources, consumers, and distribution systems. One of the challenges in these networks is the increasing influences of DGs, renewables, and the severe uncertainty of existing loads. Operation of these networks is very close to the border of network stability [6, 7]. On the other hand, some of the electricity is supplied by gas power plants, and part of the network load includes homes and industries heating loads. Renewable uncertainties and the uncertain conditions of the gas network pose many challenges during operation and situation prediction [8]. Given above, an independent reviews of electrical network stability and security without considering other energy networks impact and conditions will not be accurate and acceptable.

- Thermal energy networks
It is important to note that a large portion of the network loads is allocated to heat and thermal generation, and a high volume of greenhouse gas emissions is due to this process. Conventional thermal networks transfer the heat of gas power plants losses or underground thermal energy for consumers. These networks are usually constructed in industrial and population centers as the primary heat and power consumption. However, these networks may not be easy to control, but they effectively reduce losses and increase system efficiency.

  Thermal energy networks are created because of renewables and widespread heat loss not only for integrating high-efficiency centralized power plants. Two examples of low-temperature networks are Otokos and Lystrup. Large regional cooling networks can be found in Hong Kong, Helsinki and Munich [9].

  As reported in some sources, more than 50% of large network heat demand provided by cogenerations. This type of plants for heating of large complexes significantly reduces the initial energy requirements and carbon emissions. The trend of new approaches toward low supply temperatures (30–70 °C), i.e. the new locally heating generation, which is fully explained in some studies. It is possible to better energy management by renewable heat generation (like solar or wind) or at low temperatures in power plants. It is the main cause of change in energy infrastructure due to guaranteed heat providing. Large pipelines will be required to transmit higher currents at the request of energy consumption. Various methods have been proposed to overcome the existing network limitations due to the pipeline length to increase the number of users connected to the network without layout major changes in the pipeline and structure.

  Another trend is the increasing use of measuring devices. The ability to monitor different values of network, proposes much benefits such as real-time condition monitoring of energy performance, implementation of advanced control strategies, and detection of possible operational abnormalities (leakage, malfunction, or sediment). Data-driven models can usually be used for purposes [9].

  Regional cooling also has higher performance ability (based on initial energy) due to the new technology of chiller units. Paris is the first city that these facilities are located in large scale in Europe. Also located in smaller scale in Vienna, Helsinki, Lisbon, and London.

  The absorption chillers application in complexes and regional networks is well established, accounting for about 4% of the installed cooling capacity. Efficiency of absorption chillers is improving at these days. Overall, it can be hoped that these networks will become wider and more commercial with the advancement of heat transfer technology.

- Natural gas energy networks
Many consumers in the world are significantly dependent on natural gas energy. This dependence has led to severe effects on society's economy and security.

Also, a large part of electrical and thermal energy generation is still dependent on natural gas. The main challenges of these networks are production constraints in some cold seasons and limitations in the operation of the networks. Also, changes in the operation points of these networks have occurred with a delay, and its effects on other energy networks are inevitable.

On the other hand, these networks have affected severe interactions from the power grid due to electrical impact on compressors with electric drivers or other electrical equipment. Investigating the interactions of electricity and gas networks is one of the most critical issues that has been considered in recent years. However, the study of these networks and the creation of integrated energy management will effectively reduce challenges and implement better operation and decision making.

As mentioned above, natural gas is the main source of thermal generation in the world. It is also most flexible and affordable and main option for electricity generation. More than 3 million kilometers of natural gas pipelines have been installed in the world. The United States, Russia, and Iran, due to the vastness of the region, have the highest amount. Gas networks are divided into three main zones, production (gas field to pre-processing plants), transmission (from pre-processing plants to distribution systems), and distribution network.

Planning and operation of transmission lines are essential to achieve proper availability and reliability. Every 120–160 km in main pipelines, large stations are needed for gas compression. Optimal compression and consumption reduction with limited pressure constraints summarize the main operation for simultaneous monitoring to ensure economical and safe operation.

In various works, optimization methods for the operation of fixed current gas pipelines have been proposed for minimizing power consumption, include the installed compressor number and discharge pressure in stations. Future researches aims to increase compression stations' performance and reliability. For example, a three-generator application (cooling, heating, and combined power) is designed to achieve 30% initial energy saving when meet the station's heating, cooling, and electricity requirements.

4.3 Infrastructure of Modern Multi-Dimensional Energy

Multi-dimensional networks are created from different parts, including local/wide area networks and interfaces. It can imagine by inputs, outputs, and internal connections between components and sectors. Generations are the inputs for integrated energy systems, and all demands make outputs for the infrastructure. Some forms of energy considered input to this set would meet the energy demand

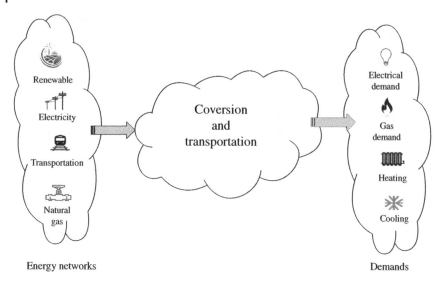

Energy networks   Demands

Figure 4.1 Structure of multi-dimensional energy network.

on the output side. Figure 4.1 provides an overview of the energy system infrastructure and sources.

It should be noted that energy conversion may occur within the network, and one type of energy may be converted to another. Therefore, electrical energy does not necessarily meet the electricity demand and may provide a part of the required cooling by converting it centrally. Increasing the efficiency and integration of energy conversions and adjusting based on the actual demand on output will be significant. In other words, this concept is designed for the integrated use of different types of energy to meet a wide range of demands. It optimizes the whole set by optimizing energy paths and conversions.

- Energy Sources
  Energy Generation in modern energy systems is extended in the wide-area network to achieve a high level of stability and reliability. It is generally in the form of electricity to provide constraints such as controllability, online measurement capability, and continuous monitoring. Some energy sources, such as wind and solar, have a severe uncertain nature. It causes energy generation changes at different times. Also, uncertainty in the unit commitment of other generators like gas or hydro. It is necessary to make more accurate predictions of the next state of the sources situation. Given above, the controllability of the generation is essential in the network structure.

As mentioned, electricity sources play a leading role in the energy network, but natural gas should not be overlooked. The usage of these resources creates some problems in controlling and monitoring gas resources, which can be attributed to delays in change and the need for high safety. On the other hand, electrical generation by gas sources makes a major dependency between two independent electricity and gas networks. Security and other operating parameters are also affected by this concept [10]. It is an issue that has been addressed in recent researches.

- Energy Consumers
  In contrast to energy sources, consumers play an essential role in an energy grid. Consumer limitations and constraints determine the type of final energy deliverable. If the final energy form is easy to use, the consumer challenges will be less and more economical. Conversely, the consumer may need to convert the delivered energy, increasing costs.

- Energy Converters
  After generating from different sources, transfer or convert to other forms is considered. Conversion is used for optimal use in an energy system or more effortless energy transfer. As mentioned, network input can include a variety of energies. Some of these types can be transferred to the consumers without conversion. Electric machines, gas turbines, heat exchangers, inverters, transformers, and other similar equipment perform energy conversions in the grid. Although most converters change energy, several energy carriers need to convert to a specific type or other. In this way, different energy demands can be met.

- Energy Storages
  Storages in an extensive energy system are a convenient solution to meet the current or over-demands at the different situations of grid operation. An optimized managed storage can store excess energy when the generation is more than consumption. These devices, which are available in different types, are also helpful in reducing transmission losses and optimal consumption. Thermal, electrical, cooling, gas, and hydrogen are different types of storage that are used depending on the application and location in the network. Electric vehicles can also play a storage role in the network. Figure 4.2 provides a list of conventional storage devices in energy networks.

- Energy Hub
  A research team introduced the concept of the energy hub and its initial frameworks at the University of Zurich. They were working on a future energy grid project. The concept was defined as an interface between energy sources, storage, transmitters, and consumers. The critical point in defining an energy hub is to connect the different types of energy [11]. Figure 4.3 shows the principal model of the energy hub.

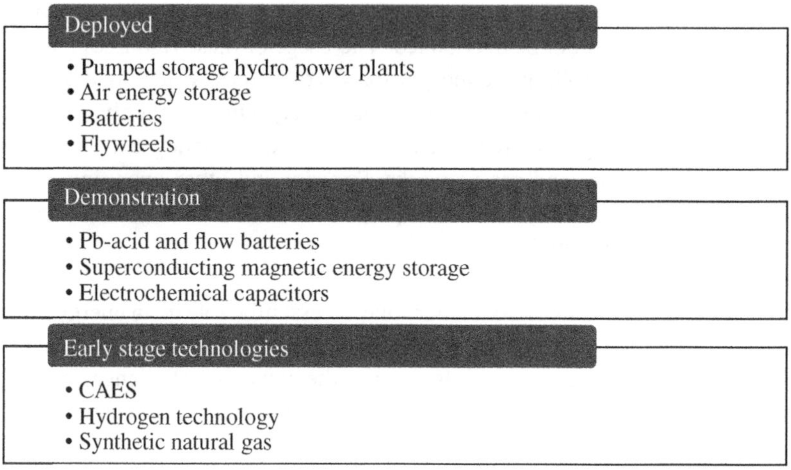

Deployed

- Pumped storage hydro power plants
- Air energy storage
- Batteries
- Flywheels

Demonstration

- Pb-acid and flow batteries
- Superconducting magnetic energy storage
- Electrochemical capacitors

Early stage technologies

- CAES
- Hydrogen technology
- Synthetic natural gas

Figure 4.2 Energy storage technologies.

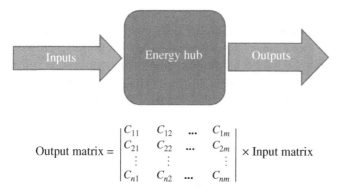

$$\text{Output matrix} = \begin{vmatrix} C_{11} & C_{12} & \dots & C_{1m} \\ C_{21} & C_{22} & \dots & C_{2m} \\ \vdots & \vdots & & \vdots \\ C_{n1} & C_{n2} & \dots & C_{nm} \end{vmatrix} \times \text{Input matrix}$$

Figure 4.3 Modeling of energy transformation.

4.4 Modeling Review

With the increasing penetration of low carbon multi-energy technologies, CHP, heat pumps, gas networks, multi-carrier systems, and energy hubs received more attention [11]. For this reason, many studies have been conducted in the field of modeling new concepts with an integrated energy systems approach [12]. Other research was done on system smartening and intelligent operation [13]. The proposed models became more competitive than previous models by increasing the accuracy of calculations and comprehensiveness in the results [14].

Interactions between gas, electricity, and heating systems are considered as a modern integrated multi-energy network that are shown in Figure 4.4. The

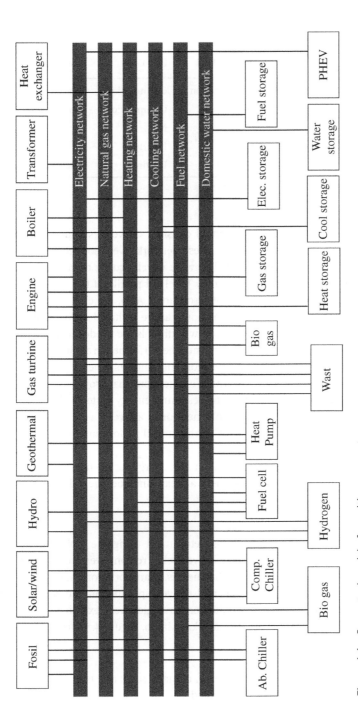

Figure 4.4 Conceptual model of a multi-energy system.

Table 4.1 Energy software applications.

	RETScreen	**EnergyPLAN**	**DER-CAM**	**Transport**
Operation	Yes	Optimization	Optimization	No
Planning	Yes	No	Optimization	Optimization
Network analyze	Yes	No	No	Optimization
Resolution	Monthly	Hourly	Hourly	Hourly
Time scale	Annual	Annual	Lifetime	Lifetime

Source: Mancarella [15]/with permission of Elsevier.

proposed model for these interactions should consider converters and evaluate their effects on energy exchanges. Initially, these networks were connected separately by an equation, but integrated models were developed by more complex equations such as Newton-Raphson methods. The load profile of the integrated network can also be calculated in different resolutions. Also, the energy generation, transfer conditions, and consumption should all be included in the model.

Due to the complexity of energy systems, powerful tools are needed to simulate the network and its components. Simulations must be performed in all dimensions of efficiency, environmental, technical, and economic impacts. Energy-Plan, Ret-Screen, DER-CAM, and E-transport are typical applications that provide operators and researchers with tools for exploiting or designing energy systems. Table 4.1 provides the application of the various tools mentioned above.

Using suitable methods or indicators for correct model performance evaluation is crucial.

Bellow items review the research conducted using energy supply models used in different parts of the world. These models at least include the following in summary:

- Markal model
  In 2003, the domestic energy price was determined using the global multi-regional model of Markal. In this model, five global regions were considered. Consumers include industry, homes, businesses, transportation, and so on. Manufacturers are also heat generators and network production and distribution chains. Then, carbon emission reduction strategies were examined using three models: Markal, Markal-ED, and Markal-Macro. Figure 4.5 shows a typical Markal-Macro block diagram.
- TIMES model
  In this model, the role of fission energies in climatic scenarios is investigated. Figure 4.6 shows the generalities of this model.

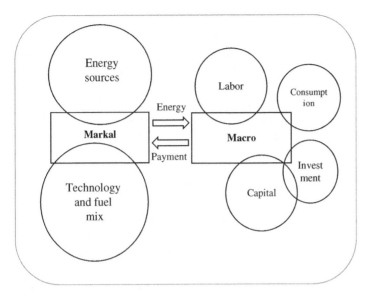

Figure 4.5 Markal-Macro model.

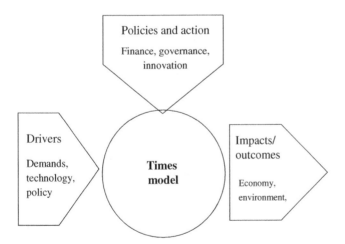

Figure 4.6 Times model.

- EFOM model

  This model was developed in 2003 for Italy renewable energy planning. The objective function in this model includes the total cost of energy conversion over the time under study. Various constraints can also be considered in this model. Finally, energy efficiency was studied with this model.

- WASP model

  Wind energy potential was studied in a single model in 2011 [16]. The relationship between energy and weather conditions was examined from WASP software. The amount of produced energy was also calculated by this model.

- JASP model

  In 2004, the generation planning model was developed. This study defined energy production planning as a high-level decision-making problem. Power plants investment and operational plans were also defined as low-level decisions.

- MESSAGE model

  Between 1999 and 2000, researchers linked an energy supply model to a macroeconomic model and solved it. The reference energy system consisted of primary and alternative sources. Energy exchanges, emissions, electrical and non-electrical demand, and related economic needs were studied. Finally, this model was developed and implemented in 2010.

- RETScreen model

  In 2007, renewable energy production and its economic impact were examined using RETScreen software. Initial studies had three main components: estimation of energy production potential, economic assessments, and emission assessment of greenhouse gases.

- LEAP model

  In 2003, a reduction in greenhouse gas emissions from biomass technology was investigated using this model. In this model, different scenarios can be considered. This model has also been used to study energy security in Asia.

- ENPEP model

  In 1997, the impact of different energy diversification strategies on the supply system was investigated. Gas exchanges and effects on demand forecasting were also studied. Integrated energy grids were also studied in 2008 with ENPP software, and market operation simulations were also studied. In various studies with this model, energy planning based on different emission and price scenarios over many years has been investigated.

- MESAP model

  In 2002, an energy supply and demand model was developed. Its purpose was to develop the electricity industry in the energy network. Fuel and energy supply simulation was performed with the MESAP model. This model categorized four groups: industrial, household, transportation, and other consumers. This method became the basis for developing a sustainable energy vision model in 2009. This scenario aims to reduce greenhouse gas emissions by 2050 [17]. Also, additional scenarios were simulated in a global energy system model in 10 regions in the MESAP environment.

- NEMS model
  In 2007, the impact of renewable energy portfolio production standards on the US energy market was examined. Different scenarios with current situations were simulated in the NEMZ model, and the results were examined.
- ENERGY2020 model
  California Environmental Protection Agency updated the economic analysis of the weather conditions in the Energy Vision Program in 2010. In this study, using the 2020 energy model, the rate of reduction of carbon emissions, changes in the cost of energy carriers, investment made, and deviations from previous plans were examined. The project thoroughly examined the assumptions and input data required by all departments in developing the reference scenario for the various government departments and the data sources required. The potential effects of proposed US policies were also examined.

4.5 Integrated Energy Management System

In many cases, each energy network type may operate independently and affect the other. Most obvious and example is the influence of e natural gas network on electricity generation. Some operations in the network occur with a delay but will lead to immediate effects in another network. On the other hand, sometimes, the type of exploitation depending on network situation can have a double impact on another network.

Examining different operating conditions, including transient situations or accidents, requires accurate modeling and knowledge of conditions and parameters extraction. One of the most critical issues discussed in integrated energy systems is the simultaneous peak of two or more networks. This situation will reduce the capacity of production or transmission and increase the probability of a trip of part or the whole energy network.

Considering all these cases, creating an integrated management and operation structure for monitoring and decision-making in different situations is one of the future requirements of complex energy networks. World-class energy infrastructures included of system operators, technical and research institutes, and the private sector with advanced operation, forecasting, management, gas networks, buildings and complexes, transportation system, industrials, and interfaces. These systems may include support for:

- State-of-the-art forecasting
- Parameter's regulation
- Grid codes

- Transmission and distribution system design
- Balancing area regulation
- Storage systems
- Smart technologies

Widespread integrated energy planning and operation are carried out via a holistic approach for large-scale and local infrastructures. It includes all energy carriers (electricity, gas, blended, hydrogen, CO_2, heating, district energy, and liquid fuels) and would cover a more extended timeframe than the ten years required by the network planning. Different parameters of each network and its changes affect the other network. The constraints that depend on the whole network or different networks under study should be considered. Therefore, the objective function may take a more complex form in each case and have many constraints. It can make optimization difficult.

For this reason, one of the appropriate methods for operation management is to use model-free optimization or intelligent methods such as machine learning, deep learning, and neural networks [18]. However, due to the multiplicity of conditions and number of data generated increasing, the users of these networks face challenges in decisions or forecasting. Mega companies like Siemens, ABB, General Electric, Hyundai, and Mapna Group offer various infrastructures and intelligent management solutions for structure and equipment. These capabilities have also been accelerated by investing in ongoing research and studies. Construction of energy generators, converters, storage devices, and required software are among the activities carried out and developed with significant investments. It is predicted that in the future, with equipment dimension and cost reduction, the possibility of a broader investment in these technologies will be more than possible.

As mentioned, integrated energy networks used effectively new modern tools and methods. Big data, Internet of Things technology, blockchain, smart contracts, and other new concepts are practical and innovative tools in energy. However, the final target is to improve and meet supply and demand. Sub-targets are optimized consumption, reduced losses, reduced carbon emissions, and saving money. In general, integrated energy management methods and systems are significant issues in current and future network design.

Managing wide-area resources is one of the problems that energy systems have faced [19]. Due to the diversity of these sources in different types and uncertainties, much data needed to be analyzed. In this regard, blockchain can solve this problem correctly [20]. This concept can widely manage the required resource parameters to achieve the appropriate network values. For example, a direct load control framework based on the blockchain can establish a good relationship between the generation and the distribution companies. This relationship makes it possibly better to manage peak load or sudden changes in the network. The network interaction management development is also enhanced by using smart contract models that can work on a blockchain basis. In these interactions, energy and financial exchanges will be monitored in the context of a contract model provisions automatically enforced. These exchanges may be done peer-to-peer or multi-faceted [21]. Smart contracts are binding, and changes are challenging to make [22]. The information required to monitor the dimensions of the contract is also received from the field using the Internet of Things and stored and analyzed in an integrated management center. Due to the high volume of exchanged data, new data mining and data analysis methods will be widely used in this field. Checking the accuracy of the data and deleting the data with noise is also done in this process. However, the deep penetration of new technologies in energy is significant. Developing these principles is in line with better management of wide energy networks and solving existing challenges.

As energy systems become more sophisticated and more involved players, the issue of maintaining cyber security becomes even more critical [23]. For example, integrating wireless systems and control equipment into communication systems increases accessibility from outside the system. Also, due to the sharing of subscribers' data in the network, third parties' possibility of information retrieval increases. It can lead to sabotage or criminal operations. So, with each development, a new challenge arises. Although network security will be a challenge for its operators, intrusion monitoring systems and smart tools have also been developed in this field. Algorithms that prevent cyber intrusion or detect it are also widely used. However, the use of complex communication networks always increases the malicious cyber challenge that must be managed.

4.6 Energy Conversion

It is important to be able to convert energy from one type to another. The most common converters using for natural gas to heat and power, electricity to heating and cooling, and heat into cooling conversion. Some of these exchanges aim to convert the excess electricity of renewables into gas or even other types. In this way, additional electrical energy is predicted or delayed at different time intervals depending on the solution adopted. The more significant number of energy conversion units (input and output energy vectors and performance curves) leads to the more significant overall flexibility of the integrated energy system. It is due to the more significant number of independent variables that make it possible to achieve a more efficient set and improve the overall objective function.

Below are the different types of these conversions.

- Gas to thermal, cooling, and electricity.
- Electricity to gas.

- Electricity to heat.
- Electricity-to-commodities.

The main application of these transformations will be in multi-dimensional energy networks and future wide area networks. In these networks, energy savings, cost reduction of energy generation, transmission and conversion, and storage will be of particular importance. In macro-policies, moving towards a cleaner environment and zero emissions is also presented as an ideal.

4.7 Economic and Environmental Impact

One-dimensional energy systems were more prevalent in the past, and retail companies played an essential role as intermediaries between energy producers and consumers. It made of more challenging to manage the energy market and its effects on other issues such as the economy or the environment [24]. The island operation of intermediary companies also complicated the process. By creating integrated energy networks such as hubs, macro-management and network operators at different levels will assume this role. Therefore, increasing the system's profitability was one of the operators' goals.

On the other hand, creating new revenues such as presence in the carbon market [25], increasing the impact on other systems according to the declared requirement, exporting the required services to off-grid sets, and using other financial instruments were considered. On the other hand, there were some contradictions. For example, increasing energy efficiency in a multi-dimensional energy network is considered. However, sometimes, consumers may be encouraged to consume more to increase the profitability of intermediaries. These issues and the requirements for economic exploitation of energy networks led to much research in the new field of integrated Energy-Economy-Environment (3E). Figure 4.7 shows a schematic of it. In this model, integrated energy networks will be developed and operated with a full view of climate change while maintaining the profitability of players in the energy field. This issue leads to the sustainable development of countries in energy and all related fields.

Numerous studies have examined the interaction of economics, energy, and the environment. Some of these studies are regional and specific to a country or region, and some are universal. In 2011, it made a tripartite connection to China. This study showed that a stable balance is created between energy consumption, economic growth, and a clean environment in the long run. With the system out of balance, greenhouse gas emissions and energy consumption will be less than the economy's growth. Therefore, creating this balance seems necessary. The Kuznets environmental hypothesis was studied in different countries and regions in later

Figure 4.7 3E model.

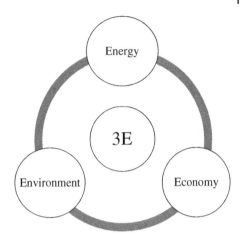

years. This model examines the cause-and-effect relationship between energy consumption, carbon emissions, economic growth, free trade, and urbanization. In a study conducted in Iran, the causal relationship between energy consumption, national income and carbon emissions, and other factors such as human resources and capital in the economy of Iran has been studied. No causal relationship was found between revenue and carbon emissions in this study. However, a one-way relationship was found between national income and energy consumption. In another study that examined the effect of economic growth, urban population, and incomes on carbon emissions, it was found that greenhouse gas emissions are a function of increasing the Gini coefficient.

However, all these studies indicate that the need for sustainable development in each region is to create a balance in the 3E model and monitor the environmental and economic effects of energy exchanges.

4.8 Future Energy Systems

The multi-energy systems discussed earlier face significant challenges in optimal managing and planning in large-scale and whole energy vectors. To meet all demand types, energy conversions must be adopted and modeled with installed energy transmission and distribution networks. Energy supply chain must be simulated for all primary energy types and/or optimizing related economy problems while maintaining the system's stability and meet all integrity constraints. Objective functions are defined such as energy balance accounting, and characteristic energy conversion curves are introduced as network constraints. The balance

Table 4.2 Energy balancing and duration.

	Seconds	Minutes	Hours	Days	Weeks	Seasons
Battery	MA	MA	MA	NA	NA	NA
Hydro pumped storage Power plant	LA	MA	MA	NA	NA	NA
Load management	LA	MA	MA	LA	NA	NA
Dams	LA	MA	MA	MA	MA	MA
Carbon capture	NA	LA	LA	MA	MA	MA
Gas reservoir	NA	NA	NA	LA	LA	MA
Re-scheduling	LA	MA	MA	MA	NA	NA

LA: less applicable, MA: most applicable, NA: not applicable.

and constraints of the power grid must also be formalized to system stabilizing and prevent uncertain conditions like overload in other network types. In addition for management optimization, there may be required additional restrictions on international policies, electricity and gas prices, and financial necessities [9]. Long-term storage combination helps to solve some significant nonlinear problems without focusing on a single step. These problems are often nonlinear because the performance curves of complex energy system and conversions with regulatory and financial constraints have a nonlinear time dependence nature. In addition, future fuel cost uncertainties play an essential role in any attempt to identify the optimal solution. Table 4.2 provides the energy balancing management in the network and the applications of each resource in a specified time period.

It also faces many energy challenges in the future. Figure 4.8 shows some of these challenges. There will be many solutions to deal with them. In this regard, many research topics have been defined in this regard.

According to some studies and reports, one of the main policies in the field of energy is the development of renewables and reducing carbon emissions. This sustenance will cause extensive changes in the field of energy generation and consumption. With increasing economic growth of countries and consumption, faster declines in energy intensity relative to history are a critical factor in mitigating in carbon emission. If energy intensity improved at the same rate as in the past 20 years, CO_2 emissions by the next 30 years would be more than 50 times. However, macro-planning in the field of energy structures and management of greenhouse gas emissions and global warming will be the main challenges in future

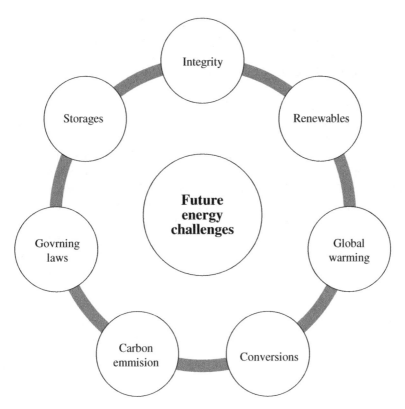

Figure 4.8 Future energy challenges.

energy networks. Most countries are developing short-term and long-term plans in this direction, and huge investments are being made [26].

4.9 Conclusion

Evolution of Multi-Energy Systems for Completion of de-carbonation goals is a highly multidisciplinary research topic that needs to define a common concept for achieving a main solution. Effective research processes are used and show opportunities to increase results at all scales. Analyzes show that electricity, heat, and gas networks have different natures and challenges despite different similarities. Multi-energy modeling should focus on methods that effectively address transient problems and accuracy. It is one of the main subjects for future research and practically technical investment for industries and complexes. Some research lines also emphasize that interconnected energy infrastructures make it possible to increase productivity by energy consumption reduction and pollutant/carbon emission.

Different research lines studies showed that future networks must be analyzes in multi-energy models and mentioned to understand the effects of dynamics on the sudden conversion of an energy form to another. However, energy issues will be one of the most important topics discussed and challenged in future engineering studies. Moving toward non-diffusion networks, optimizing energy consumption, and reducing the challenges in energy systems is one of the main important goals of multi-dimensional energy networks, which was discussed generally in this chapter. However, the challenges in modeling and prediction can be solved by smartening and using new methods. But there is still research competition in providing indicators and computational algorithms for managing complex energy systems. This chapter tries to provide a broad view of multi-dimensional energy networks by examining the generalities of the subject and to review its challenges and benefits based on recent researches. The study conducted in this chapter has been done in the direction of power and heat intercorporate networks and has provided an overview of these modern infrastructures for further studies.

References

1 Treude, S. (2022). *European Economic Policy and the European Green Deal: An Institutionalist Analysis*, Wissenschaftliche Schriften des Fachbereichs Wirtschaftswissenschaften. Econstor.eu.

2 Breyer, C. (2020). A global overview of future energy. *Future Energy* 727–756.

3 He, J., Yuan, Z., Yang, X. et al. (2020). Reliability modeling and evaluation of urban multi-energy systems: a review of the state of the art and future challenges. *IEEE Access* 8: 98887–98909.

4 Bao, J., He, D., Luo, M., and Choo, K.-K.R. (2021). A survey of blockchain applications in the energy sector. *IEEE Systems Journal* 15 (3): 3370–3381. https://doi.org/10.1109/JSYST.2020.2998791.

5 Daneshvar, M., Asadi, S., and Mohammadi-Ivatloo, B. (2021). *Grid Modernization – Future Energy Network Infrastructure: Overview, Uncertainties, Modelling, Optimization, and Analysis.*

6 Gurusinghe, D.R. and Rajapakse, A.D. (2015). Post-disturbance transient stability status prediction using synchrophasor measurements. *IEEE Transactions on Power Systems* 31 (5): 3656–3664.

7 Farrokhifar, M., Nie, Y., and Pozo, D. (2020). Energy systems planning: a survey on models for integrated power and natural gas networks coordination. *Applied Energy* 262: 114567.

8 Sardou, I.G., Khodayar, M.E., and Ameli, M.T. (2016). Coordinated operation of natural gas and electricity networks with microgrid aggregators. *IEEE Transactions on Smart Grid* 9 (1): 199–210.

9 Guelpa, E., Bischi, A., Verda, V. et al. (2019). Towards future infrastructures for sustainable multi-energy systems: a review. *Energy* 184: 2–21.

10 Ameli, H., Qadrdan, M., and Strbac, G. (2020). Coordinated operation of gas and electricity systems for flexibility study. *Frontiers in Energy Research* 8: 120.

11 Liu, T., Zhang, D., Dai, H., and Wu, T. (2019). Intelligent modeling and optimization for smart energy hub. *IEEE Transactions on Industrial Electronics* 66 (12): 9898–9908.

12 Wang, D., Liu, L., Jia, H. et al. (2018). Review of key problems related to integrated energy distribution systems. *CSEE Journal of Power and Energy Systems* 4 (2): 130–145.

13 Sechilariu, M. (2020). *Intelligent Energy Management of Electrical Power Systems*, vol. 10, 2951. Multidisciplinary Digital Publishing Institute.

14 Hilpert, S., Günther, S., Kaldemeyer, C. (2017). *Addressing Energy System Modelling Challenges: The Contribution of the Open Energy Modelling Framework (oemof)*. Preprints. https://doi.org/10.20944/preprints201702.0055.v1

15 Mancarella, P. (2014). MES (multi-energy systems): an overview of concepts and evaluation models. *Energy* 65: 1–17.

16 Mortensen, N.G., Heathfield, D.N., Rathmann, O., and Nielsen, M. (2011). *Wind Atlas Analysis and Application Program: WAsP 10 Help Facility*. Roskilde: DTU Wind Energy.

17 Tsiropoulos, I., Nijs, W., Tarvydas, D., and Ruiz, P. (2020). *Towards Net-Zero Emissions in the EU Energy System by 2050, Insights from Scenarios in Line with the 2030 and 2050 Ambitions of the European Green Deal*. Luxembourg Publications Office of the European Union.

18 Musbah, H., Aly, H.H., and Little, T.A. (2021). Energy management of hybrid energy system sources based on machine learning classification algorithms. *Electric Power Systems Research* 199: 107436.

19 Sinsel, S.R., Riemke, R.L., and Hoffmann, V.H. (2020). Challenges and solution technologies for the integration of variable renewable energy sources—a review. *Renewable Energy* 145: 2271–2285.

20 Brilliantova, V. and Thurner, T.W. (2019). Blockchain and the future of energy. *Technology in Society* 57: 38–45.

21 Alt, R. and Wende, E. (2020). Blockchain technology in energy markets–an interview with the European energy exchange. *Electronic Markets* 30 (2): 325–330.

22 Hahn, A., Singh, R., Liu, C.-C., and Chen, S. (2017). Smart contract-based campus demonstration of decentralized transactive energy auctions. *2017 IEEE Power & Energy Society Innovative Smart Grid Technologies Conference (ISGT)*. IEEE, pp. 1–5. https://doi.org/10.1109/ISGT.2017.8086092.

23 Tvaronavičienė, M., Plėta, T., Casa, S., and Latvys, J. (2020). Cyber security management of critical energy infrastructure in national cybersecurity strategies: cases of USA, UK, France, Estonia and Lithuania. *Insights into Regional Development* 2 (4): 802–813.

24 Wei, X., Qiu, R., Liang, Y. et al. (2022). Roadmap to carbon emissions neutral industrial parks: energy, economic and environmental analysis. *Energy* 238 (Part A): 121732. ISSN 0360-5442.

25 Lovcha, Y., Perez-Laborda, A., and Sikora, I. (2022). The determinants of CO_2 prices in the EU emission trading system. *Applied Energy* 305: 117903.

26 BP (2020). Energy outlook report 2020 edition. British Petrolium.

5

Overview of the Optimal Operation of Heat and Electricity Incorporated Networks

Sobhan Dorahaki[1], Sahar Mobasheri[1], Seyedeh Soudabeh Zadsar[1], Masoud Rashidinejad[1], and Mohammad Reza Salehizadeh[2]

[1] Department of Electrical Engineering, Shahid Bahonar University of Kerman, Kerman, Iran
[2] Department of Electrical Engineering, Marvdasht Branch, Islamic Azad University, Marvdasht, Iran

Abbreviations

CAES	Compressed air energy storage
CHP	Combined heat and power
DER	Distributed energy resources
DSM	Demand side management
DRPs	Demand response programs
EEA	European Environment Agency
EH	Energy Hub
EHMS	Energy Hub Management System
EENS	Expected energy not served
IC	Internal combustion
IEA	International Energy Association
IGDT	Information gap decision theory
LDC	Local distribution company
LOLE	Loss of load expectation
LOLP	Loss of load probability
MES	Multi-energy system
OILM	Optimal industrial load management
PHEV	Plug-in hybrid electric vehicles

Coordinated Operation and Planning of Modern Heat and Electricity Incorporated Networks,
First Edition. Edited by Mohammadreza Daneshvar, Behnam Mohammadi-Ivatloo, and Kazem Zare.
© 2023 The Institute of Electrical and Electronics Engineers, Inc.
Published 2023 by John Wiley & Sons, Inc.

P2H   Power to hydrogen
P2G   Power-to-gas
P2P   Peer to peer
SNG   Synthetic natural gas

5.1 Introduction

The greenhouse gas emissions, as well as the dramatic growth of energy consumption, are the main global concerns. Based on the estimations, the global electricity consumption in 2035 will be increased by 70% in comparison to 2010 [1]. These challenges encourage system operators to set environmental-friendly operation regulations and integrate different forms of energy sources so, a new concept known as "Energy Hub" (EH) was born. An EH is a multi-carrier energy system involving various energy conversion, generation, and storage, that is a promising option for integrated management. A variety of energy carriers such as electricity, heat, synthetic natural gas (SNG), wood chip, hydrogen, and renewable energy sources like solar energy, biomass, and wind power are included in the EH. Moreover, various technologies such as combined heat and power (CHP), boiler, chiller, heat storage, solar heat storage, batteries, compressed air energy storage (CAES), pumped hydro plants, steam/gas turbine, power-to-gas (P2G), fuel cell, and electric power storage can be used in EH. EHs could be enlarged on a variety of spatial scales, from the amount of a building to a bigger geographic area. Different types of applications can be imaged, e.g virtual power plants, industrial factories, big buildings, limited geographical areas, and isolated power systems. Energy conversions between different energy carriers in an EH, when merged with energy storage, increase flexibility in energy provision.

The utilization of an EH has many advantages, like reliability and flexibility improvement, optimization potential, synergistic effects, cost reduction, and emissions reduction. To sum up, the challenges are examined from technical, economic, social, and also environmental points of view. Energy management has a special place in EH owing to the important role of end-users. Indeed, demand side management (DSM) includes concepts like energy efficiency, energy-saving, and demand response programs (DRPs).

This chapter focuses on an overview of the optimal operation of combined electricity and heat networks. In the first part, the EH system is proposed as a solution for the integration of heat and electrical systems. In the second part, the multi-energy carriers of the EH system are discussed, then benefits, challenges, opportunities, and the role of DSM programs are expressed. In the next part, management methods of EH systems like the self-scheduling approach, bilateral energy trading, centralized and decentralized approaches are presented, and finally, the conclusion is summarized.

5.2 Integration of Electrical and Heat Energy Systems: The EH Solution

The electrical and heat are the main energy forms from the end-user as well as the system operator's point of view. The metropolis cities are facing environmental problems, so the water and space heating devices are superseded by greener heat resources such as electric boilers and heat pumps. Furthermore, centralized heat energy suppliers such as district heating systems are becoming popular in the world. The produced heat energy by the district heating system with a higher degree of efficiency is distributed through a pipeline network in the city, where this is an appropriate environment for implementing integrated central electrical and heat energy systems. The integration of electricity and heat distribution systems provides an extra degree of flexibility to the system operation and improves the energy productivity level.

In this regard, the EH as a key concept can play an important role in energy production, conversion, and storage. This concept opens a new window in the optimal operation of integrated energy systems by considering various primary and secondary energy forms. Given the interdependence across different physical infrastructures, significant efforts by researchers have been devoted to the investigation of modeling, planning, and operation of multi-energy systems (MESs) with EHs. Studies show that the operation cost of the integrated energy system is significantly lower than the separated ones [2]. Moreover, [3] claims that the integrated energy system is much greener than the separated energy systems. Besides the aforementioned advantages, the EH can provide a reliable and flexible environment for electrical and heat loads [4]. In this regard, the operation and planning issues of the EH system are very important aspects of energy system studies, where the operational aspects of the EHs are addressed by the current chapter.

5.3 Energy Carriers and Elements of EH

Today, owing to the development of natural gas and electricity grids, the use of multiple energy carriers in the management and operation of energy systems has been increased. An EH is a promising option for integrated management of the MES. In this regard, the development of distributed energy resources

(DER), in specific, multi-generation systems and renewable energy sources change the concept of energy production from large centralized power plants to distributed and local generation in the future. An MES ensures the efficiency and reliability of energy supply by continuously providing energy from numerous sources.

The EH is an energy conversion and storage system that converts input energy variables $(I_1, I_2, ..., I_n)$ into output energy variables $(L_1, L_2, ..., L_n)$ for various energy types in an optimal manner [2]. The conversion matrix is shown in Figure 5.1, which uses the coupling matrix to communicate various energy carriers at the input and output. Indeed, each inner element of the matrix contains the connection, the internal components, the efficiency of the converter, and the transformation coefficients.

These carriers act together using hub elements such as storage facilities, conversion, and connectors [5]. Figure 5.2 shows the components and elements, which consist of four general parts like electrical, heat, water, and gas.

Energy carriers can contain natural gas, electricity, heat, wood chips, hydrogen, seawater, SNG, drinking water, and DERs. Moreover, different technologies like CHP, fuel cell, boiler, chillers, photovoltaic, wind turbine, heat storage, hybrid electric vehicles, batteries (e.g. Li-ion battery), pumped hydro plants, and steam/gas turbine can be used. In recent years, conversion facilities and modern energy storage systems have played a key role in achieving different customer demands by providing an optimal connection between gas and electricity grids. For this reason, the use of advanced technologies, for example, P2G systems, water desalination units, hydrogen energy, batteries, pumped-storage hydroelectricity, solar heat storage, and tree-state CAES systems are required. Table 5.1 lists several studies that take technology and energy into account [3, 6].

The common energy conversion and storage technologies in the EH system are explained as follows:

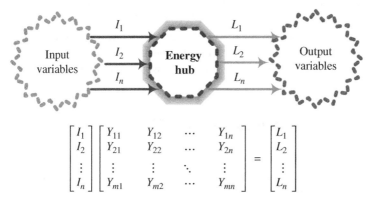

$$\begin{bmatrix} I_1 \\ I_2 \\ \vdots \\ I_n \end{bmatrix} \begin{bmatrix} Y_{11} & Y_{12} & \cdots & Y_{1n} \\ Y_{21} & Y_{22} & \cdots & Y_{2n} \\ \vdots & \vdots & \ddots & \vdots \\ Y_{m1} & Y_{m2} & \cdots & Y_{mn} \end{bmatrix} = \begin{bmatrix} L_1 \\ L_2 \\ \vdots \\ L_n \end{bmatrix}$$

Figure 5.1 The input and output matrix of the energy hub.

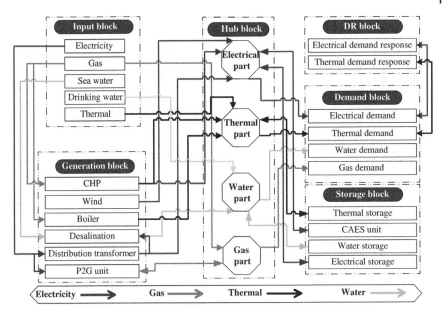

Figure 5.2 Energy hub system.

Table 5.1 Summarizes of the published energy hub papers.

	Component				Energy forms				
References	CHP	Solar	Wind	Elec	NG	Heat	H$_2$	Storage	Year
[7]	✗	✗	✗	✓	✓	✓	✗	✗	2008
[8]	✓	✓	✓	✓	✓	✓	✗	✗	2008
[9]	✓	✓	✗	✓	✓	✓	✗	✗	2009
[10]	✓	✓	✗	✓	✓	✓	✗	✗	2009
[11]	✓	✗	✗	✓	✓	✓	✗	✓	2009
[12]	✓	✗	✗	✓	✓	✓	✗	✓	2009
[13]	✓	✗	✗	✓	✓	✓	✗	✓	2009
[14]	✗	✗	✗	✓	✓	✓	✗	✓	2010
[3]	✗	✗	✓	✓	✓	✓	✗	✓	2011
[15]	✓	✓	✗	✓	✓	✓	✗	✗	2011
[16]	✓	✗	✗	✓	✓	✓	✗	✓	2012
[17]	✗	✗	✗	✓	✓	✓	✗	✓	2012
[18]	✓	✓	✓	✓	✓	✓	✗	✓	2013
[19]	✓	✗	✗	✓	✓	✓	✗	✓	2014

(Continued)

Table 5.1 (Continued)

References	Component				Energy forms			Storage	Year
	CHP	Solar	Wind	Elec	NG	Heat	H₂		
[20]	✓	✗	✓	✓	✓	✗	✗	✓	2014
[21]	✓	✗	✗	✓	✓	✓	✗	✗	2014
[22]	✗	✓	✓	✓	✗	✗	✗	✓	2015
[23]	✓	✗	✗	✓	✓	✓	✗	✓	2015
[24]	✓	✗	✗	✓	✓	✓	✗	✓	2015
[25]	✓	✗	✓	✓	✓	✓	✗	✓	2016
[26]	✓	✓	✓	✓	✓	✓	✗	✓	2016
[27]	✓	✗	✓	✓	✓	✗	✗	✓	2016
[28]	✗	✓	✓	✓	✗	✗	✓	✓	2017
[29]	✓	✓	✓	✓	✗	✓	✗	✓	2020
[30]	✓	✗	✗	✓	✓	✓	✗	✓	2021
[31]	✓	✓	✓	✓	✓	✓	✗	✓	2021
[32]	✗	✓	✗	✓	✓	✓	✗	✓	2022

5.3.1 Combined Heat and Power Technology

In recent decades, with the rapid expansion of distributed energy supply systems, developing energy efficiency and reducing emissions have been made possible by the use of CHP. The primary role of a CHP system is the combination of electricity and heat systems to achieve higher flexibility in the incorporated energy systems.

The CHP system has complex technologies that can use kinds of fuels to generate power, allowing the heat that would normally be lost in the power generation process to be used in making steam and boiling water. Figure 5.3 shows the elements and technology of CHP.

The CHP unit consists of several discrete elements:

- Reciprocating internal combustion (IC) engines
- Fuel cells
- Microturbines
- Steam turbines
- Combustion turbines

According to the report made by US Environmental Protection Agency, reciprocating IC engines make up the largest share of CHP systems over the US region [33].

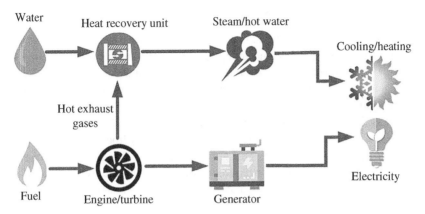

Water   Heat recovery unit   Steam/hot water   Cooling/heating

Hot exhaust gases

Electricity

Fuel   Engine/turbine   Generator

Figure 5.3 CHP technology. *Source:* Joachim/Adobe Stock and fandy/Adobe Stock.

Moreover, the benefits from CHP deployment in the electric industry, electricity customers, and society include [34]:

- Increased energy efficiency
- Economic development value
- Reduced emissions of criteria air pollutants (SO_2, NOx, and Hg)
- Resource adequacy
- Critical electrical and thermal loads can be empowered during grid power outages.
- The use of CHP units as local generation units can decrease the possibility of congestion in the power and heat networks.

Besides the significant advantages of the CHP unit, the challenges are:

- Natural gas and electricity prices can have a significant impact on the economic efficiency of CHP systems.
- In comparison to an IC engine of the same size, the particular power output is low.
- The high cost of capital is a major factor limiting growth.

Referring to the aforementioned advantage aspects, the CHP units are numerously addressed in the EH studies. For example, in [35], the role of the CHP unit in an EH system incorporating various energy inputs like electrical, heat, hydrogen, and gas are modeled. In [36, 37], natural gas is used as input energy, and the gas turbine generates electricity while the heat generated as waste heat is supplied as heating energy through a heat recovery system. An optimal combination of the electric heat pump, CHP, and different temperatures in the district heating network was studied by [38]. In [39], the economic impact of using a micro-CHP system with an electric boiler at the home level is investigated. In [40], a new model was adopted to promote the efficiency of the CHP system using the Kalina cycle process.

5.3.2 Power to Gas Technology

Natural gas plays an essential role in the transition to a green economy, constituting a natural alternative to coal and acting as a reserve resource to the intermittent nature of DERs. Power and gas energies can be structurally connected by P2G technologies because of their interaction to improve energy productivity [41]. P2G is a technology under development in the field of energy management in smart EH systems. Figure 5.4 illustrates the elements and technology of P2G.

P2G uses renewable or surplus electricity to produce hydrogen through water electrolysis ($2H_2O \longrightarrow 2H_2 + O_2$). The growing number of power to hydrogen (P2H) pilot plants producing hydrogen from a variety of renewable energy sources shows that P2H has a lot of potential as an energy storage system. P2H can also participate in facilities for power grids due to the quick resilience of hydrogen electrolyzers. Supplying extra services creates increased income, which can be used to enhance the P2H value proposition as a storage device. Storage capacity and flexibility to manage variability in renewables could be reduced by hydrogen-based products. Moreover, the flexibility of the hydrogen gas turbine and electrolyzer allows the EH to provide ancillary services to the power networks.

Recently, hydrogen as a new energy carrier in the EH system is considered. The utilization of hydrogen in the energy systems decreases natural gas consumption, especially by gasfired power plants and emissions [42]. It can be used directly as a final energy vector or converted to electricity, synthetic gas, methane, liquid fuels, or chemicals [43]. Moreover, hydrogen can serve to store energy for several months [44] while the major usage is related to uncertainties associated with renewable energy sources [45]. Reference [46] explores how P2G units can influence optimal operational strategies of integrated gas and electricity networks. In this context, it

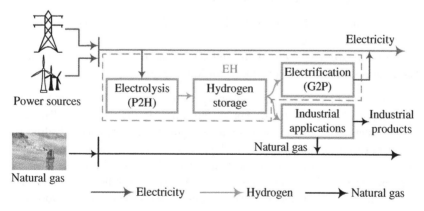

Figure 5.4 Power to gas technology. *Source:* caracterdesign/E+/Getty Images, engel.ac/ Adobe Stock, and pichitstocker/Adobe Stock.

is pointed out the benefits of using the down-regulated wind to produce hydrogen for blending to the natural gas at the terminal, as well as the SNG production that can be achieved with little disruption to gas grid operations in terms of gas flows when the P2G are placed close to gas grid terminals.

On the other hand, hydrogen storage is used to capture the wind peaks that occur during P2G production and to reduce conversion losses. In [42], a security-constrained time-based energy management system is proposed to investigate the effects of P2H units in the presence of large-scale renewable energy sources. The provided method indicates the P2H demand flexibility characterized like reactive power support, electrolyzer behavior, and hydrogen storage capability.

The key challenges of the P2G technology that need to be investigated are:

- System balancing technology requires establishing an adequate policy framework.
- Due to the lack of particular data, it's not possible to get reliable results.
- Case study analyses (involving social and economic analyses for implementation of P2G system) are required.

5.3.3 Compressed Air Energy Storage Technology

The CAES technologies are one of the efficient types of electrical storage systems. Remarkable benefits of the CAES technologies are (i) unlike pumped hydroelectric storage, it does not need a specific location for installation, (ii) storing a large amount of energy, (iii) high rate of flexibility, (iv) agility-wise (quick response to possible changes in pressure times), (v) lower investment cost compared to the pumped storage plants, and (vi) works in three modes including charging, discharging, and simple cycle [47].

Figure 5.5 shows the operation process of the CAES unit. Researchers have recently been able to store excess energy by compressing air from electricity generated using renewable energy (or other energy resources). The CAES unit is an efficient tool for energy arbitrage in smart energy systems. In this regard, [48] investigates the role of the CAES unit in the restructured electricity markets associated with electricity price uncertainties.

Although CAES can play a significant part in the future power system, the lack of operational CAES facilities that certain challenges must be overcome to make CAES a viable ES technology. Low efficiency, unavailable shelf turbomachinery for large-scale CAES, and high initial capital investment prices are major challenges. Furthermore, financial incentives and tax credits are required for supplying large energy management network services.

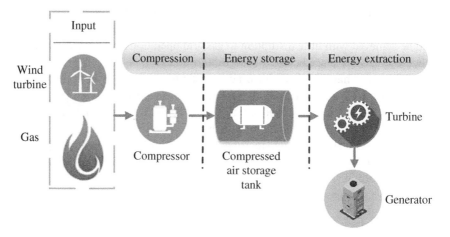

Figure 5.5 CAES technology.

5.3.4 Water Desalination Unit

DERs have a high penetration rate in the EH system, so it is a rational choice for the system operator of the energy systems to use the electrical desalination unit at low price hours. The produced freshwater by the desalination unit is stored in the water storage unit to import in the high-water demand hours. Therefore, considering the desalination unit and the water storage unit in the smart EH system increases the operational efficiency of the integrated MES. A water desalination unit is shown in Figure 5.6.

The water desalination unit is generally categorized by the technology used and the type of feed water. The reverse osmosis desalination was first applied at the end of the 1960s [49]. With further developments, its application was extended to seawater desalination and became an important participant in the desalination market in the 1980s.

Heat desalination is a kind of phase changer desalination that mimics natural water. The water is vaporized into steam and then condensed by the application of heat energy. Therefore, vapor and return phase transitions occur [50]. Reference [51] develops the role of the water desalination unit in the EH. They propose a new smart EH structure, which is made up of different sections including the smart transportation systems consisting of electric vehicles and the subway. Reference [52] proposes EH consisting of the wind turbine, photovoltaic and CHP, energy storage systems (heat energy storage system, ice storage conditioner, and solar-powered CAES), and seawater desalination with reverse osmosis technology as a flexible load.

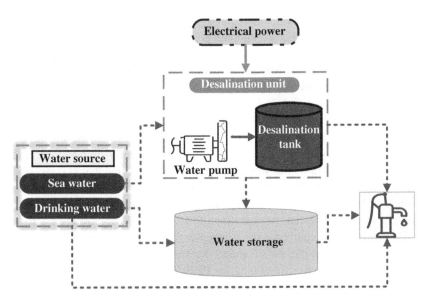

Figure 5.6 Water desalination technology.

The basic challenges of desalination technology are drought, climate change, the high planned cost of desalination devices and facilities, and the lack of research and development for improving this technology.

5.3.5 Plug-in Hybrid Electric Vehicles

The plug-in hybrid electric vehicles (PHEV) are the best solution to replace fossil fuels with clean energy in the smart EH system. Statistics indicate that in 2010 only about 17000 PHEVs are used in the world, and this number dramatically increased to 7.2 million in 2019 [53]. This willingness to use the PHEVs indicates the important role of decarbonization issues from the people's perspective. The results of the European Environment Agency (EEA) studies show that if the share of PHEVs increases to 80% by 2050, Europe's electricity consumption would probably face only about a 10% increase [54].

PHEVs get electrical power from an EH or grid and store it, also the vehicles can be used in a vehicle-to-grid or grid-to-vehicle model. When PHEVs are connected to an EH, the batteries can play a role as a storage system that is compiled. Looking at the issue from a positive privilege, this increases the potential for combining variable renewable energy sources in power grids [55]. When a power outage occurs, hybrid storage technologies allow the grid operator to rapidly access the grid and prevent a frequency drop in the grid.

Some papers describe the modeling and optimization research performed on PHEVs, charging infrastructure and integration into the power system using EHs. A novel framework for the optimal operation of an EH that includes renewable energy sources, PHEVs, fuel cell vehicles, a fuel cell, an electrolyzer, a hydrogen tank, a boiler, an inverter, a rectifier, and a heat storage system is presented [55]. Uncertainty modeling in risk-seeking and risk-averse strategies is studied by information gap decision theory (IGDT).

References [56, 57], a flexible modeling technique for PHEV based on a multi-energy carrier approach is presented in EH. Different PHEV architectures and energy management systems while driving, during additional grid-connected usage modes, and regulation services of the power grid can be easily modeled using the suggested structure. Reference [58] planned the parking lot as a storage system where the results show that the operation of EHs is very flexible and has a manageable load when the purchase price is higher than the upstream grid. Furthermore, [59] presents a residential EH model that includes a CHP and PHEV and can determine optimal charge scheduling for PHEVs at home using the EH model based on time-differentiated electricity prices.

The major challenges in using batteries for electrical storage are to make them both affordable and long. But there are many disadvantages of hybrid storage equipment such as high cost, complex maintenance, lack of standardization, low power, low efficiencies, and safety.

5.4 Advantages of the EH System

The integrated EH has advantage potentials compared to separated energy systems which are expressed as follows:

5.4.1 Reliability Improvement

The reliability of the energy system is the most critical aspect of energy satisfaction from the consumer perspective. The use of DRPs, energy storage, and DERs improve the reliability indices like expected energy not served (EENS), loss of load expectation (LOLE), and loss of load probability (LOLP) [4]. The EH provides an efficient platform for the mentioned high-tech energy conversions and generation resources. Moreover, the energy conversion in the EH system can make a more reliable energy system than the separated EH operation manner. Some novel researches address the reliability issues in the EH system. In [60] the multi-objective optimal optimization of EH considering reliability and risk indices is proposed. Also, the reliability of generation, conversion, and storage units in the EH has been imposed based on the size of components in [61]. The reliability of the EH due to the N−1 event caused by the outage is discussed in [62].

5.4.2   Flexibility Improvement

A definition for flexibility by the International Energy Association (IEA) mentions that "Power system flexibility is the ability of a power system to reliably and cost-effectively manage the variability and uncertainty of demand and supply across all relevant timescales" [63]. Referring to the above flexibility definition, the operation of DERs with high uncertainty is risky for the flexibility of the EH system. Nowadays, the penetration rate of DERs in the energy system is increasing. The uncertain nature of DERs and the dynamic nature of the loads can violate the flexibility of the energy systems. Therefore, the fast ramp resources play an important role in energy systems to cover the flexibility problems.

DRPs and energy storage are known as very important fast ramp resources which have a high penetration rate in the EH systems. Moreover, the use of a fast real-time local energy management system can improve the flexibility of the EH. Therefore, considering the flexibility index in the EH operation is a very challenging and important issue.

In this regard, [4] proposes a flexible-reliable optimization model for transactive networked EHs. Furthermore, a mathematical flexibility index for the networked EHs has been provided to investigate the role of DER, electrical energy storage, and DRPs on the flexibility of the system. The results of the mentioned study verify the positive impacts of electrical energy storage and DRPs on the flexibility level of the EH.

5.4.3   Operation Cost Reduction

The EH system provides an appropriate platform for local renewable energy generation systems such as wind and photovoltaic. The mentioned energy resources have a high capital cost with a negligible operation cost. Moreover, the use of high-efficiency generation, conversion, and storage systems in the EH framework decreases the operation cost. The results of [64] show that the EH can improve the operation cost, exergy efficiency, and carbon dioxide emissions by 30%, 28%, and 16%, respectively. In [65], a transactive multi-carrier EH as a fully decentralized system is provided. The alternating direction method of multipliers (ADMM) method is used to convert the centralized system into a decentralized control manner. The results show that the fully decentralized EH system is more cost-effective than the centralized one.

5.4.4   Emissions Mitigation

Nowadays, DERs as cost-competitive assets are well-known in the EH system. The most important dominance point of using green energies over conventional energy resources is the emission mitigation impact [66]. Although some papers such as

[67] climes that all of the renewable energy resources are not entirely "clean." However, based on the IEA reports, renewable energy sources are the most cleanest energy resource for the future of the world [68]. EHs have a high penetration rate of DER. Therefore, an EH can reduce emissions of the energy generation and transmission systems.

In [69], the distribution system with DER is modeled as an interconnected EH system to reduce the emission and operation cost. The IGDT is used to model the uncertainty of the wind power generation system. Also, the objective function is modeled by the MINLP. The results show that the networked EHs can help the system operator to decrease the emission and operation cost of the system.

5.5 Applications of the EH System

The idea of an EH system was raised by an urban utility in Switzerland [70]. This company built an EH system containing wood and methane chip gasification and a system for generating heat and electricity at the same time. The idea was to generate SNG and heat from wood chips, which are an accessible source in the company area. Then, the produced SNG is injected directly into the public natural gas system or converted into electricity through a unit of CHP and injected into the electricity distribution network. The heat dissipation attributed to both can be absorbed by the local heating network, so the whole system can be considered as an EH system [70].

EH systems are not limited in scale, so they can be used in different parts. Due to the many applications of energy systems, the real facilities that can be modeled as EHs are [71]:

➤ Residential buildings
➤ Commercial buildings
➤ Industrial factories
➤ Agricultural sector

5.5.1 Residential Buildings

The residential sector consumes a large share of energy in the power system. The average energy consumption in residential buildings was about 26.1% [72]. As a result, changes must be made in the residential sector to reduce these casualties. Moreover, a large share of energy is wasted due to the long-distance transmission network. An EH system as an appropriate solution to avoid energy loss in a long-distance transmission system is proposed for a residential building as shown in Figure 5.7. The addressed EH system includes a heat pump, a gas boiler, and a heat solar system along with the electricity and natural gas network [73]. As another

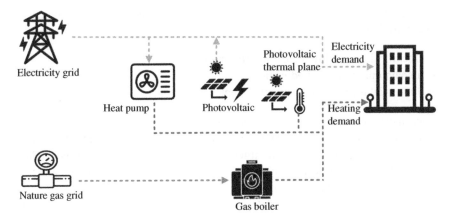

Figure 5.7 Residential energy hub.

residential EH system example, [74] provides an EH in the residential building level, which includes various parts like electric chillers, boilers, heat pumps, solar panels, and solar collectors.

5.5.2 Commercial Buildings

Large commercial buildings have the highest energy consumption, which should be given much attention. Reference [15] provides a combined cooling, heating, and power that examines the results on the costs of operating and emitting greenhouse gases in large commercial buildings. After the EH system is optimally planned, the performance of the hub in commercial buildings must be optimized. In order to reduce energy consumption and reduce environmental impact on commercial buildings, we must increase productivity in these buildings [75]. There are several ways to improve commercial buildings energy efficiency as follows [76]:

- Technological ways
- Behavioral education of residents
- Building energy management system.

5.5.3 Industrial Factories

The industrial section is the important energy consumption section. Also, the greenhouse gas emission in the industrial section is more than in other energy sections [71]. According to [72], 54% of the world's electricity is consumed in the industrial sector, with an average annual energy consumption of 1.2% from 2012 to 2040.

In general, the following can be done to improve productivity in the industry, which are [76]:

➤ Technical and technological improvements
➤ Behavioral policy and education
➤ Industrial energy management system

As an example of an industrial EH system, Figure 5.8 shows a proposed Energy Hub Management System (EHMS) for an industrial customer integrated with a distribution network as part of an intelligent network [77]. The proposed EHMS includes two essential sectors: the first sector is the central hub controller, and the second sector is an optimization part.

The market operator collects data such as energy prices, greenhouse gas emissions, and DR incentive signals. Afterward, in the smart grids, real-time information systems and communication infrastructures allow the central controller to access this information.

The central hub controller has received data from the industrial processes and their operating preferences from an industrial process operator, exchanging information such as aggregated load profiles with a local distribution control (LDC) system operator for proper coordination with the other components. The optimization part includes the desired industrial load management model which optimal schedules yield for the industrial processes to be relayed to the processes by the central hub controller for optimal dispatch purposes.

The optimal industrial load management (OILM) model uses available distribution system information and control expected in smart grids, considering load control and voltage optimization approaches together. The mentioned industrial process in Figure 5.8 depends on some variables such as applied voltage, material flow rates, and the other variables to modulate the customer's demand for the benefits of both the customers and the grid.

5.5.4 Agricultural Sector

The energy consumption of the agricultural sector has less amount than other sectors. However, in recent years some new challenges such as the lack of efficient technological assets and the growth of popularity are caused to increase the energy consumption by the agricultural sector. To increase the operational efficiency in the agricultural sector some potential challenges are [78]:

➤ Energy efficiency and consumption
➤ Costs imposed on agricultural products
➤ Emission impacts

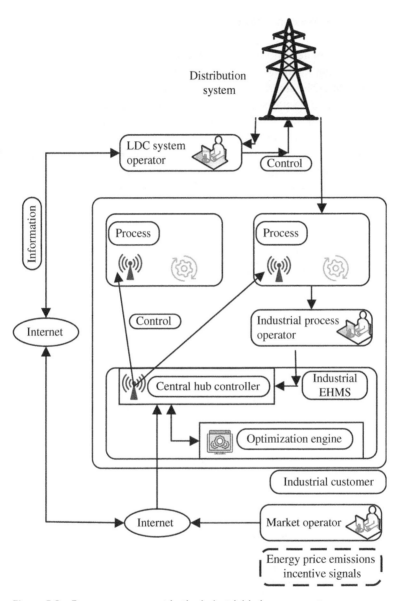

Figure 5.8 Energy management in the industrial hub energy system.

The share of each renewable resource in the agricultural sector is shown in Figure 5.9.

Renewable energy in agriculture includes the sun, geothermal, biomass, wind, and hydropower. Solar energy is used to generate heat, cold, electricity, and dry agricultural products. The smart farm uses geothermal energy to heat the greenhouse, heat the soil, and dry the crops. Biomass is used to supply energy in the form of agricultural waste. In traditional water pumps and mills, wind power is used. Hydropower is used to pump turbines for energy [71].

In [79], a fuzzy method is used to study farms in terms of energy, environmental impact, and cost. Due to the high potential of renewable energy for electricity generation, the agricultural sector still has the largest share of electricity generation

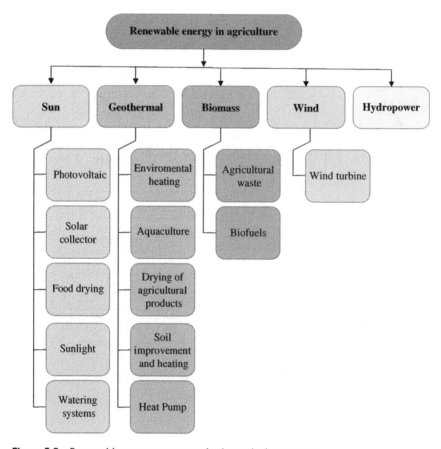

Figure 5.9 Renewable energy resources in the agriculture sector.

related to fossil fuels [71]. As a result, due to the high potential of the agricultural sector in the integration of renewable energy, there is a wide ability to operate the EH system in this sector.

5.6 Challenges and Opportunities

The investigation of challenges and opportunities of the multi-carrier energy systems is a very important subject to achieve a comprehensive understanding of the EH system. This section addresses the challenges and opportunities of EH systems from the technical, economic, social, and environmental points of view.

5.6.1 Technical Point of View

The novel energy revolution by considering various carriers, not simply electricity, represents a chance to enhance the system and can have a revolutionary impact on future electrical networks [80]. The optimization of the integrated EH system has significant technical advantages such as reliability [81] and flexibility [82] improvement. However, each energy carrier in the EH systems is susceptible to uncertainties [83]. For example, sometimes the input gas pressure of the EH system has uncertainty so, referring to the key role of the gas energy carrier, the operation of all of the EH system assets are faced with uncertainty [84]. Moreover, renewable energy resources and demands are another uncertainty source in the EH system. Therefore, the need for advanced communication and controller platforms to ensure an effectively coordinated operation in an uncertain environment is very important.

Referring to the comprehensive energy conversions, the operational flexibility of the EH systems is greater than separated electrical, heat, and gas frameworks [85]. However, operating the integrated energy system has some complicated aspects which should be considered by the system operators as a major technical challenge.

5.6.2 Economic Point of View

The EH operator has a main and key role to achieve the most optimal solution from the economic perspective [85]. In this regard, the EH operator should participate in the different energy markets such as heat, water, electrical, and gas markets, while price uncertainty is a well-known feature of the energy markets [86]. Therefore, this is a significant well-known potential challenge for EH systems from an economic perspective. To overcome the mentioned challenge, the system operator models the uncertainty of the market price by various uncertainty modeling approaches such as robust optimization, scenario-based approaches, IGDT, chance constraints method, and fuzzy method. In [87], the energy demand, market price and the output power of the renewable energy resources are modeled by the scenario-based uncertainty modeling approach. Also, in [88] the market price uncertainty is modeled by the robust approach in the EH system.

Considering various energy carriers in the EH systems provides an appropriate potential for decreasing the operation cost and risks. However, the economic modeling of the integrated energy system is very complicated than the separated one. Therefore, the mentioned multi-carrier integration can be raised as an economic challenge.

Furthermore, EH systems can be linked to each other and constitute a transactive networked EH framework. The operational and economic flexibility of the transactive networked EH system is greater than the self-scheduling optimization manner. However, the economic interactions between EH operators have an important challenge potential. A decentralized control approach is used to model the interconnected EHs in [89]. This paper uses the ADMM to model the networked EH system. In [90], the centralized multi-objective optimization approach is used to address the networked energy and water hubs. In [91], the game theory approach is used to illustrate the economic interaction between EH system operators in the transactive networked EH system.

Some novel papers show that the EH system can be considered as an appropriate framework for small energy communities. The economical interactions between small energy community operators as well as the EH small energy community operators and system operators are caused to increase the complexity of the EH framework. In [92], the bilevel optimization approach is used to indicate economic interactions between the EH operator and the market operator.

5.6.3 Environment Point of View

The emission mitigation issue is the most important preference of modern smart cities citizens, while the environmental protection and energy security problems attract much attention global with the gradual depletion of non-renewable energy and deterioration of the environment [93]. Current energy systems are operated independently, which will lead to high operation cost, low robustness, and low energy efficiency.

The EH provides an appropriate structure for green DER such as photovoltaic and wind turbine systems. However, the use of renewable energy resources such as wind turbines has some challenges like challenges to wildlife and habitat and land use issues.

Also, demand-side management programs in the EH systems are useful for emission mitigation issues. In this regard, the Department of Energy reports that the active participation of consumers, through peak demand reduction or demand shifting, could be an important solution for future decarbonized energy systems.

5.6.4 Social Point of View

Nowadays, the penetration rate of consumer-centric problems like the DRP and energy efficiency programs in the energy systems has been increased. The social behavior of end-users in the EH system has significant impacts on the output operation variables of the system. The utility function is the main basis of end-user's decision-making process. The utility function introduces how a rational EH consumer would make consumption decisions. This function models the risk-averse and risk-taker behaviors of end-users in the EH system. The utility function is a non-linear and non-convex function that is obtained by the psychological tests. Referring to the psychological studies each end-user has a different utility function where the determination of the end-user's parameters of the utility function is a significant challenge. In [94], two models for DRPs are provided based on the well-known Cobb-Douglas utility function. The provided method in [95] maximizes the individual customer's welfare based on the utility function theory under incentive-based DRPs.

5.7 The Role of DSM Programs in the EH System

End-users are the heart of future energy transition from the current conventional energy system to future energy community-based energy systems. The DSM programs as prosumer-centric actions play an important role in the mentioned energy transition. The DSM programs consist of DRP and energy efficiency programs which can work complimentary in the demand side of EH systems. In the following, the role of DSM programs in the EH system is investigated.

5.7.1 Demand Response Programs

DRPs effectively increase the technical, economic, environmental, and social roles of end-users within the energy sector. This issue facilitates the energy transition from the current conventional energy system to the community energy concept [96]. From the grid operator point of view, DRPs can postpone the investment cost of the generation [97] and transmission [98] expansion problems. Based on the previous studies, the DRPs are categorized into Incentive-Based Programs (IBP) and Time Based Rates (TBR) that shown in Figure 5.10 [99].

Moreover, the electrical, heat, and cold DRP can be considered in the EH systems [2]. In [30], the role of electrical DRPs in the served load maximization problem is

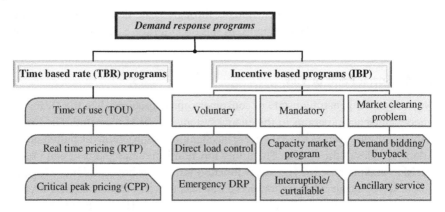

Figure 5.10 Demand response program.

studied. Some paper investigates the impacts of electrical DRPs on the operation cost of EHs such as [90, 100–102]. Results of Ref. [4] show that electrical DRPs can increase the reliability of the EH system. Furthermore, the electrical DRPs can be an effective solution for flexibility improvement in the EH system [103].

5.7.2 Energy Efficiency Programs

The energy efficiency programs have many advantages for an EH system as an energy community such as operation cost reduction, emission mitigation, and job creation solutions. From an economic point of view, improving energy efficiency is caused to stabilize electricity prices and volatility. Moreover, the loss cost of the system is decreased by the energy efficiency programs. Also, the energy efficiency programs decrease the operational risk of the EH system subjected to resource uncertainties. From the EH system operator perspective, the energy efficiency programs have long advantages by lowering overall electricity demand, therefore reducing the requirement to speculate in new electricity generation and transmission infrastructure.

The modeling of multi-carrier energy systems is the most important step in evaluating energy integration systems. The variable energy efficiencies of the storage components and energy conversion in the EH system cause nonlinearity to the structure. In this regard, [104] provides a standardized matrix method to model the energy efficiency of complicated multi-carrier energy systems. Furthermore, consider constant efficiency coefficients for energy storage and convertors in the EH modeling have a high potential for erroneous predictions. In this regard, [66] models the efficiency of EH components by the non-linear part-load efficiency curves.

5.8 Management Methods of the EH System

The smart grids consist of several transactive networked EHs. In this regard, the multi-carriers EH systems can be modeled and managed in different manners. Each approach has some advantages and disadvantages that should be investigated to achieve a practical and comprehensive approach. Figure 5.11 shows the comparison of various multi-hub control structures, which are described as follows:

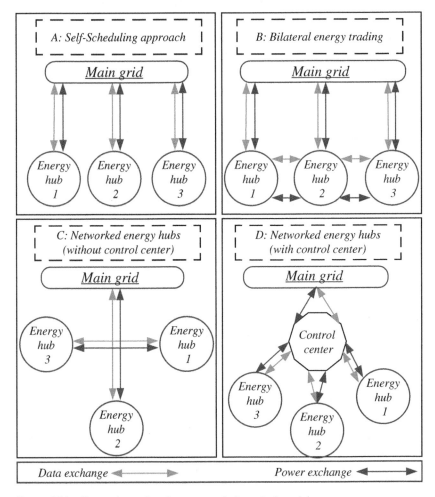

Figure 5.11 Comparison of various energy hub control models.

(A): Self-Scheduling approach: In this approach, EHs only trade with the upstream heat, electrical, gas, and water grid [31, 100, 105]. This approach does not need a very complicated communication infrastructure. However, this approach has low levels of flexibility and reliability.

(B): Bilateral energy trading: In this approach, each EH can buy/sell needed/extra energy from/to the upstream grid or paired EH [106, 107]. The Peer to Peer (P2P) energy trading between EHs can be classified into bilateral energy trading management classes [32, 108]. The possibility of P2P energy trading between the EHs provides a new opportunity for EH system operators to trade electricity in a secure and least-cost model. Therefore, the clients can participate in a P2P energy market and benefit from this possibility to decrease their energy cost by satisfying their demands from other EHs and increasing their benefits by trading their surplus power to other hubs while avoiding paying extra tax or grid service costs [65]. The communication infrastructure of approach *B* is more complicated than approach *A* but not pervasive between all of the EHs. The optimal solution of this approach for each EH is not a globally optimal solution, but it is more efficient than approach *A*.

(C): Networked EHs (Without a central controller): In this approach, EH operators can independently and directly negotiate with each other to decide on the energy trading problem without any centralized controller [109, 110]. This approach needs a more complex communication infrastructure between all EHs and also has a better optimal solution than approach *B*.

(D): Networked EHs (With a central controller): A community-based transactive networked EHs can readily be applied to community neighboring EHs, in which the members of the networked EHs have a common goal even though they are not at the same location. Usually, the members of the networked EHs work collaboratively. In a community-based EHs market, each participant generally trades its energy within the community through a central controller. Thus, the central controller manages the networked EHs to minimize the operation cost of the system (collaboratively). The obtained solution of this approach is much more efficient than other approaches.

5.9 Conclusion

This chapter presents an overview of the integrated multi-carrier energy systems as the well-known concept namely EH. In this regard, the energy carriers, components, and programs of the EH system are investigated in this chapter. Based on the conducted literature review, novel high technologies such as P2G and CAES are very effective tools for cost and emission reduction. Moreover, electric vehicles are mobile energy storage that can be useful in the operational flexibility of smart EH systems. Also, the challenges and opportunities of the EH system from the system operator and consumer's point of view are evaluated. From the technical point of view, the integrated generation, conversion, and storage system have some operational complexity such as sensing, operating, and managing processes that should be considered in the operational studies. Moreover, the economical interactions between the system operator and various energy markets associated with uncertainties can be stated as an important challenge in the smart EHs. To overcome the mentioned challenge, various uncertainty modeling approaches like robust optimization, scenario-based approaches, and IGDT are used. Moreover, the DMS programs such as DRPs and energy efficiency programs are effective tools to eliminate the negative effects of prices and resources uncertainties in the EH systems. Referring to the consumer-based programs like DRPs and energy efficiency programs in the EH system, it can be mentioned that the end-user has a key role in the operation of the EH system. Therefore, the end-user behaviors can change the operational variables, so social-perspective studies have an important station in the future EH system. Indeed, the social and behavioral studies in the EH framework have great research potential for future work.

References

1 Nozari, M.H., Yaghoubi, M., Jafarpur, K., and Mansoori, G.A. (2022). Development of dynamic energy storage hub concept: a comprehensive literature review of multi storage systems. *J. Energy Storage* 48: 103972. https://doi.org/10.1016/j.est.2022.103972.

2 Mohammadi, M., Noorollahi, Y., Mohammadi-ivatloo, B., and Yousefi, H. (2017). Energy hub: from a model to a concept – a review. *Renew. Sust. Energ. Rev.* 80: 1512–1527. https://doi.org/10.1016/j.rser.2017.07.030.

3 Maroufmashat, A., Taqvi, S.T., Miragha, A. et al. (2019). Modeling and optimization of energy hubs: a comprehensive review. *Invent* 4: 50. https://doi.org/10.3390/INVENTIONS4030050.

4 Dini, A., Hassankashi, A., Pirouzi, S. et al. (2022). A flexible-reliable operation optimization model of the networked energy hubs with distributed generations, energy storage systems and demand response. *Energy* 239: 121923. https://doi.org/10.1016/j.energy.2021.121923.

5 Geidl, M., Koeppel, G., and Favre-Perrod, P. (2007). The energy hub – a powerful concept for future energy systems. *3rd Annual Carnegie Mellon Conference on the Electricity Industry*, Pittsburgh, PA, USA (13–14 March 2007). Electrical & Computer Engineering, pp. 13–14. https://www.research-collection.ethz.ch/handle/20.500.11850/3133.

6 Divya, K.C. and Østergaard, J. (2009). Battery energy storage technology for power systems – an overview. *Electr. Power Syst. Res.* 79 (4): 511–520. https://doi.org/10.1016/j.epsr.2008.09.017.

7 Arnold, M., Negenborn, R.R., Andersson, G. and De Schutter, B. (2008). Distributed control applied to combined electricity and natural gas infrastructures. *2008 1st International Conference on Infrastructure Systems and Services: Building Networks for a Brighter Future (INFRA)*, Rotterdam, Netherlands (10–12 November 2008), vol. 2008. IEEE. doi: 10.1109/INFRA.2008.5439653.

8 Adamek, F. (2008). Optimal multi energy supply for regions with increasing use of renewable resources. *2008 IEEE Energy 2030 Conference, ENERGY*, Atlanta, GA, USA (17–18 November 2008). IEEE. doi: 10.1109/ENERGY.2008.4781045.

9 Carradore, L. and Turri, R. (2009). Modeling and simulation of multi-vector energy systems. *2009 IEEE Bucharest PowerTech Innovative Ideas Toward the Electrical Grid of the Future*, Bucharest, Romania (28 June–2 July 2009). IEEE. doi: 10.1109/PTC.2009.5281933.

10 Ramirez-Elizondo, L.M. and Paap, G.C. (2009). Unit commitment in multiple energy carrier systems. *41st North American Power Symposium*, Starkville, MS, USA (4–6 October 2009). IEEE. doi: 10.1109/NAPS.2009.5484065.

11 Arnold, M., Negenborn, R.R., Andersson, G., and De Schutter, B. (2009). Model-based predictive control applied to multi-carrier energy systems. *2009 IEEE Power & Energy Society General Meeting*, pp. 1–8. doi: 10.1109/PES.2009.5275230.

12 Barsali, S., Poli, D., Scalari, S. et al. (2009). Integration of process-side energy storage and active distribution networks: technical and economical optimisation. IET Conference Publications, no. 550 CP, Prague, Czech Republic (8–11 June 2009). IEEE. doi: 10.1049/CP.2009.0760.

13 Carradore, L. and Turri, R. (2009). Optimal co-ordinated operation of distributed multi-generation in active distribution networks. *2009 44th International Universities Power Engineering Conference (UPEC)*, Glasgow, UK (1–4 September 2009). IEEE, pp. 1–5.

14 Arnold, M., Negenborn, R.R., Andersson, G., and De Schutter, B. (2010). Distributed predictive control for energy hub coordination in coupled electricity and gas networks. *Intell. Infrastructures* 235–273. https://doi.org/10.1007/978-90-481-3598-1_10.

15 Sheikhi, A., Ranjbar, A.M., and Safe, F. (2011). Optimal dispatch of a multiple energy carrier system equipped with a CCHP. *Renew. Energy Power Qual. J.* 1413–1418. https://doi.org/10.24084/repqj09.675.

16 Akgun, E. and Cakmakci, M. (2012). Development of a supervisory controller for residential energy management problems. *Proceeding of the American Control Conference*, Montreal, QC, Canada (27–29 June 2012). IEEE, pp. 1482–1487. doi: 10.1109/ACC.2012.6315097.

17 Galus, M.D., Waraich, R.A., Noembrini, F. et al. (2012). Integrating power systems, transport systems and vehicle technology for electric mobility impact assessment and efficient control. *IEEE Trans. Smart Grid.* 3 (2): 934–949. https://doi.org/10.1109/TSG.2012.2190628.

18 Pazouki, S., Haghifam, M., and Olamaei, J. (2013). Effect of wind turbine, solar cells and storages in short term operation of coupled electricity and gas infrastructures in different climates. *Int. J. Smart Electr. Eng.* 2 (3): 159–165.

19 Yu, D., Lian, B., Dunn, R., and Le, S. (2014). Using control methods to model energy hub systems. *Proceeding of Universities Power Engineering Conference*, Cluj-Napoca, Romania (2–5 September 2014). IEEE. doi: 10.1109/UPEC.2014.6934602.

20 Soroudi, A., Mohammadi-Ivatloo, B., and Rabiee, A. (2014). Energy hub management with intermittent wind power. *Green Energy Technol.* 413–438. https://doi.org/10.1007/978-981-4585-30-9_16.

21 Kampouropoulos, K., Andrade, F., Sala, E., and Romeral, L. (2014). Optimal control of energy hub systems by use of SQP algorithm and energy prediction. *IECON 2014 – 40th Annual Conference of the IEEE Industrial Electronics Society*, Dallas, TX, USA (29 October–1 November 2014). IEEE, pp. 221–227, doi: 10.1109/IECON.2014.7048503.

22 Moazeni, S., Powell, W.B., and Hajimiragha, A.H. (2015). Mean-conditional value-at-risk optimal energy storage operation in the presence of transaction costs. *IEEE Trans. Power Syst.* 30 (3): 1222–1232. https://doi.org/10.1109/TPWRS.2014.2341642.

23 Rastegar, M., Fotuhi-Firuzabad, M., and Lehtonen, M. (2015). Home load management in a residential energy hub. *Electr. Power Syst. Res.* 119: 322–328. https://doi.org/10.1016/J.EPSR.2014.10.011.

24 Salimi, M., Ghasemi, H., Adelpour, M., and Vaez-ZAdeh, S. (2015). Optimal planning of energy hubs in interconnected energy systems: a case study for natural gas and electricity. *IET Gener. Transm. Distrib.* 9 (8): 695–707. https://doi.org/10.1049/IET-GTD.2014.0607.

25 Pazouki, S., Haghifam, M.-R., and Moser, A. (2014). Uncertainty modeling in optimal operation of energy hub in presence of wind, storage and demand response. *Int. J. Electr. Power Energy Syst.* 61: 335–345. https://doi.org/10.1016/j.ijepes.2014.03.038.

26 Skarvelis-Kazakos, S., Papadopoulos, P., Grau Unda, I. et al. (2016). Multiple energy carrier optimisation with intelligent agents. *Appl. Energy* 167: 323–335. https://doi.org/10.1016/j.apenergy.2015.10.130.

27 Hashemi, Z., Ramezani, A., and Moghaddam, M.P. (2016). Energy hub management by using decentralized robust model predictive control. *2016 4th International Conference on Control, Instrumentation, and Automation ICCIA*, Qazvin, Iran (27–28 January 2016). IEEE, pp. 105–110, doi: 10.1109/ICCIAUTOM.2016.7483144.

28 Liu, Y., Gao, S., Zhao, X. et al. (2017). Coordinated operation and control of combined electricity and natural gas systems with thermal storage. *Energies* 10 (7): 917. https://doi.org/10.3390/EN10070917.

29 Dorahaki, S., Dashti, R., and Shaker, H.R. (2020). Optimal energy management in the smart microgrid considering the electrical energy storage system and the demand-side energy efficiency program. *J. Energy Storage* 28: 101229. https://doi.org/10.1016/j.est.2020.101229.

30 Dorahaki, S., Rashidinejad, M., Farahmand, H. et al. (2021). An optimal flexible partitioning of smart distribution system considering electrical and gas infrastructure. *2021 IEEE Madrid PowerTech*, Madrid, Spain (28 June–2 July 2021). IEEE, pp. 1–6, doi: 10.1109/PowerTech46648.2021.9494956.

31 Dorahaki, S., Abdollahi, A., Rashidinejad, M., and Moghbeli, M. (2020). The role of energy storage and demand response as energy democracy policies in the energy productivity of hybrid hub system considering social inconvenience cost. *J. Energy Storage* 33: 102022. https://doi.org/10.1016/j.est.2020.102022.

32 Dorahaki, S., Rashidinejad, M., Fatemi Ardestani, S.F. et al. (2022). A home energy management model considering energy storage and smart flexible appliances: a modified time-driven prospect theory approach. *J. Energy Storage* 48: 104049. https://doi.org/10.1016/j.est.2022.104049.

33 Zhang, J., Cho, H., and Mago, P.J. (2021). Energy conversion systems and energy storage systems. In: *Energy Services Fundamentals and Financing* (ed. D. Borge-Diez and E. Rosales-Asensio), 155–179. Elsevier.

34 U.S. Environmental Protection Agency, C. Heat, and P. Partnership, "*Catalog of CHP Technologies, Section 1. Introduction*," 2015.

35 Hajimiragha, A., Canizares, C., Fowler, M. et al. (2007). Optimal energy flow of integrated energy systems with hydrogen economy considerations. *2007 iREP Symposium – Bulk Power System Dynamics and Control – VII. Revitalizing Operational Reliability*, Charleston, SC, USA (19–24 August 2007). IEEE, doi: 10.1109/IREP.2007.4410517.

36 Martínez-Mares, A. and Fuerte-Esquivel, C.R. (2011). Integrated energy flow analysis in natural gas and electricity coupled systems. *NAPS 2011 – 43rd North American Power Symposium*, Boston, MA, USA (4–6 August 2011). IEEE, doi: 10.1109/NAPS.2011.6024880.

37 Orehounig, K., Evins, R., and Dorer, V. (2015). Integration of decentralized energy systems in neighbourhoods using the energy hub approach. *Appl. Energy* 154: 277–289. https://doi.org/10.1016/J.APENERGY.2015.04.114.

38 Ommen, T., Markussen, W.B., and Elmegaard, B. (2014). Heat pumps in combined heat and power systems. *Energy* 76: 989–1000. https://doi.org/10.1016/J.ENERGY.2014.09.016.

39 Dentice d'Accadia, M., Sasso, M., Sibilio, S., and Vanoli, L. (2003). Micro-combined heat and power in residential and light commercial applications. *Appl. Therm. Eng.* 23 (10): 1247–1259. https://doi.org/10.1016/S1359-4311(03)00030-9.

40 Ogriseck, S. (2009). Integration of Kalina cycle in a combined heat and power plant, a case study. *Appl. Therm. Eng.* 29 (14–15): 2843–2848. https://doi.org/10.1016/J.APPLTHERMALENG.2009.02.006.

41 Salehi, J., Namvar, A., Gazijahani, F.S. et al. (2022). Effect of power-to-gas technology in energy hub optimal operation and gas network congestion reduction. *Energy* 240: 122835. https://doi.org/10.1016/j.energy.2021.122835.

42 Rabiee, A., Keane, A., and Soroudi, A. (2021). Green hydrogen: a new flexibility source for security constrained scheduling of power systems with renewable energies. *Int. J. Hydrog. Energy* 46 (37): 19270–19284. https://doi.org/10.1016/j.ijhydene.2021.03.080.

43 Wulf, C., Linssen, J., and Zapp, P. (2018). Power-to-gas – concepts, demonstration, and prospects. *Hydrog. Supply Chain Des. Deploy. Oper.* 309–345. https://doi.org/10.1016/B978-0-12-811197-0.00009-9.

44 Jiao, J., Chen, C., and Bai, Y. (2020). Is green technology vertical spillovers more significant in mitigating carbon intensity? Evidence from Chinese industries. *J. Clean. Prod.* 257: 120354. https://doi.org/10.1016/J.JCLEPRO.2020.120354.

45 Sadeghi, H., Rashidinejad, M., Moeini-Aghtaie, M., and Abdollahi, A. (2019). The energy hub: an extensive survey on the state-of-the-art. *Appl. Therm. Eng.* 161: 114071. https://doi.org/10.1016/j.applthermaleng.2019.114071.

46 Clegg, S. and Mancarella, P. (2016). Storing renewables in the gas network: modelling of power-to-gas seasonal storage flexibility in low-carbon power systems. *IET Gener. Transm. Distrib.* 10 (3): 566–575. https://doi.org/10.1049/IET-GTD.2015.0439.

47 Aliasghari, P., Zamani-Gargari, M., and Mohammadi-Ivatloo, B. (2018). Look-ahead risk-constrained scheduling of wind power integrated system with compressed air energy storage (CAES) plant. *Energy* 160: 668–677. https://doi.org/10.1016/J.ENERGY.2018.06.215.

48 Cai, W., Mohammaditab, R., Fathi, G. et al. (2019). Optimal bidding and offering strategies of compressed air energy storage: a hybrid robust-stochastic approach. *Renew. Energy* 143: 1–8. https://doi.org/10.1016/J.RENENE.2019.05.008.

49 Van der Bruggen, B. and Vandecasteele, C. (2002). Distillation vs. membrane filtration: overview of process evolutions in seawater desalination. *Desalination* 143 (3): 207–218. https://doi.org/10.1016/S0011-9164(02)00259-X.

50 Elsaid, K., Kamil, M., Sayed, E.T. et al. (2020). Environmental impact of desalination technologies: a review. *Sci. Total Environ.* 748: 141528. https://doi.org/10.1016/j.scitotenv.2020.141528.

51 Mohamed, M.A., Almalaq, A., Mahrous Awwad, E. et al. (2020). An effective energy management approach within a smart Island considering water-energy hub. *IEEE Trans. Ind. Appl.* https://doi.org/10.1109/TIA.2020.3000704.

52 Jalili, M., Sedighizadeh, M., and Sheikhi Fini, A. (2021). Optimal operation of the coastal energy hub considering seawater desalination and compressed air energy storage system. *Therm. Sci. Eng. Prog.* 25. https://doi.org/10.1016/J.TSEP.2021.101020.

53 IEA (2021). *Global EV Outlook 2021*. Paris: IEA https://www.iea.org/reports/global-ev-outlook-2021.

54 Andreas Unterstaller (2022). Electric vehicles: a smart choice for the environment. https://www.eea.europa.eu/articles/electric-vehicles-a-smart.

55 Moghaddas-Tafreshi, S.M., Jafari, M., Mohseni, S., and Kelly, S. (2019). Optimal operation of an energy hub considering the uncertainty associated with the power consumption of plug-in hybrid electric vehicles using information gap decision theory. *Int. J. Electr. Power Energy Syst.* 112 (October 2018): 92–108. https://doi.org/10.1016/j.ijepes.2019.04.040.

56 Galus, M.D. and Andersson, G. (2009). Integration of plug-in hybrid electric vehicles into energy networks. *2009 IEEE Bucharest PowerTech Innovative Ideas Toward the Electrical Grid of the Future*, Bucharest, Romania (28 June–2 July 2009). IEEE, doi: 10.1109/PTC.2009.5282135.

57 Galus, M.D. and Andersson, G. (2009). Power system considerations of plug-in hybrid electric vehicles based on a multi energy carrier model. *2009 IEEE Power & Energy Society General Meeting*, Calgary, AB, Canada (26–30 July 2009). IEEE. doi: 10.1109/PES.2009.5275574.

58 Damavandi, M.Y., Moghaddam, M.P., Haghifam, M.R. et al. (2014). Stochastic modeling of plug-in electric vehicles' parking lot in smart multi-energy system. *IFIP Adv. Inf. Commun. Technol.* 423: 332–342. https://doi.org/10.1007/978-3-642-54734-8_37.

59 Rastegar, M. and Fotuhi-Firuzabad, M. (2014). Optimal charge scheduling of PHEV in a multi-carrier energy home. *2014 14th International Conference on Environment and Electrical Engineering*, pp. 199–203, doi: 10.1109/EEEIC.2014.6835863.

60 Mokaramian, E., Shayeghi, H., Sedaghati, F., and Safari, A. (2021). Four-objective optimal scheduling of energy hub using a novel energy storage, considering reliability and risk indices. *J. Energy Storage* 40: 102731. https://doi.org/10.1016/j.est.2021.102731.

61 Zamani Gargari, M. and Ghaffarpour, R. (2020). Reliability evaluation of multi-carrier energy system with different level of demands under various weather situation. *Energy* 196: 117091. https://doi.org/10.1016/j.energy.2020.117091.

62 Faraji, J., Hashemi-Dezaki, H., and Ketabi, A. (2021). Stochastic operation and scheduling of energy hub considering renewable energy sources' uncertainty and

N–1 contingency. *Sustain. Cities Soc.* 65: 102578. https://doi.org/10.1016/j.scs.2020.102578.

63 Babatunde, O.M., Munda, J.L., and Hamam, Y. (2020). Power system flexibility: a review. *Energy Rep.* 6: 101–106. https://doi.org/10.1016/j.egyr.2019.11.048.

64 Mostafavi Sani, M., Mostafavi Sani, H., Fowler, M. et al. (2022). Optimal energy hub development to supply heating, cooling, electricity and freshwater for a coastal urban area taking into account economic and environmental factors. *Energy* 238: 121743. https://doi.org/10.1016/j.energy.2021.121743.

65 Javadi, M.S., Esmaeel Nezhad, A., Jordehi, A.R. et al. (2022). Transactive energy framework in multi-carrier energy hubs: a fully decentralized model. *Energy* 238: 121717. https://doi.org/10.1016/j.energy.2021.121717.

66 Ahmadisedigh, H. and Gosselin, L. (2022). Combined heating and cooling networks with part-load efficiency curves: optimization based on energy hub concept. *Appl. Energy* 307: 118245. https://doi.org/10.1016/j.apenergy.2021.118245.

67 Abbasi, S. and Abbasi, N. (2000). The likely adverse environmental impacts of renewable energy sources. *Appl. Energy* 65 (1–4): 121–144. https://doi.org/10.1016/S0306-2619(99)00077-X.

68 International Energy Agency (IEA) (2021). Renewables 2021-analysis and forecasts to 2026. https://iea.blob.core.windows.net/assets/5ae32253-7409-4f9a-a91d-1493ffb9777a/Renewables2021-Analysisandforecastto2026.pdf.

69 Zare Oskouei, M., Mohammadi-Ivatloo, B., Abapour, M. et al. (2021). Techno-economic and environmental assessment of the coordinated operation of regional grid-connected energy hubs considering high penetration of wind power. *J. Clean. Prod.* 280: 124275. https://doi.org/10.1016/j.jclepro.2020.124275.

70 Geidl, M., Koeppel, G., Favre-Perrod, P. et al. (2007). Energy hubs for the future. *IEEE Power Energy Mag.* 5 (1): 24–30. https://doi.org/10.1109/MPAE.2007.264850.

71 Azar, B.M., Kazemzadeh, R., and Baherifard, M.A. (2020). Energy hub: modeling and technology – a review. *2020 28th Iranian Conference on Electrical Engineering (ICEE)*, Tabriz, Iran (4–6 August 2020). IEEE, pp. 1–6, doi: 10.1109/ICEE50131.2020.9260955.

72 Energy Commission (2020). Eurostat: energy consumption and use by households. https://ec.europa.eu/eurostat/web/products-eurostat-news/-/DDN-20200626-1 (accessed 21 July 2022).

73 Yuan, Y., Bayod-Rújula, A.A., Chen, H. et al. (2019). An advanced multicarrier residential energy hub system based on mixed integer linear programming. *Int. J. Photoenergy* 2019: 1–12. https://doi.org/10.1155/2019/1384985.

74 Fabrizio, E., Corrado, V., and Filippi, M. (2010). A model to design and optimize multi-energy systems in buildings at the design concept stage. *Renew. Energy* 35 (3): 644–655. https://doi.org/10.1016/j.renene.2009.08.012.

75 Ruparathna, R., Hewage, K., and Sadiq, R. (2016). Improving the energy efficiency of the existing building stock: a critical review of commercial and institutional buildings. *Renew. Sust. Energ. Rev.* 53: 1032–1045. https://doi.org/10.1016/j.rser.2015.09.084.

76 Mohammadi, M., Noorollahi, Y., Mohammadi-ivatloo, B. et al. (2018). Optimal management of energy hubs and smart energy hubs – a review. *Renew. Sust. Energ. Rev.* 89: 33–50. https://doi.org/10.1016/J.RSER.2018.02.035.

77 Paudyal, S., Canizares, C.A., and Bhattacharya, K. (2015). Optimal operation of industrial energy hubs in smart grids. *IEEE Trans. Smart Grid.* 6 (2): 684–694. https://doi.org/10.1109/TSG.2014.2373271.

78 Vadiee, A. and Martin, V. (2014). Energy management strategies for commercial greenhouses. *Appl. Energy* 114: 880–888. https://doi.org/10.1016/j.apenergy.2013.08.089.

79 Shamshirband, S., Khoshnevisan, B., Yousefi, M. et al. (2015). A multi-objective evolutionary algorithm for energy management of agricultural systems – a case study in Iran. *Renew. Sust. Energ. Rev.* 44: 457–465. https://doi.org/10.1016/j.rser.2014.12.038.

80 Eladl, A.A., El-Afifi, M.E., and El-Saadawi, M.M. (2019). Communication technologies requirement for energy hubs: a survey. *2019 21st International Middle East Power Systems Conference (MEPCON)*, Cairo, Egypt (17–19 December 2019). IEEE, pp. 821–827, doi: 10.1109/MEPCON47431.2019.9008006.

81 AkbaiZadeh, M., Niknam, T., and Kavousi-Fard, A. (2021). Adaptive robust optimization for the energy management of the grid-connected energy hubs based on hybrid meta-heuristic algorithm. *Energy* 235: 121171. https://doi.org/10.1016/j.energy.2021.121171.

82 Wang, Y., Cheng, J., Zhang, N., and Kang, C. (2018). Automatic and linearized modeling of energy hub and its flexibility analysis. *Appl. Energy* 211: 705–714. https://doi.org/10.1016/j.apenergy.2017.10.125.

83 Davatgaran, V., Saniei, M., and Mortazavi, S.S. (2019). Smart distribution system management considering electrical and thermal demand response of energy hubs. *Energy* 169: 38–49. https://doi.org/10.1016/j.energy.2018.12.005.

84 Ficco, G., Dell'Isola, M., Vigo, P., and Celenza, L. (2015). Uncertainty analysis of energy measurements in natural gas transmission networks. *Flow Meas. Instrum.* 42: 58–68. https://doi.org/10.1016/j.flowmeasinst.2015.01.006.

85 Ghanbari, A., Karimi, H., and Jadid, S. (2020). Optimal planning and operation of multi-carrier networked microgrids considering multi-energy hubs in distribution networks. *Energy* 204: 117936. https://doi.org/10.1016/j.energy.2020.117936.

86 Heidari, A. and Bansal, R.C. (2021). Probabilistic correlation of renewable energies within energy hubs for cooperative games in integrated energy markets. *Electr. Power Syst. Res.* 199: 107397. https://doi.org/10.1016/j.epsr.2021.107397.

87 Khorasany, M., Najafi-Ghalelou, A., Razzaghi, R., and Mohammadi-Ivatloo, B. (2021). Transactive energy framework for optimal energy management of multi-carrier energy hubs under local electrical, thermal, and cooling market constraints. *Int. J. Electr. Power Energy Syst.* 129: 106803. https://doi.org/10.1016/j.ijepes.2021.106803.

88 Najafi-Ghalelou, A., Nojavan, S., Zare, K., and Mohammadi-Ivatloo, B. (2019). Robust scheduling of thermal, cooling and electrical hub energy system under market price uncertainty. *Appl. Therm. Eng.* 149: 862–880. https://doi.org/10.1016/j.applthermaleng.2018.12.108.

89 Nikmehr, N. (2020). Distributed robust operational optimization of networked microgrids embedded interconnected energy hubs. *Energy* 199: 117440. https://doi.org/10.1016/j.energy.2020.117440.

90 Vahid Pakdel, M.J., Sohrabi, F., and Mohammadi-Ivatloo, B. (2020). Multi-objective optimization of energy and water management in networked hubs considering transactive energy. *J. Clean. Prod.* 266: 121936. https://doi.org/10.1016/j.jclepro.2020.121936.

91 Sobhani, S.O., Sheykhha, S., and Madlener, R. (2020). An integrated two-level demand-side management game applied to smart energy hubs with storage. *Energy* 206: 118017. https://doi.org/10.1016/j.energy.2020.118017.

92 Chen, Y., Wei, W., Liu, F. et al. (2018). Analyzing and validating the economic efficiency of managing a cluster of energy hubs in multi-carrier energy systems. *Appl. Energy* 230: 403–416. https://doi.org/10.1016/j.apenergy.2018.08.112.

93 Saberi, K., Pashaei-Didani, H., Nourollahi, R. et al. (2019). Optimal performance of CCHP based microgrid considering environmental issue in the presence of real time demand response. *Sustain. Cities Soc.* 45: 596–606. https://doi.org/10.1016/j.scs.2018.12.023.

94 Vidyamani, T. and Swarup, K.S. (2019). Demand response based on utility function maximization considering time-of-use price. *2019 IEEE PES Innovative Smart Grid Technologies Europe (ISGT-Europe)*,Bucharest, Romania (29 September–2 October 2019). IEEE, pp. 1–5, doi: 10.1109/ISGTEurope.2019.8905475.

95 Niromandfam, A., Yazdankhah, A.S., and Kazemzadeh, R. (2020). Modeling demand response based on utility function considering wind profit maximization in the day-ahead market. *J. Clean. Prod.* 251: 119317. https://doi.org/10.1016/j.jclepro.2019.119317.

96 Vand, B., Ruusu, R., Hasan, A., and Manrique Delgado, B. (2021). Optimal management of energy sharing in a community of buildings using a model predictive control. *Energy Convers. Manag.* 239: 114178. https://doi.org/10.1016/j.enconman.2021.114178.

97 Muller, G. (2020). Impact of demand response on generation expansion planning in the brazilian interconnected power system. *2020 IEEE PES Transmission &*

Distribution Conference and Exhibition – Latin America (T&D LA), Montevideo, Uruguay (28 September–2 October 2020). IEEE, pp. 1–6, doi: 10.1109/ TDLA47668.2020.9326129.

98 Zakeri, A. (2017). Transmission expansion planning using TLBO algorithm in the presence of demand response resources. *Energies* 10 (9): 1376. https://doi.org/ 10.3390/en10091376.

99 Moghaddam, M.P., Abdollahi, A., and Rashidinejad, M. (2011). Flexible demand response programs modeling in competitive electricity markets. *Appl. Energy* 88 (9): 3257–3269. https://doi.org/10.1016/j.apenergy.2011.02.039.

100 Jadidbonab, M., Mohammadi-Ivatloo, B., Marzband, M., and Siano, P. (2021). Short-term self-scheduling of virtual energy hub plant within thermal energy market. *IEEE Trans. Ind. Electron.* 68 (4): 3124–3136. https://doi.org/10.1109/ TIE.2020.2978707.

101 Dezfouli, M.M.S., Dorahaki, S., Rashidinejad, M. et al. (2019). A new energy hub scheduling model considering energy efficiency and demand response programs as energy democracy policy. *2019 IEEE 6th International Conference on Engineering Technologies and Applied Sciences (ICETAS)*, Kuala Lumpur, Malaysia (20–21 December 2019). IEEE, pp. 1–6, doi:10.1109/ ICETAS48360.2019.9117472.

102 Bostan, A., Nazar, M.S., Shafie-khah, M., and Catalão, J.P.S. (2020). Optimal scheduling of distribution systems considering multiple downward energy hubs and demand response programs. *Energy* 190: 116349. https://doi.org/10.1016/ j.energy.2019.116349.

103 Wang, A., Liu, J., and Wang, W. (2018). Flexibility-based improved model of multi-energy hubs using linear weighted sum algorithm. *J. Renew. Sustain. Energy* 10 (1): 015901. https://doi.org/10.1063/1.5005571.

104 Huang, W., Zhang, N., Wang, Y. et al. (2020). Matrix modeling of energy hub with variable energy efficiencies. *Int. J. Electr. Power Energy Syst.* 119: 105876. https:// doi.org/10.1016/j.ijepes.2020.105876.

105 Cao, Y., Wang, Q., Du, J. et al. (2019). Optimal operation of CCHP and renewable generation-based energy hub considering environmental perspective: an epsilon constraint and fuzzy methods. *Sustain. Energy, Grids Networks* 20: 100274. https:// doi.org/10.1016/j.segan.2019.100274.

106 Yang, J., Huang, Y., and Jiang, T. (2020). Optimal strategy of RIES based on bilateral stackelberg game. *2020 IEEE Sustainable Power and Energy Conference (iSPEC)*, Chengdu, China (23–25 November 2020). IEEE, pp. 1456–1461, 10.1109/ iSPEC50848.2020.9351148.

107 Wang, Y., Huang, Z., Li, Z. et al. (2020). Transactive energy trading in reconfigurable multi-carrier energy systems. *J. Mod. Power Syst. Clean Energy* 8 (1): 67–76. https://doi.org/10.35833/MPCE.2018.000832.

108 Dorahaki, S., Rashidinejad, M., Ardestani, S.F.F. et al. (2021). A peer-to-peer energy trading market model based on time-driven prospect theory in a smart and sustainable energy community. *Sustain. Energy, Grids Networks* 28: 100542. https://doi.org/10.1016/J.SEGAN.2021.100542.

109 Zhong, W., Yang, C., Xie, K. et al. (2018). ADMM-based distributed auction mechanism for energy hub scheduling in smart buildings. *IEEE Access* 6: 45635–45645. https://doi.org/10.1109/ACCESS.2018.2865625.

110 Feng, C., Wen, F., Zhang, L. et al. (2018). Decentralized energy management of networked microgrid based on alternating-direction multiplier method. *Energies* 11 (10): 2555. https://doi.org/10.3390/en11102555.

6

Modern Heat and Electricity Incorporated Networks Targeted by Coordinated Cyberattacks for Congestion and Cascading Outages

Arash Asrari[1], Ehsan Naderi[1], Javad Khazaei[2], Poria Fajri[3], and Valentina Cecchi[4]

[1] School of Electrical, Computer, and Biomedical Engineering, Southern Illinois University, Carbondale, IL, USA
[2] Department of Electrical and Computer Engineering, Lehigh University, Bethlehem, PA, USA
[3] Department of Electrical and Biomedical Engineering, University of Nevada, Reno, NV, USA
[4] Department of Electrical and Computer Engineering, University of North Carolina, Charlotte, NC, USA

Abbreviations

ACSE	AC state estimation
BDD	Bad data detection
CHP	Combined heat and power
CTR	Current transformer ratio
DCSE	DC state estimation
DG	Distributed generation
FDI	False data injection
FOR	Feasible operating region
LOI	Line outage index
LMP	Local marginal price
OT	Operating time
PMU	Phasor measurement unit
PV	Photovoltaic
RTU	Remote terminal unit
TD	Time dial
WT	Wind turbine

Coordinated Operation and Planning of Modern Heat and Electricity Incorporated Networks,
First Edition. Edited by Mohammadreza Daneshvar, Behnam Mohammadi-Ivatloo, and Kazem Zare.
© 2023 The Institute of Electrical and Electronics Engineers, Inc.
Published 2023 by John Wiley & Sons, Inc.

6.1 Introduction

6.1.1 Scope of the Chapter

The power outage for a quarter of a million Ukrainian people in 2015 was due to an organized cyberattack, which *simultaneously* targeted 60 substations of Ukraine's power grid [1], and is considered as one of the real-world examples motivating engineers and researchers to address the following question: "how to mitigate the consequences of future cyberattacks on power systems?". In fact, modernization of heat and electricity incorporated networks leads to a higher complexity in systems operation resulting in a more challenging task for operators to handle such attacks. According to the SANS (SysAdmin, Audit, Network, and Security) Institute report, "the effects of the attack on Ukraine's power grid were largely mitigated because grid operations there could be returned to manual control; however, the current *complexity* of grid operations in the U.S. would make a switch to manual operations difficult" [2]. Therefore, to accurately address the indicated question (i.e. how to mitigate the consequences of future cyberattacks?), researchers have scrutinized the following three research roadmaps:

1) *Modeling* of complicated cyberattacks from different standpoints targeting energy/power systems [3, 4],
2) *Detection* of complex cyberattacks [5, 6], and
3) *Protection* of power systems against cyberattacks via performing remedial actions [7, 8].

The scope of this chapter is associated with the first research line, where system operator tries to be in *attackers' shoe* in modeling of a coordinated cyberattack that can potentially target transmission and distribution sectors of the corresponding system. More specifically, the following research question will be addressed in this chapter: "How to design a coordinated cyberattack to result in system congestions and potentially cascading outages on a modern heat and electricity incorporated network including distributed generations (DGs) as well as combined heat and power (CHP) units?". More specifically, the following research goals are targeted in this chapter.

- Investigation of cyberattacks targeting modern heat and electricity incorporated networks upgraded by CHP units and renewable distributed energy resources.
- Examining coordinated cyberattacks targeting transmission and distribution sectors of smart grids simultaneously to cause cascading power outages.

6.1.2 Literature Review

Due to the indicated significance of research in this field, researchers have investigated the modeling of cyberattacks targeting energy/power systems. For example, in [9], the link between data attack and physical consequences was scrutinized, where an attacker constructed an FDI attack to trigger a targeted branch *outage* in order to trip multiple downstream branches leading to subsequent failures. In [10], a coordinated bilevel cyberattack model was proposed to result in undetectable line *outage* in transmission level. Moreover, a proper reaction mechanism was also introduced in [10] to address the attack's impacts. In [11], integrity cyberattacks against substation networks equipped with firewall were studied to cause unnecessary tripping of breakers leading to cascading *outages*. In [12], a learning-based attack model, termed as Q-learning algorithm, was implemented on a closed-loop automatic voltage control of a power transmission network resulting in local/system-scale *outages*. An attack model based on contingency analysis of system components was developed in [13] to significantly increase the occurrence of cascading *outages* by overloading key tie-lines in a typical transmission system. In [14], a malicious FDI cyberattack was studied to target flexible AC transmission system devices of a smart transmission network via bypassing the bad data detection (BDD) in DC state estimation (DCSE) leading to *congestion* of transmission lines. In [15], an FDI cyberattack was examined to result in overloading and initiating cascading failures with extreme consequences. In addition, a relevant reaction mechanism was elaborated in [15] to alleviate the experienced physical overloads caused by the launched cyberattack. In [16], a cyberattack model was designed to compromise internet of things (IoT)-controllable high wattage loads and distributed energy resources to negatively affect grid stability resulting in power *outages* and even large-scale *blackouts*. In [17], *outage*-related cyberattack models were developed to target the vulnerable transmission lines through bypassing the security constraints. Based on the proposed attack strategy in [17], a typical tie-line was physically disconnected as the initial step. Then, the corresponding *outage* was masked via misleading the system operator into detecting it as an actual outage at a different location of the power grid. Topological cyberattacks were elucidated in [18] via targeting modern distribution systems leading to *cascading power outages* in different timescales, where other types of cyberattack models including false data injection attack and false command injection attack were taken into account to analyze their impacts on the intended outages.

6.1.3 Research Gap and Contributions of This Chapter

Despite extensive research efforts in the existing literature on modeling of attacks leading to congestions/outages, a framework is not yet proposed to investigate the impacts of a coordinated attack on transmission and distribution sectors. In other words, the following question is yet to be addressed: "How to model a coordinated attack by a team of attackers simultaneously targeting transmission and distribution sectors of a smart grid leading to system congestions and potentially cascading outages?". As was indicated earlier, it is essential for system operators to be in attackers' shoe and investigate different possible approaches of cyberattacks to be able to proactively cope with them if such cyberattacks are ever experienced in the future. The motivation of this chapter is to address the indicated research gap. More specifically, the contributions of this chapter are itemized as follows:

1) A cyberattack framework to optimize the effect of congestions (as its *primary* consequence) via maximizing LMPs (local marginal prices) and to optimize the severity of potential outages (as its *secondary* consequence) via targeting the critical lines with the most sensitive protective devices,
2) A formulation to coordinate between lead attacker and co-attackers in targeting the *main system* (transmission network) and the corresponding *subsystems* (distribution networks) containing DGs and CHPs.

6.1.4 Organization of the Chapter

Section 6.2 presents a detailed description about the proposed framework in this chapter. Section 6.3 provides the problem formulation including the objective functions to be optimized in each level of the framework and the related technical constraints. The case study, containing a transmission system and eleven distribution systems, as well as the obtained results and relevant discussions are elaborated in Section 6.4. A comprehensive analysis is performed in this section to validate the superiority of the proposed *coordinated attack on distribution and transmission sectors, simultaneously,* compared to attacking *only the transmission network* or targeting *only the distribution systems.* Section 6.5 outlines the conclusions and indicates the future step of this research.

6.2 Proposed Framework

6.2.1 Illustration of the Proposed Framework

Figure 6.1 illustrates the proposed framework. The main system (i.e., transmission network) and the subsystems (distribution networks) are equipped with phasor measurement units (PMUs). PMU is employed to directly monitor and precisely measure the state of the system (i.e., voltage magnitude and voltage phase angle) [19, 20]. Remote terminal units (RTUs) are also used to interface sensors of network to supervisory control and data acquisition system for a more effective communication mechanism [14]. For the sake of clarity of presentation, only three attackers are depicted in Figure 6.1 as the team of attackers. As can be perceived from the figure, lead attacker has hacked/manipulated the corresponding PMUs leading to congestion of lines $i-j$ and $k-l$. In addition, co-attackers #1 and #2 have compromised the relevant PMUs resulting in congestion of lines $m-n$ and $o-p$.

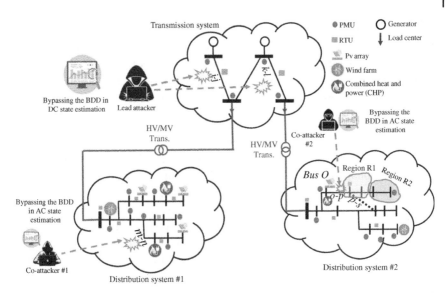

Figure 6.1 The proposed framework.

According to the figure, the lead attacker coordinates with the co-attackers in identification of the (to be) targeted lines. The primary intention of the team of attackers is to result in a more destructive system congestion via targeting more vulnerable lines with the criterion of maximizing LMP (local marginal price). The secondary goal of the team of attackers is to cause cascading outages via targeting the critical branches with the most sensitive protective devices. To further clarify the concept, let us concentrate on the second subsystem in the example illustrated in Figure 6.1. The breaker of line $o-p$ is equipped with a sensitive relay having a *lower trip current* and a *less time dial setting*. In addition, it is considered as the upstream branch for regions R_1 & R_2 of the second subsystem. Hence, it is considered as a suitable target for the indicated attack. The lead attacker and co-attacker #2 simultaneously target lines $k-l$ and $o-p$. Since regions R_1 & R_2 will be disconnected from the upstream PV and CHP of Bus O after the attack (see Figure 6.1), a possible solution to the severe congestion of line $o-p$ is distribution network reconfiguration; i.e., closing $p-s$ after opening $o-p$ such that *extra power supply* can be requested from the utility grid to feed regions R_1 & R_2. However, this strategy will not be an effective solution for this attack since the upstream line itself (i.e., see line $k-l$ in Figure 6.1) is also targeted by the lead attacker at the same time (i.e., when co-attacker #2 targets line $o-p$). Thus, the continuation of the experienced congestion (i.e., ongoing increase of LMP at the location of attacked line $o-p$) will lead to tripping of the corresponding breaker (with *sensitive relay settings; i.e., less time dial*), resulting in the cascading outages for regions R_1 & R_2.

6.2.2 Assumptions of the Attack Framework

There are a set of realistic and typical assumptions in the literature for targeting power systems in different levels of transmission and distribution sectors. The following explains the adopted assumptions in development of the described framework in order to ensure that the coordinated cyberattacks do not trigger the security thresholds.

- Set of measurements (e.g., z), which encompass active and reactive power injections to the buses, active and reactive power flow through tie-lines and branches, and also voltage magnitudes are generally redundant; hence, the number of measurements is more than the state variables (e.g., x). According to [19–21], the redundancy in the number of measurements stems from the fact that many measurement tools are located in a power grid to precisely estimate the state of power networks for different purposes such as billing services.
- Due to uncertainty of measurements, a measurement error (e.g., e) is considered in the problem formulation.
- FDI cyberattacks targeting different assets (e.g., load centers, generation units, etc.) are limited to a certain range of the rated values to ensure that the launched attacks do not violate the rated limits. The formulated attacks (i.e., manipulations) in this study are limited to up to 20% of the corresponding rated values [22].
- Attackers have gained access to the detailed configuration of the targeted smart grid. This assumption is based on what was experienced in the cyberattack on the Ukrainian power grid in 2015, when attackers sent superfishing emails, hacked the supervisory control and data acquisition system to isolate the substations, and attacked the Master boot files on stations via a KillDisk malware [23].
- The lead attacker is assigned to the transmission sector to supervise the attack process and ensure the coordinated cyberattack by the co-attackers targeting the downstream distribution sectors can successfully result in the intended congestions and outages.
- Although attacks on distribution systems should bypass AC state estimation (ACSE) [3], it is practical to assume that successful attacks on transmission systems should bypass DCSE-based BDD [10]. Hence, in the proposed framework, lead attacker tries to bypass DCSE BDD while co-attackers need to bypass ACSE BDD.

6.3 Problem Formulation

6.3.1 Objective Functions of the Attack Framework

The optimization problem is generally presented in (6.1), where OF_1 and OF_2 are, respectively, the objective functions to be addressed in the upper-level (i.e., congestion) and lower-level (i.e., outage) of the proposed attack framework. In the following bi-level optimization problem, attackers maximize OF$_1$ to identify the most vulnerable transmission/distribution branches as the primary targets of the attack leading to maximum system congestion. Then, considering the identified branches, they will minimize OF$_2$ to find the most vulnerable ones whose congestion has a higher possibility to result in cascading power outages (i.e., minimization of overcurrent relay's operation time over the severity of intentional outage (IO)).

$$
\left\{
\begin{array}{l}
\max \mathrm{OF}_1^{t,d,c} \\
\min \mathrm{OF}_2^{t,d,c}
\end{array}
\right\}
t \in T, d \in D, c \in C \;\; \text{Subject to} \;\;
\begin{array}{l}
g(x) = 0 \\
h(x) \leq 0
\end{array}
\tag{6.1}
$$

where t, d, and c, respectively, refer to transmission system lines, distribution system branches, and the counter of distribution networks being fed by the corresponding transmission system; g and h are, respectively, the set of equality and inequality constraints; and x denotes the decision variables.

Equations (6.2) and (6.3), respectively, formulate the optimization (i.e., maximization and minimization) problems related to OF$_1$ and OF$_2$ in (6.1).

$$
\max \mathrm{OF}_1^{t,d,c} = \max \left(\sum_{t=1}^{T} \xi^t \times \mathrm{LMP}^t \right) + \max \left(\sum_{c=1}^{C} \sum_{d=1}^{D} \varphi^{d,c} \times \mathrm{LMP}^{d,c} \right)
\tag{6.2}
$$

$$
\min \mathrm{OF}_2^{t,d,c} = \min \left(\sum_{t \in \{\xi^t = 1\}}^{T} \frac{\mathrm{OT}^t}{\mathrm{IO}^t} \right) + \min \left(\sum_{c=1}^{C} \sum_{d \in \{\varphi^{d,c} = 1\}} \frac{\mathrm{OT}^{d,c}}{\mathrm{IO}^{d,c}} \right)
\tag{6.3}
$$

In (6.2) and (6.3), ξ^t and $\varphi^{d,c}$, respectively, denote the binary variables of attack for tth transmission line and dth distribution branch associated with cth distribution system. It is noted that the numerical values of ξ and φ will be equal to one if the corresponding transmission line and distribution branch are targeted for a cyberattack; otherwise, they will be equal to zero. In addition, OT^t and IO^t are, respectively, the operating time (OT) of overcurrent relay and IO associated with tth transmission tie-line. Similarly, OT^{d,c} and IO^{d,c} are, respectively, the OT and IO related to dth distribution branch located in cth distribution system. According to (6.2), the main objective of the coordinated cyberattack is to maximize the LMP in the transmission system (i.e., the first term of (6.2)) and the downstream distribution systems (i.e., the second term of (6.2)). Referring to (6.3), the lead attacker and the corresponding co-attackers can identify the most vulnerable sections of system for congestion which can result in cascading outages. In (6.3), the ratio of OT over IO is minimized in both transmission system (i.e., the first term of (6.3)) and relevant distribution systems (i.e., the second term of (6.3)). As an example, if a co-attacker identifies two branches with the same LMP (i.e., same level of destructive impact for congestion according to (6.2)) and the same IO, the one with the *lower* OT will be selected so that successful implementation of such a cyberattack on the congested line will most likely lead to tripping of the more sensitive protective device resulting in the intended outages. As another example, if the co-attacker identifies two branches with the same LMP and the same OT, the one with the *higher* IO will be selected so that a more noticeable power outage can be experienced as the consequence of severe congestion caused by the attack.

According to [24], LMP can be divided into three parts: the first part represents the marginal energy price (i.e., the first term of (6.4)), the second part denotes the marginal congestion price (i.e., the second term of (6.4)), and the third part indicates the marginal loss price (i.e., the third term of (6.4)).

$$\text{LMP} = \mu + \left(\sum_{b=1}^{L} \Xi^{l-b} \times \mu^l \right) + \left(\mu \times \left(\Delta^b - 1 \right) \right) \tag{6.4}$$

where μ and μ^l are the Lagrange multipliers of the equality and inequality constraints related to the security constraint economic dispatch covered in Section 6.3.2; Ξ^{l-b} denotes the generation shift factor from bus b to line l by considering the line's power flow limit; L is the total number of lines in the system; and Δ^b is the delivery factor at bth bus of the system. Interested readers are directed to [24] for more information about formulation of LMP.

With respect to OT in the second objective function (i.e., (6.3)), it is noted that a typical inverse time directional overcurrent relay comprise two operating units namely *instantaneous* part, which is independent from the time, and inverse *overcurrent* part, which depends on the time. The instantaneous part can operate without any delay as soon as the magnitude of current is greater than the threshold (i.e., instantaneous pickup current). However, the overcurrent part is associated with LT/ST (long-time/short-time) modality of relay to take into account currents higher than 1.15 times the rated current according to NEC (national electric code) [25]. To precisely determine the OT of relay for different current magnitudes (refer to (6.3)), time dial (TD) setting is employed as one of the common settings for overcurrent relays. The OT of a relay has a nonlinear relationship with the TD setting. In other words, OT is a function of TD, pick-up current, and fault current (i.e., the compromised current due to the launched cyberattack). Thus, the OT associated with a directional overcurrent relay can be calculated as (6.5) [26], where CTR denotes the current transformer ratio; $I_t^{\text{pick-up}}$ and $I_{d,c}^{\text{pick-up}}$ are, respectively, the pick-up currents associated with tth transmission tie-line and dth distribution branch located in cth distribution system; I_t^{attack} and $I_{d,c}^{\text{attack}}$ are the abnormal currents due to the attack related to tth transmission tie-line and dth distribution branch located in cth distribution system; and TD^t and $\text{TD}^{d,c}$ are, respectively, the TDs of the relays located at tth tie-line and dth distribution branch located in cth distribution system.

$$
\begin{bmatrix} OT^t \\ OT^{d,c} \end{bmatrix} = \begin{bmatrix} TD^{d,c} \times \left(\dfrac{0.14}{\left(\dfrac{I_{d,c}^{attack}}{CTR \times I_{d,c}^{pick\text{-}up}} \right)^{0.02} - 1} \right) \\ TD^t \times \left(\dfrac{0.14}{\left(\dfrac{I_t^{attack}}{CTR \times I_t^{pick\text{-}up}} \right)^{0.02} - 1} \right) \end{bmatrix}
\tag{6.5}
$$

With respect to IO in the second objective function (i.e., (6.3)), it is noted that the line outage index (LOI) is implemented to measure how a manipulation in the status of a tie-line/branch in the system affects the power flow of other tie-lines/branches. Thus, LOI, adopted from [27], is presented in (6.6) and has been reflected on (6.7), where l refers to both transmission tie-lines and distribution branches.

$$
LOI_{Normalized}^l = \frac{AVE(|LOI^l|, \forall l)}{STD(|LOI^l|, \forall l)}
\tag{6.6}
$$

$$
IO^l = LOI_{Normalized}^l \times PF_{AA}^l
\tag{6.7}
$$

In (6.6) and (6.7), LOI^l denotes line outage index associated with lth line; AVE and STD are, respectively, the average and standard deviation functions of LOI (refer to [27] for background about AVE and STD); $LOI_{Normalized}^l$ is the corresponding normalized value for lth line, which is in range of [0–1]. Moreover, PF_{AA}^l is the electric power flowing into lth tie-line/branch after the experienced congestion. Therefore, multiplication of normalized LOI and power flow of the congested line indicates the IO in (6.3). To sum up, the formulated LMP in (6.4) is reflected on (6.2) to identify the vulnerable transmission/distribution lines to *congestion*. In addition, the formulated OT and IO in (6.5)–(6.7) are reflected on (6.3) to attack the most critical lines with sensitive settings having higher impact on downstream regions leading to outages. Therefore, (6.2) and (6.3) are solved to result in identification of transmission/distribution lines leading to most destructive congestions with the highest possibility of causing *cascading outages*. Hence, the solution of solving (6.2) and (6.3) is a set of FDI vectors on the electric load data leading to: (i) severe congestion of the most vulnerable transmission/distribution lines to congestion (greater LMP) and (ii) cascading outages initiated by the attacked *critical* lines (greater PF_{AA}^l & IO) equipped with the most *sensitive* protection setting devices (less TD & OT).

6.3.2 Technical Constraints

6.3.2.1 Constraints Related to Bypassing DCSE BDD and ACSE BDD

In general, the measurement model for state estimation is presented as (6.8), where e denotes the measurement error vector and $h(x)$ signifies the relationship between state variables x and measurements z [28]. Nevertheless, for transmission-level systems analysis, related to the lead attacker (see Figure 6.1), DCSE is used, in which the resistive component of transmission line impedances is ignored, and the voltage magnitudes are considered to be 1.00 p.u. Hence, (6.8) can be simplified to (6.9) for transmission systems [29]. The logic of DCSE and the conditions based on which the lead attacker in Figure 6.1 can retain the cyberattack stealthy are formulated in (6.10)–(6.17). Referring to (6.17), since the transmission-level cyberattacks are designed to ensure $\Delta D_G - \Delta D_L = H\Delta\Theta$, the cyberattack can successfully bypass the DCSE BDD.

$$z = h(x) + e \tag{6.8}$$

$$z = Hx + e \tag{6.9}$$

$$\begin{bmatrix} z_1 \\ z_2 \\ \vdots \\ z_r \end{bmatrix} = \begin{bmatrix} H_{11} & H_{12} & \cdots & H_{1s} \\ H_{21} & H_{22} & \cdots & H_{2s} \\ \vdots & \vdots & \ddots & \vdots \\ H_{r1} & H_{r2} & \cdots & H_{rs} \end{bmatrix} \begin{bmatrix} x_1 \\ x_2 \\ \vdots \\ x_s \end{bmatrix} + \begin{bmatrix} \Omega_1 \\ \Omega_2 \\ \vdots \\ \Omega_r \end{bmatrix} \tag{6.10}$$

$$W = E(NN^T) = \begin{bmatrix} v_1^2 & 0 & \cdots & 0 \\ 0 & v_2^2 & \cdots & 0 \\ \vdots & \vdots & \ddots & \vdots \\ 0 & 0 & \cdots & v_r^2 \end{bmatrix} \tag{6.11}$$

$$\hat{x} = \left(H^T W^{-1} H \right)^{-1} H^T W^{-1} z \tag{6.12}$$

$$\text{if} \begin{cases} \|z - H\hat{x}\| \leq \varepsilon & \text{No bad data in the system} \\ \|z - H\hat{x}\| \geq \varepsilon & \text{Bad data exist in the system} \end{cases} \tag{6.13}$$

$$\|z_{\text{bad}} - H\hat{x}_{\text{bad}}\| = \|z + \Delta z + H\hat{x}_{\text{bad}}\| \tag{6.14}$$

$$\hat{x}_{\text{bad}} = \left(H^T W^{-1} H \right)^{-1} H^T W^{-1} z_{\text{bad}} = \hat{x} + \left(H^T W^{-1} H \right) H^T W^{-1} \Delta z \tag{6.15}$$

$$\|z_{\text{bad}} - H\hat{x}_{\text{bad}}\| = \left\| z + \Delta z - H\left(\hat{x} + \left(H^T W^{-1} H \right) H^T W^{-1} \Delta z \right) \right\| = \|z - H\hat{x}\| \leq \varepsilon \tag{6.16}$$

$$\|z_{\text{Clean}} + \Delta D_G - \Delta D_L - Hx_{\text{Clean}} + H\Delta\Phi\| \geq \varepsilon \tag{6.17}$$

where r and s are the number of measurements and state variables in the transmission system; Ω is the measurement noise; H is the function that relates measurements to the state variables and depends on transmission line reactance in DCSE formulation; $\hat{x}$ is the estimated state variable vector achieved by minimizing the weighted sum of squares of measurement residuals; ε is the threshold of BDD algorithm embedded in DCSE; Δz is the vector of false data injected into the system by the lead attacker. It is noted that if $Hc = \Delta z$, then the cyberattack remains stealthy.

On the other hand, ACSE is utilized in order to estimate the unknown parameters of distribution systems [30]. Since the magnitude of voltage at different nodes of a distribution system cannot be considered constant, the DCSE can no longer be utilized. The principles of ACSE and the conditions based on which co-attackers can retain the cyberattacks stealthy are presented in (6.18)–(6.26) [30]. It is noted that the cooperation of lead attacker (bypassing DCSE BDD) with co-attackers (bypassing ACSE BDD) will result in the intended congestions and cascading outages within the transmission and distribution sectors of the power system.

$$\|z - h(\hat{x})\| \geq \varepsilon \tag{6.18}$$

$$z_{bad} = \begin{bmatrix} z_c^u \\ z_c^t + FDI_t \end{bmatrix} \tag{6.19}$$

$$\Delta z_{bad} = [0 \;\; FDI_t] \tag{6.20}$$

$$\|z_{cm} - h(\hat{x}_{bad})\| = \|z + cm - h(\hat{x} + cs)\| = \left\| \begin{pmatrix} z_1 \\ z_2 + cm_2 \end{pmatrix} - \begin{pmatrix} h_1(\hat{x}_1) \\ h_2(\hat{x}_1, \hat{x}_1 + cs) \end{pmatrix} \right\|$$

$$= \left\| \begin{pmatrix} z_1 \\ z_2 \end{pmatrix} - \begin{pmatrix} h_1(\hat{x}_1) \\ h_2(\hat{x}_1, \hat{x}_1) \end{pmatrix} \right\| = \|z - h(\hat{x})\| \leq \varepsilon \tag{6.21}$$

$$FDI_t = h(\hat{x} + c) - h(\hat{x}) \tag{6.22}$$

$$\text{Re}\{h(\hat{x} + c) - h(\hat{x})\} = \Delta P_{G_n} - \Delta P_{D_n} \tag{6.23}$$

$$\text{Im}\{h(\hat{x} + c) - h(\hat{x})\} = \Delta Q_{G_n} - \Delta Q_{D_n} \tag{6.24}$$

$$P_{ij} = V_i^2 \times g_{ij} - V_i \times V_j \times g_{ij} \cos(\theta_i - \theta_j) - V_i \times V_j \times b_{ij} \sin(\theta_i - \theta_j) \tag{6.25}$$

$$Q_{ij} = -V_i^2 \times b_{ij} + V_i \times V_j \times b_{ij} \cos(\theta_i - \theta_j) - V_i \times V_j \times g_{ij} \sin(\theta_i - \theta_j) \tag{6.26}$$

where ε is the threshold for BDD; z is the set of measurements; $h(\hat{x})$ is the function relating state variables to measured values; $\hat{x}$ is the vector of estimated variables; z_c^u is the clean data associated with the non-targeted buses; z_c^t is the clean data related to the targeted buses; and FDI_t is the vector of false data to be injected into the targeted buses. Moreover, P_{ij}, Q_{ij}, g_{ij}, b_{ij}, V_i, and θ_i, respectively, stand for active and reactive powers flowing into line ij, real and imaginary parameters of the indicated line, magnitude, and angle of voltage of bus i. Readers are referred to [30] for more details about formulation of $h(\hat{x})$ in bypassing ACSE.

6.3.2.2 Constraints Related to Thermal Units and CHP Units

The cost functions associated with thermal generation units located at different buses of the transmission system and combined heat and power (CHP) units feeding different buses of distribution systems are, respectively, presented in (6.27)–(6.28), where C_m denotes the generation cost associated with the mth thermal unit in terms of \$/h; a_m, b_m, and c_m are the cost coefficients for mth thermal unit; C_n signifies the generation cost associated with the nth CHP unit in terms of \$/h; α_n, β_n, δ_n, η_n, λ_n, and ω_n are the cost coefficients related to nth CHP; P_m^{power} represents the output power for mth thermal unit; and P_n^{CHP} and H_n^{CHP} are, respectively, the generated power and heat associated with nth CHP [31].

$$C_m\left(P_m^{\text{power}}\right) = a_m + b_m \times \left(P_m^{\text{power}}\right) + c_m \times \left(P_m^{\text{power}}\right)^2 \tag{6.27}$$

$$C_n\left(P_n^{\text{CHP}}, H_n^{\text{CHP}}\right) = \begin{pmatrix} \alpha_n + \beta_n \times \left(P_n^{\text{CHP}}\right) + \delta_n \times \left(P_n^{\text{CHP}}\right)^2 + \\ \eta_n \times \left(H_n^{\text{CHP}} \times P_n^{\text{CHP}}\right) + \lambda_n \times \left(H_n^{\text{CHP}}\right) + \omega_n \times \left(H_n^{\text{CHP}}\right)^2 \end{pmatrix} \tag{6.28}$$

The capacity constraints for thermal units and CHP units are presented in (6.29)–(6.32), where $P_{m,\text{Min}}^{\text{power}}$ and $P_{m,\text{Max}}^{\text{power}}$ are, respectively, the minimum and maximum boundaries of output active power associated with mth thermal unit; $Q_{m,\text{Min}}^{\text{power}}$ and $Q_{m,\text{Max}}^{\text{power}}$ are the minimum and maximum limits of output reactive power for mth thermal unit, respectively; and $P_{n,\text{Min}}^{\text{CHP}}\left(H_n^{\text{CHP}}\right)$ and $P_{n,\text{Max}}^{\text{CHP}}\left(H_n^{\text{CHP}}\right)$ are, respectively, the minimum and maximum boundaries of output power related to nth CHP. It is noted that $P_{n,\text{Min}}^{\text{CHP}}\left(H_n^{\text{CHP}}\right)$ and $P_{n,\text{Max}}^{\text{CHP}}\left(H_n^{\text{CHP}}\right)$ are functions of the output heat (i.e.,H_n^{CHP}). Likewise, $H_{n,\text{Min}}^{\text{CHP}}\left(P_n^{\text{CHP}}\right)$ and $H_{n,\text{Max}}^{\text{CHP}}\left(P_n^{\text{CHP}}\right)$ are, respectively, minimum and maximum boundaries of output heat associated with nth CHP unit, which are functions of the output power of the CHP (i.e., P_n^{CHP}).

$$P_{m,\text{Min}}^{\text{power}} \leq P_m^{\text{power}} \leq P_{m,\text{Max}}^{\text{power}} \tag{6.29}$$

$$Q_{m,\text{Min}}^{\text{power}} \leq Q_m^{\text{power}} \leq Q_{m,\text{Max}}^{\text{power}} \tag{6.30}$$

$$P_{n,\text{Min}}^{\text{CHP}}\left(H_n^{\text{CHP}}\right) \leq P_n^{\text{CHP}} \leq P_{n,\text{Max}}^{\text{CHP}}\left(H_n^{\text{CHP}}\right) \tag{6.31}$$

$$H_{n,\text{Min}}^{\text{CHP}}\left(P_n^{\text{CHP}}\right) \leq H_n^{\text{CHP}} \leq H_{n,\text{Max}}^{\text{CHP}}\left(P_n^{\text{CHP}}\right) \tag{6.32}$$

According to (6.31) and (6.32), one can infer that the produced *power* and *heat* in a typical CHP unit are related. Herein, the concept of feasible operating region

Figure 6.2 Feasible operating region (FOR) associated with a typical convex-type CHP. *Source:* Adapted from Moradi-Dalvand et al. [31].

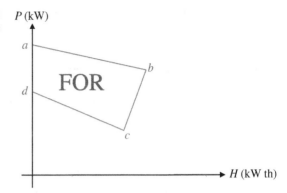

(FOR) is employed to model the boundaries associated with CHP units (see Figure 6.2). It is noted that only convex-type CHP units are considered in this work; however, the presented model can be extended to a nonconvex FOR [31, 32].

The mathematical model of a typical convex CHP depicted in Figure 6.2 is provided in (6.33)–(6.37).

$$P_n^{CHP} - P_a^{CHP} - \frac{\left(P_a^{CHP} - P_b^{CHP}\right)}{\left(H_a^{CHP} - H_b^{CHP}\right)} \times \left(H_n^{CHP} - H_a^{CHP}\right) \leq 0 \tag{6.33}$$

$$P_n^{CHP} - P_b^{CHP} - \frac{\left(P_b^{CHP} - P_c^{CHP}\right)}{\left(H_b^{CHP} - H_c^{CHP}\right)} \times \left(H_n^{CHP} - H_b^{CHP}\right) \geq -L \times \left(1 - CS_n^{CHP}\right) \tag{6.34}$$

$$P_n^{CHP} - P_c^{CHP} - \frac{\left(P_c^{CHP} - P_d^{CHP}\right)}{\left(H_c^{CHP} - H_d^{CHP}\right)} \times \left(H_n^{CHP} - H_c^{CHP}\right) \geq -L \times \left(1 - CS_n^{CHP}\right) \tag{6.35}$$

$$0 \leq P_n^{CHP} \leq P_a^{CHP} \times CS_n^{CHP} \tag{6.36}$$

$$0 \leq H_n^{CHP} \leq H_b^{CHP} \times CS_n^{CHP} \tag{6.37}$$

where CS_n^{CHP} is a binary variable presenting the commitment status of nth CHP unit; and L is an arbitrary large number. Regarding the presented constraints, (6.33) formulates the line from point a to point b presented in Figure 6.2, (6.34) reflects the formula of line connecting point b to point c (see Figure 6.2), (6.35) is included to consider the line from point c to point d depicted in Figure 6.2, and (6.36) and (6.37) take into account the minimum and maximum limits associated with output power and heat in the proposed convex model for CHP units. Reflection of Constraints (6.33) and (6.37) ensures that the operating point of the CHP units is inside the feasible region. As an example, constraint (6.33) considers the operating points *below* the line connecting point a to point b. Likewise, considering (6.34) and (6.35) confines the operating points to be, respectively, in the *left* side of the line connecting point b to point c and on *top* of the line connecting point c to point d.

6.3.2.3 Constraints Related to Wind Turbines

The output power of a typical wind turbine (WT) depends on the wind speed, which can be mathematically presented as (6.38) considering (6.39)–(6.41), where v is wind speed in terms of m/s; P_k^{WT} is the output active power associated with kth WT; P_k^{rated} is the rated power for kth WT's generator; Φ_1–Φ_3 are relevant coefficients for WTs; and v_{ci}, v_{co}, and v_{rated} are, respectively, the cut in, cut out, and rated wind speeds in terms of m/s. According to (6.42), κ is a coefficient confining the level of wind power penetration (i.e., contribution) into the system with respect to the total demand, which can be altered from zero (i.e., 0% contribution in supplying the system's demand) to one (i.e., fully serving the load); and K is the number of WTs [33]. The presented equations are reflected based on the Weibull probability density function. Interested readers are directed to [33] for further details about modeling of WTs.

$$P_k^{WT} = \begin{cases} 0 & 0 \leq v \leq v_{ci} \\ P_k^{rated} & v_{rated} \leq v \leq v_{co} \\ P_k^{rated} \times (\Phi_1 v^2 + \Phi_2 v + \Phi_3) & v_{ci} \leq v \leq v_{rated} \\ 0 & v \geq v_{co} \end{cases} \tag{6.38}$$

$$\Phi_1 = \frac{1}{(v_{ci} - v_{rated})^2} \times \left(v_{ci}^2 + v_{ci} v_{rated} - 4 v_{ci} v_{rated} \left(\frac{v_{ci} + v_{rated}}{2 v_{rated}} \right)^3 \right) \tag{6.39}$$

$$\Phi_2 = \frac{1}{(v_{ci} - v_{rated})^2} \times \left(4 v_{ci} v_{rated} \left(\frac{v_{ci} + v_{rated}}{2 v_{rated}} \right)^3 - 3 v_{ci} - v_{rated} \right) \tag{6.40}$$

$$\Phi_3 = \frac{1}{(v_{ci} - v_{rated})^2} \times \left(2 - 4 \left(\frac{v_{ci} + v_{rated}}{2 v_{rated}} \right)^3 \right) \tag{6.41}$$

$$\sum_{k=1}^{K} P_k^{WT} \leq \kappa \times P^{Demand} \tag{6.42}$$

6.3.2.4 Constraints Related to PV Modules

The power generated by yth PV is presented in (6.43) and (6.44), where I_{std} and I_C are, respectively, the solar irradiances in the standard environment and at a certain point; P_{rated}^{PV} is the rated power generated by module; Ψ is a coefficient indicating the level of PV's penetration, which can be altered from zero (i.e., 0% contribution in supplying the demand) to one (i.e., fully serving the load); and Y is the number of modules [34, 35].

$$
P_y^{PV} = \begin{cases} \left(\dfrac{I^2}{I_{std} \times I_C}\right) \times P_{rated}^{PV} \rightarrow I \leq I_C \\[4mm] \left(\dfrac{I^2}{I_{std}}\right) \times P_{rated}^{PV} \rightarrow I_C \leq I \end{cases}
\tag{6.43}
$$

$$
\sum_{y=1}^{Y} P_y^{PV} \leq \Psi \times P^{Demand}
\tag{6.44}
$$

In order to handle the uncertainties associated with solar irradiance for PV modules, Beta probability density function is utilized to model the stochastic nature of solar irradiance. Interested readers are directed to [35, 36] for more information about formulation of Beta function as is reflected on (6.43) and (6.44).

6.3.2.5 Power and Heat Balance Constraints
The generated power and heat associated with thermal units and CHP units need to meet the electric and thermal demand within the interconnected system. The relevant power and heat equilibriums are mathematically presented in (6.45)–(6.47), where M, N, K, and H are, respectively, the number of thermal units, CHP units, WTs, and PV modules; P^{Demand} and H^{Demand} are, respectively, the electric power and thermal demands of the entire system; P^{Loss} and H^{Loss} denote the summation of power loss and heat loss in the entire system, respectively; Q_m^{power} and $Q_{k,c}^{WT}$ are, respectively, the reactive power produced by mth thermal unit and kth WT located in cth distribution system.

$$
\sum_{m=1}^{M} P_m^{power} + \sum_{c=1}^{C}\sum_{n=1}^{N} P_{n,c}^{CHP} + \sum_{c=1}^{C}\sum_{k=1}^{K} P_{k,c}^{WT} + \sum_{c=1}^{C}\sum_{y=1}^{Y} P_{y,c}^{PV} = P^{Demand} + P^{Loss}
$$

$$
\tag{6.45}
$$

$$
\sum_{m=1}^{M} Q_m^{power} + \sum_{c=1}^{C}\sum_{k=1}^{K} Q_{k,c}^{WT} = Q^{Demand} + Q^{Loss}
\tag{6.46}
$$

$$
\sum_{c=1}^{C}\sum_{n=1}^{N} H_{n,c}^{CHP} = H^{Demand} + H^{Loss}
\tag{6.47}
$$

In addition, each distribution system needs to satisfy majority of its local demand. Moreover, power and heat balance should be reflected in each distribution system besides the equilibriums in the entire system (i.e., (6.45)–(6.47)). Herein, (6.48)–(6.50) present the electric power and heat balance constraints in a typical distribution network (e.g., the first subsystem when $c = 1$).

$$P^{\text{Grid}} + \sum_{n=1}^{N} P_n^{\text{CHP}} + \sum_{k=1}^{K} P_k^{\text{WT}} + \sum_{y=1}^{Y} P_y^{\text{PV}} = P_{\text{ss}}^{\text{Demand}} + P_{\text{ss}}^{\text{Loss}} \tag{6.48}$$

$$Q^{\text{Grid}} + \sum_{k=1}^{K} Q_k^{\text{WT}} = Q_{\text{ss}}^{\text{Demand}} + Q_{\text{ss}}^{\text{Loss}} \tag{6.49}$$

$$\sum_{n=1}^{N} H_n^{\text{CHP}} = H_{\text{ss}}^{\text{Demand}} + H_{\text{ss}}^{\text{Loss}} \tag{6.50}$$

where $P_{\text{ss}}^{\text{Demand}}$ and $H_{\text{ss}}^{\text{Demand}}$ are, respectively, the local active power and thermal demands in a typical distribution subsystem; $P_{\text{ss}}^{\text{Loss}}$ and $H_{\text{ss}}^{\text{Loss}}$ refer to the local active power loss and local heat loss in a typical distribution system, respectively; Q^{Grid} signifies the reactive power provided by the main transmission system, which is injected to the relevant bus (i.e., load center); Q_k^{WT} denotes the reactive power generated by kth WT; $Q_{\text{ss}}^{\text{Demand}}$ and $Q_{\text{ss}}^{\text{Loss}}$, respectively, represent the local reactive power demand and reactive power loss in a typical distribution system.

6.3.2.6 Rest of System's Constraints
The rest of inequality constraints reflecting system's technical limits are presented in (6.51)–(6.54).

$$I_\tau^{\text{Trans}} \leq I_{\tau,\max}^{\text{Trans}} \tag{6.51}$$

$$I_f^{\text{Feeder}} \leq I_{f,\max}^{\text{Feeder}} \tag{6.52}$$

$$V_b^{\min} \leq V_b \leq V_b^{\max} \tag{6.53}$$

$$|S_l| \leq S_l^{\max} \tag{6.54}$$

where I_τ^{Trans} and $I_{\tau,\max}^{\text{Trans}}$ are, respectively, the current magnitude and maximum current limit of τth transformer; I_f^{Feeder} and $I_{f,\max}^{\text{Feeder}}$, respectively, refer to the current magnitude and maximum current of fth feeder; V_b, $V_b^{\min}$, and $V_b^{\max}$, respectively, denote the voltage magnitude and its minimum/maximum limits at bth bus in the entire system; and S_l and $S_l^{\max}$, respectively, signify the apparent power flow and its maximum limit in lth tie-line/branch.

6.4 Case Study and Simulation Results

6.4.1 Utilized Solver

The presented formulations in (6.1)–(6.54) are reflected on the proposed framework in Figure 6.1, where MATLAB [37] and general algebraic modeling system (GAMS) environments are utilized for the programming purpose. GAMS CPLEX solver [38] has been utilized to solve the optimization problems. It is noted that we have used the well-known "piecewise linear approximation" in order to convert the nonlinear functions to mixed-integer linear programming (MILP) to be solved by the CPLEX solver. Interested readers are referred to our previous work [39], where it is elaborated how the piecewise linear approximation can be implemented for conversion of a nonlinear problem into MILP. In addition, GDX-MATLAB Read-Write (GDX-MRW) functions [40] have been implemented to link MATLAB and GAMS to take advantage of exchanging data between them. In other words, GAMS-related scripts are embedded in the body of different loops of MATLAB M-files to solve the optimization problems. It is noted that the programming and simulations have been performed on a computer with a processor of Intel(R) Core(TM) CPU i5-7500 @ 3.4 GHz.

6.4.2 Case Study

One of the most commonly used test systems in the existing literature to study different aspects of transmission networks is the IEEE 14-bus system [41–43]. Moreover, IEEE 33-bus system is one of the test systems commonly utilized to analyze the distribution networks from cybersecurity point of view [44–46]. As a result, the proposed framework has been implemented on the IEEE 14-bus system feeding the modified IEEE 33-bus systems as the case study, which is illustrated in Figure 6.3, where the main system is the IEEE 14-bus transmission network containing different subsystems (i.e., different modified versions of IEEE 33-bus distribution network). The subsystems are modified to contain renewable-based distributed generation (DG) units such as WT, photovoltaic (PV), and CHP units.

It is noted that the location of PV units, WTs, and CHP units is randomly altered in eleven distribution networks to result in different subsystems. In addition, all of the distribution systems connected to the load centers indicated by purple arrows in Figure 6.3 have approximately the same amount of electric demand of the corresponding buses in IEEE 14-bus system, which are extracted from the MATPOWER package [47].

The following explains how the IEEE 33-bus network is modified to form 11 different subsystems for 11 buses of the transmission system. The rated power associated with all PV units is 50 kW. In addition, the solar irradiance at standard test conditions and the certain irradiance point are, respectively, equal to 1000 W/m^2 and 120 W/m^2. The rated power and diameter of WTs manufactured by Wind World [48] are, respectively, 250 kW and 29.2 m. The v_{ci}, v_{cout}, and v_{rated} associated with the mentioned WT are, respectively, 2 m/s, 12 m/s, and 25 m/s. However, the average wind speed (i.e., v) related to each distribution system is provided in Table 6.1. The characteristics of different types of CHP units (i.e., small-scale and large-scale units) integrated into distribution systems are provided in

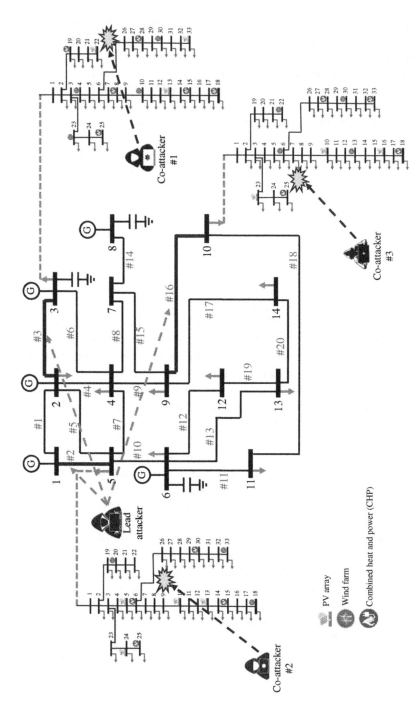

Figure 6.3 Transmission/distribution systems as the case study.

Table 6.1 Average wind speeds associated with eleven distribution systems.

Subsys. #	1	2	3	4	5	6	7	8	9	10	11
v (m/s)	8.4	6.1	5.9	7.5	9.2	7.0	8.6	6.5	7.9	8.0	4.7

Table 6.2 Descriptions of CHP units integrated into the distribution systems.

Feasible regions [P^{CHP}, H^{CHP}]				α_n	β_n	δ_n	η_n	λ_n	ω_n
[20,0]	[10,40]	[45,55]	[60,0]	2650	34.5000	0.1035	0.0510	2.2030	0.0250
[98.8,0]	[81,104.8]	[215,180]	[247,0]	2650	14.5000	0.0345	0.0310	4.2000	0.0300

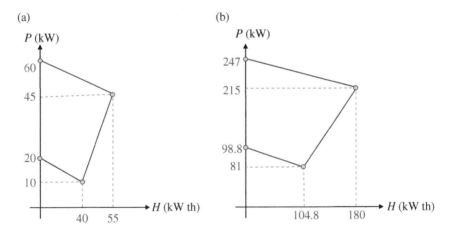

(a)   (b)

Figure 6.4 FORs associated with CHP units: (a) Small-scale unit and (b) large-scale unit. *Source:* Mohammadi-Ivatloo et al. [32]/ with permission of Elsevier.

Table 6.2. To have a better perspective, Figure 6.4 shows the feasible regions associated with two types of CHP units integrated into the case study. Herein, the CHP units embedded in different subsystems are a combination of these two types of CHPs. For example, the subsystem connected to bus #10 of the transmission system includes two small-scale CHP units (refer to Figure 6.4a) and two large-scale CHP units (see Figure 6.4b). It is noted that the generation units in each distribution system are sized in such a way that each subsystem can supply up to 80% of its local demand (i.e., both electric and thermal demands). Therefore, all of the generation units in the case study (i.e., combination of transmission and distribution levels) need to contribute to serve the total demand and meet the system's losses (i.e., both electric and thermal losses).

The main system (i.e., IEEE 14-bus transmission network) and the eleven subsystems (i.e., IEEE 33-bus distribution networks) are equipped with overcurrent relays, where in *downstream* tie-lines of IEEE 14-bus system and branches of IEEE 33-bus systems, the current flow is bidirectional; however, in *upstream* tie-lines and branches, the flow of current is unidirectional. To this end, each bidirectional tie-line/branch is secured through two overcurrent relays, which are located at both ends. However, each unidirectional tie-line/branch is protected by only one overcurrent relay, which is considered at the sending end. To have a better perspective, Figure 6.5 illustrates a section of IEEE 14-bus test system and a section of IEEE 33-bus test system depicting the described relays configuration throughout the case study (see the squares in Figure 6.5). In addition, Tables 6.3 and 6.4, respectively, display the TDs associated with protection systems embedded in the transmission system and the distribution systems. It is noted that Table 6.4 presents a general case of relays configuration considering two relays per distribution branch meaning that all of the branches are bidirectional. The CTR is equal to 400/5 for all of the overcurrent relays in the system. As was indicated, the subsystems are modified versions of the IEEE 33-bus network with different type/ number of DG units. For example, in the first distribution system, buses #19 and #22 have generation units (i.e., CHP and PV, respectively, refer to Figure 6.3); therefore, two relays per line is considered since there will be currents flowing in the *reverse* direction. However, in another distribution system, there is no need

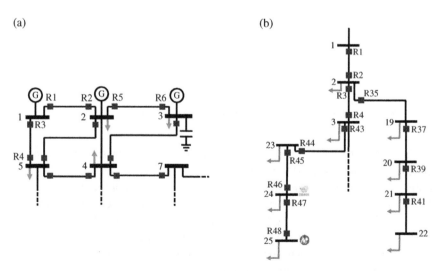

Figure 6.5 Selected sections of systems showing the installed relays: (a) A part of IEEE 14-bus system and (b) a part of IEEE 33-bus subsystem.

Table 6.3 TDs associated with the overcurrent relays embedded in IEEE 14-bus system.

Tie-line #	Relay #	TD (s)	Tie-line #	Relay #	TD (s)
1	1	0.3038	11	21	0.1534
	2	0.1211		22	0.1473
2	3	0.1418	12	23	0.2554
	4	0.1184		24	0.1437
3	5	0.1184	13	25	0.3358
	6	0.1020		26	0.2024
4	7	0.1718	14	27	0.1641
	8	0.1452		28	0.1779
5	9	0.1258	15	29	0.2033
	10	0.1523		30	0.2124
6	11	0.1006	16	31	0.2053
	12	0.1414		32	0.1001
7	13	0.1307	17	33	0.1511
	14	0.1250		34	0.2285
8	15	0.1872	18	35	0.3681
	16	0.2302		36	0.1300
9	17	0.3258	19	37	0.3143
	18	0.1301		38	0.1892
10	19	0.3402	20	39	0.1596
	20	0.1057		40	0.1485

Source: Adapted from Liu and Yang [49].

to consider two relays per line since there is no generation unit located at buses #19–#22 (Figure 6.5b as an example). The rest of data associated with the IEEE 14-bus and 33-bus test systems including thermal generators' cost coefficients, impedance of transmission lines, and distribution branches are extracted from the MATPOWER package [47].

According to Figure 6.3, the lead attacker coordinates with co-attackers assigned to different subsystems including but not limited to the ones connected to buses #3, #5, and #10. As was indicated earlier, there are eleven subsystems connected to buses #2, #3, #4, #5, #6, #9, #10, #11, #12, #13, and #14. For the sake of cleanness and clarity of the figure, the rest of the nine subsystems have not been included in Figure 6.3. Referring to the case study, Bus #3 of the transmission system feeds the first distribution system, which will be targeted by co-attacker #1.

Table 6.4 TDs associated with the overcurrent relays embedded in IEEE 33-bus systems.

Branch #	Relay #	TD (s)	Branch #	Relay #	TD (s)
1	1	1.005	17	33	0.050
	2	0.050		34	0.166
2	3	0.406	18	35	5.000
	4	0.325		36	5.000
3	5	0.353	19	37	2.537
	6	5.000		38	0.072
4	7	1.216	20	39	0.899
	8	0.050		40	0.132
5	9	0.747	21	41	0.050
	10	0.108		42	0.090
6	11	0.464	22	43	5.000
	12	0.111		44	5.000
7	13	1.008	23	45	3.856
	14	0.052		46	0.053
8	15	1.028	24	47	1.967
	16	0.068		48	0.096
9	17	0.272	25	49	0.186
	18	0.099		50	2.500
10	19	0.110	26	51	1.299
	20	0.058		52	0.050
11	21	0.642	27	53	1.077
	22	0.050		54	0.058
12	23	0.458	28	55	0.378
	24	0.070		56	0.112
13	25	0.252	29	57	0.539
	26	0.127		58	0.066
14	27	0.238	30	59	0.326
	28	0.077		60	0.078
15	29	0.490	31	61	0.200
	30	0.065		62	0.091
16	31	0.331	32	63	0.050
	32	0.137		64	0.058

Source: Data from Balyith et al. [50].

Similarly, buses #5 and #10 of the transmission system, respectively, feed the second and third distribution systems, which will be targeted by co-attackers #2 and #3, respectively. The cooperation of the lead attacker with co-attackers #1 to #3 and also the rest of nine co-attackers (not shown in Figure 6.3) will result in increasing the LMPs in (i) the transmission system (e.g., tie-lines #2, #3, and #16 indicated by red lines in Figure 6.3) and (ii) the targeted distribution systems based on the upper-level's objective function (i.e., OF_1) presented in (6.2).

According to the lower-level objective function (i.e., OF_2) formulated in (6.3), the team of co-attackers solve another optimization problem to (i) minimize the overall OTs associated with the overcurrent relays (refer to (6.3)) and (ii) identify a set of distribution branches in the vulnerable subsystems with higher IOs and lower OTs, which have been equipped with faster overcurrent relays; i.e., relays with less TDs provided in Tables 6.3 and 6.4. Cooperating with the co-attackers, the lead attacker ensures to attack the corresponding upstream buses (e.g., buses #3, #5, and #10 demonstrated in Figure 6.3) via targeting the lines having the highest contribution in feeding the (to be targeted) distribution networks. The intention of this cooperation is to push the entire system toward severe congestions and potentially cascading power outages if there is no proper remedial action against the coordinated attack.

6.4.3 Investigated Scenarios of Cyberattacks

This section provides three scenarios of cyberattacks targeting the power system illustrated in Figure 6.3 to cause congestions and possible cascading outages. The attack scenarios are described as follows.

- **Scenario I: Transmission level cyberattack**
  In this scenario, only the transmission system is the target of cyberattacks performed by the lead attacker. Therefore, there is no coordination between the lead attacker and co-attackers. However, the impacts of cyberattacks on the LMPs and line flows will be scrutinized from different points of view.
- **Scenario II: Distribution level cyberattack**
  In this scenario, several co-attackers target the corresponding subsystems to cause different levels of congestions or possible power outages by increasing the LMPs in the targeted distribution systems. The negative impacts of such cyberattacks will be investigated on the rest of the systems.
- **Scenario III: Coordinated cyberattacks on transmission and distribution levels concurrently**
  In this scenario, which is the main contribution of this chapter, both sectors (i.e., transmission network as the main system and distribution networks as the subsystems) are the target of a coordinated cyberattack performed by the lead attacker and the co-attackers at the same time (see Figure 6.3). The consequences of such coordinated attacks on the operation of the entire system will be analyzed.

6.4.4 Numerical Results and Analysis

6.4.4.1 Elaboration of Results Associated with Scenario I

Referring to the transmission level of the case study presented in Figure 6.3 and the proposed framework depicted in Figure 6.1, the lead attacker maximizes the upper-level objective function OF_1 in (6.2), where only the first term of (6.2) is considered in Scenario I. The LMPs associated with buses #1–#14 in IEEE 14-bus test system are illustrated in Figure 6.6.

As can be inferred from Figure 6.6, the lead attacker's action has led to an overall enhancement in LMPs of the targeted transmission system (compare the red curve indicated by black stars with the green curve indicated by blue squares). However, it is noted that buses #4, #6, and #14 experience the highest rates of changes regarding LMPs before and after the cyberattack. Therefore, the tie-lines associated with these three buses are more vulnerable to congestions. Since the main objective of the lead attacker in this scenario is to optimally push the transmission system toward congestion in different tie-lines, the second objective function (i.e., the first term of (6.3)) needs to be minimized to recognize a set of tie-lines: (i) which are equipped with faster overcurrent relays and (ii) if tripped, they have a more noticeable impact on the rate of cascading outages. In other words, according to (iii), the lead attacker minimizes the rate of OTs of the overcurrent relays over the IOs (i.e., min {OT/IO}). Toward this end, the numerical values of OT/IO for all of the twenty tie-lines in IEEE 14-bus system are depicted in Figure 6.7. As can be perceived from the figure, tie-lines #4, #5, #6, #7, #9, #11, #12, #13, #14, #16, #18, and #20 are not among suitable candidates (from

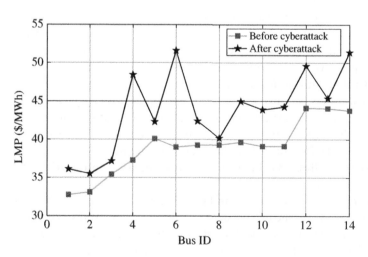

Figure 6.6 Comparison of LMPs before and after cyberattack associated with Scenario I.

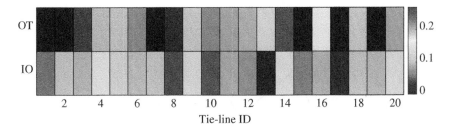

Figure 6.7 The importance of tie-lines to be targeted in Scenario I.

the lead attacker's standpoint) to be targeted by FDI cyberattacks leading to congestions since these tie-lines do not have either fast enough (i.e., sufficiently sensitive) protections operation or sufficiently high rates of IO. However, tie-lines #8, #10, and #17 have faster overcurrent relays (see the dark blue blocks in the first row indicating lower OTs) and higher rates of negative effects on the rest of the system due to possible outages (see the dark red and red blocks in the second row indicating higher IOs). The results of these two upper-level and lower-level optimization problems (i.e., Figures 6.6 and 6.7) provide the lead attacker with a set of tie-lines (to be targeted by FDIs) which have the following characteristics at the same time: (i) the recognized tie-lines have higher impacts on the overall LMPs in the transmission system, (ii) they are equipped with more sensitive protection operation (i.e., faster overcurrent relays), and (iii) they have a noticeable impact on the downstream subsystems. Hence, the indicated tie-lines #8, #10, and #17 meet all of the mentioned conditions based on which they are targeted by the lead attacker. These tie-lines are connected to buses #4, #6, and #14, which have the highest rate of changes in their corresponding LMPs before and after the cyberattack (see Figure 6.6).

Figure 6.8 displays the optimal false data injections in terms of kW to result in congestions in tie-lines #8, #10, and #17. It is noted that the positive sign in Figure 6.8 means the load of the bus is increased and negative sign is associated with reducing the load. According to Figure 6.8, one can infer that the injected false data have a unique pattern to maintain the cyberattack undetectable by meeting (6.10)–(6.17) to bypass the DCSE. The FDIs associated with bus #7 is equal to zero since this bus does not have any load center (see Figure 6.3). In addition, Figure 6.9 illustrates the power flow associated with the targeted tie-lines before and after injecting the false data into the load centers. It is noted that if a tie-line experiences a current more than 80% of its rated current, it is considered as congested. However, if a current higher than 110% of rated value flows into it, the corresponding breaker will trip leading to possible outages for downstream buses. Hence, according to Figure 6.9, although the three targeted tie-lines have experienced different rates of congestions, the power flow of tie-line #17 increases by 198

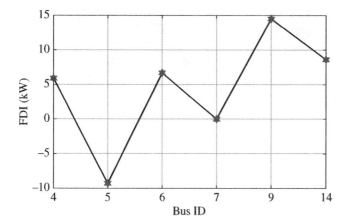

Figure 6.8 FDIs on load centers to result in congestions of lines #8, #10, and #17 in Scenario I.

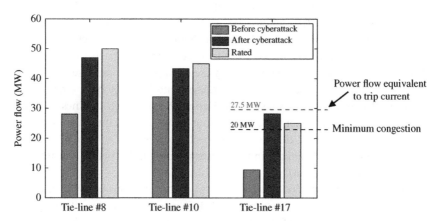

Figure 6.9 Power flow of targeted lines before/after cyberattack associated with Scenario I.

percent confirming that the consequence of this cyberattack is an outage since the tie-line is loaded about 113% (i.e., higher than the rated value).

6.4.4.2 Elaboration of Results Associated with Scenario II

In Scenario II, the focus of launched cyberattacks is only on the distribution level of the case study presented in Figure 6.3. In other words, only the second terms of the objective functions (6.2) and (6.3) are optimized in this scenario. As a result, co-attackers #1 to #11 separately solve (6.2) and (6.3) and target the corresponding distribution systems to increase the LMPs and cause possible congestions. Figure 6.10–6.12 show the LMPs before and after maximizing the second term of the upper-level's objective function provided in (6.2) for the three distribution systems connected to buses #4, #6, and #14 from the IEEE 14-bus test system.

As can be observed from Figures 6.10–6.12, the patterns of LMPs before and after the cyberattacks are different, which stems from the difference between the

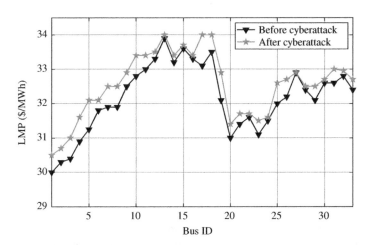

Figure 6.10 Comparison of LMPs on subsystem connected to bus #4 in Scenario II.

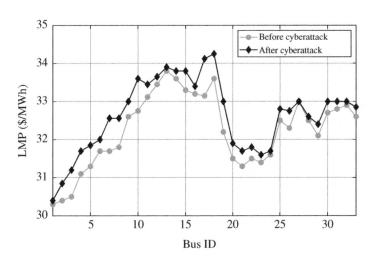

Figure 6.11 Comparison of LMPs on subsystem connected to bus #6 in Scenario II.

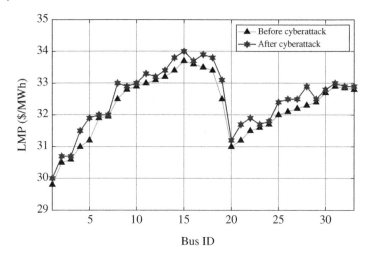

Figure 6.12 Comparison of LMPs on subsystem connected to bus #14 in Scenario II.

characteristics of subsystems. In addition, it can be inferred that distribution buses #17, #10, and #5, respectively, associated with distribution systems connected to transmission buses #4, #6, and #14 have the highest rate of changes in their corresponding LMPs compared to the rest of buses.

The numerical values of OT/IO for all of the 32 branches in all of the 11 subsystems are depicted in Figure 6.13. According to this figure, each co-attacker has a set of distribution branches that have higher impacts on the overall LMPs in the corresponding distribution system and are also equipped with more sensitive protection systems (i.e., fast enough overcurrent relays). For example, in the distribution system connected to bus #10 in the IEEE 14-bus system, the first two branches have the highest rate of IOs (see the dark red blocks depicted in Figure 6.13a). In addition, the first two branches in the same subsystem (i.e., distribution system connected to bus #10 in the IEEE 14-bus system) have fast enough protection systems (see the first two dark blue blocks demonstrated in Figure 6.13b). A similar explanation can be provided for branch #21 in the distribution system connected to bus #10. Moreover, branches #27 and #19 in the same distribution system (i.e., distribution system connected to bus #10 in the IEEE 14-bus system) have lower priorities (i.e., fourth and fifth, respectively) to be targeted since these two branches are equipped with slower relays compared to branches #1, #2, and #21. Therefore, the co-attacker assigned to this subsystem has the set of identified vulnerable branches as {branches #1, #2, #19, #21, and #27}.

Bypassing the ACSE (i.e., (6.18)–(6.26)), Figure 6.14 illustrates FDIs in terms of kW injected to distribution buses to result in congestions in the most vulnerable branches associated with the distribution system connected to bus #10 of the main

(a)

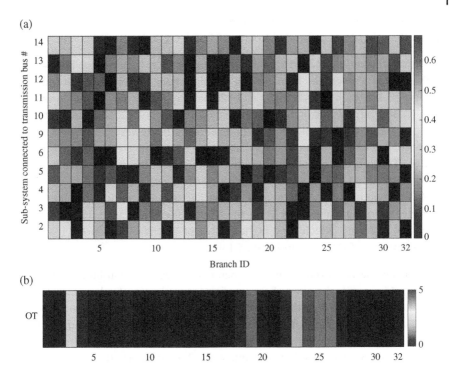

(b)

Figure 6.13 Importance of lines to be targeted in Scenario II: (a) IOs and (b) OTs.

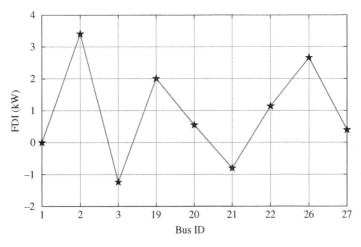

Figure 6.14 FDIs on the distribution buses to result in congestions in the most vulnerable branches in the subsystem connected to bus #10 of the transmission system related to Scenario II.

transmission system shown in Figure 6.13. The following provides an analysis about the subsystems connected to buses #4 and #12 in the main system.

- The co-attacker assigned to the subsystem connected to transmission bus #4 has the set of vulnerable branches as {branch #26}; however, according to Figure 6.13a, both branches #22 and #26 have approximately the same values of IO (see the only two orange blocks related to branches #22 and #26). According to Figure 6.13b, branch #22 is equipped with the slowest overcurrent relay; hence, this branch cannot be a suitable target from the co-attacker's point of view. This is because the main objective of the cyberattack is to optimally push the distribution system toward congestion and potential power outages.
- Referring to the same strategy, the co-attacker assigned to the subsystem connected to transmission bus #12 has the set of vulnerable branches as {branches #5, #9, and #25}, which have higher values of IO (refer to Figure 6.13a) and lower values of OT (refer to Figure 6.13b).

Verifying the effectiveness of the cyberattacks targeting distribution systems, Figure 6.15 presents the power flow associated with the targeted branches before and after injecting the false data into the targeted buses. It is necessary to note that if a distribution line experiences a current more than 80% of its rated current, it is considered as a congested branch. However, in this distribution system, if a current higher than 105% of rated value flows into a distribution branch, the corresponding

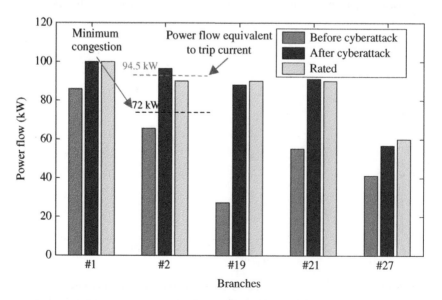

Figure 6.15 Power flows of the subsystem connected to bus #10 in Scenario II.

breaker will trip leading to possible outages for downstream buses. Therefore, targeting line #2 can result in its tripping, which will lead to power outages for downstream buses. Thus, according to Figure 6.15, although the indicated five distribution branches have experienced different rates of congestions, the power flows of branches #2 and #21, respectively, increase by almost 50 and 68% confirming that the consequence of this cyberattack is an outage for branch #2 since the indicated branches are loaded about 107% and 101%, respectively. Hence, the obtained results presented in Figure 6.15 verify that branch #2 can be the optimal target from the corresponding co-attacker's standpoint.

6.4.4.3 Elaboration of Results Associated with Scenario III

This section reflects the coordination between lead attacker and co-attackers to concurrently target the main system and subsystems to cause congestions leading to potential power outages. Toward this end, Figure 6.16 presents the LMPs related to buses #1 to #14 in the main system, when the case study is targeted by a coordinated cyberattack to maximize the LMPs in the transmission system and the relevant distribution systems. As can be perceived from this figure, after the coordinated cyberattack on the system (i.e., Scenario III), the transmission buses experience higher values of LMPs compared to Scenario I, where only the transmission system was the main target of the cyberattack (i.e., compare Figures 6.6 and 6.16). As can be inferred from Figure 6.16, transmission buses #7, #8, #10, and #14 experience the highest rate of alterations in LMPs before and after the coordinated cyberattack.

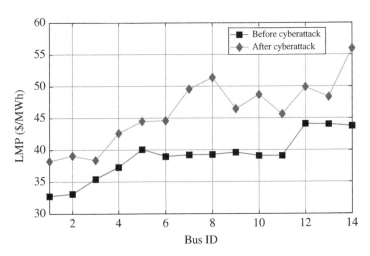

Figure 6.16 Comparison of LMPs on main system before/after cyberattack in Scenario III.

Table 6.5 presents the most vulnerable tie-lines/branches in the main system (i.e., IEEE 14-bus system) and the subsystems (i.e., 11 versions of the IEEE 33-bus system).

It is noted that the reported tie-lines/branches in Table 6.5 are obtained via applying the same strategy elaborated in Scenarios I and II. In other words, the team of attackers need to look into the numerical values of IOs and OTs to recognize the targeted tie-lines/branches, which: (i) are equipped with faster

Table 6.5 Sets of tie-lines/branches to be simultaneously targeted in Scenario III.

Level	Subsys. connected to trans. bus #	System	Buses with highest LMPs	Set of candidate tie-lines/branches
Transmission	–	IEEE 14-bus	#7, #8, #10, and #14	#8, #14, #15, **#16**, #17, **#18**, and **#20**
Distribution	2	IEEE 33-bus	#4, #5, #18, #25, and #30	**#3**, #4, #5, #17, **#24**, #29, and #30
	3		#2, #6, #13, #20, and #24	#1, **#3**, #5, #6, #12, #13, **#19**, #20, **#23**, and **#24**
	4		#6, #16, #21, #23, #29, and #30	#5, #6, #15, #16, #20, #21, **#22**, **#23**, #28, #29, and #30
	5		#4, #7, #13, #21, and #33	**#3**, #4, #6, #7, #12, #13, #20, #21, and #32
	6		#9, #10, and #20	#8, #9, **#19**, and #20
	9		#5, #6, #10, and #25	#4, #5, #6, #9, #10, and **#24**
	10		#3, #9, #13, #18, #24, and #33	#2, **#3**, #8, #9, #12, #13, #17, **#23**, **#24**, and #32
	11		#8, #15, #16, #19, and #26	#7, #8, #14, #15, #16, **#18**, **#19**, **#25**, and **#26**
	12		#6, #11, #14, and #23	#5, #6, #10, #11, #13, #14, **#22**, and **#23**
	13		#2, #5, #13, #18, and #24	#1, #2, #4, #5, #12, #13, #17, **#23**, and **#24**
	14		#7, #9, #17, #19, #24, and #33	#6, #7, #8, #9, #16, #17, **#18**, **#19**, **#23**, **#24**, and #32

overcurrent relays and (ii) have a noticeable impact on the rate of congestions or possible outages. It is also noted that those tie-lines/branches which do not meet the mentioned attack conditions are **boldfaced** (condition #1: faster overcurrent relays) and indicated by *italic* font (condition #2: having a significant impact on congestions/outages) in Table 6.5. For example, regarding the subsystem connected to Bus #2 of the main system (i.e., second row of Table 6.5), branch #3 and branch #30 are, respectively, **boldfaced** and indicated by *italic* font meaning that (i) branch #3 is not equipped with fast enough overcurrent relay and (ii) branch #30 is not vulnerable to severe congestion. Therefore, attackers can exclude these two branches from the list of the "to be targeted" branches. Table 6.6 provides the FDI vectors injected to transmission and distribution buses resulting in congestions or power outages in the targeted tie-lines/branches (see the last column in Table 6.5). The pattern of FDI vectors are different to keep the coordinated attack stealthy via bypassing DCSE and ACSE, respectively, by lead attacker and co-attackers.

After concurrently injecting the FDI vectors into the main system and subsystems by the corresponding lead attacker and the co-attackers, the set of candidate

Table 6.6 FDIs on the transmission and distribution buses related to Scenario III.

System/subsystem connected to trans. bus #	FDI vectors
Main transmission	$[0.00, 0.00, 22.69, -14.11]$
2	$[-0.55, -0.72, 2.11, 0.88, -1.32]$
3	$[1.29, 3.47, -1.06, -0.04, 0.93]$
4	$[0.29, 1.21, -2.19, -0.56, -0.60, 2.00]$
5	$[-2.43, -0.05, 3.16, -1.28, 0.89]$
6	$[-3.77, 1.69, 2.45]$
9	$[0.99, 0.34, -1.67, 0.00]$
10	$[3.63, -1.52, -0.61, -2.32, 2.74, 0.08]$
11	$[1.00, 0.66, 0.11, 0.40, 1.03]$
12	$[-2.99, -1.18, -0.53, 2.95]$
13	$[-0.33, -1.28, 1.49, 1.16, 0.33]$
14	$[1.73, 1.02, -2.78, -1.80, 0.64, 1.96]$

tie-lines/branches presented in Table 6.5 experience noticeable rates of congestions or possible power outages compared to separately targeting the transmission system and distribution systems (i.e., Scenario III vs. Scenarios I and II). In this regard, Figure 6.17 displays the power flows associated with the targeted tie-lines before and after injecting the malicious data in Scenario III in the main system.

In Figure 6.17, although tie-lines #8, #14, and #17 have experienced different rates of congestions, only one tie-line can experience tripping of its breaker. The power flow of tie-line #17 increases by 133% confirming that the consequence of the coordinated cyberattack can be a power outage in this tie-line since the tie-line is loaded about 128%. Comparing Figures 6.9 and 6.17, one can infer that the intensity of imposed congestion in tie-line #17 increases by almost 14% when *both* the transmission and distribution levels of the proposed case study are the target of a *coordinated* cyberattack (i.e., Scenario III vs. Scenario I).

To obtain a better perspective about the impacts of targeting both transmission system and distribution systems at the same time, Figure 6.18 illustrates the power flow of the set of candidate branches associated with distribution system connected to transmission bus #5 (i.e., targeted distribution lines #4, #6, #7, #13, #21, and #32). According to this figure, although the six targeted distribution branches have experienced different rates of congestions, the power flow of branches #4, #21, and #32, respectively, increase by almost 182%, 225%, and 191% meaning that the mentioned branches are loaded about 103.3%, 115.6, and 107%, respectively. Given that the trip current settings of the entire breakers in this subsystem (i.e., the distribution network fed from bus #5 of the

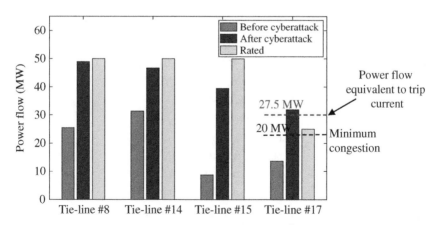

Figure 6.17 Power flows of targeted lines of main system before/after attack in Scenario III.

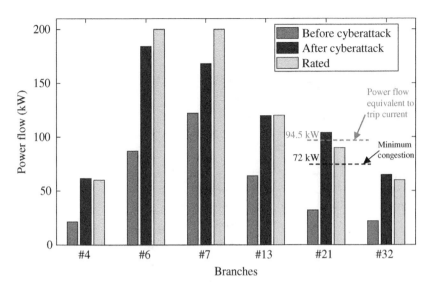

Figure 6.18 Power flows of the subsystem connected to bus #5 in Scenario III.

transmission network) are assumed to be 103% of the corresponding rated currents, the breakers of branches #4, #21, and #32 will trip after the indicated coordinated attack.

According to Figure 6.19, tripping of breaker related to branch #21 will lead to the power outage of bus #22. However, tripping of breaker associated with branch #32 will not result in any power outage for bus #33 since the local DG can sufficiently supply its demand. On the other hand, the attack targeted on branch #4 will result in disconnection of three downstream regions from the utility source as indicated in Figure 6.19. As can be inferred from the figure, although region R2 is disconnected from the utility source due to the coordinated cyberattack, it will not experience any power outage thanks to having sufficient contributions from WT/PV units and CHP in the region. However, the other disconnected regions from the power grid (i.e., regions R1 & R3) will experience outages as the consequence of severe congestion on critical line #4 (having higher IO) with more sensitive breaker (having less OT). This implies that, when the lead attacker coordinates with the co-attacker assigned to the subsystem connected to transmission bus #5, their major concentration is to enhance the LMPs of transmission bus #5 and distribution bus #4 so that successful tripping of breaker on distribution line #4, having a *more sensitive relay setting*, can result in the intended cascading outages for two downstream regions.

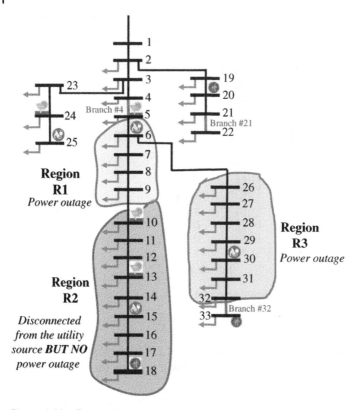

Figure 6.19 Targeted branches/regions in subsystem connected to bus #5 in Scenario III.

6.5 Conclusions and Future Work

In this chapter, a coordinated attack framework was proposed as a platform for a team of attackers simultaneously targeting the vulnerable lines of a transmission network (i.e., main system) and the corresponding distribution networks (i.e., subsystems), leading to severe congestions to cause potential cascading outages. The effectiveness of the proposed framework was validated on the IEEE 14-bus transmission network (as the main system) and 11 modified versions of IEEE 33-bus distribution network (as 11 subsystems of the main system).

The following indicates the main achievements of this chapter. The obtained simulation results verified that targeting the vulnerable sections of the case study with higher LMPs leads to more severe congestions. The results also validated that targeting the more critical lines with greater IOs and more sensitive protective devices with less OTs results in more noticeable cascading outages. In addition, the results demonstrated that the proposed coordinated attack (i.e., simultaneous attack on transmission/distribution networks via the lead attacker and co-attackers) causes more significant congestions/outages compared to the scenarios of *only* targeting the transmission network or *only* attacking the distribution networks. According to the results, when only the transmission system was the target of the cyberattack, the power flow of tie-line #17 was increased by 113% compared to the normal condition. However, in case of launching the coordinated attack targeting both transmission and distribution levels, tie-line #17 was loaded about 128% with respect to the normal condition. Furthermore, the LMP of bus #14 was increased by 17.9% in targeting only the transmission system, whereas the same LMP was increased by 28.3% when both transmission and distribution levels were the targets of the coordinated attack.

In the next step of this research, we will propose a coordinated remedial action scheme against the attack model investigated in this chapter. Our remedial action scheme will be based on a coordinated transmission/distribution networks reconfiguration so that the transmission system operator and the distribution system operators effectively mitigate the consequences of the examined coordinated attack caused by the lead attacker (on the main system) and the co-attackers (on the subsystems).

References

1 Liang, G., Weller, S.R., Zhao, J. et al. (2017). The 2015 Ukraine blackout: implications for false data injection attacks. *IEEE Transactions on Power Systems* 32 (4): 3317–3318. http://dx.doi.org/10.1109/TPWRS.2016.2631891.

2 Lee, R.M., Assante, M.J., and Conway, T. (2016). Analysis of the cyber attack on the Ukrainian power grid. SysAdmin, Audit, Network and Security (SANS). https://media.kasperskycontenthub.com/wp-content/uploads/sites/43/2016/05/20081514/E-ISAC_SANS_Ukraine_DUC_5.pdf.

3 Liu, C., Liang, H., and Chen, T. (2021). Network parameter coordinated false data injection attacks against power system AC state estimation. *IEEE Transactions on Smart Grid* 12 (2): 1626–1639. http://dx.doi.org/10.1109/TSG.2020.3033520.

4 Tu, H., Xia, Y., Tse, C.K., and Chen, X. (2020). A hybrid cyber attack model for cyber-physical power systems. *IEEE Access.* 8: 114876–114883. http://dx.doi.org/10.1109/ACCESS.2020.3003323.

5 Cui, M., Wang, J., and Chen, B. (2020). Flexible machine learning-based cyberattack detection using spatiotemporal patterns for distribution systems. *IEEE Transactions on Smart Grid* 11 (2): 1805–1808. http://dx.doi.org/10.1109/TSG.2020.2965797.

6 Li, B., Ding, T., Huang, C. et al. (2019). Detecting false data injection attacks against power system state estimation with fast Go-Decomposition approach. *IEEE Transactions on Industrial Informatics* 15 (5): 2892–2904. http://dx.doi.org/1109/TII.2018.2875529.

7 Wang, P. and Govindarasu, M. (2020). Multi-agent based attack-resilient system integrity protection for smart grid. *IEEE Transactions on Smart Grid* 11 (4): 3447–3456. http://dx.doi.org/10.1109/TSG.2020.2970755.

8 Li, Y., Wang, Y., and Hu, S. (2020). Online generative adversary network based measurement recovery in false data injection attacks: a cyber-physical approach. *IEEE Transactions on Industrial Informatics* 16 (3): 2031–2043. http://dx.doi.org/10.1109/TII.2019.2921106.

9 Che, L., Liu, X., Li, Z., and Wen, Y. (2019). False data injection attacks induced sequential outages in power systems. *IEEE Transactions on Power Systems* 34 (2): 1513–1523. http://dx.doi.org/10.1109/TPWRS.2018.2871345.

10 Li, Z., Shahidehpour, M., Alabdulwahab, A., and Abusorrah, A. (2016). Bilevel model for analyzing coordinated cyber-physical attacks on power systems. *IEEE Transactions on Smart Grid* 7 (5): 2260–2272. http://dx.doi.org/10.1109/TSG.2015.2456107.

11 Bahrami, M., Fotuhi-Firuzabad, M., and Farzin, H. (2020). Reliability evaluation of power grids considering integrity attacks against substation protective IEDs. *IEEE Transactions on Industrial Informatics* 16 (2): 1035–1044. http://dx.doi.org/10.1109/TII.2019.2926557.

12 Chen, Y., Huang, S., Liu, F. et al. (2019). Evaluation of reinforcement learning-based false data injection attack to automatic voltage control. *IEEE Transactions on Smart Grid* 10 (2): 2158–2169. http://dx.doi.org/10.1109/TSG.2018.2790704.

13 Che, L., Liu, X., Ding, T., and Li, Z. (2019). Revealing impacts of cyber attacks on power grids vulnerability to cascading failures. *IEEE Transactions on Circuits and Systems II: Express Briefs* 66 (6): 1058–1062. http://dx.doi.org/10.1109/TCSII.2018.2869941.

14 Naderi, E., Pazouki, S., and Asrari, A. (2021). A remedial action scheme against false data injection cyberattacks in smart transmission systems: application of thyristor controlled series capacitor (TCSC). *IEEE Transactions on Industrial Informatics* 18 (4): 2297–2309. Early Access. http://dx.doi.org/10.1109/TII.2021.3092341.

15 Che, L., Liu, X., and Li, Z. (2019). Mitigating false data attacks induced overloads using a corrective dispatch scheme. *IEEE Transactions on Smart Grid* 10 (3): 3081–3091. http://dx.doi.org/10.1109/TSG.2018.2817515.

16 Soltan, S., Mittal, P., and Poor, H.V. (2018). BlackIoT: IoT botnet of high wattage devices can disrupt the power grid. *27th USENIX Security Symposium, Baltimore, MD, USA*. pp. 15–32. https://www.usenix.org/system/files/conference/usenixsecurity18/sec18-soltan.pdf.

17 Chung, H., Li, W., Yuen, C. et al. (2019). Local cyber-physical attack for masking line outage and topology attack in smart grid. *IEEE Transactions on Smart Grid* 10 (4): 4577–4588. http://dx.doi.org/10.1109/TSG.2018.2865316.

18 Zhang, Z., Huang, S., Liu, F., and Mei, S. (2020). Pattern analysis of topological attacks in cyber-physical power systems considering cascading outages. *IEEE Access.* 8: 134257–134267. http://dx.doi.org/10.1109/ACCESS.2020.3006555.

19 Mohammadi-Ivatloo, B. (2009). Optimal placement of PMUs for power system observability using topology based formulated algorithms. *Journal of Applied Sciences* 9 (13): 2463–2468. https://www.researchgate.net/publication/26630412.

20 Nazari-Heris, M. and Mohammadi-Ivatloo, B. (2015). Optimal placement of phasor measurement units to attain power system observability utilizing an upgraded binary harmony search algorithm. *Energy Systems* 6: 201–220. https://doi.org/10.1007/s12667-014-0135-3.

21 Chakhchoukh, Y. and Ishii, H. (2015). Coordinated cyber-attacks on the measurement function in hybrid state estimation. *IEEE Transactions on Power Systems* 30 (5): 2487–2497. http://dx.doi.org/10.1109/TPWRS.2014.2357182.

22 Cui, M. and Wang, J. (2021). Deeply hidden moving-target-defense for cybersecure unbalanced distribution systems considering voltage stability. *IEEE Transactions on Power Systems* 36 (3): 1961–1972. http://dx.doi.org/10.1109/TPWRS.2020.3031256.

23 Asrari, A. (2021). How to employ competitive smart home retailers to react to cyberattacks in smart cities? In: *Operation of Smart Homes*, ch. 3 (ed. M. Rahmani-Andebili), 63–92. Springer Nature https://doi.org/10.1007/978-3-030-64915-9_3.

24 Esmalifalak, M., Nguyen, H., Zheng, R. et al. (2018). A stealthy attack against electricity market using independent component analysis. *IEEE Systems Journal* 12 (1): 297–307. https://doi.org/10.1109/JSYST.2015.2483742.

25 National Fire Protection Association (NFPA) (2017). *National Electrical Code*, 1e, 1–1306. Independence, KY: Delmar Cengage Learn https://www.nfpa.org/codes-and-standards/all-codes-and-standards/list-of-codes-and-standards/detail?code=70.

26 Amraee, T. (2012). Coordination of directional overcurrent relays using seeker algorithm. *IEEE Transactions on Power Delivery* 27 (3): 1415–1422. https://doi.org/10.1109/TPWRD.2012.2190107.

27 Narimani, M.R., Huang, H., Umunnakwe, A. et al. (2021). Generalized contingency analysis based on graph theory and line outage distribution factor. *IEEE Systems Journal* 16 (1): Early Access. https://doi.org/10.1109/JSYST.2021.3089548.

28 Abur, A. and Exposito, A.G. (2004). *Power Systems State Estimation: Theory and Implementation*. Boca Raton: CRC Press https://doi.org/10.1201/9780203913673.

29 Bi, S. and Zhang, Y.J. (2014). Using covert topological information for defense against malicious attacks on DC state estimation. *IEEE Journal on Selected Areas in Communications* 32 (7): 1471–1485. https://doi.org/10.1109/JSAC.2014.2332051.

30 Hug, G. and Giampapa, J.A. (2012). Vulnerability assessment of AC State estimation with respect to false data injection cyber-attacks. *IEEE Transactions on Smart Grid* 3 (3): 1362–1370. https://doi.org/10.1109/TSG.2012.2195338.

31 Moradi-Dalvand, M., Nazari-Heris, M., Mohammadi-Ivatloo, B. et al. (2020). A two-stage mathematical programming approach for the solution of combined heat and power economic dispatch. *IEEE Systems Journal* 14 (2): 2873–2881. https://doi.org/10.1109/JSYST.2019.2958179.

32 Mohammadi-Ivatloo, B., Moradi-Dalvand, M., and Rabiee, A. (2013). Combined heat and power economic dispatch problem solution using particle swarm optimization with time varying acceleration coefficients. *Electric Power Systems Research* 95: 9–18. https://doi.org/10.1016/j.epsr.2012.08.005.

33 Naderi, E., Pourakbari-Kasmaei, M., and Lehtonen, M. (2020). Transmission expansion planning integrated with wind farms: a review, comparative study, and a novel profound search approach. *International Journal of Electrical Power & Energy Systems* 115: 105460. https://doi.org/10.1016/j.ijepes.2019.105460.

34 Azizivahed, A., Naderi, E., Narimani, H. et al. (2018). A new bi-objective approach to energy management in distribution networks with energy storage systems. *IEEE Transactions on Sustainable Energy* 9 (1): 56–64. https://doi.org/10.1109/TSTE.2017.2714644.

35 Elkadeem, M.R., Abd Elaziz, M., Ullah, Z. et al. (2019). Optimal planning of renewable energy-integrated distribution system considering uncertainties. *IEEE Access* 7: 164887–164907. https://doi.org/10.1109/ACCESS.2019.2947308.

36 Ullah, Z., Wang, S., Radosavljević, J., and Lai, J. (2019). A solution to the optimal power flow problem considering WT and PV generation. *IEEE Access* 7: 46763–46772. https://doi.org/10.1109/ACCESS.2019.2909561.

37 MATLAB and Simulink (2022). https://www.mathworks.com/products/matlab.html.

38 GAMS (2022). CPLEX solver manual. https://www.gams.com/latest/docs/S_MAIN.html.

39 Moazeni, F., Khazaei, J., and Asrari, A. (2021). Step towards energy-water smart microgrids; buildings thermal energy and water demand management embedded in economic dispatch. *IEEE Transactions on Smart Grid* 12: 3680–3691. https://ieeexplore.ieee.org/document/9383103.

40 GAMS-MATLAB Interface (2022). GDXMRW manual. https://www.gams.com/latest/docs/T_GDXMRW.html.

41 Liu, C., Liang, H., and Chen, T. (2021). Network parameter coordinated false data injection attacks against power system AC state estimation. *IEEE Transactions on Smart Grid* 12 (2): 1626–1639. https://doi.org/10.1109/TSG.2020.3033520.

42 Ustun, T.S., Hussain, S.M.S., Yavuz, L., and Onen, A. (2021). Artificial intelligence based intrusion detection system for IEC 61850 sampled values under symmetric and asymmetric faults. *IEEE Access.* 9: 56486–56495. https://doi.org/10.1109/ACCESS.2021.3071141.

43 Thams, F., Venzke, A., Eriksson, R., and Chatzivasileiadis, S. (2020). Efficient database generation for data-driven security assessment of power systems. *IEEE Transactions on Power Systems* 35 (1): 30–41. https://doi.org/10.1109/TPWRS.2018.2890769.

44 Verma, A., Krishan, R., and Mishra, S. (2018). A novel PV inverter control for maximization of wind power penetration. *IEEE Transactions on Industry Applications* 54 (6): 6364–6373. https://doi.org/10.1109/TIA.2018.2854875.

45 Dolatabadi, S.H., Ghorbanian, M., Siano, P., and Hatziargyriou, N.D. (2021). An enhanced IEEE 33 bus benchmark test system for distribution system studies. *IEEE Transactions on Power Systems* 36 (3): 2565–2572. https://doi.org/10.1109/TPWRS.2020.3038030.

46 Moradi, M.H., Abedini, M., and Hosseinian, S.M. (2016). A combination of evolutionary algorithm and game theory for optimal location and operation of DG from DG owner standpoints. *IEEE Transactions on Smart Grid* 7 (2): 608–616. https://doi.org/10.1109/TSG.2015.2422995.

47 Zimmerman, R.D., Murillo-Sanchez, C.E., and Thomas, R.J. (2011). MATPOWER: Steady-state operations, planning and analysis tools for power systems research and education. *IEEE Transactions on Power Systems* 26 (1): 12–19. https://doi.org/10.1109/TPWRS.2010.2051168.

48 Wind-Turbine-Models (2021). https://en.wind-turbine-models.com/ (accessed 10 October 2021).

49 Liu, A. and Yang, M.T. (2012). A new hybrid Nelder-Mead particle swarm optimization for coordination optimization of directional overcurrent relays. *Mathematical Problems in Engineering* 456047. https://doi.org/10.1155/2012/456047.

50 Balyith, A.A., Sharaf, H.M., Shaaban, M. et al. (2020). Non-communication based time-current-voltage dual setting directional overcurrent protection for radial distribution systems with DG. *IEEE Access* 8: 190572–190581. https://doi.org/10.1109/ACCESS.2020.3029818.

7

Cooperative Unmanned Aerial Vehicles for Monitoring and Maintenance of Heat and Electricity Incorporated Networks: A Learning-based Approach

Fereidoun H. Panahi and Farzad H. Panahi

Department of Electrical Engineering, University of Kurdistan, Sanandaj, Iran

Abbreviations

UAV	Unmanned aerial vehicle
GPS	Global positioning system
EH	Energy harvesting
SUAV	Searching unmanned aerial vehicle
GUAV	Ground-based unmanned aerial vehicle
RL	Reinforcement learning
VGUAV	Visiting ground-based unmanned aerial vehicle
NGUAV	Nearest ground-based unmanned aerial vehicle
SNR	Signal-to-noise ratio
QL	Q-learning

7.1 Introduction

Modern technologies, such as drones, open up plenty of new prospects for companies in the power/energy generation, transmission, and distribution sectors. Unmanned aerial vehicles (UAVs) can be used to monitor and maintain various sorts of power and energy-producing facilities [1–11]. They not only capture the present condition but also allow analysis of various aspects impacting the facility's functioning, thanks to the use of specific sensors. UAVs are nowadays equipped

Coordinated Operation and Planning of Modern Heat and Electricity Incorporated Networks,
First Edition. Edited by Mohammadreza Daneshvar, Behnam Mohammadi-Ivatloo, and Kazem Zare.
© 2023 The Institute of Electrical and Electronics Engineers, Inc.
Published 2023 by John Wiley & Sons, Inc.

with cutting-edge technologies such as photo or video cameras, multi-spectral cameras, thermal imagers, infrared cameras, GPS, and laser scanners [12]. UAVs can broadcast high-definition or infrared video in real-time, as well as capture comprehensive high-resolution pictures that are later processed into photogrammetry products. A thermal camera, for example, may aid in locating overheated infrastructure or areas that require further attention. High-resolution infrared cameras could also detect from a great distance the change in soil temperature that happens when leaked water soaks into the underground. Thermal cameras and trained drone operators can inspect and test electrical facilities such as power stations, powerlines, and voltage regulation devices from a safe distance for excessive heat spots, defective parts, and weak connections. Refinery and pipeline monitoring are two specific domains of use for drone thermal imaging in the oil and gas industry. Thus, the analysis may be done on a variety of infrastructure types, including energy towers (poles), water management infrastructure, wind farms, oil and gas industry, photovoltaic (PV) panels, etc. UAVs' potential to aid in the maintenance of power and energy networks is much larger. UAVs can conduct maintenance on assets dispersed across a large region, access to hard-to-reach infrastructure, and risky inspection operations typically performed by humans or costly helicopters or airplanes. Inspections with UAVs are less expensive, faster, and safer than with other technologies. They also enable for more accuracy and easier access to difficult-to-reach locations [7]. Most significantly, UAV-based inspections may be carried out without shutting down the energy network.

The authors in [1] discuss the details of an on-board visual-based navigation system that allows a UAV to perform power line following. In [4], a deep learning-based technique for monitoring power infrastructure elements using drone-captured images is proposed. A smart inspection system that can be mounted on a UAV for patrol inspection of transmission lines and auxiliary systems of power transmission systems has been developed in [6]. In [10], the solution of identification of pipeline leakage using UAVs is described. Their proposed system allows for identifying both large and small hydrocarbon leaks in pipelines laid above and underground. The authors in [13] developed a reliable autonomous power line tracking and inspection system based on a quadrotor helicopter. The model of the UAV was presented and evaluated in simulation and experiments performed in the real environment. Reference [14] focuses on the automatic measurement of power line 3D coordinates using UAV images to improve the automation level of power line inspections. The authors in [15] try to design feasible schemes to boost the international promotion of China's UAV transmission line inspection standardization and further development of international UAV inspection technology. In [16], UAVs are used for power infrastructure monitoring. A low-cost optimal aerial drone for surveillance and heat leakage detection in heating networks is developed in [17]. The authors in [18] propose a UAV-based thermal anomaly detection for distributed heating networks.

We propose a learning-based platform for early defect identification and repair of all types of power and energy-producing facilities. Unlike previous researchers' work, this chapter discusses an early fault detection and repair system that would not only detect the fault using an energy harvesting (EH)-assisted searching UAV (SUAV) but would also repair the fault and protect valuable lives and properties by notifying ground-based UAVs (GUAVs) via triggering an alarm with accurate position parameters. Indeed, data gathering is not the only UAV application being explored for the power and energy industries. UAVs can also be utilized to carry out perilous activities from great heights. We explicitly define the state, action space, and reward function for the learning agent (the SUAV). At each state, the learning agent is limited to a subset of its action set, which is determined by the action it selected in the previous stage. This is a novel strategy to dealing with large action spaces that relies on action elimination; that is, limiting each state's available actions to a subset of the most likely ones. Therefore, by exploring invalid actions less frequently, we can develop an algorithm that converges faster and performs better.

7.2 Application of Machine Learning in Power and Energy Networks

Energy operators can discover quality defects, malfunctions, or inventory shortages rapidly and at a reduced cost by using machine learning (ML)-based techniques. Identification can be done automatically, without the need for human intervention, using photos or a real-time video stream provided by UAVs. UAVs can already undertake autonomous inspections of power plants and transmission lines, as well as provide real-time analytics on infrastructure conditions, thanks to advancements in technology. We now describe several notable ML paradigms pertaining to different settings in which ML can be used [19, 20]:

- **Supervised learning** is the process of developing a mathematical model for the labeled training data set using learning strategies and then marking the new data using the developed mathematical model. Regression and classification algorithms are two examples of supervised learning algorithms.
- **Unsupervised learning** is the process of describing unmarked data and discovering the rules hidden within it. Unsupervised learning algorithms that are often used include single-class density estimation, single-class data dimension reduction, clustering, etc.

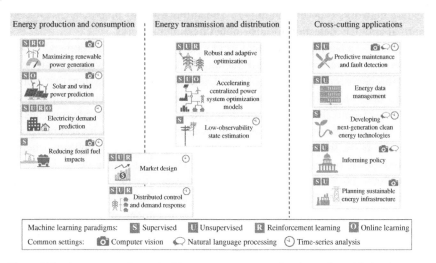

Figure 7.1 A summary of sustainable energy system applications where ML has been utilized, as well as typical ML paradigms used within each context. *Source:* Donti and Kolter [19]/Annual Reviews/CC BY-4.0.

- **Reinforcement learning (RL)** is when the system maximizes the value of an output function by learning in some situations. Unmanned driving and robot chess are two examples of when RL has proved successful.
- **Online learning**: The paradigms outlined above (with the possible exception of RL) often occur in an offline (or batch) setting, where the ML algorithm is given a whole dataset to learn over in advance. However, in the online (or streaming) setting, data points arrive one at a time, and the algorithm must make a prediction before receiving the next data point.

A summary of sustainable energy system applications where ML has been utilized, as well as typical ML paradigms used within each context, is given in Figure 7.1.

7.3 Unmanned Aerial Vehicle Applications in Energy and Electricity Incorporated Networks

UAV applications for data collection and inspection are not the only ones being explored for the power and energy industries. UAVs can also be used to do dangerous tasks at high altitudes, such as construction work, repairs, or vegetation trimming along power lines or thermal supply pipe networks [8, 11, 21].

For example, trees and growing plants that are too close to electricity lines are a serious hazard and a primary cause of power outages during severe winds and storms. In dry weather, branches contacting electricity lines can catch fire, risking human life and causing massive harm to the environment and existing infrastructure. UAVs can assist make the trimming operation more effective while also collecting data that can help predict and prevent damages from falling trees. UAVs' growing capabilities allow them to deliver construction supplies as well as assemble, weld, and connect various sorts of pieces. They also can be used as flame-throwing drones to clean rubbish of underground or aboveground electrical transmission lines and energy pipelines [22]. UAVs can complete these activities quicker than humans while lowering the danger of injury or death. The potential of UAV applications in the power and energy industries, however, is not restricted to these solutions.

UAV utilization presents both benefits and concerns regarding operational safety. UAV technology's growing application increases the number of aviation accidents and ground collisions. Human mistake, a loss of signal between the UAV and the operator, and technological flaws can all lead to events that endanger public safety. A UAV falling into power lines may cause blackouts and major damage to energy infrastructure. As UAVs collect precise data on power and energy infrastructure, there is an increasing danger that sensitive information may be hacked. To achieve effective protection, this concern necessitates the development of data security measures. Fixed-wing UAVs might replace planes, while multirotor UAVs could replace helicopters. UAVs have distinct benefits, including the capability to fly lower than other aerial vehicles, resulting in improved data quality, and the absence of a human pilot on board, which minimizes the danger of fatal accidents. The ultimate choice of UAV type and on-board equipment should be based on the company's expectations for data quality, acquisition time, and prices. A diverse set of new technologies may be employed to address the most critical challenges and improve their operations. One of the most intriguing technological applications is the use of ML in monitoring and maintenance operations [23, 24].

This chapter discusses the use of cooperative UAVs in monitoring and maintenance of all types of power and energy production facilities. We investigate an innovative monitoring and maintenance system for multi-UAV teams applicable to modern heat and electricity incorporated networks. The proposed system consists of (i) an EH-assisted SUAV that detects and reports abnormalities in any infrastructure component, or spots that require further action, and (ii) GUAVs that travel to the waypoints to perform dangerous tasks at high altitudes, such as construction work, repairs, and etc. First, the SUAV carries out gathering data and inspection with special equipment capable of collecting images and measurements of the deficiencies for evaluation. The SUAV then sends alarm signals to the GUAVs, instructing them to commence the relevant mission. The goal is to inform as many GUAVs as possible, as the required tasks can be completed more efficiently and quickly with more GUAVs. To do this, the SUAV with on-board processing capabilities makes use of the advantages of RL [25–27]. In particular, we discuss an RL-based trajectory design problem that enables the energy-constrained SUAV to effectively notify as many GUAVs as possible while considering the limited execution time. The critical issue is to identify defects and risks associated with investments as quickly as possible, allowing for quick intervention and therefore minimizing extra budget spending.

7.4 Cooperative UAVs for Monitoring and Maintenance of Heat and Electricity Incorporated Networks: A Learning-based Approach

7.4.1 Network Topology

We consider an SUAV enabled with solar EH capability [28–30] that informs the GUAVs deployed over the infrastructure area. The location of each GUAV is (x_n, y_n, z_n), where x_n and y_n are the Cartesian coordinates, and z_n is the altitude of the n-th UAV placed at its own ground charging station. The SUAV harvests energy from solar energy according to a certain EH model. More specifically, as depicted in Figure 7.2, a wireless network with downlink communications [31] is considered where an EH-assisted SUAV carries out gathering data and inspection with special

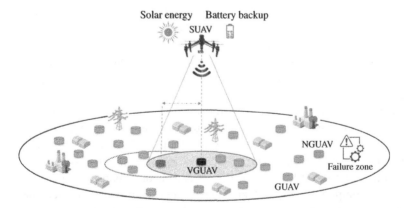

Figure 7.2 System model. The solar EH-enabled SUAV informs the GUAVs deployed over the infrastructure area about failure by starting from the nearest GUAV (NGUAV) to the failure zone. As shown, the SUAV is currently visiting the n-th GUAV (VGUAV). The next GUAV to visit by the RL-enabled SUAV is an GUAV inside the cluster (the filled-in red cylindrical shape) of which the currently VGUAV is located at its center point.

equipment capable of collecting images and measurements of the deficiencies for evaluation. The SUAV then transmits alarm signals to inform a range of GUAVs. In this model, we consider a scenario in which N GUAVs are distributed uniformly over the infrastructure area. At each time step (each SUAV's operation), only a cluster of C_n GUAVs ($C_n \leq N$) within the circular coverage region of the SUAV will receive the alarm signals from the SUAV. The GUAV located at the center point is called the visiting GUAV (VGUAV). Notice that the coverage region of the SUAV is a circular disk with radius R_c, and C_n is not a constant. Indeed, the SUAV will cover (inform) only a cluster of GUAVs at each time step (operation) in the presented model, and the SUAV then travels to wake up/inform more GUAVs after the current operation is done. Since the SUAV will fly at medium altitude, it might trigger false alarms due to the height or lack of good vision. Thus, we can conclude that in each operation, each informed GUAV within the cluster will be activated (starting to fly toward the spot where abnormalities in infrastructure components have been discovered) with probability $A_R(1-P_{FA})$, where P_{FA} is the false alarm probability of detecting the damaged spot by the SUAV, and A_R is its detection accuracy. Note that P_{FA} and A_R can take any value in the range of [0–1] depending on the detection technique employed at the SUAV [12, 32–35]. At the end, a "mission success" metric can be measured simply by counting the total number of informed GUAVs (over the entire infrastructure area) in each SUAV's tour who fly to the reported zone to start the relevant mission after being activated. Note that from the first to the last operation is one SUAV's tour or mission. The goal in each SUAV's mission is to inform as many GUAVs as possible as the required task can be completed more effectively and faster with more GUAVs.

Furthermore, the SUAV sends alarm signals to under-coverage UAVs at a fixed altitude H. The SUAV's alarm signals, which are converted into packet-based messages of size P_s, consist of two elements: (i) the estimated spot of damage and (ii) the mission residual time at time step i, i.e., T_i^R, as each relevant mission has to be completed in its entirety under a single time limit T^M, i.e., the mission or response time. As mentioned, the critical issue is to identify defects and risks associated with infrastructures as quickly as possible, allowing for quick intervention and therefore minimizing extra budget spending. It is worth noting that, if a typical GUAV is informed (about the spot where abnormalities in infrastructure components have been discovered) by the SUAV before T^M and is activated (starting to fly toward the spot of damage), this will be counted a success for the proposed RL-enabled SUAV.

7.4.2 Solar Power Harvesting Model

As previously stated, the SUAV has solar EH capability, which harvests energy using a PV cell. The amount of power harvested from solar energy is determined by the time of day and the months of the year, as well as the solar radiance and the PV system efficiency. Using the probability distribution function (PDF) of the harvested solar power P_{PV} (P_{PV}: the output power of the PV system), the cumulative distribution function (CDF) $F_{P_{PV}}(p_{PV})$ of the harvested solar power can be obtained as [28, 36]

$$F_{P_{PV}}(p_{PV}) = \begin{cases} X, & p_{PV} < \eta_c K_c \\ Y, & \text{otherwise,} \end{cases} \tag{7.1}$$

where,

$$X = \frac{1}{2}\left(\text{erf}\left[\frac{I_d}{\sqrt{2}}\right] - \text{erf}\left[\frac{I_d - \sqrt{\frac{K_c p_{PV}}{\eta_c}}}{\sqrt{2}}\right]\right) \tag{7.2}$$

and

$$Y = \frac{1}{2}\left(\text{erf}\left[\frac{I_d - K_c}{\sqrt{2}}\right] + \text{erf}\left[\frac{p_{PV} - I_d \eta_c}{\sqrt{2}\eta_c}\right]\right), \tag{7.3}$$

in which, K_c is a radiation intensity threshold beyond which the efficiency η_c can be regarded fixed. $I_d(t)$ is the deterministic component of the solar radiation intensity, which is generally affected by the time of day and the months/seasons of the year.

7.4.3 SUAV's Energy Outage

We now describe the SUAV's energy outage probability in light of solar power harvesting. If the battery energy, together with the harvested energy, is insufficient to sustain the SUAV's overall energy consumption for the whole mission period T^M, the SUAV is considered to be in an energy outage. The energy outage probability E_O is expressed mathematically as:

$$E_O = P[P_{PV} < \delta], \tag{7.4}$$

where $\delta = \left(E^l_{SUAV} + E^f_{SUAV} - P_B T^M\right)/T^M$, E^l_{SUAV} denotes the total energy consumed by the SUAV over the communication links during the whole transmission periods (while hovering), E^f_{SUAV} represents the total energy consumption of the SUAV over the whole flight periods, and P_B is the SUAV's battery power. The energy coverage probability ($E_C = 1 - E_O$), defined as the probability of the SUAV receiving enough energy to ensure a successful mission, can be computed based on the CDF of the harvested power in Eq. (7.1) and the expression of energy outage event in Eq. (7.4), as follows:

$$E_C = 1 - F_{P_{PV}}(\delta). \tag{7.5}$$

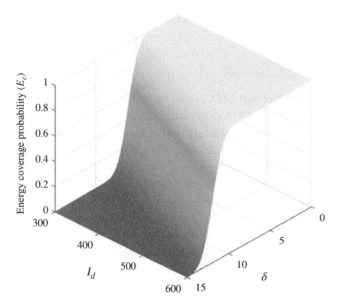

Figure 7.3 SUAV's energy coverage probability vs. the deterministic component of the solar radiation intensity (i.e., I_d) and the SUAV's energy outage threshold (i.e., δ). The values of K_c and η_c are respectively set to 150 and 0.02.

In Figure 7.3, we illustrate the SUAV's energy coverage probability versus the deterministic component of the solar radiation intensity (i.e., I_d) and the SUAV's energy outage threshold (i.e., δ). It is intuitive that when I_d increases and δ decreases, the probability of the SUAV receiving enough energy to ensure a successful mission increases.

7.4.4 Mission Success Metric

Given the provided system model, our objective is to design an optimal flight trajectory for the SUAV. By optimizing the trajectory of the SUAV, we maximize the mission success metric, which reflects the total number of GUAVs in the entire infrastructure region who not only get notification of a network behavior anomaly at an acceptable signal-to-noise ratio (SNR) level but also fly into the failure zone to do construction works, repairs, and etc. In particular, we consider an RL-based trajectory design problem that enables the energy-constrained SUAV to effectively inform as many number of GUAVs as possible while considering the limited execution time, as the critical issue is to identify defects and risks associated with investments as quickly as possible, allowing for quick intervention and therefore minimizing extra budget spending.

We now formulate the the "mission success metric" (M_s) while taking into account the P_{FA} and A_R:

$$M_s = A_R.(1 - P_{FA}).\sum_{n \in V} I^n_{\text{inf}}, \tag{7.6}$$

where I^n_{inf} is the number of informed GUAVs within the mission time limit T^M. V represents the visiting order of the GUAVs in each SUAVs tour (mission), subject to T^R_i. In other words, the sequence of visited states in each SUAV's tour is shown by V. The solution will be reached if the SUAV finds the optimal sequence of states (GUAVs) to visit, so that the total number of informed GUAVs will be maximized. Note that if T^R_i is not imposed, then the SUAV can easily find a trajectory that visits all the GUAVs through flying over the entire area. In fact, here, we discuss the problem of computing an optimal flight trajectory for the SUAV to visit a set of GUAVs under a limited mission time. Since determining the optimal trajectory necessitates evaluating all potential permutations of visit order V, which is very time-consuming, it is critical to implement a learning method to reduce the computation time for the optimal trajectory. It is worth noting that the SUAV can learn its optimal policy offline and then utilize that policy to make the optimal decisions and find the optimal trajectory. This can aid in real-time decision-making. If a typical GUAV is informed by the SUAV about the abnormalities in infrastructure component or spots that require further action before the time limit T^M (i.e., as long as $T^R_i \geq 0$) and is activated (starting to fly toward the zone of damage), this will be counted a success for the proposed RL-enabled searching UAV. As a result, not all of the informed GUAVs will fly toward the failure zone; only those who have been activated will. The activation of the informed GUAVs depends on the SUAVs false alarm probability (P_{FA}) and its detection accuracy (A_R).

7.4.5 Learning Strategy

To maximize mission success metric M_s (see Eq. (7.6)), we provide an RL framework based on double Q-learning (QL). In contrast to current RL methods such as QL, which may lead to a sub-optimal trajectory, the proposed double QL approach allows the SUAV to determine the optimal flying trajectory, maximizing the number of informed GUAVs. Furthermore, unlike the conventional QL algorithm, which normally utilizes a single Q-table to store and update the values resulting from various states and actions [37], the proposed double QL method employs two Q-tables (Q^A and Q^B) to choose and evaluate the action separately [38]. In this sense, the proposed double QL algorithm prevents the overestimation of Q values. Overestimation is common in conventional QL algorithms due to the positive feedback induced by choosing and evaluating the action in the same Q-table.

The QL and double QL algorithms used by the SUAV, are made up of four basic components: an agent, states, actions, and a reward function. These basic components are described in this chapter as follows:

- **Agent**: Agents are classified into five categories based on their degree of perceived intelligence and capability: simple reflex agent, model-based reflex agent, goal-based agent, utility-based agent, and learning agent. All these agents can improve their performance and generate better action over the time. Please refer to [39] for more detailed information. The agent in this chapter is a learning agent. Agents with the capability of learning from their previous experience are learning agents. They begin by acting on their basic knowledge, and eventually, via learning, they can act and adapt automatically. Learning agents can learn, analyze, and improve their performance. The searching UAV, i.e., the SUAV, is obviously the agent in our case study. The agent collects information or keeps track of a specific event in the environment. Here, the SUAV uses its on-board cutting-edge technologies such as photo or video cameras, thermal imagers, infrared cameras, and laser scanners to detect and locate the damaged infrastructure or areas that require further attention. The SUAV then sends out alarm signals, ordering the GUAVs to begin any relevant mission.
- **State**: The learning agent's current state can be interpreted as the currently visited ground station (at time step i) on which a GUAV is mounted. The number of states is N, which represents all of the existing ground-based stations on which the GUAVs are mounted.
- **Action**: As mentioned, at each time step, the agent observes the current state and then chooses an action using an exploration strategy [26]. The action of the learning algorithm determines which GUAV to visit next by the RL-enabled SUAV, which is from those GUAVs inside the cluster of which the currently visited GUAV is located at its center point (see Figure 7.2). In other words, it is assumed that at each time step, the RL-enabled SUAV is limited to a subset of its action set, not all of the $N-1$ possible actions.
- **Reward**: Following the execution of the action a_q, the agent obtains a reward (a measure of how good the action is in the short term) and advances to the next state. In particular, the received reward corresponding to the action taken in a state s_p is defined as the total number of GUAVs that will be inside the SUAV's circular coverage region by choosing the n-th GUAV to visit next (i.e., action $a_q = n$). The greater the number of GUAVs inside the SUAV's coverage when visiting the n-th GUAV, the greater the reward received. Therefore, $\mathcal{R}(s_p, a_q)$ (reward) can be specified as follows:

$$\mathcal{R}(s_p, a_q) = C_n, \qquad \forall n \in \{1, ..., N\}, \tag{7.7}$$

where C_n is the number of GUAVs inside the circular coverage of the SUAV when it visits the n-th GUAV.

The interested reader is referred to [38, 40, 41] for more detailed description of the double QL algorithm.

7.4.6 Convergence Analysis

It is straightforward to verify that in the proposed double QL-based framework, both Q^A and Q^B will eventually converge with probability one to the same optimum value function Q^* [40–42]. The optimum value refers to the optimal trajectory that maximizes the number of GUAVs that are informed. As a result, as the proposed framework converges, it will be able to maximize the number of informed GUAVs.

Figure 7.4 depicts the number of independent runs (each run can be considered as an offline mission) needed till convergence for the proposed double QL-based strategy. Let us consider a particular state (the n-th visited GUAV, $n \in \{1, ..., N\}$). The SUAV has to choose an action among all possible actions at that state, where the action specifies the next GUAV to visit. The maximum Q value over all possible

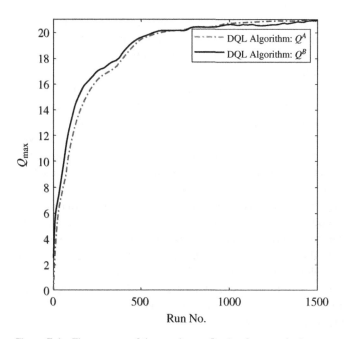

Figure 7.4 The average of the maximum Q value for a particular state and over all possible actions vs. the number of simulation runs: a convergence indicator of the double QL (DQL).

actions in that state can be obtained and considered as a convergence indicator. As time passes, we can observe that the maximum Q value for that particular state increases until convergence to its final value. From the figure, we also can see that the Q^A and Q^B tables may have different maximum Q values as time passes. Nonetheless, when the passage of time continues, both tables Q^A and Q^B will converge to the same ultimate value. This is owing to the fact that, in each run, the double QL strategy chooses an action based on the value of one Q-table and updates the actions Q value in another Q-table. As can be seen in the figure, the maximum of the Q values for a particular state (and overall possible actions) smoothly approaches convergence.

7.5 Simulation Results

We consider a system in which a single SUAV detects and reports abnormalities in any infrastructure component, or spots that require further action. The SUAV then informs several GUAVs to fly toward the location where faults in infrastructure components have been found and perform necessary actions at high altitudes, such as construction work, repairs, etc. Through designing a proper SUAV flight trajectory, a higher percentage of the GUAVs can be informed about the damaged spot. We assume that the trajectory optimization of the SUAV is implemented in a two-dimensional (2D) space using the proposed double QL-based strategy. There exist $N = 100$ uniformly distributed GUAVs within a 2D area of 10 km × 10 km. Both tables Q^A and Q^B have a size of $N \times (N-1)$. Table 7.1 contains a list of other system settings. Note that the simulation settings and the values explored in our simulation analysis are case specific. The proposed strategy is compared with two other strategies: (i) a random strategy (RND) that opts and informs the GUAVs for the relevant mission in a random order, and (ii) a traditional QL-based strategy [37].

Table 7.1 Major simulation parameters.

Parameter	Value
N: number of GUAVs	100
v: flying speed of the SUAV	50 m/s
h_{SUAV}: flying height of the SUAV	500 m
T^M: mission time limit	180 s
γ_{TH}: GUAV's SNR target threshold	18.5dB
ϵ: learning exploration parameter	0.95

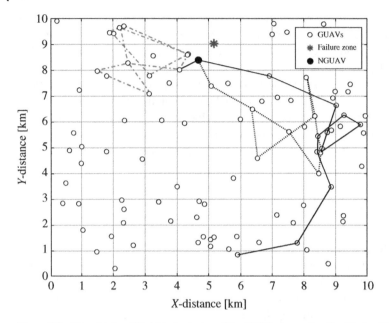

Figure 7.5 The obtained flight trajectories of the SUAV when using the QL, double QL and a random strategy. The SUAVs flight trajectories for the QL, double QL and random strategy are respectively shown by black dotted line, black solid line, and dash-dot line.

In Figure 7.5, we observe the obtained flight trajectories of the SUAV when using the QL, double QL, and a random strategy. As seen in the figure, the SUAV visits and informs the GUAVs by starting from the nearest GUAV (i.e., the NGUAV). GUAV visit repetitions may occur during the SUAVs mission. For the given scenario and within the time limit T^M, the double QL-enabled SUAV travels a greater distance to areas with a higher density of GUAVs in order to inform a greater number of GUAVs. The swarm or formation formed by multi-GUAVs offers apparent benefits over a single-GUAV system [43]. If a single GUAV is shot down during a mission in a single-GUAV system, the mission is termed a failure. In multi-GUAV systems, however, a single out-of-control GUAV is insignificant because other GUAVs will continue to operate and complete the mission. In addition, the infrastructure maintenance/repair of heat and electricity incorporated networks can be performed more efficiently and quickly with more GUAVs, as multi-GUAV systems can process tasks in parallel, thereby speeding up the time it takes to complete tasks [44]. Coordination of multiple GUAVs to complete tasks in a swarm environment is attractive because it overcomes the constraints of a single GUAV while providing more functionality [45]. Indeed, a swarm of GUAVs is a coordinated unit of GUAVs that perform a specific mission or set of activities.

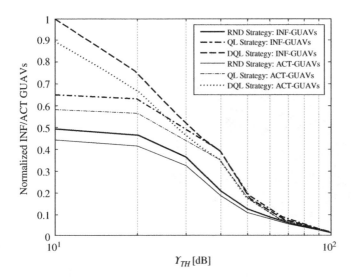

Figure 7.6 The normalized number of informed/activated GUAVs when using the Q-learning (QL), double QL (DQL) and a random strategy (RND) as a function of the SNR threshold γ_{TH} of the GUAVs.

Individual GUAVs can communicate with each other and react to the environment autonomously because they have computing, communication, and sensing capabilities locally on-board. The investigation of how to coordinate a swarm of multiple GUAVs to complete a task or set of activities is outside the scope of this study.

Figure 7.6 shows that the proposed double QL-based strategy performs better in terms of the number of informed GUAVs than the random and QL strategies, particularly for low SNR threshold γ_{TH} values. This is because the proposed strategy tries to determine the optimal flight trajectory in order to maximize the number of informed GUAVs, whereas the random strategy just provides a trail random trajectory independent of the number of informed GUAVs, and the QL strategy may lead to sub-optimal policies, resulting in a worse outcome. As depicted in the figure, the lower the value of γ_{TH} at the GUAVs, the more GUAVs receive the SUAV's alarm signal at an acceptable SNR level, which directly translates into more informed GUAVs. Each of the informed GUAVs will be activated with the probability of $A_R(1-P_{FA})$, which depends on the fault detection technique employed at the SUAV. Obviously, since the double QL-based approach informs more GUAVs, it also will outperform in terms of the number of activated GUAVs as well as the mission success metric M_s (see Eq. (7.6)), assuming that all techniques use the same detection method. The activated GUAVs will then climb and fly toward

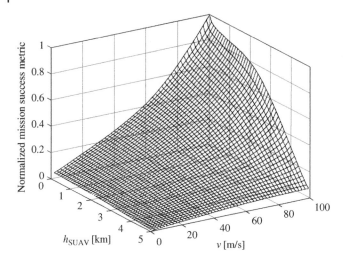

Figure 7.7 The mission success metric vs. the SUAVs flying height (h_{SUAV}) and speed (v).

the location where faults in infrastructure components have been found, by using the location information provided by the SUAV.

The mission success metric has been depicted in Figure 7.7 as a function of the SUAV's flying height (h_{SUAV}) and speed (v). It is intuitive that the mission success increases with the increase of the SUAV's speed. This is due to the fact that higher SUAV's speeds result in shorter travel times between the visiting nodes (the GUAVs). This means that the SUAV will inform a greater number of GUAVs about the presence of network failure during the time limit T^M, resulting in a higher overall mission success metric. On the other hand, when the SUAV is flying at a higher altitude, it may issue false warnings due to the altitude or a lack of good vision. Furthermore, operating at higher altitudes causes the GUAVs to receive the SUAV's alarm signal below the acceptable SNR level, reducing the number of informed GUAVs.

7.6 Conclusions

In this chapter, we discussed an intelligent monitoring and maintenance system for multi-UAV teams applicable to modern heat and electricity incorporated networks. We investigated the issue of flight trajectory optimization in a general scenario, where a solar EH-enabled SUAV, travels over randomly deployed ground stations to inform as many number of GUAVs as possible about abnormalities in any infrastructure component or spots that require further action. The optimal solution (flight trajectory) will be reached if the SUAV finds the optimal sequence of states (GUAVs) to visit within the mission time so that the total number of informed GUAVs will be maximized. To solve this optimization problem, we have introduced an RL framework based on the double QL mechanism. Simulations are conducted to show the benefits of using the proposed framework in terms of a mission success metric.

References

1 Ceron, A., Mondragon, I., and Prieto, F. (2018). Onboard visual-based navigation system for power line following with UAV. *International Journal of Advanced Robotic Systems* 15: 172988141876345.

2 Zahariadis, T., Voulkidis, A., Panagiotis, K., and Panagiotis, T. (2017). Preventive maintenance of critical infrastructures using 5G networks and drones. In: *2017 14th IEEE International Conference on Advanced Video and Signal Based Surveillance (AVSS)*, Lecce, Italy (29 August 2017–01 September 2017), 1–4. https://doi.org/10.1109/AVSS.2017.8078465. IEEE.

3 Gupta, A., Afrin, T., Scully, E., and Yodo, N. (2021). Advances of UAVs toward future transportation: the state-of-the-art, challenges, and opportunities. *Future Transportation* 1 (2): 326–350. https://www.mdpi.com/2673-7590/1/2/19.

4 Varghese, A., Gubbi, J., Sharma, H., and Balamuralidhar, P. (2017). Power infrastructure monitoring and damage detection using drone captured images. *International Joint Conference on Neural Networks (IJCNN)* 2017: 1681–1687.

5 Gouglidis, A., Green, B., Hutchison, D. et al. (2018). Surveillance and security: protecting electricity utilities and other critical infrastructures. *Energy Informatics* 1: 09.

6 Kim, S., Kim, D., Jeong, S. et al. (2020). Fault diagnosis of power transmission lines using a UAV-mounted smart inspection system. *IEEE Access* 8: 149 999–150 009.

7 Nooralishahi, P., Ibarra-Castanedo, C., Deane, S. et al. (2021). Drone-based non-destructive inspection of industrial sites: a review and case studies. *Drones* 5 (4): 106. https://www.mdpi.com/2504-446X/5/4/106.

8 Shakhatreh, H., Sawalmeh, A.H., Al-Fuqaha, A. et al. (2019). Unmanned aerial vehicles (UAVs): a survey on civil applications and key research challenges. *IEEE Access* 7: 48 572–48 634.

9 Liu, P., Chen, A., Huang, Y.-N. et al. (2014). A review of rotorcraft unmanned aerial vehicle (UAV) developments and applications in civil engineering. *Smart Structures and Systems* 13: 1065–1094.

10 Kochetkova, L. (2018). Pipeline monitoring with unmanned aerial vehicles. *Journal of Physics: Conference Series* 1015: 042021.

11 PWC. (2017). Clarity from above: leveraging drone technologies to secure utilities systems. https://www.pwc.pl/en/publikacje/2017/clarity-from-above-leveraging-drone-technologies-to-secure-utilities-systems.html (accessed 27 September 2021).

12 Colomina, I. and Molina, P. (2014). Unmanned aerial systems for photogrammetry and remote sensing: a review. *ISPRS Journal of Photogrammetry and Remote Sensing* 92: 79–97. https://www.sciencedirect.com/science/article/pii/S0924271614000501.

13 Kenta Takaya, K.S., Ohta, H., and Kroumov, V. (2021). Tracking control of unmanned aerial vehicle for power line inspection. In: *Motion Planning*. London, United Kingdom: IntechOpen. https://www.intechopen.com/chapters/78679 https://doi.org/10.5772/intechopen.100067.

14 Zhang, Y., Yuan, X., Li, W., and Chen, S. (2017). Automatic power line inspection using UAV images. *Remote Sensing* 9 (8): 824. https://www.mdpi.com/2072-4292/9/8/824.

15 Li, X., Li, Z., Wang, H., and Li, W. (2021). Unmanned aerial vehicle for transmission line inspection: status, standardization, and perspectives. *Frontiers in Energy Research* 9: 1–13. https://www.frontiersin.org/article/10.3389/fenrg.2021.713634.

16 Matikainen, L., Lehtomki, M., Ahokas, E. et al. (2016). Remote sensing methods for power line corridor surveys. *ISPRS Journal of Photogrammetry and Remote Sensing* 119: 10–31. https://www.sciencedirect.com/science/article/pii/S0924271616300697.

17 Kayan, H., Eslampanah, R., Yeganli, F., and Askar, M. (2018). Heat leakage detection and surveiallance using aerial thermography drone. In: *2018 26th Signal Processing and Communications Applications Conference (SIU)*, Izmir, Turkey (2-5 May 2018), 1–4. IEEE. https://doi.org/10.1109/SIU.2018.8404366.

18 Sledz, A., Unger, J., and Heipke, C. (2020). UAV-based thermal anomaly detection for distributed heating networks. *ISPRS - International Archives of the Photogrammetry, Remote Sensing and Spatial Information Sciences* 43B1: 499–505.

19 Donti, P.L. and Kolter, J.Z. (2021). Machine learning for sustainable energy systems. *Annual Review of Environment and Resources* 46 (1): 719–747. https://doi.org/10.1146/annurev-environ-020220-061831.

20 Jiao, J. (2020). Application and prospect of artificial intelligence in smart grid. *IOP Conference Series: Earth and Environmental Science* 510: 022012.

21 Inagaki, Y., Ikeda, H., Yato, Y. et al. (2020). Application of UAV for sewer pipe inspection. *International Journal of Water and Wastewater Treatment* 6 (2): 1–7. https://doi.org/10.16966/2381-5299.169.

22 Farquhar, P. (2017). China is using a flamethrower drone to clean rubbish off powerlines, business insider. www.businessinsider.com/china-is-using-a-flamethrowing-drone-to-clean-rubbish-off-power-lines-2017-2?IR=T (accessed 10 November 2021).

23 Bithas, P.S., Michailidis, E.T., Nomikos, N. et al. (2019). Survey on machine-learning techniques for UAV-based communications. *Sensors (Basel)* 19 (23): 5170. https://doi.org/10.3390/s19235170.

24 Jenssen, R. and Roverso, D. (2019). Intelligent monitoring and inspection of power line components powered by UAVs and deep learning. *IEEE Power and Energy Technology Systems Journal* 6: 11–21.

25 Sutton, R.S. and Barto, A.G. (1998). *Reinforcement Learning: An Introduction.* MIT Press.

26 Panahi, F.H. and Ohtsuki, T. (2014). Optimal channel-sensing scheme for cognitive radio systems based on fuzzy Q-learning. *IEICE Transactions on Communications* E97.B (2): 283–294.

27 Panahi, F.H., Panahi, F.H., Hattab, G. et al. (2018). Green heterogeneous networks via an intelligent sleep/wake-up mechanism and D2D communications. *IEEE Transactions on Green Communications and Networking* 2 (4): 915–931.

28 Sekander, S., Tabassum, H., and Hossain, E. (2019). On the performance of renewable energy-powered UAV-assisted wireless communications. https://doi.org/10.48550/arXiv.1907.07158

29 Panahi, F.H. and Panahi, F.H. (2020). *Energy Harvesting Technologies and Market Opportunities*, 1–18. Cham: Springer International Publishing. https://doi.org/10.1007/978-3-030-36979-8_1.

30 Panahi, F.H., Panahi, F.H., and Ohtsuki, T. (2021). Spectrum-aware energy efficiency analysis in k-tier 5G HetNets. *Electronics* 10 (7): 839. https://www.mdpi.com/2079-9292/10/7/839.

31 Panahi, F.H., Panahi, F.H., and Ohtsuki, T. (2020). Energy efficiency analysis in cache-enabled D2D-aided heterogeneous cellular networks. *IEEE Access* 8: 19 540–19 554.

32 Muhammad, K., Ahmad, J., Mehmood, I. et al. (2018). Convolutional neural networks based fire detection in surveillance videos. *IEEE Access* 6: 18 174–18 183.

33 Frizzi, S., Kaabi, R., Bouchouicha, M. et al. (2016). Convolutional neural network for video fire and smoke detection. *IECON 2016 - 42nd Annual Conference of the IEEE Industrial Electronics Society* 877–882. https://doi.org/10.1109/IECON.2016.7793196.

34 Stojni, V., Risojevi, V., Mutra, M. et al. (2021). A method for detection of small moving objects in UAV videos. *Remote Sensing* 13 (4): 653. https://www.mdpi.com/2072-4292/13/4/653.

35 Yao, H., Qin, R., and Chen, X. (2019). Unmanned aerial vehicle for remote sensing applications—a review. *Remote Sensing* 11 (12): 1443. https://www.mdpi.com/2072-4292/11/12/1443.

36 Liang, H., Su, J., and Liu, S. (2010). Reliability evaluation of distribution system containing microgrid. *CICED 2010 Proceedings*, Nanjing, China (13-16 September 2010). IEEE, pp. 1–7.

37 Bennis, M. and Niyato, D. (2010). A Q-learning based approach to interference avoidance in self-organized femtocell networks. In: *2010 IEEE Globecom Workshops*, 706–710. IEEE.

38 Van Hasselt, H. (2010). *Double Q-learning*, 2613–2621. The MIT Press.

39 Javapoint (2011). Types of AI agents. https://www.javatpoint.com/types-of-ai-agents (accessed 30 January 2022).

40 Liu, X., Chen, M., and Yin, C. (2018). Optimized trajectory design in UAV based cellular networks: a double Q-learning approach. *2018 IEEE International Conference on Communication Systems (ICCS)*, Chengdu, China (19–21 December 2018). IEEE, pp. 13–18.

41 Liu, X., Chen, M., and Yin, C. (2019). Optimized trajectory design in UAV based cellular networks for 3D users: a double Q-learning approach. *Journal of Communications and Information Networks* 4 (1): 24–32.

42 Abu Alsheikh, M., Hoang, D.T., Niyato, D. et al. (2015). Markov decision processes with applications in wireless sensor networks: a survey. *IEEE Communication Surveys and Tutorials* 17 (3): 1239–1267.

43 Chen, X., Tang, J., and Lao, S. (2020). Review of unmanned aerial vehicle swarm communication architectures and routing protocols. *Applied Sciences* 10 (10): 3661. https://www.mdpi.com/2076-3417/10/10/3661.

44 Yanmaz, E., Costanzo, C., Bettstetter, C., and Elmenreich, W. (2010). A discrete stochastic process for coverage analysis of autonomous UAV networks. *2010 IEEE Globecom Workshops*, Miami, FL, USA (6–10 December 2010). IEEE, pp. 1777–1782.

45 Zhu, X., Liu, Z., and Yang, J. (2015). Model of collaborative UAV swarm toward coordination and control mechanisms study. *Procedia Computer Science* 51: 493–502. International Conference on Computational Science, ICCS 2015. https://www.sciencedirect.com/science/article/pii/S1877050915010820.

8

Coordinated Operation and Planning of the Modern Heat and Electricity Incorporated Networks

Aminabbas Golshanfard and Younes Noorollahi

Energy Modelling and Sustainable Energy System (METSAP) Research Lab., Faculty of New Sciences and Technologies, University of Tehran, Tehran, Iran

Nomenclature

Abbreviation

CHP	Combined heat and power
CHPwte	Combined heat and power waste to energy
HS	Heat storage
GSHP	Ground source heat pump
PV	Photovoltaic
WT	Wind turbine
MCS	Monte Carlo simulation
PDF	Probability distribution function
MILP	Mixed integer linear programing
MINLP	Mixed integer non-linear programing
TOU	Time of use
O&M	Operation and maintenance
GA	Genetic algorithm
PSO	Particle swarm optimization
ACO	Ant colony optimization
DG	Distributed generation
COP	Coefficient of performance
EER	Energy efficiency ratio
IEA	International energy agency
HP	Heat pump

Coordinated Operation and Planning of Modern Heat and Electricity Incorporated Networks,
First Edition. Edited by Mohammadreza Daneshvar, Behnam Mohammadi-Ivatloo, and Kazem Zare.
© 2023 The Institute of Electrical and Electronics Engineers, Inc.
Published 2023 by John Wiley & Sons, Inc.

AC	Absorption chiller
EV	Electrical vehicle
CO_2	Carbon dioxide

Parameters

η_{PV}	Photovoltaic panel Efficiency
pr_{PV}	PV system performance coefficient
K_t	Solar clearness index
G	Solar radiation
G_0	Standard solar radiation
μ	Average
σ	Standard deviation
α	Beta distribution parameter
β	Beta distribution parameter
Γ	Gamma function
PPV_t	PV production at time t
P_{rated}^{PV}	Nominal power of the PV module
Y_{PV}	Environmental coefficient for PV
T_{C_t}	Cell temperature at time t
T_{ref}	Standard temperature of the PV module
v_m	Maximum wind speed
c	Weibull distribution parameter (scale)
v	wind speed
u	Random number
P_{WT}^t	Wind turbine production at time t
P_{WT}^{rated}	Nominal power of wind turbine
v_{wind}^t	Wind speed at the time t
v_{ci}	WT's cut-in speed (m/s)
v_{co}	WT's cut-out speed (m/s)
v_{rated}	WT's rated wind speed (m/s)
COP	GSHP's coefficient of performance
Q_H	Heating Generated
W	Work done to transfer energy
EER	Coefficient of performance in cooling mode
Q_C	Cooling Generated
pr_{se}	Probability of the se-th season
$pr_{se,s}$	Probability of the s-th scenarios of PV in se season

$\mathrm{pr}_{se,w}$	Probability of the w-th scenarios of WT in se season
$P_{t,w,s,se}^{\mathrm{Egrid}}$	Electricity input at the t-th time interval under the s-th and w-th scenarios of the PV and WT in se season (kW)
$\lambda_{se,t}^{\mathrm{Elec}}$	Electricity price at the t-th time interval in se season (\$/kWh)
$P_{t,w,s,se}^{\mathrm{NGgrid}}$	Natural gas input at the t-th time interval under the s-th and w-th scenarios of the PV and WT in season se (kW)
$\lambda_t^{\mathrm{Ngas}}$	Natural gas price at the t-th time interval (\$/kWh)
$P_{t,w,s,se}^{\mathrm{EExtra}}$	Excess electrical energy from renewable sources at the t-th time interval under the s-th and w-th scenarios of the PV and WT in season se (kW)
N_{PV}	PV capacity coefficient
$C_{\mathrm{daily}}^{\mathrm{PV}}$	Daily cost of PV (\$/kW)
N_{WT}	Wind turbine capacity coefficient
$C_{\mathrm{daily}}^{\mathrm{WT}}$	Daily cost of wind turbine (\$/kW)
N_{CHPwte}	CHPwte capacity coefficient
$C_{\mathrm{daily}}^{\mathrm{CHPwte}}$	Daily cost of CHPwte (\$/kW)
N_{HS}	HS capacity coefficient
$C_{\mathrm{daily}}^{\mathrm{HS}}$	Daily cost of HS (\$/kW)
N_{GSHP}	GSHP capacity coefficient
$C_{\mathrm{daily}}^{\mathrm{GSHP}}$	Daily GSHP cost (\$/kW)
N_{Boiler}	Boiler capacity coefficient
$C_{\mathrm{daily}}^{\mathrm{Boiler}}$	Daily cost of boiler (\$/kW)
OF	Objective function
$\mathrm{Em}_{\mathrm{Grid}}$	Upstream carbon dioxide emission coefficient (g/kWh)
$\mathrm{Em}_{\mathrm{Boiler}}$	Boiler carbon dioxide emission coefficient (g/kWh)
$\mathrm{Em}_{\mathrm{CHPwte}}$	CHPwte carbon dioxide emission coefficient (g/kWh)
$\lambda^{\mathrm{CO_2}}$	Carbon dioxide emission cost (\$/kg)
$P_{t,w,s,se}^{\mathrm{Egrid,enduser}}$	The portion of the electricity grid that directly supplies electricity demand at the t-th time interval under the s-th and w-th scenarios of the PV and WT in *se* season (kW)
$P_{t,w,s,se}^{\mathrm{Egrid,2GSHP}}$	The portion of the electricity grid that supplies the GSHP at the t-th time interval under the s-th and w-th scenarios of the PV and WT in se season (kW)
$P_{t,w,s,se}^{\mathrm{Boiler,in}}$	The amount of natural gas boiler input power GSHP at the t-th time interval under the s-th and w-th scenarios of the PV and WT in se season (kW)
$P_{t,se}^{\mathrm{Demand,Elec}}$	Electricity demand at the t-th time interval in se season (kW)

P_{CHPe}	CHP's electricity output at the t-th time interval under the s-th and w-th scenarios of the PV and WT in se season (kW)
$P_{t,s,se}^{\text{PV}}$	PV power generation at the t-th time interval under the s-th scenarios of the PV in *se* season (kW)
$P_{t,w,se}^{\text{WT}}$	WT power generation at the t-th time interval under the w-th scenarios of the WT in *se* season (kW)
$\text{SurRES}_{t,w,s,se}$	The excess power that remained after feeding consumers at the t-th time interval under the s-th and w-th scenarios of the PV and WT in *se* season (kW)
$P_{t,w,s,se}^{\text{CHPin}}$	Input power to CHPwte at the t-th time interval under the s-th and w-th scenarios of the PV and WT in *se* season (kW)
η_e^{CHPwte}	CHPwte electrical efficiency
η_h^{CHPwte}	CHPwte thermal efficiency
$\text{HSS}_{t,w,s,se}$	Excess heating of CHPwte that remained after feeding consumers at the t-th time interval under the s-th and w-th scenarios of the PV and WT in *se* season (kW)
$P_{t,se}^{\text{Demand,Heat}}$	Heating demand at the t-th time interval in *se* season (kW)
$\text{HNC}_{t,w,s,se}$	The amount of heating that not supplied by CHPwte at the t-th time interval under the s-th and w-th scenarios of the PV and WT in *se* season (kW)
P_{CHPh}	CHP's heating output at the t-th time interval under the s-th and w-th scenarios of the PV and WT in *se* season (kW)
$\text{Ra}_{\text{Boiler}}$	Boiler efficiency
$P_{t,w,s,se}^{\text{Boiler,out}}$	Boiler output power at the t-th time interval under the s-th and w-th scenarios of the PV and WT in *se* season (kW)
$P^{\text{Boiler,Max}}$	Boiler capacity (kW)
$P_{t,w,s,se}^{\text{Boiler,enduser}}$	The portion of the boiler that directly supplies electricity demand at the t-th time interval under the s-th and w-th scenarios of the PV and WT in *se* season(kW)
$\beta_{t,w,s,se}^{\text{Heat}}$	Dispatching coefficient for heating demand at the t-th time interval under the s-th and w-th scenarios of the PV and WT in *se* season (kW)
$\alpha_{t,w,s,se}^{\text{Heat}}$	Dispatching coefficient for heating demand at the t-th time interval under the s-th and w-th scenarios of the PV and WT in *se* season (kW)
$P_{t,w,s,se}^{\text{Boiler,2HS}}$	Share of boiler production that supplies HS at the t-th time interval under the s-th and w-th scenarios of the PV and WT in *se* season (kW)

$P^{HSin}_{t,w,s,se}$	HS input at the t-th time under the s-th and w-th scenarios of the PV and WT in se scenario (kW)
$P^{GSHPh}_{t,w,s,se}$	Heating output power by GSHP at the t-th time under the s-th and w-th scenarios of the PV and WT (kW)
$P^{GSHPh,Max}$	GSHP heating capacity (kW)
$P^{GSHP,in}_{t,w,s,se}$	GSHP input power at the t-th time under the s-th and w-th scenarios of the PV and WT in se scenario (kW)
COP_{GSHP}	GSHP performance coefficient in heating mode
EER_{GSHP}	GSHP performance coefficient in cooling mode

8.1 Introduction

One of the most critical human issues has been energy and related topics. Nowadays, with the diversity of energy components technology and the increasing dependence of human life on energy, it has received more and more attention. In the present century, fossil fuels are the most important energy source used to generate electricity in power plants [1]. In addition to electricity generated, fossil fuels have been consumed to meet heat demand. In other words, fossil fuels have been considered the most important energy source in the last decade [2]. As a result, the portion of carbon dioxide (CO_2) emitted by the heat and electricity sector is about 42%, and this section was the most important CO_2 emitter in 2019 [3]. Generally, conventional energy systems like current power systems have a hierarchical structure. Fossil fuel power plants are placed on top of the energy system to meet energy consumers by transmission and distribution lines at different distances. In this situation, grid losses and total cost of the system increase, total system efficiency decreases, and some constraints related to relays and protection can be created [1]. Recent research introduced the smart grid concept in electricity and energy to eliminate the existing problems in conventional energy networks. In addition to the usual tasks of a conventional electricity grid, a smart grid, which include production, transmission, and distribution, can create interaction between system components, storage, and optimal decision making [1]. In other words, smart grids improve the system's performance and increase the overall efficiency by analyzing information in different parts of the network [4]. The smart grid concept focuses on only one energy carrier: electricity from a macro perspective. Still, the idea of smart energy systems is expressed in a more comprehensive view, which includes different energy carriers [2]. This concept expressed a pattern in moving from single-sector systems to integrating different sectors of the energy system. Smart energy systems incorporate electricity, heat, and transportation, are intelligently controlled using the information exchanged in the system, and operate optimally [5]. In other words, all energy carriers interact with each other in this system and do not work individually [6]. Distributed generations (DG), especially renewable energy resources, play a significant role in smart energy systems. With the development of this group of producers, the generation portfolio is diversified and can be used instead of power plants; consequently, energy systems move to local DG instead of large power plant concentrations [1].

For this reason, the importance of renewable sources is increasing, so that according to the latest statistics of the International Energy Agency (IEA), the production of photovoltaic (PV) systems in the world from 0.8 TWh in 2000 reached more than 550 TWh in 2018 and production from wind energy from about 32 TWh in 2000 get to nearly 1275 TWh in 2018 [7]. With the addition of other energy carriers to the system and providing the interaction between different parts, a concept called energy hubs or incorporated energy systems were developed, which in various scales, such as a house, a residential complex, commercial, hospitals, industrial units, greenhouses, rural and urban energy systems are implemented. In general, in an incorporated energy system, goals are met simultaneously that address the challenges of the energy world, including efficiency, cost, environmental impact, sustainable development, integration and community of different systems, and commercial index [8]. A smart energy system or an incorporated system combines operation, power supply, transportation, and energy consumption with intelligent control and optimization to achieve an optimal energy system. These systems have five layers that interact with each other: production, conversion, transmission, distribution, storage, and consumers [5]. The concept of an incorporated system has been defined to provide a scientific basis for the transition from one-dimension thinking to the concept of multi-dimensional and integrated systems to more and more move toward energy systems based on sustainable development. The objectives of this chapter can be stated as follows:

1) Modeling of renewable resources according to their uncertain behavior.
2) Optimal planning and operation of a hybrid electricity and heating system using a scenario-based method.
3) Application of modern technologies to provide heating in the form of an incorporated energy system.
4) Development of a system under technical-economic and environmental criteria.

For achieving the mentioned objectives, it is necessary to model each component according to its position in the energy system and determine the optimal capacity and scheduling using the appropriate optimization method. Each of the steps is described in detail in the following sections.

8.2   Literature Review

In recent years, the study of incorporated energy systems has increased day by day. As a result, the terms used to describe such systems have evolved: energy hub, smart energy system, decentralized energy systems, virtual power plants, etc. [9]. The four main components of an energy hub include inputs, converters, storage, and outputs, which contain various technologies like gas micro-turbines, fuel cells, waste heat recovery systems, renewable sources, heat pump (HP), combined heat and power (CHP), absorption chiller (AC), etc. The inputs of this system can consist of fossil fuels, garbage, waste, renewable sources, electricity from the upstream network, etc. Natural gas and electricity are generally considered the most common inputs of these systems [10]. The research in this field investigates energy systems with various components in planning and operation views to achieve an optimal model for the studied energy system; Research [11] considered CHP, storage in both heat and electricity types, wind turbines (WTs), and PV to obtain optimal operation for the studied system. The research [12] shows that WT and gas turbines and PV systems have been used as electricity sources. In the heating and cooling section, boilers, electric and thermal storage devices, heat exchangers, solar thermal systems, absorption, and electrical chillers are operated to schedule multiple energy hubs. The proposed structure in [12] reduced energy cost and CO_2 emissions by 22 and 13%, respectively. Rahman Habib et al. [13] optimized an electricity system consisting of a diesel generator, fuel cell, battery, WT, PV, and bidirectional electrical vehicles (EVs). Incorporated energy systems play a crucial role in providing clean, efficient, cost-effective, and reliable energy services. For this reason, today's societies are moving towards this category of new energy systems [1, 14]. In research [8], implementing a smart energy system has been considered a suitable way to achieve intelligent goals. These goals include better efficiency, more incredible economic benefits, better use of production resources, better design and analysis, better energy security, and ultimately a better environment [8]. These objectives can be accessible by optimal operation and planning for current and future energy systems. In energy system modeling, research deals with many challenges, that uncertainty is one of the most important issues. Energy systems have many uncertain or complex elements that depend on different factors like weather and social features [15]. By increasing the share of renewable resources such as solar and WT systems and restructuring energy systems, the issue of uncertainties in this field has become critical [16]. The challenges arise from environmental factors and energy market interactions and directly affect power and energy systems [15]. System uncertainties must be considered and investigated to make an energy system model closer to reality [17]. Uncertainties related to the parameters of an energy system are studied in both technical and economic aspects. Technical parameters such as load curve, capacities of solar and wind systems, charging and discharging behaviors of EVs, the exit of a device from the network, etc. Economic parameters include prices in the energy market, operating costs, interest rates, etc. [17]. After performing the preliminary steps in energy system modeling, the general model of the system structure is presented and using optimization methods, planning or operation is applied. In general, optimization in energy systems can be studied in the form of two deterministic and stochastic approaches [18]. In the deterministic approach, the search for the optimal solution is based on deterministic processes, and the analytical features of the problem are used to increase the computational efficiency [18]. Ma et al. [19], using the mixed-integer linear programming (MILP) method, optimized an energy hub system to minimize the costs of the whole system and compared three different structures with different components of this system. Majidi et al., considering load uncertainties and demand response programing, optimized a CHP-based energy system with the MILP method and modeled the existing uncertainties in the research using a robust optimization technique [20]. In [21], the WTs and PV systems, due to their stochastic nature, are modeled by scenario-based technique, and finally, the studied energy system is optimized using the MILP method. In order to simplify the complex problem presented in [22], the Bender decomposition algorithm was used, and then the optimization was performed on the energy system using the mixed-integer non-linear programming (MILNP) method. In stochastic approaches, unlike the deterministic perspective, optimization methods apply under uncertainties and possible predictions. This approach has different techniques such as particle swarm optimization (PSO), genetic algorithm (GA), ant colony optimization (ACO), etc., and unlike the deterministic view, using random operators that it uses to prevent the problem does not fall into the optimal local solution [18]. This approach is also referred to in research as meta-heuristic methods [23]. The characteristics and behaviors of nature inspire these methods. These optimization algorithms usually have uncertain performance, which means that the response is provided close to the optimal response, but their computational time is better than conventional methods [24]. One of the most common methods of these algorithms is the PSO method used in [25]. Eriksson et al. investigated a hybrid system consisting of renewable resources and, in addition to technical and environmental indexes, also considered social parameters as objective functions [25]. In the study [26], the energy hub is optimized in two planning and operation stages by PSO and MINLP methods, respectively. In other words, a two-stage probabilistic model has been considered for this research in order to determine the optimal size of the equipments in the first stage and reach optimum operation in the second stage for the studied energy system. In Table 8.1, the research is reviewed and compared in accordance with what is stated in the chapter title.

Table 8.1 Relevant literature in incorporated energy systems.

References	Optimization approach		Study approach		Uncertainty element					Supply		Demand		Publish year
	Deterministic	Stochastic	Operation	Planning	WT	PV	EV	Demand	Economic	Gas	Electricity	Electricity	Heating	
[20]	✓	×	✓	×	×	×	×	✓	×	✓	✓	✓	✓	2019
[21]	✓	×	✓	×	✓	✓	×	×	×	✓	✓	✓	✓	2019
[22]	✓	×	✓	✓	✓	×	×	✓	×	✓	✓	✓	✓	2020
[26]	✓	✓	✓	✓	×	✓	×	✓	×	✓	✓	✓	✓	2020
[27]	×	✓	✓	×	✓	×	×	✓	×	✓	✓	✓	✓	2019
[28]	✓	×	✓	×	×	×	×	×	✓	×	✓	✓	✓	2019
[29]	✓	✓	×	✓	✓	×	✓	✓	×	✓	✓	✓	×	2017
[30]	✓	×	✓	×	×	✓	✓	×	×	×	✓	✓	✓	2019
[31]	✓	×	✓	×	✓	✓	✓	×	×	✓	✓	✓	×	2019
[32]	✓	×	✓	×	✓	✓	✓	✓	×	×	✓	✓	×	2020
[33]	✓	×	✓	×	✓	✓	✓	×	×	✓	✓	✓	✓	2019
[34]	✓	×	✓	×	✓	✓	✓	✓	×	✓	✓	✓	✓	2019
[13]	×	✓	✓	×	×	×	×	×	×	×	✓	✓	×	2017
[35]	×	✓	×	✓	×	✓	✓	×	✓	✓	✓	✓	✓	2020
[36]	✓	×	×	✓	×	✓	×	×	✓	×	✓	✓	×	2013
[37]	✓	×	✓	×	×	×	×	×	✓	✓	✓	✓	✓	2020

8.3 Optimal Operation and Planning

8.3.1 Optimization in Incorporated Energy Networks

According to Figure 8.1, to plan and operate the incorporated energy system, it is necessary to study and model the components of the system in the first stage. Then, the system's economic and environmental characteristics are analyzed in the second stage. Finally, each element's optimal capacities and operation are determined [38]. In this way and according to the optimization results, it was determined which component, capacity, and time should meet demand in the system. In general, the development of these systems to achieve a more economical, more efficient, and environmental-friendly energy system [8]. In this study, GA has been used to optimize the case study system, one of the proposed stochastic optimization methods. This method is inspired by the laws and behaviour of nature and is based on the biological sciences [39]. This method has three important operators, including selection, crossover, and mutation, to achieve the optimal solution. The steps of the optimization process using the GA method are shown in Figure 8.2.

8.3.2 Stochastic Modelling

In this research, the scenario-based method, one of the probabilistic methods, has been used to investigate different scenarios of renewable sources. In other words,

Figure 8.1 Conceptual model for optimal operation and planning.

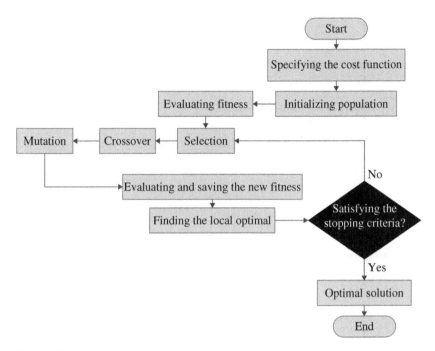

Figure 8.2 GA steps.

according to Figure 8.3, uncertainties arising from renewable sources are modeled using the Monte Carlo simulation (MCS) and k-means methods. In many research in energy systems, stochastic models such as MCS are used to model the uncertainty of behavior of renewable sources. Therefore, the MCS method has been used to generate the scenarios. MCS is a probabilistic simulation method [40] widely used due to its easy and simple implementation [41]. In general, the MCS method for generating scenarios is divided into the following four steps [40, 41]:

1) Select the best distribution density function for the variable with uncertainty.
2) Obtain the parameters related to the distribution function based on the nature and behavior of the variables.
3) Adjust the range of the probability distribution function based on the nature and behavior of the variables.
4) Perform simulation using the information obtained in the previous steps and generate random numbers according to them.

In the Scenario-based method, all scenarios are obtained based on the probability density functions (PDF). Then, the scenarios with the highest probability of occurrence are selected as the best or most probable scenarios [42]. Thus, the

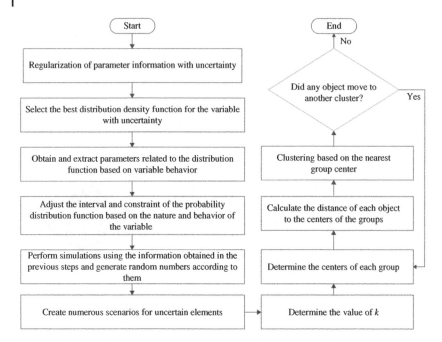

Figure 8.3 MCS and k-means procedure in scenario-based modeling.

number of studied scenarios is reduced, and the computational burden of the optimization problem is reduced [26]. Furthermore, increasing the number of selected scenarios may improve the accuracy of the final results [43]. Still, because of the mentioned point, it is necessary to limit the scenarios and complexity of the model [26]. Clustering methods have been used for this purpose in these issues, which will be examined in the following clustering concepts, emphasizing the k-means technique. One of the most popular clustering methods is the k-means clustering method, useful for data analysis. The importance of this tool is due to the simplicity and high efficiency of this method. Furthermore, grouping in this algorithm is done in such a way that the components of each cluster are the most similar to their groupmates in terms of the desired features and the least similar to the members of other clusters [44].

8.3.3 Objective Function

To develop an incorporated energy system that is economical and environmentally friendly, in addition to energy costs, investment and operation, it is necessary to consider CO_2 emissions. For this purpose, CO_2 emissions have been translated at economic costs using the carbon tax considered in the research. Therefore, it is included in the objective function to minimize CO_2 emissions as much as possible. Equation (8.1) shows the objective function of the studied system, which pr_{se}, $\text{pr}_{se,s}$, and $\text{pr}_{se,w}$ represent the probability of each season, the probability of occurrence of scenario s in the solar scenario in season se and the probability of occurrence of wind scenario w in season se, respectively. $P_{t,w,s,se}^{\text{Egrid}}$ and $P_{t,w,s,se}^{\text{NGgrid}}$ show the power of electricity and natural gas purchased from the upstream grid. $P_{t,w,s,se}^{\text{EExtra}}$ determines the surplus power of the system that is sold to the grid. In the environmental parameters section, the upstream network CO_2 emissions coefficient, boiler, and CHP are determined by the terms Em_{Grid}, $\text{Em}_{\text{Boiler}}$, and $\text{Em}_{\text{CHPwte}}$, respectively. The economic parameters considered in this objective function are expressed in three sections. The first part is the cost of investment and operation, which in terms of $C_{\text{daily}}^{\text{PV}}$, $C_{\text{daily}}^{\text{WT}}$, $C_{\text{daily}}^{\text{CHPwte}}$, $C_{\text{daily}}^{\text{HS}}$, $C_{\text{daily}}^{\text{GSHP}}$, and $C_{\text{daily}}^{\text{Boiler}}$, the second part is related to the cost of CO_2 emissions (λ^{CO_2}) and the third part is the cost of energy ($\lambda_{se,t}^{\text{Elec}}$ and λ_t^{Ngas}).

It should also be noted that the system's daily operation has been studied. As a result, all costs have been converted into a daily cost scale. In the stochastic view, this Equation indicates the probability of PV power generation and WT generation scenarios in a given season and the probability of each season, which is considered 25%.

$$\text{Min}\left\{ \begin{aligned} \text{OF} = \sum_{se}\sum_{s}\sum_{w}\sum_{t} \left[\text{pr}_{se}\cdot\text{pr}_{se,s}\cdot\text{pr}_{se,w}\cdot \left[\begin{array}{l} P_{t,w,s,se}^{\text{Egrid}}.\lambda_{se,t}^{\text{Elec}} + P_{t,w,s,se}^{\text{NGgrid}}.\lambda_t^{\text{Ngas}} - P_{t,w,s,se}^{\text{EExtra}}.\lambda_{se,t}^{\text{Elec}} \\ + \left[\begin{array}{l} P_{t,w,s,se}^{\text{Egrid}}.\text{Em}_{\text{Grid}} \\ + P_{t,w,s,se}^{\text{Boiler}}.\text{Em}_{\text{Boiler}} \\ + N_{\text{CHPwte}}.P_{\text{CHPe}}.\text{Em}_{\text{CHPwte}} \end{array} \right].\lambda^{CO_2} \end{array} \right] \right] \\ + N_{\text{PV}}.C_{\text{daily}}^{\text{PV}} + N_{\text{WT}}.C_{\text{daily}}^{\text{WT}} + N_{\text{CHPwte}}.C_{\text{daily}}^{\text{CHPwte}} + N_{\text{HS}}.C_{\text{daily}}^{\text{HS}} + N_{\text{GSHP}}.C_{\text{daily}}^{\text{GSHP}} \\ + N_{\text{Boiler}}.C_{\text{daily}}^{\text{Boiler}} \end{aligned} \right\}$$

$$(8.1)$$

8.4 Components and Constraints

8.4.1 Combined Heat and Power by Waste to Energy

Nowadays, CHPwte power plants are used to prevent the problem of waste management and reduce the cost of fuel transfer for power plants [45]. The performance of this type of power plant is evaluated using their heat and electricity production and greenhouse gas emissions. Waste used in these power plants can include: organic, wood, fabric, plastic, paper, etc. Generally, CHPwte is the incinerator paired with a heat recovery steam generator system [46]. Equations (8.2)–(8.4) model the relationship between electricity and heating in the production of CHPwte. Equations (8.5) and (8.6) calculate the extra heat and heating not supplied by CHPwte. The operation strategy in this system is such that in meeting the heating demand, CHPwte takes precedence.

$$N_{\text{CHPwte}} \cdot P_{\text{CHPe}} \geq 0 \tag{8.2}$$

$$N_{\text{CHPwte}} \cdot P_{\text{CHPh}} \geq 0 \tag{8.3}$$

$$P_{t,w,s,se}^{\text{CHPin}} \leq \frac{N_{\text{CHPwte}} \cdot P_{\text{CHPe}}}{\eta_e^{\text{CHPwte}}} + \frac{N_{\text{CHPwte}} \cdot P_{\text{CHPh}}}{\eta_h^{\text{CHPwte}}} \tag{8.4}$$

$$HSS_{t,w,s,se} = N_{\text{CHPwte}} \cdot P_{\text{CHPh}} - P_{t,se}^{\text{Demand,Heat}} \tag{8.5}$$

$$HNC_{t,w,s,se} = P_{t,se}^{\text{Demand,Heat}} - N_{\text{CHPwte}} \cdot P_{\text{CHPh}} \tag{8.6}$$

8.4.2 Photovoltaic

The PV output power is obtained using the technical information related to solar modules and weather information. Equation (8.7) is related to PV output power and is as follows [47]:

$$\text{PPV}_t = P_{\text{rated}}^{\text{PV}} \times Y_{\text{PV}} \times K_t \times \left(1 + K_t \times (T_{C_t} - T_{\text{ref}})\right) \tag{8.7}$$

In Eq. (8.7), environmental issues are also considered technical issues. It should be noted that the environmental coefficient depends on factors such as heat generated in the module, system inverter efficiency, system wiring, dirt and dust, and shadow coefficient [48].

8.4.3 Wind Turbine

To calculate the power of WT according to the wind speed in the region and the technical characteristics considered for the turbine, which is extracted from the reference [49], Eq. (8.8) is used [50]:

$$P_{\text{WT}}^t = \begin{cases} 0 & v_{\text{wind}}^t \leq v_{\text{ci}} \ or \ v_{\text{wind}}^t \geq v_{\text{co}} \\ P_{\text{WT}}^{\text{rated}} \times \left(\frac{v_{\text{wind}}^t - v_{\text{ci}}}{v_{\text{rated}} - v_{\text{ci}}}\right)^3 & v_{\text{ci}} \leq v_{\text{wind}}^t \leq v_{\text{rated}} \\ P_{\text{WT}}^{\text{rated}} & v_{\text{rated}} \leq v_{\text{wind}}^t \leq v_{\text{co}} \end{cases} \tag{8.8}$$

In Eq. (8.8), cut-in, cut-off, and rated speed of turbine are considered as the main specifications of this element in the relationship, and Figure 8.4 demonstrates each of these parameters [51].

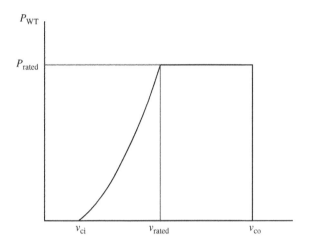

Figure 8.4 WT operation curve.

8.4.4 Ground Source Heat Pump

In recent years, the use of near-surface geothermal energy using GSHP has been considered to provide heating, cooling, and hot water to one or more buildings using these systems [52]. These systems have more advantages than conventional systems, especially regarding efficient operating and maintenance costs, distinguishing GSHP from other thermal systems. GSHP uses the earth instead of the air as an energy source because this causes the environment with less temperature fluctuation during different times to be selected as the energy source of the HP [53]. Coefficient of performance (COP) and energy efficiency ratio (EER) are related to the performance of GSHPs, which is because, in this system, electricity is not converted to heating or cooling, but electricity is a factor for transferring energy from one place to another [54]. As a result, the high efficiency of these systems saves electricity consumption and ultimately reduces the emission of environmental pollutants [53]. Equations (8.9) and (8.10) show the concept of COP and EER [53], but this study concentrated on the heating section of GSHP. Equations (8.11) and (8.12) show the input and output relationship of GSHP. In Eq. (8.13), two heat dispatching factors, which are the decision variables of the problem, determine the amount of contribution of this system in supplying the heating demand. Equation (8.14) also shows the power supplies of GSHP that, as can be seen, the incorporated system provides optional input for supplying demand to increase the reliability and flexibility of the system.

$$\text{COP} = \frac{Q_H}{W} \tag{8.9}$$

$$\text{EER} = \frac{Q_C}{W} \tag{8.10}$$

$$0 \leq P_{t,w,s,se}^{\text{GSHPh}} \leq P^{\text{GSHPh,Max}} \tag{8.11}$$

$$P_{t,w,s,se}^{\text{GSHP,in}} = \frac{P_{t,w,s,se}^{\text{GSHPh}}}{\text{COP}_{\text{GSHP}}} + \frac{P_{t,w,s,se}^{\text{GSHPc}}}{\text{EER}_{\text{GSHP}}} \tag{8.12}$$

$$P_{t,w,s,se}^{\text{GSHPh}} = \left(1 - \beta_{t,w,s,se}^{\text{Heat}}\right) \times \alpha_{t,w,s,se}^{\text{Heat}} \times \text{HNC}_{t,w,s,se} \tag{8.13}$$

$$P_{t,w,s,se}^{\text{GSHP,in}} \leq \text{SurRES}_{t,w,s,se} + P_{t,w,s,se}^{\text{Egrid,2GSHP}} \tag{8.14}$$

8.4.5 Boiler

Equations (8.15)–(8.18) describe the inputs and outputs of the boiler, which in this study used a gas-fired boiler to provide heating [20]. Equation (8.19) determines the share of boiler production that reaches the consumer directly by the heating dispatching factor. Equation (8.20) also shows that the portion of the boiler output enters the heat storage (HS), and the other part meets the consumers directly.

$$P_{t,w,s,se}^{\text{NGgrid}} \geq 0 \tag{8.15}$$

$$P_{t,w,s,se}^{\text{NGgrid}} = P_{t,w,s,se}^{\text{Boiler,in}} \tag{8.16}$$

$$P_{t,w,s,se}^{\text{Boiler,in}} = \frac{P_{t,w,s,se}^{\text{Boiler,out}}}{\text{Ra}_{\text{Boiler}}} \tag{8.17}$$

$$0 \leq P_{t,w,s,se}^{\text{Boiler,out}} \leq P^{\text{Boiler,Max}} \tag{8.18}$$

$$P_{t,w,s,se}^{\text{Boiler,enduser}} = \beta_{t,w,s,se}^{\text{Heat}} \cdot \text{HNC}_{t,w,s,se} \tag{8.19}$$

$$P_{t,w,s,se}^{\text{Boiler,out}} = P_{t,w,s,se}^{\text{Boiler,enduser}} + P_{t,w,s,se}^{\text{Boiler,HS}} \tag{8.20}$$

8.4.6 Heat Storage

HS inputs are provided using heating system equipment such as boilers, CHPs, etc., to enter the circuit at appropriate times. Equations (8.21) and (8.22) indicate the charge and discharge of the HS and the range of the storage amount [20]. Equations (8.23) and (8.24) calculate the share of heating demand supplied by the HS and the inputs of this equipment, respectively.

$$\text{AH}_{t,w,s,se}^{\text{HS}} = \text{AH}_{t-1,w,s,se}^{\text{HS}} + P_{t,w,s,se}^{\text{HSin}} \cdot \eta_{\text{ch}} - \frac{P_{t,w,s,se}^{\text{HSout}}}{\eta_{\text{disch}}} \tag{8.21}$$

$$0 \leq \text{AH}_{t,w,s,se}^{\text{HS}} \leq \text{AH}^{\text{HS, max}} \tag{8.22}$$

$$P_{t,w,s,se}^{\text{HSout}} = \left(1 - \beta_{t,w,s,se}^{\text{Heat}}\right) \times \left(1 - \alpha_{t,w,s,se}^{\text{Heat}}\right) \times \text{HNC}_{t,w,s,se} \tag{8.23}$$

$$P_{t,w,s,se}^{\text{HSin}} = \text{HSS}_{t,w,s,se} + P_{t,w,s,se}^{\text{Boiler,HS}} \tag{8.24}$$

8.4.7 Heat and Electricity Demand

Equations (8.25) and (8.26) determine the amount of electricity that must enter this energy hub system. Its division is done to supply the demand for electricity and the inlet of the GSHP. Equation (8.27) shows the amount of excess renewable energy after supplying the electricity demand of the system. Equations (8.28) and (8.29) also justify balancing load and output in the electrical and heating sections.

$$P_{t,w,s,se}^{\text{Egrid,enduser}} = P_{t,se}^{\text{Demand,Elec}} - N_{\text{CHPwte}}.P_{\text{CHPe}} - N_{\text{PV}}.P_{t,s,se}^{\text{PV}} - N_{\text{WT}}.P_{t,w,se}^{\text{WT}}$$
$$(8.25)$$

$$P_{t,w,s,se}^{\text{Egrid}} = P_{t,w,s,se}^{\text{Egrid,enduser}} + P_{t,w,s,se}^{\text{Egrid,2GSHP}} \tag{8.26}$$

$$\text{SurRES}_{t,w,s,se} = N_{\text{CHPwte}}.P_{\text{CHPe}} + N_{\text{PV}}.P_{t,s,se}^{\text{PV}} + N_{\text{WT}}.P_{t,w,se}^{\text{WT}} - P_{t,se}^{\text{Demand,Elec}}$$
$$(8.27)$$

$$P_{t,se}^{\text{Demand,Elec}} \leq N_{\text{CHPwte}}.P_{\text{CHPe}} + N_{\text{PV}}.P_{t,s,se}^{\text{PV}} + N_{\text{WT}}.P_{t,w,se}^{\text{WT}} + P_{t,w,s,se}^{\text{Egrid,enduser}}$$
$$(8.28)$$

$$P_{t,se}^{\text{Demand,Heat}} \leq N_{\text{CHPwte}}.P_{\text{CHPh}} + P_{t,w,s,se}^{\text{Boiler,enduser}} + P_{t,w,s,se}^{\text{GSHPh}} + P_{t,w,s,se}^{\text{HSout}} - \text{HSS}_{t,w,s,se}$$
$$(8.29)$$

8.5 Incorporated Heat and Electricity Structure

Various components are used in modern heat and electricity incorporated networks, and this study has included the elements mentioned above in its modeling. Figure 8.5 exhibits the energy system structure and how the various components of the system connect. In this system, renewable resources are considered a priority to meet demand.

8.6 Case Study

The available information for an industrial zone is used to implement the studied model. Constraints for renewable resources planning in this system for wind and PV power plants are mainly related to the available area of the industrial zone, and constraints for the CHPwte power plant are related to the amount of waste generated in that area, which is about 214 tons is per day for the case study. These constraints determine the search interval for decision variables so that the objective function converges to the optimal response. Required information for this study is expressed in two sections: demand and techno-economical, which will be explained in the next section.

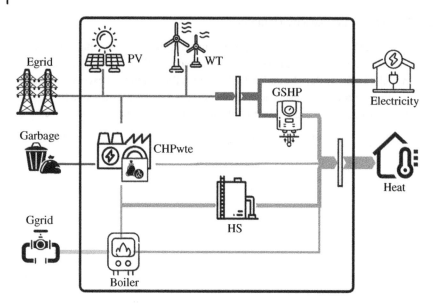

Figure 8.5 Structure and components of the proposed model.

8.7 Demand Profile

Figures 8.6 and 8.7 show the daily electricity and heating demand profiles for spring, summer, fall, and winter, respectively.

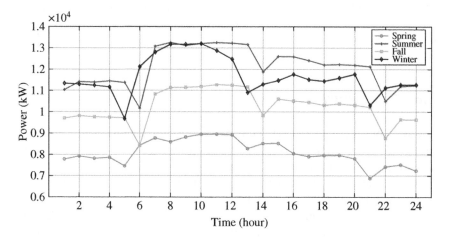

Figure 8.6 Daily electricity demand for the studied energy system.

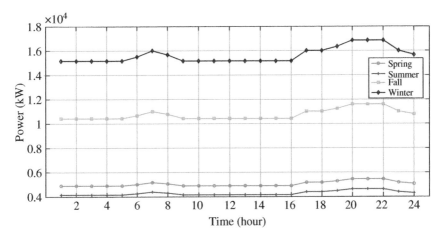

Figure 8.7 Daily heating demand for the studied energy system.

8.8 Economic and Environmental Features

In order to study solar and wind systems, the most important indicators are solar radiation and wind speed, respectively, which are shown in Figures 8.8 and 8.9. In economic terms, in this research, time of use pricing (TOU) programing is implemented for electricity to manage consumers' side; as a result, electricity and

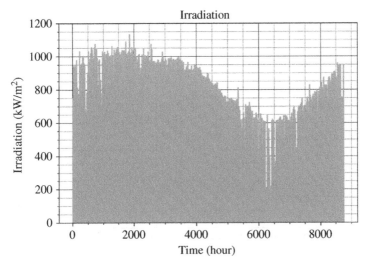

Figure 8.8 Solar irradiation for the case study during one year.

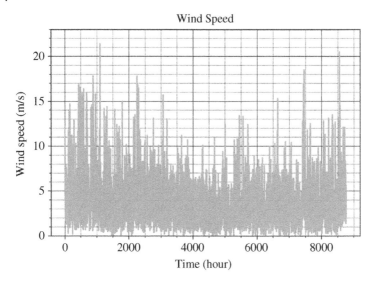

Figure 8.9 Wind speed for the case study during one year.

natural gas cost are shown in Figure 8.10 [55]. Table 8.2 shows the investment, operation and maintenance (O&M) costs and lifetime of each component used in the incorporated energy network of the studied industrial zone.

Figure 8.11 shows the 1000 scenarios generated from PV power generation for a 100-kW system and selected clusters. In other words, the two clusters shown in the figure represent PV power production behavior for each season. The probability of each cluster in each season is specified in Table 8.3. The mentioned method for WT

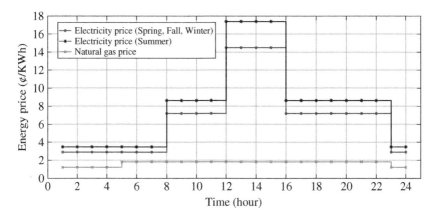

Figure 8.10 Electricity and natural gas price.

Table 8.2 Detailed cost for the energy system's components.

Component	Capital cost ($/kW)	O&M cost ($/kWh)	Life (year)	References
PV	995	16.7	25	[56]
WT	1473	13.6	25	[56]
CHPwte	3057	122	20	[57, 58]
GSHP	373	7.4	30	[59]
HS	56	0	10	[60]
Boiler	50	0.5	20	[59, 61]

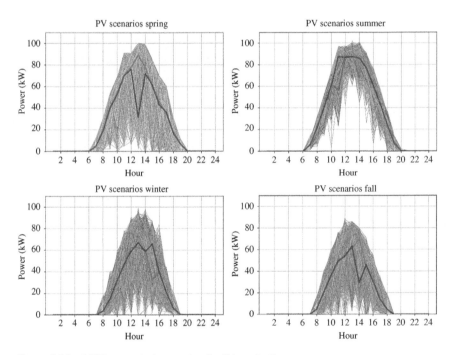

Figure 8.11 1000 generated scenarios for PV production.

power generation is also repeated, shown in Figure 8.12 and Table 8.4 of power generation behavior in this case study and the probability of each cluster.

Due to environmental constraints, the imposition of a carbon tax can force the industry to reduce CO_2 emissions. According to research [62], the cost of carbon emissions is $6 02086/kg of CO_2. The emission coefficients for the upstream network, CHPwte, and gas boiler are 975.2, 418.5, and 468.2 g/kWh, respectively [62, 63].

Table 8.3 PV clusters and their probability.

Cluster	Seasons			
	Spring	Summer	Fall	Winter
Cluster 1	0.597	0.3	0.647	0.443
Cluster 2	0.403	0.7	0.353	0.557

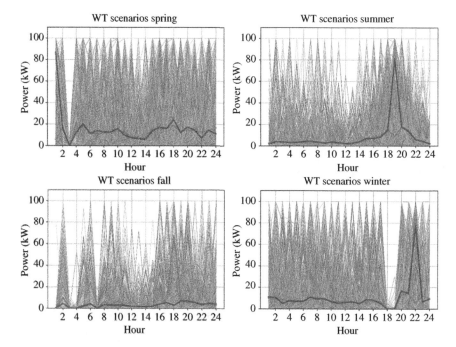

Figure 8.12 1000 generated scenarios for WT production.

Table 8.4 WT clusters and their probability.

Cluster	Seasons			
	Spring	Summer	Fall	Winter
Cluster 1	0.847	0.784	0.078	0.868
Cluster 2	0.153	0.216	0.922	0.132

8.9 Result and Discussion

The mathematical model based on MINLP for the proposed energy system has been obtained, and the GA has been used in MATLAB software to solve the optimization problem. In order to mathematically model, a computer system with 1.8 GHz Core-i7 and 16 GB of RAM is employed. In this research, to achieve optimal operation and planning, heating dispatching factors, WT, PV, CHPwte, boiler, HS, and GSHP capacity have been analyzed as decision variables. The heating demand in the studied system is equal to 120, 104, 257, and 374 MWh, respectively, and the electricity demand is 192, 290, 245, and 280 MWh for a day in spring, summer, fall, and winter. Also, in addition to the mentioned electricity consumption, about 1, 0.8, 16, and 26 MWh of electricity for each season of the year, respectively, is used to meet the needs of the GSHP to satisfy the heating demand at specific time hours. The strategy for this system is to meet the demand of the industrial zone using renewable sources as much as possible and to sell the excess renewable energy to the upstream network. After applying the process of optimizing the capacity of electrical components were determined for PV and CHPwte 16 and 2 MW, respectively. Due to the wind speed curve of this region and the high investment cost of this element, WT is not the optimal source for this energy system.

On the other hand, in the heating section, the optimal capacity of GSHP, boiler, and HS is calculated as 8154, 12876, and 17200 kWh, respectively. As shown in Figures 8.13–8.16, PV output generation is different in various seasons, so one of the most apparent differences in these four seasons is the amount of electricity purchased and sold to the grid. Since according to the energy price curve, from 12:00 to 16:00, the highest price level has been proposed for electricity, and on the other hand, the production of PV output power reaches its peak at the same time, so in these hours, the industrial zone not only do not impose on the network but by injecting electricity into the upstream network, helps the system in peak times. This pattern of electricity supply is shown in Figure 8.17 for each season. The amount of this electricity injection in spring and summer is significantly higher than in the fall and winter seasons due to the favorable weather conditions in the region. When there is no sunlight, only the CHPwte remind in the circuit, and the rest of the energy is satisfied by the upstream network. These times, buying electricity from the upstream network is at low price levels. In terms of heating, CHPwte plays a key role, and in spring and summer, it provides about 76 and 88% of the heating required by the system, respectively. However, with the increase of heating demand in fall and winter, the percentage of heating supply by CHPwte decreases by 35 and 24%, respectively, and the GSHP, boiler, and HS satisfy the rest of the heating demand. As can be seen in the systems, various natural gas, waste, and electricity have been used to produce heating. Providing such diversity in the production basket increases the flexibility and reliability of

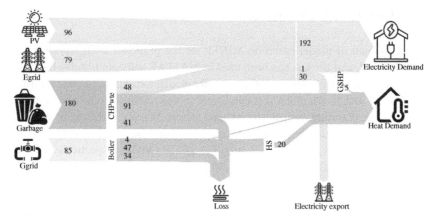

Figure 8.13 Optimal operation of the proposed incorporated energy network in a day of spring (Energy unit: MWh).

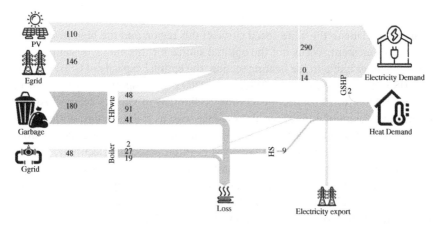

Figure 8.14 Optimal operation of the proposed incorporated energy network in a day of summer (Energy unit: MWh).

the energy system, and the overall costs of the system are decreased. The GSHP plays a key role in this system, as it is the input of this component. As a result, when natural gas or waste is unavailable or has a higher cost, or there is not enough production when renewable sources are in the circuit, and free electricity is prepared for GSHP, this component can meet the heating demand. This equipment enters the circuit due to its high efficiency and cooperates with other heating equipment. GSHP in winter is the largest share in the supply of heating load with a contribution of about 34% and fall with about 30% after CHPwte is the most important producer of heating in the system. But in other seasons, this importance has diminished, and the CHPwte and HS operate with a larger share.

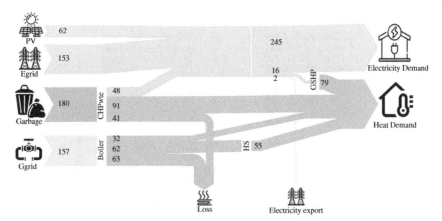

Figure 8.15 Optimal operation of the proposed incorporated energy network in a day of fall (Energy unit: MWh).

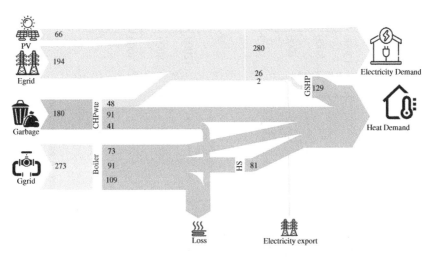

Figure 8.16 Optimal operation of the proposed incorporated energy network in a day of winter (Energy unit: MWh).

Another component used in this section is a natural gas boiler that can be examined directly and indirectly. In direct mode, this equipment satisfies 3, 2, 12, and 19% of the heating load in spring, summer, fall, and winter, and a large share of boiler production acts as an HS input. HS input values are 47, 27, 62, and 91 MWh for spring, summer, fall, and winter. Figure 8.18 shows each heating element's contribution to providing heating per hour for different seasons. As shown in Figures 8.17 and 8.18, every hour of the day, a specific contribution of sources of electricity and heating satisfies energy consumption. This means that by

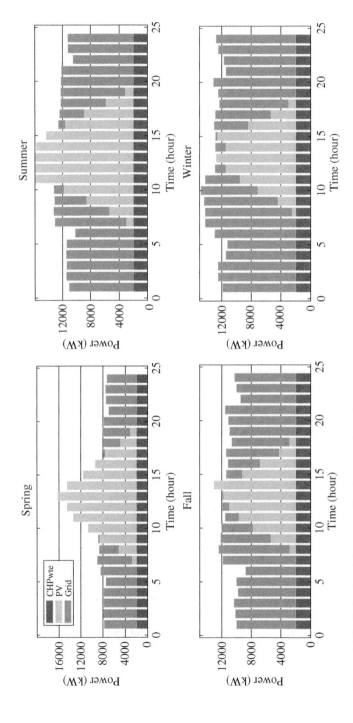

Figure 8.17 Daily electricity generation by source in each season.

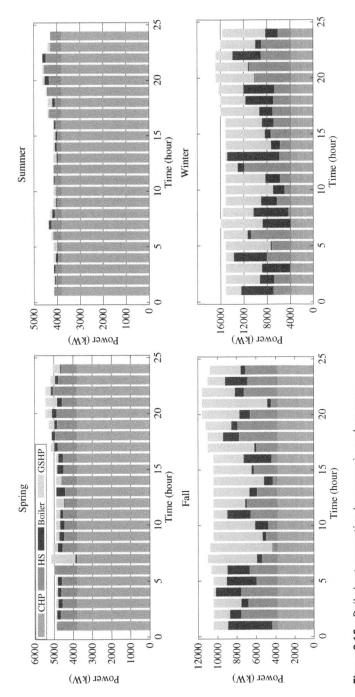

Figure 8.18 Daily heat generation by source in each season.

creating interaction between different parts of the energy system, the energy management system has various options per hour to meet the needs of consumers. This selection is based on economic and environmental parameters, and resources are chosen in each hour of the day.

8.10 Conclusion

In these years, a large portion of energy consumption is used to generate electricity and heat. As a result, the focus and study on incorporated energy systems have increased. Switching to hybrid heat and electricity systems is one of the effective solutions in this field. In addition to improving the overall efficiency of the system and related economic issues, it is also environmentally effective. In incorporated power and heating systems, the product portfolio is diverse, so different options can contribute to the supply of energy demand. But the main question is which component and with what capacity should be used in the energy system when that component should be operated in the system, and when should it be taken out of the circuit to optimize the system's objective function. Therefore, to plan and operate the incorporated heat and electricity system, it is necessary to study and model the elements of the system in the first step. Then, the system's technical, economic, and environmental features are analyzed in the second step. Finally, each element's optimal capacities and operation are determined using computer systems and algorithms. In general, in this research, considering the uncertainty of the variables affecting the operation of an energy system in a scenario-based method and using modern technologies in heating, an interconnected energy system based on the technical-economic and environmental aspects were developed.

References

1 Mohammadi, M., Noorollahi, Y., Mohammadi-Ivatloo, B. et al. (2018). Optimal management of energy hubs and smart energy hubs – a review. *Renew. Sust. Energ. Rev.* 89 (March): 33–50.

2 Mohammadi, M., Noorollahi, Y., and Mohammadi-Ivatloo, B. (2018). Impacts of energy storage technologies and renewable energy sources on energy hub systems. In: *Operation, Planning, and Analysis of Energy Storage Systems in Smart Energy Hubs* (ed. B. Mohammadi-Ivatloo and F. Jaberi), 1–456. Springer International Publishing. https://doi.org/10.1007/978-3-319-75097-2.

3 IEA (2020). CO2 Emissions by Sector. https://www.iea.org/data-and-statistics/data-browser/?country=WORLD&fuel=CO2%20emissions&indicator=CO2BySector (accessed June 2020).

4 Lund, H., Østergaard, P.A., Connolly, D., and Mathiesen, B.V. (2017). Smart energy and smart energy systems. *Energy* 137: 556–565.

5 Haitao, L., Yiming, L., Ling, W. et al. (2017). Research on the conceptual model of smart energy system. *2017 IEEE Conference on Energy Internet and Energy System Integration (EI2)*, Beijing, China. IEEE, pp. 1–6. https://doi.org/10.1109/EI2.2017.8245690

6 Lund, H., Vad Mathiesen, B., Connolly, D., and Østergaarda, P.A. (2014). Renewable energy systems – a smart energy systems approach to the choice and modelling of 100% renewable solutions. *Chem. Eng. Trans.* 39 (Special issue): 1–6.

7 IEA (2020). Electricity generation from renewable energy. https://www.iea.org/data-and-statistics country=WORLD&fuel=Renewables%20and%20waste&indicator=RenewGenBySource (accessed December 2020).

8 Dincer, I. and Acar, C. (2017). Smart energy systems for a sustainable future. *Appl. Energy* 194: 225–235.

9 Rae, C., Kerr, S., and Maroto-Valer, M.M. (2020). Upscaling smart local energy systems: a review of technical barriers. *Renew. Sust. Energ. Rev.* 131 (July): 110020.

10 Mohammadi, M., Noorollahi, Y., Mohammadi-Ivatloo, B., and Yousefi, H. (2017). Energy hub: from a model to a concept – a review. *Renew. Sust. Energ. Rev.* 80 (December 2016): 1512–1527.

11 Rahmatian, M.R., Shamim, A.G., and Bahramara, S. (2021). Optimal operation of the energy hubs in the islanded multi-carrier energy system using cournot model. *Appl. Therm. Eng.* 191 (January): 116837.

12 Khorasany, M., Najafi-Ghalelou, A., Razzaghi, R., and Mohammadi-Ivatloo, B. (2021). Transactive energy framework for optimal energy management of multi-carrier energy hubs under local electrical, thermal, and cooling market constraints. *Int. J. Electr. Power Energy Syst.* 129 (February): 106803.

13 Habib, H.U.R., Subramaniam, U., Waqar, A. et al. (2020). Energy cost optimization of hybrid renewables based V2G microgrid considering multi objective function by using artificial bee colony optimization. *IEEE Access* 8: 62076–62093.

14 Sadeghi, H., Rashidinejad, M., Moeini-Aghtaie, M., and Abdollahi, A. (2019). The energy hub: an extensive survey on the state-of-the-art. *Appl. Therm. Eng.* 161 (July): 1–26. https://doi.org/10.1016/j.applthermaleng.2019.114071.

15 Zhu, J. (2015). *Optimization of Power System Operation*, 2e. Wiley.

16 Aien, M., Hajebrahimi, A., and Fotuhi-Firuzabad, M. (2016). A comprehensive review on uncertainty modeling techniques in power system studies. *Renew. Sust. Energ. Rev.* 57: 1077–1089.

17 Jordehi, A.R. (2018). How to deal with uncertainties in electric power systems? A review. *Renew. Sust. Energ. Rev.* 96 (June 2017): 145–155.

18 Alzahrani, A.M., Zohdy, M., and Yan, B. (2021). An overview of optimization approaches for operation of hybrid distributed energy systems with photovoltaic and diesel turbine generator. *Electr. Power Syst. Res.* 191 (October 2020): 106877.

19 Ma, T., Wu, J., Hao, L. et al. (2018). The optimal structure planning and energy management strategies of smart multi energy systems. *Energy* 160: 122–141.

20 Majidi, M., Mohammadi-Ivatloo, B., and Anvari-Moghaddam, A. (2019). Optimal robust operation of combined heat and power systems with demand response programs. *Appl. Therm. Eng.* 149 (December 2018): 1359–1369.

21 Ata, M., Erenoğlu, A.K., Şengör, İ. et al. (2019). Optimal operation of a multi-energy system considering renewable energy sources stochasticity and impacts of electric vehicles. *Energy* 186 (2019): 115841.

22 Mansouri, S.A., Ahmarinejad, A., Ansarian, M. et al. (2020). Stochastic planning and operation of energy hubs considering demand response programs using Benders decomposition approach. *Int. J. Electr. Power Energy Syst.* 120 (March): 106030.

23 Twaha, S. and Ramli, M.A.M. (2018). A review of optimization approaches for hybrid distributed energy generation systems: Off-grid and grid-connected systems. *Sustain. Cities Soc.* 41: 320–331.

24 De Leon-Aldaco, S.E., Calleja, H., and Aguayo Alquicira, J. (2015). Metaheuristic optimization methods applied to power converters: a review. *IEEE Trans. Power Electron.* 30 (12): 6791–6803.

25 Eriksson, E.L.V. and Gray, E.M.A. (2019). Optimization of renewable hybrid energy systems – a multi-objective approach. *Renew. Energy* 133: 971–999.

26 Mansouri, S.A., Ahmarinejad, A., Javadi, M.S., and Catalão, J.P.S. (2020). Two-stage stochastic framework for energy hubs planning considering demand response programs. *Energy* 206: 118124. https://doi.org/10.1016/j.energy.2020.118124.

27 Moghaddas-tafreshi, S.M., Mohseni, S., Karami, M.E., and Kelly, S. (2019). Optimal energy management of a grid-connected multiple energy carrier. *Appl. Therm. Eng.* 152 (January): 796–806.

28 Najafi-Ghalelou, A., Nojavan, S., Zare, K., and Mohammadi-Ivatloo, B. (2019). Robust scheduling of thermal, cooling and electrical hub energy system under market price uncertainty. *Appl. Therm. Eng.* 149 (April 2018): 862–880.

29 Rabiee, A. and Mohseni-bonab, S.M. (2017). Maximizing hosting capacity of renewable energy sources in distribution networks: a multi-objective and scenario-based approach. *Energy* 120: 417–430.

30 Moghaddas-Tafreshi, S.M., Jafari, M., Mohseni, S., and Kelly, S. (2019). Optimal operation of an energy hub considering the uncertainty associated with the power consumption of plug-in hybrid electric vehicles using information gap decision theory. *Int. J. Electr. Power Energy Syst.* 112 (April): 92–108.

31 Bagher Sadati, S.M., Moshtagh, J., Shafie-khah, M. et al. (2019). Operational scheduling of a smart distribution system considering electric vehicles parking lot: a bi-level approach. *Int. J. Electr. Power Energy Syst.* 105 (June 2018): 159–178.

32 Noorollahi, Y., Golshanfard, A., Aligholian, A. et al. (2020). Sustainable energy system planning for an industrial zone by integrating electric vehicles as energy storage. *J. Energy Storage* 30 (January): 101553.

33 Dini, A., Pirouzi, S., Norouzi, M., and Lehtonen, M. (2019). Grid-connected energy hubs in the coordinated multi-energy management based on day-ahead market framework. *Energy* 188: 116055.

34 Dong, J., Nie, S., Huang, H. et al. (2019). Research on economic operation strategy of CHP microgrid considering renewable energy sources and integrated energy demand response. *Sustainability* 11 (18): 1–22.

35 Emrani-Rahaghi, P. and Hashemi-Dezaki, H. (2020). Optimal scenario-based operation and scheduling of residential energy hubs including plug-in hybrid electric vehicle and heat storage system considering the uncertainties of electricity price and renewable distributed generations. *J. Energy Storage* 33 (April): 102038. https://doi.org/10.1016/j.est.2020.102038.

36 Dehghan, S., Kazemi, A., and Amjady, N. (2014). Multi-objective robust transmission expansion planning using information-gap decision theory and augmented ε-constraint method. *IET Gener. Transm. Distrib.* 8 (5): 828–840.

37 Mansour-Saatloo, A., Agabalaye-Rahvar, M., Mirzaei, M.A. et al. (2020). Robust scheduling of hydrogen based smart micro energy hub with integrated demand response. *J. Clean. Prod.* 267: 122041.

38 Golshanfard, A., Noorollahi, Y., and Hashemi-Dezaki, H. (2021). *Energy Modeling for Sustainable Smart Energy System Planning for Industrial Zones.* University of Tehran.

39 Li, G., Jin, Y., Akram, M.W. et al. (2018). Application of bio-inspired algorithms in maximum power point tracking for PV systems under partial shading conditions – a review. *Renew. Sust. Energ. Rev.* 81 (July 2017): 840–873.

40 Kim, S. and Hur, J. (2020). A probabilistic modeling based on Monte Carlo simulation of wind powered ev charging stations for steady-states security analysis. *Energies* 13: 5260. https://doi.org/10.3390/en13205260.

41 Uwineza, L., Kim, H.G., and Kim, C.K. (2021). Feasibilty study of integrating the renewable energy system in Popova Island using the Monte Carlo model and HOMER. *Energy Strateg. Rev.* 33: 100607.

42 Roustaei, M., Niknam, T., Salari, S. et al. (2020). A scenario-based approach for the design of smart energy and water hub. *Energy* 195: 116931. https://doi.org/10.1016/j.energy.2020.116931.

43 Biswas, P.P., Suganthan, P.N., Mallipeddi, R., and Amaratunga, G.A.J. (2019). Optimal reactive power dispatch with uncertainties in load demand and renewable energy sources adopting scenario-based approach. *Appl. Soft Comput. J.* 75: 616–632.

44 Liu, G., Yang, J., Hao, Y., and Zhang, Y. (2018). Big data-informed energy efficiency assessment of China industry sectors based on K-means clustering. *J. Clean. Prod.* 183: 304–314.

45 Zsigraiová, Z., Tavares, G., Semiao, V., and de Carvalho, M.G. (2009). Integrated waste-to-energy conversion and waste transportation within island communities. *Energy* 34 (5): 623–635.

46 Singh, K. and Hachem-vermette, C. (2019). Influence of mixed-use neighborhood developments on the performance of waste-to-energy CHP plant. *Energy* 189: 116172.

47 Nyeche, E.N. and Diemuodeke, E.O. (2020). Modelling and optimisation of a hybrid PV-wind turbine-pumped hydro storage energy system for mini-grid application in coastline communities. *J. Clean. Prod.* 250: 119578.

48 Idoko, L., Anaya-Lara, O., and McDonald, A. (2018). Enhancing PV modules efficiency and power output using multi-concept cooling technique. *Energy Rep.* 4: 357–369.

49 You, C. and Kim, J. (2020). Optimal design and global sensitivity analysis of a 100% renewable energy sources based smart energy network for electrified and hydrogen cities. *Energy Convers. Manag.* 223 (July): 113252.

50 Nosratabadi, S.M., Jahandide, M., and Nejad, R.K. (2020). Simultaneous planning of energy carriers by employing efficient storages within main and auxiliary energy hubs via a comprehensive MILP modeling in distribution network. *J. Energy Storage* 30 (March): 101585.

51 Sohoni, V., Gupta, S.C., and Nema, R.K. (2016). A critical review on wind turbine power curve modelling techniques and their applications in wind based energy systems. *J. Energy* 2016 (region 4): 1–18.

52 David, I., Ştefănescu, C., and Vlad, I. (2015). Efficiency assessment of ground-source heat pumps in comparison with classical heating syastem. *Int. Multidiscip. Sci. GeoConference Surv. Geol. Min. Ecol. Manag. SGEM* 1 (4): 191–198.

53 Rees, S.J. (2016). *An Introduction to Ground-Source Heat Pump Technology.* Elsevier Ltd.

54 Noorollahi, Y., Saeidi, R., Mohammadi, M. et al. (2018). The effects of ground heat exchanger parameters changes on geothermal heat pump performance – a review. *Appl. Therm. Eng.* 129: 1645–1658.

55 Mansour-saatloo, A., Agabalaye-rahvar, M., and Amin, M. (2020). Robust scheduling of hydrogen based smart micro energy hub with integrated demand response. *J. Clean. Prod.* 267: 122041.

56 IRENA (2019). *Renewable Power Generation Costs in 2019.* Abu Dhabi.

57 Escamilla-García, P.E., Camarillo-López, R.H., Carrasco-Hernández, R. et al. (2020). Technical and economic analysis of energy generation from waste incineration in Mexico. *Energy Strateg. Rev.* 31: 100542. https://doi.org/10.1016/j.esr.2020.100542.

58 Cost Information for Energy Component (2020). https://www.energyplan.eu/useful_resources/costdatabase/ (accessed December 2020).

59 Zhang, G., Wang, J., Ren, F. et al. (2021). Collaborative optimization for multiple energy stations in distributed energy network based on electricity and heat interchanges. *Energy* 222: 119987.

60 Ma, W., Fang, S., and Liu, G. (2017). Hybrid optimization method and seasonal operation strategy for distributed energy system integrating CCHP, photovoltaic and ground source heat pump. *Energy* 141: 1439–1455.

61 Yang, X., Chen, Z., Huang, X. et al. (2021). Robust capacity optimization methods for integrated energy systems considering demand response and thermal comfort. *Energy* 221: 119727.

62 Chamandoust, H., Derakhshan, G., Hakimi, S.M., and Bahramara, S. (2019). Tri-objective optimal scheduling of smart energy hub system with schedulable loads. *J. Clean. Prod.* 236: 117584.

63 Münster, M. and Meibom, P. (2011). Optimization of use of waste in the future energy system. *Energy* 36 (3): 1612–1622.

9

Optimal Coordinated Operation of Heat and Electricity Incorporated Networks

S. Mahdi Kazemi-Razi and Hamed Nafisi

Department of Electrical Engineering, Amirkabir University of Technology (Tehran Polytechnic), Tehran, Iran

Nomenclature

A. Acronyms

ASHRAE	American society of heating, refrigerating and air-conditioning engineers
BSUoS	Balancing system use of system
BT	Back-pressure turbine
CAES	Compressed air energy storage
CCHP	Combined cool, heat and power
CHP	Combined heat and power
CSP	Concentrated solar power
DHW	Domestic hot water
DNUoS	Distribution network use of system
EFR	Enhanced frequency response
EHP	Electric heat pump
EB	Electrical boiler
EP	Electrical parameter
ERPS	Enhanced reactive power service
FACTS	Flexible AC transmission system
FFR	Firm frequency response
GB	Gas boiler
HEIN	Heat and electricity incorporated network
ISO	Independent system operator
KKT	Karush–Kuhn–Tucker

Coordinated Operation and Planning of Modern Heat and Electricity Incorporated Networks,
First Edition. Edited by Mohammadreza Daneshvar, Behnam Mohammadi-Ivatloo, and Kazem Zare.
© 2023 The Institute of Electrical and Electronics Engineers, Inc.
Published 2023 by John Wiley & Sons, Inc.

PDF	Probabilistic distribution function
P2G	Power-to-gas
PCM	Phase change material
STOR	Shot-term operating reserve
SEL	Stable export limit
TSUoS	Transmission system use of system
TES	Thermal energy storage

B. Indices

s	Scenario index
t	Time step index
b	Bus index
e	Power transmission equipment index
i	Generation/storage unit index
Δt	Time interval (h)

C. Parameters

B_b^d / B_b^u	Min/Max ES capacity (kWh)
$H_b^{CHP\_d} / H_b^{CHP\_u}$	Min/Max CHP thermal power (kW)
$H_b^{EB\_d} / H_b^{EB\_u}$	Min/Max EB thermal power (kW)
$H_b^{GB\_d} / H_b^{GB\_u}$	Min/Max GB thermal power (kW)
$P_b^{ES\_d} / P_b^{ES\_u}$	Min/ Max ES charging rate (kW)
$P_b^{EHP\_d} / P_b^{EHP\_u}$	Min/ Max EHP electrical power (kW)
X_b^d / X_b^u	Min/Max temperature of TES (°C)
$v_{t,b}^d / v_{t,b}^u$	Max down/up temperature variation (°C)
$\eta_b^{CHP\_e} / \eta_b^{CHP\_t}$	CHP electrical/thermal efficiency (%)
$\eta_b^{GB} / \eta_b^{EB}$	GB/EB efficiency (%)
η_b^{ES}	Round-trip efficiency of ES (%)
$CP_{s,t,b}^{EHP}$	EHP coefficient of performance (%)
C^{RES_u}	Max call length of reserve (h)
P^{RES}	Probability of reserve call (%)
$X_{s,t,b}^{DHW}$	DHW load (kWh)
$I\_G_{s,t,b} / S\_G_{s,t,b}$	Internal/PV heat gains (kWh)
$O_{s,t,b}$	Binary occupancy ($\in (0,1)$)
C_b^B / C_b^{TES}	Build/TES thermal capacitance (kWh/°C)
R_b^B / R_b^{TES}	Build/TES thermal resistance (°C/kW)
$T_{t,b}^A$	Adjusted temperature (°C)

p_s	Scenario probability (%)
$T_{s,t}^{ENV}$	Environmental temperature (°C)
$\epsilon_t^I/\epsilon_t^E$	Day-ahead import/export price (£/kWh)
$\pi_{s,t}^I/\pi_{s,t}^E$	Imbalance import/export price (£/kWh)
δ_t/σ_t^d	Gas/down reserve availability price (£/kWh)
ξ_t^d/ξ_t^s	Temperature deficit/surplus penalties (£/°Ch)
$R_{s,t,i}^{solar}$	Solar radiation percent (-)
P_i^{max}	Maximum installed power of solar field (kW)
$P_{s,t}^{dissipated}$	Heat loss of BT start up (kW)
η^{BT}	BT efficiency (-)
η^{TES+}/η^{TES-}	TES thermal charging/discharging power efficiency (-)
η^{CSP}	Thermal power transfer efficiency of CSP
k	BT thermoelectric ratio (-)
α/β	Wind farm power coefficient ()
v	Wind speed (m/s)
P_{rated}/V_{rated}	Wind farm rated power/speed (kW)
V_{CI}/V_{CO}	Low/up cut speed of wind farm
$\overline{P^{CAES+}}/\overline{P^{CAES-}}$	Max CAES charging/discharging power (kW)
$\eta^{CAES+}/\eta^{CAES-}$	CAES charging/discharging power efficiency (-)
$E_{min}^{CAES}/\eta_{max}^{CAES}$	Min/Max CAES stored energy (kWh)

D. Variables

$P_{s,t}^{BT}/Q_{s,t}^{BT}$	BT output electrical/thermal power (kW)
$R_{s,t}^{TES+}/R_{s,t}^{TES-}$	TES thermal charging/discharging power (kW)
$P_{s,t,i}^{Wind}$	Wind farm output power (kW)
u_t^{CAES+}/u_t^{CAES-}	Binary charging/discharging indicator of CAES ($\in (0,1)$)
P_t^{CAES+}/P_t^{CAES-}	CAES charging/discharging power (kW)
E_t^{CAES}	CAES stored energy (kWh)
$B_{s,t,b}/X_{s,t,b}$	ES/TES energy level (kWh)
$F_{s,t,b}^B/H_{s,t,b}^B$	Build thermal storage footroom/headroom (kW)
D_t^I/D_t^E	Day-ahead energy import/export (kW)
$G_{s,t,b}^{GB}$	Gas consumed by GB (kW)
$H_{s,t,b}^{GB}/H_{s,t,b}^{CHP}$	GB/CHP thermal power (kW)
$X_{s,t,b}^I/X_{s,t,b}^{LOSS}$	TES heat import/loss (kWh)
$X_{s,t,b}^{SH}$	Space heating demand (kWh)
$I_{s,t}^I/I_{s,t}^E$	Imbalance energy import/export (kW)
$P_{s,t,b}^{ES}/P_{s,t,b}^{EB}$	ES/EB electrical power (kW)

$P_{s,t,b}^{CHP}/P_{s,t,b}^{EHP}$	CHP/EHP electrical power (kW)
$R_{s,t,b}^{ES\_d}/R_{s,t,b}^{CHP\_d}$	ES/CHP down reserve (kW)
$R_{s,t,b}^{EB\_d}/R_{s,t,b}^{EHP\_d}$	EB/EHP down reserve (kW)
$R_{s,t}^{dis\_a}$	Down reserve district level (kW)
$T_{s,t,b}$	Build temperature (°C)
$T_{s,t,b}^{d}/T_{s,t,b}^{s}$	Temperature deficit/surplus (°C)
$T_{s,t,b}^{R\_d}/T_{s,t,b}^{R\_s}$	Temperature deficit/surplus of reserve (°C)
$\Gamma_{s,t,b}$	Internal/PV gain vent (%)
$P_{s,t,i}^{photovoltaic}$	Power produced by solar field (kW)

9.1 Introduction

HEINs including electricity, gas, heat and cool have drawn huge attention to themselves due to their high features in minimizing the cost and pollutants emissions and also maximizing some important technical features of HEINs such as flexibility, reliability, resiliency, and sustainability. Therefore, in addition to low operational cost and high end-user satisfaction due to maximum flexibility, it also can operate optimally over critical conditions such as natural disasters (including floods) or failures/maintenance periods of equipment due to high reliability and resiliency. Also, they can operate permanently due to high sustainability, and their equipment would not be damaged. Therefore, the maintenance cost would also decrease.

Because HEINs can integrate a huge amount of renewable units, thus, CO_2 emission will be reduced significantly and results in climate change abatement. Exploiting more renewable units, which have more accessibility even in far places compared to traditional power plants also leads to rural electrification. The new strategies achieved for HEINs power management, such as machine learning, allow us to model the uncertainties of renewable generators more accurately.

The first step to optimize the performance of HEINs is modeling all parts of HEINs exactly to be able to schedule the resources and loads of HEINs optimally. These parts include demand, equipment (such as resources and storages) and the available markets for HEINs to participate. The mentioned parts are elaborated and modeled in Sections 9.2 and 9.5. Section 9.5 also defines all types of markets such as energy market for electricity (including day-ahead and unbalanced market), the wholesale gas market and local heat markets are explained. Furthermore, the ancillary services markets, such as the reserve market are defined. Also, there are other incentives, such as tax or use-of-system charges, which has a significant impact on HEINs operation. These incentives are completely explained in Section 9.5. Also, the existing uncertainties in power management problem of HEINs are explained in Section 9.3, and then the available strategies to overcome these uncertainties are suggested. Afterward, Section 9.4 defines the optimal operation of HEINs to determine which targets must be considered through HEINs optimal operation. According to these targets, the available potentials of HEINs (such as demand response) which lead to optimal operation are elaborated. According to this section, through the optimal operation of HEINs, three major goals are considered, as follows:

- Technical issues such as congestion and power losses.
- Economic issues such as cost reduction.
- Environmental issues such as carbon emissions reduction.

Reference [1] proposes an optimal power management method for HEINs to increase flexibility by power scheduling; however, it does not consider thermal comfort. The authors in [2] also enhance the flexibility using energy vector substitution of multi-carrier resources and storages. But reserve market and selling energy in market participation is not modeled in [2]. The role of power and reserve scheduling in optimal operation of HEINs is evaluated by [3], however it missed thermal comfort investigation. Also, Refs. [4, 5] model the new technologies including power to gas devices and electric vehicles to optimize the HEINs operation, but they do not investigate the impact of thermal comfort, voltage regulation and congestion. As an interesting research, the flexibility enhancement in unbalanced networks are investigated in [6], but it only consider the electricity network and does not consider to HEINs. The potential of electric vehicles as mobile storage in enhancing the flexibility in the level of distribution networks are investigated by [7]. The robustness of this model allows to reach the optimal operation in presence of electric vehicles uncertainties. As a module to regulate voltage and avoid the congestion in models that optimize the operation of HEINs without these two properties, the authors in [8] propose a model to set the voltage and current profile in the standard range while it minimizes the players participation. The potential of thermal energy storage (TES) in increasing the flexibility using time consumption shift that provides more demand response is demonstrated by [9] in multi-time scale. The authors in [10] use power scheduling, demand response, and market participation to increase the HEINs flexibility, while it also maximizes the end-users thermal comfort in a way that is robust even during a reserve call. This model uses a software (OpenDSS) linked to optimization process (MATLAB) to model the voltage/current profile for voltage regulation and congestion avoidance, while it enhances the calculation speed and accuracy. Furthermore, using this model, the power system development can be added easily. The model proposed by [11] augments the HEINs flexibility and thermal comfort using market participation and power scheduling, but it does not consider power flow calculations for voltage and current. Reference [12] demonstrates the impact of TES and demand response in HEINs optimal operation including thermal comfort, but it does not model voltage/current profiles. The thermal demand response achieved from thermal loads are addressed by [13, 14], while also reserve market is modeled in [14]. In these two papers, the temperature provided for end-users is allowed to be degraded in a predefined range to minimize the HEIN operational cost, however, EPs are set into their standard range. The research done by [15] illustrates the impact of electric vehicles, as a source of added loads and also as a mobile storage, on optimal operation of smart cities. Also, the authors in [16] have reduced the emissions of pollutants by 3.5% by the flexible management of electric vehicles, while it uses ac power flow and Wohler curve through the electric vehicles scheduling. The control of residential heating, ventilation and air conditioning units as a source of flexibility enhancement is investigated by [17], while also it models the impact of storage units to increase the end-users comfort by minimizing their participation in this model. The authors in [18] use a roust optimization for modeling the uncertainties through network power management. Also another strategy in modeling the uncertainties is stochastic programing that is used in [19] for modeling the uncertainties of renewable sources and electric vehicles, while it aims to do phase voltage balancing. Therefore, it results in service voltage, decreasing the voltage unbalance that leads to more lifetime of transformers. The study done by [20] models the impact of porous pavement and plant transpiration and proves that higher buildings increase the relationship between provided temperature and the ambient temperature. Therefore, tall buildings do not provide thermal comfort, while the building with more space improve end-users thermal comfort. A model of competitive markets that HEINs operator can offer their own power/ price quantities as a price-maker is presented in [21], while it also considers the power/price offers of other players in market clearing process. But this model does not model their thermal comfort and power flow calculations. As a strategy to aggregate reactive power from loads and renewables for ancillary services markets, such as modifying the voltage profile, the authors in [22] present an index showing the flexibility achieved by flexible loads and inverter-based photovoltaics. As an external tool for maximizing the flexibility in power systems, the authors in [23] quantify the flexibility achieved from water distribution system. Reference [24] proposes a decentralized approach for optimal managing of HEINs, while it also models the gas pipes using linepack model and allows the HEINs to participate in imbalanced and reserve market. It proves decentralized model results in more optimal operation than centralized models. In the level of smart cities, the proposed framework for optimal operation of HEINs by [25], minimizes the power loss and costs alongside the voltage regulation. The authors in [26] compare a centralized and decentralized strategy in HEINs operation management which use the potentials of power scheduling and demand response. Although, the centralized strategy results in less carbon emissions, but it has more value of cost. To propose an economically and computationally method for optimal operation of networks, the authors in [27] provide a strategy that is distributed between some systems that it wants to aggregate their flexibility. The four-layer model proposed by [28] can aggregate the flexibility in level of end-users, local systems, aggregators, and market, while this model also reduces the operational risk. Therefore, based on previous studies, the research gaps in this field are as follows:

1) There is no work to model all the existing potential of HEINs exactly. Therefore, it can lead to the non-optimal results in optimizing HEINs operation.
2) The available markets for HEINs operators are not modeled completely, therefore, this important part is missed in HEINs power management, which prevents from achieving the maximum revenue.
3) There is no model to consider the power management of HEINs, while the demand response is completely modeled.
4) A model that optimizes several HEINs which are participating in a competitive market simultaneously is not proposed that leads to unfair results for some HEINs and reduces their participation in the markets.
5) The different uncertainties affecting on optimal operation of several HEINs participating in a competitive market is not modeled completely and the previous models often use approximate methods.
6) The optimal energy trading through HEINs is not modeled, which can take away the HEINs from their possible optimal setpoints.

This chapter investigates the mentioned research gaps and provides proper solutions for each part. Thus, the innovations of this chapter are as follows:

1) To define all the potentials of HEINs for achieving the optimal operation, all four parts of HEINs including operator, markets, equipment, and loads/services are completely modeled.
2) All of the available energy markets and ancillary services, where HEINs can participate and apply their offers to maximize their revenue are defined. Also, the tax and incentives affecting on HEINs operation are determined.
3) Electrical and thermal demand response potentials as two important parts of demand response are defined to use electrical and thermal load respectively for optimizing the HEINs operation.
4) To reach the fair results for all of the HEINs participating in a competitive market, the fair power management of a community of microgrids is investigated. Therefore, in addition to optimal results for whole community, all of microgrids increase their participation in available market that results in more competitive environment of these markets.

5) The available methods to consider the uncertainties in HEINs power management are discussed to propose the best method. Also, as a new and accurate method, machine learning-based algorithms are proposed to model the uncertainties of HEINs parameters.

6) The peer-to-peer energy trading within HEINs or among several HEINs are proposed to achieve optimal energy trading, while the players can apply their power/price offers freely, and also, they can purchase power from other players with the minimum price.

9.2 Heat and Electricity Incorporated Networks Components and Their Modeling

The structure of each HEIN includes several important parts, which are loads or services provided for end-users in buildings, equipment to produce/store the energy for supplying the loads and network operator to optimize the performance of the mentioned two parts. These parts are illustrated in Figure 9.1.

The following sections discuss the suitable model for each component including loads and equipment to be modeled properly in the problem of HEINs optimal power management. To reach the optimal operation, firstly, it is needed to have an accurate mathematical model for different parts of HEINs (demonstrated in Figure 9.1). Thus, the following parts have an high importance that need to be studied carefully.

9.2.1 Loads/Services

As Figure 9.1 shows, one of the most important parts of HEINs is their loads, which need to be modeled exactly. Therefore, based on this exact modeling, it is possible to supply them completely and augment the end-users satisfaction. Figure 9.2 shows what factors are effective on amounts of HEINs loads.

According to Figure 9.2, there are two major loads which are electrical and thermal loads. These loads are elaborated and modeled in the following sections.

9.2.1.1 Electrical Loads
The electrical loads within a HEINs can be listed as follows:

- Lighting which needs to consider all types of light bulbs and their characteristics.
- Cooking devices (such as oven and microwave).
- Home devices (such as washing machines) and other devices including electrical machines.
- Electrical machines in commercial buildings.

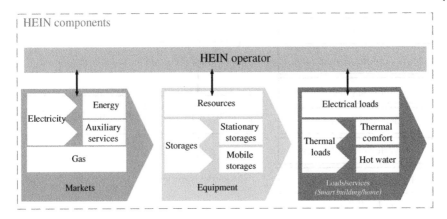

Figure 9.1 The framework of a HEIN.

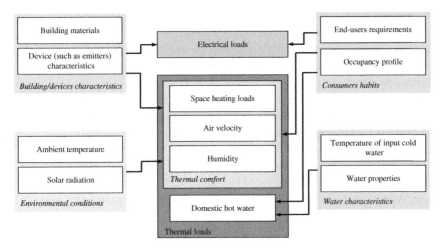

Figure 9.2 The effective factors on different loads of HEINs.

- Industrial electrical machines.
- Inverter-based devices (such as computers and laptops), which also needs to consider their effects on power quality.
- Heating/Cooling devices and air conditioners in domestic/commercial/ industrial section.

These loads must be supplied to end-users completely to maximize their satisfaction. The considered model for these loads should also model their participation in demand response to help HEIN reduce its costs. There are different methods to

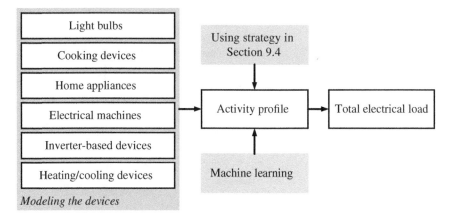

Figure 9.3 The effective factors on the different loads of HEINs.

model these loads. After modeling each device based on its characteristics, it is necessary to determine how much end-users uses that device. Therefore, in modeling electrical devices, there are two major factors, named activity profile and device characteristics. Different factors affecting on the electrical loads of HEINs are elaborated in Figure 9.3.

The amount of activity profile can be predicted based on previous data using strategies explained in Section 9.3. Also, machine learning-based algorithms can be used to exploit the values measured by smart meters and determine the amount of loads.

The mentioned loads above exclude the loads and devices generating heat because these loads will be explained in Section 9.2.1.2 separately. There are some data sources such as [29] which provides electricity consumption for standard networks.

9.2.1.2 Thermal Loads

Thermal loads, based on Figure 9.2, are categorized into two major groups which are:

- Air quality (including space heating, air velocity, and humidity).
- Domestic hot water.

These parts are elaborated in Figure 9.4.

In modeling the space heating loads, devices (such as heat emitters) characteristics, dwelling physical parameters such as the characteristics of building walls and ceiling (how much thermal loss or thermal insulation they have) should be

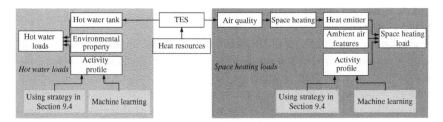

Figure 9.4 The overall structure of HEINs thermal loads.

quantified. After modeling all the devices, the activity profile of consumers must be modeled (similar to electrical loads, it can be determined by strategies in Section 9.3 or machine learning-based algorithms). By applying the activity profiles to models achieved for devices, the space heating loads are calculated. The devices in this type of thermal load include radiator systems, heat emitters, storage heater systems, etc.

For hot water loads, the characteristic of devices such as hot water tank/buffer tank (the characteristics can be parameters such as thermal loss or hot water flow rate) must be modeled. Also, the environmental properties such as the in-feed water features to the tank (for instance, the temperature difference between the in-feed water and delivered water as the hot water) should be modeled. Then, the activity profile must be modeled and applied to the models achieved for devices to obtain hot water loads. Hot water in HEINs is usually used in four parts which modeling each part must consider the special features of that part. These parts are (i) bath or shower units, (ii) cooking, (iii) laundry, and (iv) washing (such as washing the dishes). Some researches such as [30, 31], model the domestic hot water profile. Also, the data of hot water for some standard HEINs are provided in [29].

It is noteworthy some types of loads have interactions with each other that need to be considered. For example, cooking devices such as ovens consume electricity, but they also warm up the internal air of buildings; therefore, they have interactions with thermal loads. The space heating loads and hot water loads also have some interactions with each other that is important to be modeled. For instance, hot water pipes transferring hot water to different parts of the building are not completely insulated and have an amount of losses that leads to warming up the building air.

One of the most important parts in both space heating and hot water is heating system controllers that enable HEINs operators to adjust thermal loads through some processes such as failures or thermal demand response. Also, this additional equipment can also add an amount of losses. As a result, modeling this equipment has a significant impact on the accuracy of the achieved models.

9.2.1.3 Thermal Comfort
9.2.1.3.1 Definition
To get the optimal operation of HEINs, one of the most important factors, especially in modern smart grids/cities, is to supply the end-users requirements completely to maximize their satisfaction. One of these requirements is thermal comfort, which is to supply the target temperature (adjusted by consumers) for them within each building. This is the simplest form of thermal comfort used in some researches such as [10, 11]. If the amount of supplied temperature is closer to the adjusted temperature, it increases the thermal comfort of consumers and vice versa.

However, the more exact models of thermal comfort, also consider two other factors alongside the temperature, which are air velocity and humidity [32, 33]. Thus, these models also manage the amount of air circulation and moisture within smart building that definitely affect occupants comfort.

But in last researches, more factors are considered in thermal comfort evaluation. These factors include [33–37]:

- Clothing insulation;
- Metabolic rate;
- Daily average outdoor temperature and mean radiant temperature;
- Age and gender of end-users;
- Insulation with vapor pressure;
- Containing globe temperature.

In [34], the American society of heating, refrigerating and air-conditioning engineers (ASHRAE) RP-884 thermal comfort data were used to drive a thermal comfort model. ASHRAE is a standard that helps to model the thermal comfort exactly including all of the mentioned factors and will be solved by artificial intelligence approaches such as deep learning methods.

9.2.1.3.2 Robust Thermal Comfort
Although maximizing the thermal comfort of end-users is a key point in the optimized operation of HEINs, also maintaining it within the allowed range through each operational condition of HEINs is important. If the amount of thermal comfort is within the allowed range for each possible operational condition, then the thermal comfort will be robust. One group of HEINs operational conditions affecting on thermal comfort is created by different uncertainties, which are defined in Section 9.3. To evaluate the impact of uncertainties on thermal comfort, the amount of thermal comfort must be within the allowed range for each state of uncertainties [11], which are modeled by a method such as scenarios generation or probabilistic distribution function (PDF), as defined in Section 9.3. There is

another group of critical conditions occurred when the ISO (or upstream grid) call for reserve to be supplied by HEIN. To produce the required reserve, the HEINs resources must produce more electrical power, which results in changes in scheduled power. Thus, the amount of electrical power and thermal power in HEINs changes and can lead to amounts outside the allowed range for thermal comfort. In this case, if only the air temperature is considered, robust thermal comfort means the framework used for optimizing the operation of HEIN must be able to provide adjusted temperature for the consumers even during a reserve call. For example, for a HEIN including combined heat and power (CHP), electric heat pump (EHP), electrical boiler (EB), and TES, to have a robust thermal comfort during a reserve call, there must be enough energy in TES or building fabric or heat sources must have enough energy capacity to deliver the required energy immediately. Equation (9.1) applies this fact; therefore, it prevents the temperature reduction in buildings and end-users discomfort [10].

$$O_{s,t,b}\left(R_{s,t,b}^{\text{EHP}\_d}CP_{s,t,b}^{\text{EHP}} + R_{s,t,b}^{\text{EB}\_d}\eta_b^{\text{EB}} - R_{s,t,b}^{\text{CHP}\_d}\eta_b^{\text{CHP}\_t}/\eta_b^{\text{CHP}\_e}\right) \leq$$

$$O_{s,t,b}\left(\left(\frac{\left(\frac{X_{s,t,b}}{C_b^{\text{TES}}} + T_{s,t,b} - X_b^d\right)C_b^{\text{TES}} + F_{s,t,b}^B}{C^{\text{RES}\_u}}\right) + \left(H_b^{\text{GB}\_u} - H_{s,t,b}^{\text{GB}}\right)\right) \forall_{s,t,b}$$

$$(9.1)$$

where $R_{s,t,b}^{\text{EHP}\_d}$, $R_{s,t,b}^{\text{EB}\_d}$, and $R_{s,t,b}^{\text{CHP}\_d}$ are the produced reserve by EHP, EB, and CHP, respectively. The subscript s, t, and b stand for scenario (Section 9.3), time-step, and bus, respectively. C^{RES_u} shows the maximum call length of possible reserve calls. $O_{s,t,b}$ which is proposed in [10], shows the occupancy profile in each building, as a result, it applies Eq. (9.1) only in buildings that are occupied at that hour because when the consumers are outside their buildings, changes in temperature can occur freely with no constraints. Therefore, when an amount of reserve is called, $R_{s,t,b}^{\text{EHP}\_d}$ and $R_{s,t,b}^{\text{EB}\_d}$ show respectively how much input power of EHP and EB is reduced to be used as reserve that show how much output thermal power of them is reduced. Because electrical power is an output for CHP and has a direct relation with output heat, therefore, CHP has a minus sign in Eq. (9.1). The first term in the right side of Eq. (9.1) shows the highest thermal power that can generated by storages including TES and building fabric to be replaced with reduced power of EHP, EB, and CHP. Also, the second last term shows the maximum heat power produced by GB to be replaced. Because of the reverse behavior of CHP compared to EHP and EB (the output heat of CHP increases when its reserve increases), Eq. (9.2) is necessary to ensure the heat produced by CHP can be stored in TES or replace the gas boiler (GB) heat ($H_{s,t,b}^{\text{GB}}$) and does not increase the buildings temperature.

$$O_{s,t,b} R_{s,t,b}^{\text{CHP}\_d} \eta_b^{\text{CHP}\_t} / \eta_b^{\text{CHP}\_e} \leq$$

$$O_{s,t,b} \left(\frac{\left(X_b^u - \frac{X_{s,t,b}}{C_b^{\text{TES}}} - T_{s,t,b} \right) C_b^{\text{TES}} + H_{s,t,b}^B}{C^{\text{RES}\_u}} + H_{s,t,b}^{\text{GB}} \right) \forall_{s,t,b} \tag{9.2}$$

As defined in Section 9.2.3, the building structure can also store a limited amount of heat. In Eq. (9.1) and (9.2), $F_{s,t,b}^B$ and $H_{s,t,b}^B$ respectively model the footroom and headroom amount of building fabric storages and are determined by Eqs. (9.3) and (9.4).

$$0 \leq F_{s,t,b}^B = O_{s,t,b} \left(T_{s,t,b} - \left(T_{t,b}^A - v_{t,b}^d \right) + T_{s,t,b}^{R\_d} \right) C_b^B \, \forall_{s,t,b} \tag{9.3}$$

$$0 \leq H_{s,t,b}^B = O_{s,t,b} \left(\left(T_{t,b}^A + v_{t,b}^u \right) - T_{s,t,b} + T_{s,t,b}^{R\_s} \right) C_b^B \, \forall_{s,t,b} \tag{9.4}$$

The mentioned model for robust thermal comfort can be developed to involve other factors of thermal comfort, which are listed in Section 9.2.1.3.1.

9.2.2 Equipment

According to Figure 9.2, another important part of each HEIN is equipment including resources to produce energy and storages to store the electricity/heat during the energy surplus or release energy during the energy deficit. The Sections 9.2.2.1 and 9.2.2.2 define these equipment completely. Figure 9.5 shows the relationship between these equipment using the energy carriers of HEINs. Therefore, it can be understood which carriers are transferred between each two equipment and how the equipment supply different loads. As Figure 9.5 illustrates, the produced heat by heat resources must be stored in TES at first, and then it can be consumed by thermal loads.

9.2.2.1 Resources

The resources are a group of equipment to produce power for loads of networks using their special input carrier. Because within a HEINs, the existing carriers include electricity, gas, heat, and cool, therefore the available resources of HEINs include CHP, combined cool, heat and power (CCHP), power-to-gas (P2G), EHP, EB, GB, solar power including photovoltaic and solar thermal, wind power (including onshore and offshore), tidal power, geothermal power, and traditional power plants utilizing fossil fuels. Table 9.1 illustrates the input/output carriers of each of these resources.

In Table 9.1, the multi-carrier is assigned to the devices, which ones include more than one carrier and can convert one carrier to another one. Although

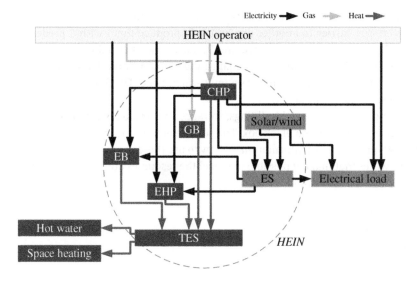

Figure 9.5 The relationship amongst different equipment used in HEINs.

Table 9.1 The input/output carriers of HEINs resources.

Equipment		Input carriers	Output carriers	Multi-carrier	Feature
CHP		Gas	Electricity-heat	Yes	—
CCHP		Gas	Electricity-heat-cool	Yes	—
Power-to-gas		Electricity	Gas	Yes	—
EHP		Electricity	Heat	Yes	—
EB		Electricity	Heat	Yes	Mostly as a support
GB		Gas	Heat	Yes	Mostly as a support
Solar power	Photovoltaic	—	Electricity	No	—
	Solar thermal	—	Electricity-heat	Yes	—
Wind power	Onshore	—	Electricity	No	—
	Offshore	—	Electricity	No	—
Tidal power		—	Electricity	No	—
Geothermal power		—	Electricity-heat	No	—
Traditional power plant		—	Electricity-heat	No	—

geothermal and traditional power plants have more than one carrier, but they cannot convert one carrier to another one and only produce them with a predetermined efficiency based on their structure. Because solar thermal can includes a back-pressure CHP in its output and change the amount of produced electricity and heat [38], therefore it is defined as a multi-carrier one. The following sections model these resources in detail.

According to [30], to model buildings, storages and resources can be modeled by physical approaches as a black box. To get more details on such physical models, [39, 40] provide more information on how to use these models. There are also some softwares that make it possible to model the resources and storages used in HEINs. These softwares are:

- OpenDSS/OpenDSS-G [41]
- MATLAB [42]
- EnergyPlus [43]
- esp-r [44]
- IES [45]
- NetLOGO [46]

Amongst these softwares to model the equipment, OpenDSS and OpenDSS-G (which uses a graphical panel instead of using scripting used in OpenDSS) has the exact model of equipment and allow to model HEINs accurately. It has the model of multi-energy technologies such as CHP, renewables such as solar panels and their inverters, storages such as batteries and electric vehicles. MATLAB, as a powerful known tool in modeling, needs to make the mathematical model of each equipment at first and then model it in software environment. Also, NetLogo, as a new tool, makes it possible to model HEINs at the level of a city considering its equipment, loads, and renewable units.

Based on the above discussion, mathematical modeling of HEINs equipment is important. Therefore, the following sections determine the model of each component.

9.2.2.1.1 Combined Heat and Power

The multi-carrier devices such as CHP, EHP, and EB allow using the flexibility of energy carrier substitution, which is a potential that can reduce the cost or improve the technical constraints such as congestion. Therefore, modeling of these devices has significant importance in case of optimizing the operation of HEINs. CHP as a multi-carrier device has good efficiency, while its cost and pollutants emissions are low. As a result, it is a typical device in HEINs, and usually, it is supported by agreement and international or governmental programs [47].

Table 9.2 The characteristics of some typical types of CHPs.

CHP type	Rated thermal power [kW]	Thermal efficiency [–]	Electrical efficiency [–]
Senertec	12.50	0.61	0.27
Ecopower	12.50	0.65	0.25
Solo	8.00–26.00	0.72	0.24
Whispertech	4.90–8.00	0.80	0.12
Idatech	9.00	0.55	0.25

Source: Adapted from Tookanlou et al. [48].

CHP gets the gas as the input and produces the heat and electricity with a fixed efficiency simultaneously. Table 9.2 shows the characteristics of some typical CHP devices.

Also, CHPs can be classified based on their internal structures. This classification includes four classes including [30]:

- Gas turbine;
- Stirling engine;
- Organic Rankine cycle engine;
- Fuel cell.

The mathematical model of CHP can be determined by Eq. (9.5) [10].

$$H_b^{\text{CHP}\_d} \leq H_{s,t,b}^{\text{CHP}} \leq H_b^{\text{CHP}\_u} \; \forall_{s,t,b}, \; H_{s,t,b}^{\text{CHP}} = \frac{P_{s,t,b}^{\text{CHP}} \eta_b^{\text{CHP}\_t}}{\eta_b^{\text{CHP}\_e}} \tag{9.5}$$

The heat produced by CHP ($H_{s,t,b}^{\text{CHP}}$) is limited by a lower/upper limit based on Eq. (9.5). CHPs also include periods of unit start-up/shut-down, in which electricity is consumed while heat production is not done.

9.2.2.1.2 Power-to-Gas Technology

Power-to-gas technology is a device that converts electricity to gas. The function of this technology is the reverse form of CHP or gas turbine process, which converts gas to electricity. The authors in [49, 50] model P2G technology through power management problems.

9.2.2.1.3 Electric Heat Pump (EHP)

CHP as a resource for producing the thermal power, uses electricity as its input. Therefore, with the ability to convert electricity to heat, it is a multi-energy device and enables the energy vector substitution within HEINs. Also, as a low carbon resource for thermal loads compared to CHP and GB and due to its high efficiency, lots of research such as [11] have been done to integrate more EHPs within the HEINs.

The operational model of EHP includes a limitation on operational power of EHP, as stated in Eq. (9.6).

$$P_b^{EHP\_d} \leq P_{s,t,b}^{EHP} \leq P_b^{EHP\_u} \ \forall_{s,t,b} \tag{9.6}$$

Equation (9.6) determines the input electrical power of EHP ($P_{s,t,b}^{EHP}$) should be within an allowed range, which is determined by $P_b^{EHP\_d}/P_b^{EHP\_u}$ [31]. The output thermal power ($H_{s,t,b}^{EHP}$) is achieved by multiplying the performance coefficient of EHP ($CP_{s,t,b}^{EHP}$) in input electrical power, as stated in Eq. (9.7).

$$H_{s,t,b}^{EHP} = P_{s,t,b}^{EHP} CP_{s,t,b}^{EHP} \ \forall_{s,t,b} \tag{9.7}$$

9.2.2.1.4 Electrical Boiler

EB is a multi-carrier technology which consumes electricity to produce heat in HEINs. To model the EB operation, Eq. (9.8) stating the operational constraint of EB is necessary. Equation (9.8) says the output thermal power of EB (achieved by multiplying the EB efficiency (η_b^{EB}) in its input electrical power ($P_{s,t,b}^{EB}$)) must be within an allowed range.

$$H_b^{EB\_d} \leq P_{s,t,b}^{EB} \eta_b^{EB} \leq H_b^{EB\_u} \ \forall_{s,t,b} \tag{9.8}$$

9.2.2.1.5 Gas Boiler

GB is a technology that consumes gas and produces heat. It is a suitable device during the hours when the electricity price is high; therefore, GB helps to consume gas for producing heat. In other words, it is a CHP that only produces heat. GB, similar to EB, is usually used as a supportive heat source because, in addition to production during peak prices of electricity, also it can produce heat when electricity is not supplied to be used by EHP and EB for heat production [51].

The output thermal power of GB ($H_{s,t,b}^{GB}$) can be only within an allowed range, as shown in Eq. (9.9). The relationship between input gas and output heat of GB is determined by its efficiency (η_b^{GB}).

$$H_b^{GB\_d} \leq H_{s,t,b}^{GB} \leq H_b^{GB\_u} \; \forall_{s,t,b}, \; H_{s,t,b}^{GB} = G_{s,t,b}^{GB} \eta_b^{GB} \tag{9.9}$$

9.2.2.1.6 Renewables

Due to the importance of being low carbon, the HEINs integrate more renewable sources than before. Therefore, this type of resource needs to be modeled exactly to help the HEIN operate more optimally. There are several types of renewables which are:

- Solar power including photovoltaic and solar thermal;
- Wind power including onshore and offshore;
- Geothermal power plant;
- Tidal power plant.

There are also other types of renewables that can be used, but the mentioned ones are the most typical cases that are more used to generate electricity. The features of renewables are:

- Low carbon power generation;
- High availability/accessibility;
- Low operation cost, because it does not include any cost for supplying its input.

Solar power can be used by solar panels (that is also called photovoltaic) that directly generates electricity. A lot of research is implemented to increase the efficiency of solar panels. The output power of a solar panel ($P_{s,t,i}^{photovoltaic}$) can be achieved by Eq. (9.10) for a solar farm. P_i^{max} is the rated power of each panel. $R_{s,t,i}^{solar}$ states the amount of solar radiation in the location of solar panel and can be determined by previous data (Section 9.3). i is an index to display the generation/storage unit number. The solar radiation for each location on the Earth can be exported from some sites such as [29, 52].

$$P_{s,t,i}^{photovoltaic} = R_{s,t,i}^{solar} P_i^{max} \; \forall_{s,t} \tag{9.10}$$

Another strategy to use solar power is solar thermal power plants which convert the solar radiation into heat. One of the most typical methods of solar thermal is concentrating solar power. The overall structure of concentrating solar power is shown in Figure 9.6.

As Figure 9.6 shows, all of the solar radiation is reflected to one point in the receiver to heat a fluid (which is usually molten salt). Then, superheated liquid

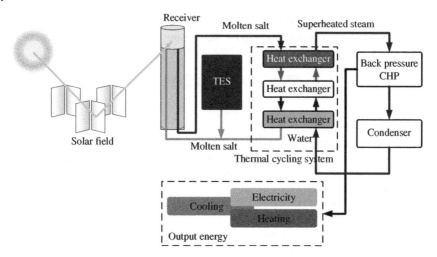

Figure 9.6 Overall structure of concentrating solar power. *Source:* Adapted from Sun et al. [21].

which has a temperature of about 535 °C goes to the thermal cycling system. In this system, the heat is transferred from molten salt to water to produce superheated steam. As it is shown, solar radiation is converted to thermal energy stored in steam. This steam can empower a turbo generator to generate electricity or can be applied to a back pressure CHP unit to generate electricity and heat at the same time. The output electrical ($P_{s,t}^{\mathrm{BT}}$) and thermal ($Q_{s,t}^{\mathrm{BT}}$) power of concentrating solar power is achieved by Eqs. (9.11) and (9.12), respectively. $P_{s,t}^{\mathrm{dissipated}}$ is the dissipated power of back pressure turbine. The coefficient between electrical and thermal power of back pressure turbine is stated by k [21].

$$P_{s,t}^{\mathrm{dissipated}} + \frac{P_{s,t}^{\mathrm{BT}}}{\eta^{\mathrm{BT}}} - R_{s,t}^{\mathrm{TES}-}\eta^{\mathrm{TES}-} + \frac{P_{s,t}^{\mathrm{TES}+}}{\eta^{\mathrm{TES}+}} = R_{s,t}^{\mathrm{solar}}\eta^{\mathrm{CSP}} \ \forall_{s,t} \tag{9.11}$$

$$Q_{s,t}^{\mathrm{BT}} = kP_{s,t}^{\mathrm{BT}} \ \forall_{s,t} \tag{9.12}$$

Wind turbine converts the energy of wind into electricity using blades rotating to power a generator. Wind power as low carbon and high-efficiency technology has attracted the attention of governments to be invested in lots of countries. Spain and the United Kingdom with 2319 MW and 1764 MW, have the most wind capacity in onshore and offshore types, respectively. The data of wind speed can also be exported from [52]. Equation (9.13) shows the operational equation of wind turbines, where V_{CI} and V_{CO} are the lower/upper limits of wind turbine, respectively. V_r and P_{rated} are respectively the nominal speed/power. The amount of α and β are

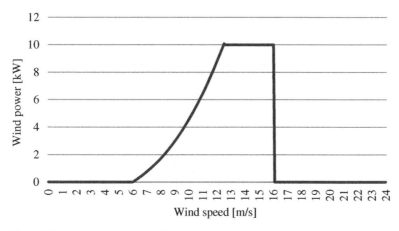

Figure 9.7 A typical operational curve of wind turbines.

calculated by Eqs. (9.14) and (9.15), respectively [53–55]. Figure 9.7 shows a typical curve for the output power of wind turbines. This figure illustrates how the power is associated with wind speed rotating the blades.

$$P_{s,t,i}^{\text{Wind}} = \begin{cases} 0, v < V_{\text{CI}} \\ \alpha v^3 - \beta P_{\text{rated}}, V_{\text{CI}} < v < V_r \\ P_{\text{rated}}, V_r < v < V_{\text{CO}} \\ 0, V_{\text{CO}} < v \end{cases} \forall_{s,t,i} \tag{9.13}$$

$$\alpha = \frac{P_{\text{rated}}}{V_{\text{rated}}^3 - V_{\text{CI}}^3} \; \forall_{s,t} \tag{9.14}$$

$$\beta = \frac{V_{\text{CI}}^3}{V_{\text{rated}}^3 - V_{\text{CI}}^3} \; \forall_{s,t} \tag{9.15}$$

Solar power and wind power are the most important renewable resources that are researched in lots of studies. However, other renewable resources such as geothermal power and tidal power can also be modeled and optimized in HEINs.

9.2.2.1.7 Traditional Power Plant

Traditional power plants include the typical power plants consuming fossil fuels to generate electricity and heat. These power plants have different types such as thermal (which uses superheated steam to rotate the turbo generator), gas turbine (burning gas to rotate the generator) and etc. These power plants emit a huge amount of environmental pollutants and have a large initial investment cost, while they are not suitable for decentralizing the HEINs, as it is an important strategy nowadays to improve the efficiency, reliability, and resiliency of HEINs. Although these power plants are less focused compared to renewable resources nowadays, they exist in lots of networks, and it is essential to be modeled to optimize the

Table 9.3 Different storages used in HEINs.

Type	Stored carrier	Status
Battery	Electricity	Stationary
TES	Heat	Stationary
PCM	Heat	Stationary
Building fabric	Heat	Stationary
Electric vehicle	Electricity	Mobile
Compressed air energy storage	Electricity	Stationary

HEINs performance. Reference [56] models the thermal power plant completely that can be used in modeling the HEINS.

9.2.2.2 Storages

Because of different carriers in HEINs, there are different storages used in HEINs. Table 9.3 shows these storages alongside their features.

9.2.2.2.1 Stationary Storages

Battery, as a stationary storage, store the electricity to be discharged during electricity deficit. But battery has a high initial investment cost. The mathematical model of battery can be stated by Eqs. (9.16)–(9.21). Equation (9.16) shows how the battery energy is calculated at each hour, while Eqs. (9.17) and (9.18) respectively show the allowed range of battery energy and power. Equation (9.19) is usually used to prevent the results distortion. The reserve produced by battery in HEINs can be calculated by Eq. (9.20). To ensure the battery can supply the expected reserve, Eq. (9.21) is added. This equation ensures the battery has sufficient stored energy to supply reserve even over the maximum call length of the reserve.

$$B_{s,(t+1),b} = B_{s,t,b} + \left(\frac{P_{s,t,b}^{ES}}{\eta_b^{ES}}\right) \Delta t \ \forall_{s,t,b} \tag{9.16}$$

$$B_b^d \leq B_{s,t,b} \leq B_b^u \ \forall_{s,t,b} \tag{9.17}$$

$$P_b^{ES\_d} \leq P_{s,t,b}^{ES} \leq P_b^{ES\_u} \ \forall_{s,t,b} \tag{9.18}$$

$$B_{s,0,b} = B_{s,N_t,b} \ \forall_{s,b} \tag{9.19}$$

$$0 \leq R_{s,t,b}^{ES\_d} \leq P_b^{ES\_u} + P_{s,t,b}^{ES} \ \forall_{s,t,b} \tag{9.20}$$

$$R_{s,t,b}^{ES\_d} \leq B_{s,t,b}/C^{RES\_u} \ \forall_{s,t,b} \tag{9.21}$$

Another important stationary storage is TES which stores heat. The efficiency of TES is higher and has a lower initial investment cost compared to the battery. As a result, it is more focused on new researches. Based on recent researches, utilizing TES can enhance the efficiency and flexibility of HEINs and reduce their dependency on the upstream grid, while it leads to lower costs than older networks with batteries. Two different strategies to store heat in TES are:

1) Sensible TES: The fluids used in sensible type of TES are usually water, solid media or other types of materials including concrete/brick/rock. Table 9.4 shows the parameters of these materials.
2) Latent TES: One of the most popular technologies of this type are phase change materials (PCMs). Table 9.5 shows the characteristics of PCM to be compared with Table 9.4.
3) Sorption: This type of TES stores the heat and releases it to HEINs using the control of the absorption process [30].

The operational equations of TES within HEINs can be stated as Eqs. (9.22)–(9.26). Equation (9.22) shows the lower/upper limit of thermal energy stored in TES ($X_{s,\ t,\ b}$), while TES losses ($X_{s,t,b}^{\text{LOSS}}$) are modeled by Eq. (9.23). Equation (9.24) shows how the TES energy changes at each hour, while its input energy at each hour is calculated by Eq. (9.25), which is supplied by GB, CHP, EHP, and EB. To prevent optimization results distortion in HEINs power management process at each hour, Eq. (9.26) is usually applied [10].

$$\left(X_b^d - T_{s,t,b}\right)C_b^{\text{TES}} \leq X_{s,t,b} \leq \left(X_b^u - T_{s,t,b}\right)C_b^{\text{TES}} \ \forall_{s,t,b} \tag{9.22}$$

Table 9.4 Characteristics of materials used in sensible-based TESs.

	Density [kg/m³]	Heat capacity [kJ/kg.K]	Usable temperature range [°C]
Water	1000	4.18	35–80
Bricks	2400	1.09	680

Source: Good [30]/Nicholas Paul Good.

Table 9.5 The characteristics of PCM-based TES.

	Density [kg/m³]	Heat capacity [kJ/kg.K]	Usable temperature range [°C]
PCM	1375	2.5	58

$$X_{s,t,b}^{\text{LOSS}} = \frac{\left(\frac{X_{s,t,b}}{C_b^{\text{TES}}} - T_{s,t,b}\right)\Delta t}{R_b^{\text{TES}}} \ \forall_{s,t,b} \tag{9.23}$$

$$X_{s,(t+1),b} = X_{s,t,b} + X_{s,t,b}^I - X_{s,t,b}^{\text{LOSS}} - X_{s,t,b}^{\text{SH}} - X_{s,t,b}^{\text{DHW}} \ \forall_{s,t,b} \tag{9.24}$$

$$\left(H_{s,t,b}^{\text{GB}} + H_{s,t,b}^{\text{CHP}} + P_{s,t,b}^{\text{EHP}}\text{CP}_{s,t,b}^{\text{EHP}} + P_{s,t,b}^{\text{EB}}\eta_b^{\text{EB}}\right)\Delta t = X_{s,t,b}^I \ \forall_{s,t,b} \tag{9.25}$$

$$X_{s,0,b} = X_{s,N_t,b} \ \forall_{s,b} \tag{9.26}$$

Another heat storage is provided by material used in building construction, named building fabric storage. Because it exists, there is no need for additional investment. But this storage is limited and can degrade the thermal comfort of building occupants. The more heat it is stored in building fabric, the higher temperature the building occupants experience and vice versa. Therefore, it is essential to model this storage exactly and exploit it in a way that does not set the air temperature outside the allowed range. The equations of building fabric can be stated as Eqs. (9.27) and (9.28), where $T_{s,t,b}$ is the supplied temperature and $T_{t,b}^A$ is the target temperature adjusted by consumers. $v_{t,b}^d/v_{t,b}^u$ are allowed amount for temperature degradation during a reserve call, respectively. Also, $F_{s,t,b}^B/H_{s,t,b}^B$ respectively define the footroom ad headroom of building fabric storage [31].

$$0 \le F_{s,t,b}^B = O_{s,t,b}\left(T_{s,t,b} - \left(T_{t,b}^A - v_{t,b}^d\right) + T_{s,t,b}^{R_d}\right)C_b^B \ \forall_{s,t,b} \tag{9.27}$$

$$0 \le H_{s,t,b}^B = O_{s,t,b}\left(\left(T_{t,b}^A + v_{t,b}^u\right) - T_{s,t,b} + T_{s,t,b}^{R_s}\right)C_b^B \ \forall_{s,t,b} \tag{9.28}$$

Compressed air energy storage is one the most important electricity storage which has drawn attentions to itself recently. It is very suitable for storing energy on a large scale. Compressed air energy storage, as an emerging bulk energy storage, stores the electricity during the off-peak hour and then discharges to the HEIN during the on-peak hours. In charging mode, compressed air energy storage imports power to compress air inside its tank and then in discharging mode, the compressed air is used for powering the gas turbine and electricity production. To reduce the investment cost of compressed air energy storage, it is possible to use underground salt cavities or similar cavities such as mines instead of tanks to be able to store more energy. The compressed air energy storage applications are as follows:

- Tackling the load at each hour in the presence of renewable, loads and market prices uncertainties.
- Decoupling electricity and heat produced by multi-carrier devices such as CHP to improve the power dispatching inside HEINs [21].

- More optimal performance of HEINs participating in electricity and heat markets;
- Enhancing flexibility, reliability, and resiliency.
- Reusing cavities created underground after finishing the mine project or similar project.
- Needing less space on the ground to create the storage.

So far, a few projects have used compressed air energy storage, but the project using CAES are penetrating in more HEINs. For example, the most important projects using compressed air energy storage are [57]:

- McIntosh in Alabama, United States with the capacity of 110 MW.
- Huntorf in Germany, with the capacity of 321 MW.

Compressed air energy storage can also be combined with other technologies to create better characteristics in HEINs. For example, Ref. [58] integrates compressed air energy hstorage with a thermochemical technology, using Cobalt monoxide, to increase the round-trip and exergy efficiency by 56.4% and 75.6%, respectively.

A typical set of operational equations of compressed air energy storage can be stated by Eqs. (9.29)–(9.33), where $P_i^{CAES\,+}$ and $P_i^{CAES\,-}$ are charging and discharging power, respectively. Equation (9.32) shows how the CAES energy at each hour (E_t^{CAES}) is calculated, while Eq. (9.33) illustrates the allowed range of E_t^{CAES}.

$$u_t^{CAES\,+} + u_t^{CAES\,-} \leq 1 \tag{9.29}$$

$$0 \leq P_t^{CAES\,+} \leq u_t^{CAES\,+} \overline{P^{CAES\,+}} \tag{9.30}$$

$$0 \leq P_t^{CAES\,-} \leq u_t^{CAES\,-} \overline{P^{CAES\,-}} \tag{9.31}$$

$$E_t^{CAES} = E_{t-1}^{CAES} + P_t^{CAES\,+} \eta^{CAES\,+} - \frac{P_t^{CAES\,-}}{\eta^{CAES\,-}} \tag{9.32}$$

$$E_{min}^{CAES} \leq E_t^{CAES} \leq E_{max}^{CAES} \tag{9.33}$$

9.2.2.2.2 Mobile Storages

Another important type of storage is mobile storage which includes electric vehicles. Because of the mobility of these storages, they can be scheduled in a way to be located in the appropriate location of the HEINs at each hour. It is noteworthy through these scheduling, the drivers' preferences are also considered with the highest priority. Because modeling of electric vehicles is outside the scope of this chapter, they are not modeled here. To achieve the complete models for electric vehicles, [59] has introduced a complete structure.

9.2.3 Buildings/Smart Homes

Buildings are one of the most important parts of each HEINs that can have two types of domestic and commercial. There are four types of modeling for buildings that are:

- Flats;
- Terraced;
- Semi-detached;
- Detached.

These models are completely explained in [30]. The buildings equations for the type of detached are as Eqs. (9.34)–(9.38). $T^s_{s,t,b}/T^d_{s,t,b}$ show the surplus/deficit amount of temperature, which is allowed due to heat storage of building fabric. The parameters $v^u_{t,b}$ and $v^d_{t,b}$ are set to zero and just can be equal to a non-zero amount, when the optimization model cannot be converted for a special condition. Equations (9.34) and (9.35) limit the temperature of building to preserve the thermal comfort of end-users. Equation (9.36) determines the building temperature at each hour. R^B_b and C^B_b are parameters for modeling the building that are building thermal resistance/capacitance, respectively. To avoid result distortion, Eq. (9.38) is also applied.

$$O_{s,t,b}\left(T_{s,t,b} - T^s_{s,t,b}\right) \leq O_{s,t,b}\left(T^A_{t,b} + v^u_{t,b}\right) \forall_{s,t,b} \tag{9.34}$$

$$O_{s,t,b}\left(T^A_{t,b} - v^d_{t,b}\right) \leq O_{s,t,b}\left(T_{s,t,b} + T^d_{s,t,b}\right) \forall_{s,t,b} \tag{9.35}$$

$$T_{s,(t+1),b} = T_{s,t,b} + \left(X^{SH}_{s,t,b} + (1-\Gamma_{s,t,b})(I\_G_{s,t,b} + S\_G_{s,t,b})\right.$$
$$\left. - \left(T_{s,t,b} - T^{ENV}_{t,b}\right)\Delta t R_b^{B^{-1}} + X^{LOSS}_{s,t,b}\right)C_b^{B^{-1}} \forall_{s,t,b} \tag{9.36}$$

$$X^{SH}_{s,t,b} \geq 0 \forall_{s,t,b} \tag{9.37}$$

$$T_{s,0,b} = T_{s,N_t,b} \forall_{s,b} \tag{9.38}$$

Equations (9.34)–(9.38) are stating the thermal structure/loads of each building. A building in HEIN also has several electrical loads that are explained in Section 9.2.1.

9.2.4 Heat and Electricity Incorporated Network Operator

As Figure 9.1 shows, each HEIN has a unit that dispatches the HEIN loads and equipment and also determines how much energy must be imported/exported from/to the upstream grid. Totally, the responsibility of this unit is to manage the HEIN performance. All the proposed centralized approaches for HEINs power management are run by this unit and also this unit determines the power/price offering strategies of HEIN to participate in competitive markets.

Although, in new HEINs which are managed in a decentralized way or have peer-to-peer transactions, the responsibility of HEIN operator is reduced and some parts of its tasks are done by local operators or and then these local operators all send the results of their decisions to HEIN operator to coordinates all of the local operators in whole HEIN. The important point is that in three cases, HEINs are essential, which are:

- Whenever a HEIN must be managed centrally;
- When it needs to coordinate local operators;
- Whenever the HEIN power management problem includes power flow calculations, because power flow calculation needs to be done for all parts of HEIN.

9.2.5 Different Layers/Networks and Their Connection

As stated in Section 9.1, there are three carriers including electricity, gas, and heat in HEINs. Sometimes cool is also considered in HEINs. In this chapter, the carrier of cool has not been addressed. One layer for each carrier can be considered, and the equipment and loads are categorized into these groups. Figure 9.8 also shows each equipment and loads presented in Sections 9.2.1–9.2.3 is related to which layer.

9.3 Uncertainties

In the problem of HEINs power management, there are several probabilistic parameters leading to uncertainties. These probabilistic parameters are as Table 9.6. As Table 9.6 shows, there are three major groups of uncertainties that are environmental parameters, market-related parameters such as the prices in an unbalanced market and uncertainties of consumers such as the amount of load demand.

There are several methods to model the uncertainties in HEINs power management, as follows:

- Stochastic programming which uses probabilistic distribution function or scenario generation (such as Monte Carlo [60] or frameworks that use previous data and then reduce the number of scenarios to improve the speed and volume of computations (there are different methods to reduce the number of scenarios such as simultaneous backward reduction algorithm [10, 61] or Taguchi

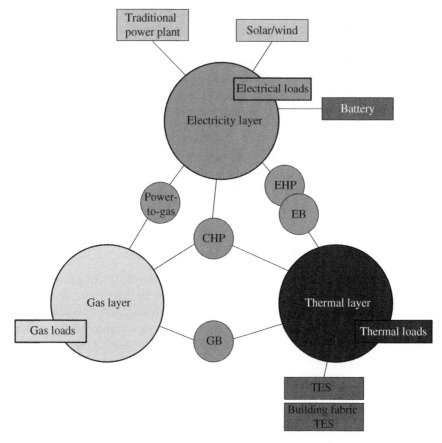

Figure 9.8 The connection between different layers of HEINs.

algorithm [62])). This strategy is suitable when the behavior of the parameter can be predicted by its previous data such as load demands or the paths which consumers select to drive.

- Robust optimization which is used for parameters that cannot be predicted by their previous data exactly such as wind speed.
- Another strategy uses robust optimization and stochastic programming simultaneously.
- Point estimate method is another strategy to consider the uncertainties [63].

Table 9.6 shows each probabilistic parameter and suggests the strategies the probabilistic parameter can be modeled by which at its last column.

Table 9.6 Existing probabilities in HEIN power management.

Type	Parameter	Suggested strategy to model
Environmental	Solar radiation-wind speed	Probabilistic method-Robust optimization
	Temperature	Probabilistic method-Robust optimization
Market	Price	Probabilistic method
	The amount of reserve	Probabilistic method
Consumers	Electricity load	Probabilistic method-Robust optimization
	Adjusted temperature	Probabilistic method-Robust optimization
	Domestic hot water	Probabilistic method-Robust optimization
	Occupant behavior (such as occupancy or the selected path to drive)	Probabilistic method-Robust optimization

9.4 Optimal Operation of Heat and Electricity Incorporated Networks

The major target of this chapter is to define what is the optimal operation of HEINs and how it can reach to this operation. Also, it is important to mention what tools are available in HEINs to have an optimal operation. The following sections discuss these concepts in detail.

9.4.1 Definition of Optimal Operation

The optimal operation of HEINs is when it provides the features below:

1) It provides maximum flexibility [64]. The maximum flexibility means the HEIN operates with the minimum cost, while it provides the end-users the services completely so that they have the maximum possible satisfaction. Thermal comfort is one of the most important services in modern smart grid/cities that has drawn the researchers' attention to be addressed.
2) Also, HEIN has maximum sustainability, which means it can proceed with its performance without any problem or stop. Some problems that can interrupt the HEIN operation or stop it includes congestion or voltage amounts outside the permitted range. It is needed to prevent congestion occurrence and unallowed voltage values, which can reduce the sustainability and damage the

HEINs equipment such as transformers. Improving the amount of voltage and current also improves the services for end-users and enhances the flexibility.

3) Some searches also consider other factors such as power loss reduction or unserved loads reduction to augment end-users satisfaction.

4) One of the most important results of optimal operation is minimum environmental pollutant emissions. Therefore, HEIN must be scheduled in a way that has minimum greenhouse gas emissions. Because of recent international agreement or governmental assignment, to reduce the climate change and Earth warming rate, it is essential to consider this feature in HEINs power management.

The first factor mentioned above considers economic and technical factors, while the second and third ones consider technical issues. The fourth factor improves environmental issues. Therefore, the optimal operation of HEINs has a techno-economic-environmental effect.

9.4.2 Different Goals in Heat and Electricity Incorporated Networks Exploitation

As explained in Section 9.4.1, the important goals in HEINs optimal power management are as shown in Figure 9.9. These goals are categorized into three parts named technical, economic and environmental.

As Figure 9.9 shows, the goals in HEINs power management are:

- *Economic Goals:*
  This group of goals includes cost minimization that tries to minimize the HEINs operations cost. The researches in [10, 11] have an objective function of cost minimization. For example, Eq. (9.39) shows the objective function of [10], which considers all costs of energy purchasing from day-ahead market ($\epsilon_t^I D_t^I \Delta t$) and energy selling to day-ahead market ($\epsilon_t^E D_t^E \Delta t$), energy purchasing from imbalanced market ($\pi_{s,t}^I I_{s,t}^I \Delta t$) and energy selling to imbalanced market ($\pi_{s,t}^E I_{s,t}^E \Delta t$) and also the penalties ($\xi_t^s T_{s,t,b}^s + \xi_t^d T_{s,t,b}^d + P^{RES} \xi_t^s T_{s,t,b}^{R\_s} + P^{RES} \xi_t^d T_{s,t,b}^{R\_d}$

), which must be paid to the end-users because their adjusted temperature is not supplied exactly. Furthermore, the cost of gas purchasing ($\delta_t G_{s,t}^I \Delta t$) and reserve selling ($\sigma_t^d R_{s,t}^{dis\_a} \Delta t$) to upstream network is also modeled.

$$
\begin{aligned}
\text{Min}\Big\{ \sum_{s=1}^{N_s} \Big[p_s \sum_{t=1}^{N_t} \big(\epsilon_t^I D_t^I \Delta t - \epsilon_t^E D_t^E \Delta t \\
+ \pi_{s,t}^I I_{s,t}^I \Delta t - \pi_{s,t}^E I_{s,t}^E \Delta t + \delta_t G_{s,t}^I \Delta t - \sigma_t^d R_{s,t}^{dis_a} \Delta t \\
+ \sum_{b=1}^{N_b} \big(\xi_t^s T_{s,t,b}^s + \xi_t^d T_{s,t,b}^d + P^{RES} \xi_t^s T_{s,t,b}^{R_s} + P^{RES} \xi_t^d T_{s,t,b}^{R_d} \big) \big) \Big] \Big\}
\end{aligned}
$$
(9.39)

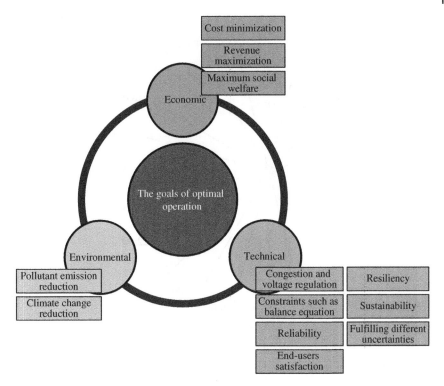

Figure 9.9 The goals in HEINs exploitation through optimal operation.

Also, there are some studies that try to maximize the revenue of HEINs. Another objective function that is very applicable in relevant literature is social welfare that considers both revenue and cost of all players participating in power management even other competitors in the market [65–68].

- *Technical Goals:*
  Technical goals usually are applied in most studies to provide a techno-economic analysis. One of the most important factors in technical goals is to prevent congestion and set voltage amount in the allowed range [1], which is usually between −0.95 per unit and +0.95 per unit (based on standards on voltage regulation). Also, the authors in [8] present a strategy to set the amount of voltage and current in the standard range. Figures 9.10 and 9.11 respectively show how a framework (done in [10]) that has considered voltage regulation and congestion avoidance as the technical goals improve the amount of voltage and current.

Other works usually try to minimize power loss [60] or maximize renewable power exploitation [69]. The other factor as a technical goal is maximizing the

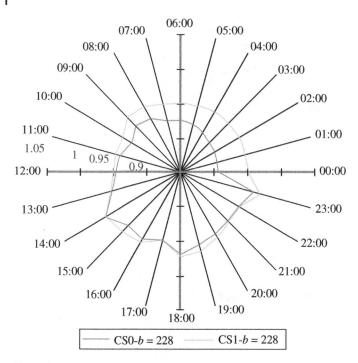

Figure 9.10 Voltage regulation within a model with power flow.

end-users satisfaction [12]. The authors in [70, 71] maximize the resiliency of HEINs to ensure the HEINs can continue their performance even under critical situations.

- *Environmental Goals:*
  There are also some works that try to mitigate the greenhouse gas emissions [72] as a goal through the HEIN optimal operation. Usually, this type of goal is accompanied with two previous types of goals.

9.4.3 Different Levels of Heat and Electricity Incorporated Networks Exploitation

HEINs, as multi-carrier networks including electricity, gas, heat, and cool, can have different sizes. They can be classified into three classes, as follows:

- Small HEINs which has a small scale in size of a home or building. Therefore, some researches also name this HEINs as the smart-home or the smart-building. Reference [73] investigates HEINs in level of homes.
- Large HEINs which are in size of a city or a part of a city, so that some researches name them smart grid or smart city. The authors in [1, 2, 74] evaluate the available tools to optimize the large-scale HEINs performance.

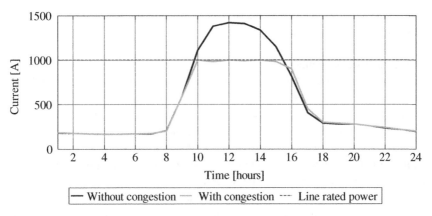

Figure 9.11 Congestion elimination within a model with power flow.

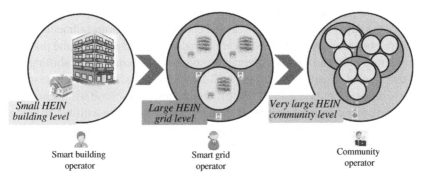

Figure 9.12 The connection between different levels of HEINs.

- The third group are very large HEINs which are a number of large-scale HEINs that combine with each other and create a community of HEINs. These HEINs can be in size of a large city or several small cities [75].

Some researches optimize two or three levels of these classes simultaneously. Figure 9.12 shows how these levels are related to each other.

9.4.4 Existing Potential of Heat and Electricity Incorporated Networks for Optimizing Their Operation

HEINs operators can use some tools to optimize the HEIN performance. In fact, these tools which are available in HEINs, make HEIN attractive to be more exploited. These tools can be categorized into two groups, which are investigated in Sections 9.4.4.1 and 9.4.4.2.

9.4.4.1 Internal Potential

The internal structure of HEINs is in a way that has high potential to help HEINs operate more optimally. Because HEINs manage the operation of the heat and electricity networks simultaneously, thus, the ability of energy vector substitution provides more optimized results. Moreover, hybrid energy storages (such as electric vehicles and TES) make it possible to shift the time of consumption. Also, managing the reactive power alongside the active power can manipulate the power factor in a way that leads to optimal operation.

Because the HEINs usually have a well-structured communication system, therefore, the prices and other data can be exchanged with end-users. As a result, demand response is another tool that can be implemented in such networks. Dispatchable loads (the loads that can be shifted to consume in off-peak hours) can make a decision based on price data received from market to consume at which hours of the day. This ability is also named end-users services curtailment. There are two types of demand response that are electrical demand response and thermal demand response. The electrical demand response is implemented by shifting the consumption of dispatchable loads (such as washing machines) from on-peak hours to off-peak hours [12, 76, 77]. The thermal demand response is implemented by building fabric thermal storage. If the supplied temperature within a building can be located in a range around the target temperature instead of a point, then the building fabric can store an amount of exergy. This mechanism is named thermal demand response [14].

Figure 9.13 is drawn based on the results of [14]. In this figure, Case 1 does not include any amount of thermal demand response, and HEIN operator must provide end-users the exact amount set for temperature. But in cases 2–5, there is a variation range of 4 °C, 2 °C (in both sides of target temperature), 2 °C (just one side of target temperature) and 1 °C for temperature, respectively. As Figure 9.13 shows, if the variation range of temperature is larger (which means more demand response capacity is available), the operational cost of the HEIN decreases.

The other tool available in HEINs is operating in grid-connected or islanding mode, which can improve HEIN cost, as well as enhance its reliability and sustainability.

Connecting to other HEINs to create a microgrid community can also result in more optimal operation. Because in this case, HEINs can exchange power with other HEINs and send/receive better power/price offering that helps to decrease its cost. Also, it can receive power during the unexpected events such as earthquakes or floods or when its equipment such as its resources have a failure and cannot supply its loads.

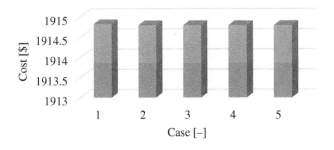

■ Imported electricity cost ■ Imported gas cost

Figure 9.13 Cost per building in a HEIN including 720 buildings with thermal demand response.

Because HEINs have their own resources and they can often supply their loads independently, this special structure of HEINs makes them be able to manage themselves completely without any dependency on the upstream grid. Therefore, HEINs can operate as a virtual power plant and schedule their own power resources independently, which helps them reach the most optimal level of operation. Then, they can be as an independent load from the viewpoint of the upstream grid and send their own power/price offering to the upstream grid to improve their operation [78]. This structure also makes them able to act as a price-maker in the competitive market and can affect the market-clearing process in a way that leads to more benefits for HEIN [21].

Many studies are done to integrate more different devices into HEINs to evaluate the impact of each one on optimal operation of HEINs. The authors in [2, 11] evaluate the impact of each device such as CHP on flexibility enhancement. Because HEINs can integrate more types of devices, therefore, they have more optimal operation compared to network having just the electricity layer.

The concept of peer-to-peer energy transaction has been focused on recently. It allows to HEINs to send each other power/price offers and decide individually the offers of which HEINs they must accept. Also, peer-to-peer transaction within a HEIN, allows its internal players (which are usually buildings) to propose their own power/price offers to other players. This ability allows them to select their best-received offer and, therefore, decrease their own cost. Because their operation strategy at each hour can be determined just based on their conditions and is not affected by the conditions of other players like the centralized power management. As a result, their cost is just associated with their own condition and is in the possible minimum amount. Figure 9.14 shows a brief demonstration of peer-to-peer transactions in HEINs or amongst HEINs.

Therefore, the internal potentials of HEINs to optimize their performance are similar to Figure 9.15.

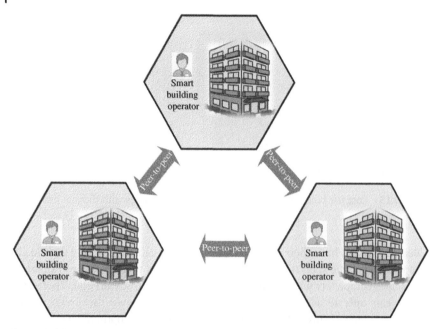

Figure 9.14 Peer-to-peer energy transactions in a HEIN or amongst HEINs.

9.4.4.2 External Potential
In addition to internal potentials in Section 9.4.4.1, which are created due to the structure of HEINs, also HEINs can use some external potentials to achieve a more optimal operation.

The first external tool is participating in hybrid energy markets, including electricity, gas, and heat market. Based on the prices of gas and electricity market, HEINs can decide to supply the heat load by a gas-consuming device (such as CHP or GB) or an electricity-consuming device (such as EHP or EB). Also, the HEIN operator can decide to supply electrical loads by CHP or supply them correctly from the electricity market. This selection based on the prices of the electricity or the gas market makes the HEIN able to operate more optimally. The mentioned process also can occur when a local heat market exists and the HEIN can participate in it. The HEIN operator also can decide to import electricity and store it in storages such as a battery or a compressed air energy storage during the low-price hours and then sells the stored energy to the market over high-price hours. The evaluation of the hybrid market impact on HEINs operation is addressed in [30, 79].

In addition to the energy market mentioned above, the ancillary services market can also decrease the HEINs cost and also improve their operation technically.

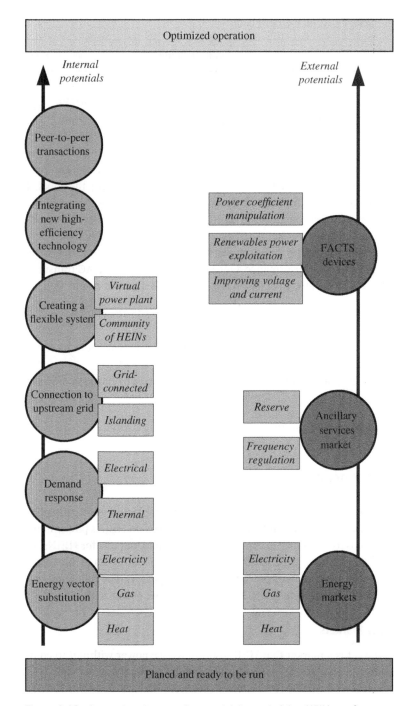

Figure 9.15 Internal and external potential for optimizing HEINs performance.

Table 9.7 Reserve impact on the HEIN cost.

	Without reserve	With reserve
Reserve revenue [Euro/day]	0	331
Electricity revenue [Euro/day]	52721	62759
HEIN revenue [Euro/day]	52607	62986

One of the most important ancillary services is reserve provision which can provide revenue by selling reserve to the upstream grid. According to [10], the effect of reserve on HEINs operation includes thermal comfort improvement. Also, reserve provision reduces the operational cost of HEINs. Table 9.7 achieved by the results of [10] shows how reserve market participation decreases the HEINs cost.

Furthermore, there are other tools that can be installed in HEINs to help HEINs operators improve HEINs performance. One of these tools is flexible AC transmission system (FACTS) devices. FACTS devices can lead to different results [80], as follows:

- Maximizing renewable power exploitation [69].
- Improving the amount of voltage and reducing the congestion amount by changing the impedance of HEINs lines.
- Enhancing the flexibility and reducing the operational cost by power coefficient manipulation.

Figure 9.15 summarizes external potentials available in HEINs.

9.5 Market/Incentives

As discussed in Section 9.4.4.2, market participation is an effective strategy to optimize HEINs performance. Thus, it has a significant impact on HEINs power management and needs to be studied completely. The available markets for HEINs, in addition to operation stage, are also important through planning stage. It is noteworthy to ask what types of market HEINs can participate in. The following sections discuss available markets that HEINs can exchange power with.

9.5.1 Energy Markets

The energy market is a market that HEINs sell/purchase power within it to compensate its power deficit or sells its power surplus during supplying its loads [81]. This type of market is available for carriers of electricity and gas. Sometimes, there is a local heat market for HEINs to trade the produced heat in a market mechanism, but it is local and, unlike the electricity, is not on a large scale. The gas market, which is often named wholesale gas market, usually have a flat price over 24 hours of the day. The energy markets of electricity can be divided into two groups. The first type is day-ahead market which its price profile is clear on the day before the power delivery day. Thus, this profile will be used in power management and leads to exact results without uncertainties. But there is a balancing mechanism which is also called the unbalanced electricity market. It makes HEINs able to compensate their deficit/surplus power in the form of real-time (up to five minutes before the power delivery time). But the price profile of this market is not determined before the power delivery [82]. As a result, it is essential to predict its prices and consider the uncertainties of its prices by methods introduced in Section 9.3.

9.5.2 Ancillary Services Market

There are different services that HEINs can provide them in ancillary services market and by selling these services, make additional revenue for themselves. Based on [83], ancillary services can be listed as elaborated in Table 9.8. These services can

Table 9.8 Ancillary services markets available for HEINs.

Group	Ancillary services
Reserve	Fast reserve
	STOR
	Balancing market start-up
	Demand-side response
	Demand turn up
Power system security	Black start
	Intertrips
	Super SEL
	System operator to system operator
	Transmission constraint management
Frequency response	EFR
	FFR
	Mandatory response services
Reactive power	ERPS
	Obligatory reactive power services

Source: Kazemi-Razi and Nafisi [83]/Springer Nature.

be signed between HEIN operators and the power system operator. For each signed service, HEINs offer their own prices, then power system operator based on received offers of whole HEINs and network conditions, decides which HEINs should supply the required services and pay them an amount of money which is calculated based on the agreement between the HEIN operator and the power system operator. The strategy used for each ancillary service to calculate its revenue is elaborated in [83].

9.5.3 Tax/Incentives Impact on Heat and Electricity Incorporated Networks Operation

In addition to energy markets and ancillary services markets and their impact on HEINs operation, there are other incentives and price signals having an important role on optimizing HEINs operation. As a result, it is essential to identify all of these taxes for each HEIN and model these prices in power management problems. These incentives for the electricity carrier can be listed as follows [83]:

- Capacity market;
- Local balancing services;
- Use of system charges which includes three parts named DNUoS charges for distribution network, TSUoS charges for transmission network and BSUoS charges for balancing the system;
- Environmental and social obligations to control the amounts of pollutants emitted or the effects of energy system on society.

Also, there are similar incentives for the gas carrier, which can be stated as follows [83, 84]:

- Imbalance charges;
- Use of system charges which includes two separate charges for both transmission network and distribution network;
- Environmental and social obligations to control the effects of activities related to the gas carrier on nature and society.

It is noteworthy both electricity and gas have an amount of tax due to using the facilities, and it is considered to be used for developing the gas and electricity network in the future [30].

9.5.4 Offering Strategy

Many researches include some of the mentioned markets in Sections 9.5.1–9.5.3. But these studies use a price profile for each of these markets, which is predicted based on past data of prices using strategies in Section 9.3. But the other studies which also models competitive behavior of HEINs within competitive markets, consider the power/price offering of HEINs in the markets. By these studies, there is a mechanism in problem of HEINs power management that makes HEINs operator able to determine the optimal power/price offering strategy at each hour of the day. In some of these studies, HEINs which propose their power/price offering in the markets, act as a price taker [85, 86].

But more developed researches have proposed some strategies in which the HEIN proposing its power/price offering, acts as a price maker in the competitive market that can affect the market-clearing process and changes the cleared prices in a way to improve its operation [21].

More developed works including power/price offering models also consider the market-clearing process to model the competitive market, which has a significant role in optimal operation of HEINs and their operational cost. To simplify the utilized models including market clearing process and to increase the calculation speed, some approaches such as Karush–Kuhn–Tucker (KKT) are used to replace the market clearing process equations with their simple converted forms in the problem of HEINs power management [21].

9.6 Main Achievements on Heat and Electricity Incorporated Networks Operation

As mentioned in the previous section, the optimal integration of heat and electricity networks results in more optimal operation of these two networks. The benefits of these two networks operation can be summarized as follows:

- Enhancing the flexibility of these networks by achieving the ability of energy vector substitution based on the price of heat and electricity in the markets which results in less cost and more end-users comfort.
- Enhancing the sustainability of these networks that results in less equipment failure and more lifetime as well as sustainable service delivery to the end-users that increase their satisfaction.
- Using multi-carrier sources and storages that can supply loads by different energy carriers increases the resiliency during the unpredicted events.
- HEINs allow using new technologies such as P2G technology and EHP that reduces CO_2 emissions.
- There are different markets where HEINs with more flexibility than single-carrier networks can optimally participate in such as ancillary services market or participating as a price-maker with more optimal power/price offers.
- Equipment used in HEINs and also flexible power trading with different available markets makes it possible to integrates more small-size high-efficient technologies and electric vehicles. Also, HEINs is able to be used as a smart-building

or a smart city. These equipment allow HEINs to be scheduled more optimally so that the voltage and current can be set in their standard range.

- The flexible structure of HEINs participating in the markets also allows to be integrated with other HEINs and make a community of HEINs that participate in the markets fairly.
- The structure of HEINs is able to allows end-users to offer their power/price as a prosumer in peer-to-peer transactions.
- The high flexibility of HEINs permits more demand response ability to be controlled based on heat and electricity price.

Although integration of heat and electricity results in mentioned benefits, but there are some challenges through their implementation, which are:

- A high-speed communication network is needed to be invested that can results in a high investment cost. This communication network needs a high bandwidth to transmit the big data of HEINs between operator, power generators and consumers.
- Some devices installed in some electricity or heat networks cannot be used in flexible HEINs and need to be replaced. Also, installing the new high-efficiency technologies can cost a lot.
- Some of network operator do not know how control an incorporated network that results in non-optimal operation of HEINs.
- There is not a complete model that can predict loads and prices exactly, then schedule the HEINs operation optimally. The existing models have missed some parts of prediction or optimization stages or do not work optimally. For example, there are some models that optimize the HEINs operation, but they do not model the thermal comfort.
- Some equipment needed in HEINs are not made optimally yet, or example the existing electric vehicles are not as optimal as they are modeled in researches. Therefore, it causes unreal results of researches and reduces end-users satisfaction.

The most important principle through the heat and electricity network integration is to model and quantify the different parts of these networks (including operator, markets, equipment, and loads/services) exactly. Then, it is necessary to use high-efficient multi-carrier devices such as CHP that can serve loads using both heat and electricity. Also, the structure of existing market needs some updates to support incorporated operation of these networks simultaneously such as more competitive market to get power/price offers and clears the market based on received offers. Furthermore, new markets such as ancillary services need to be more concentrated to increase the flexibility of HEINs. A crucial part through heat and electricity integration is to train operators so that they can optimize the heat and electricity networks as an incorporated network.

Based on mentioned benefits and needed principles, this chapter models and quantifies the different parts of HEINs at first to help the optimal results achievements through the HEINs power management. Then, the optimal operation of HEINs power management including technical, economic, and environmental parts are defined. To reach the defined optimal operation, it is necessary to know what tools are available for optimal operation of HEINs. Therefore, the internal potentials of HEINs such as peer-to-peer transactions and external available tools such as ancillary services market are elaborated. Because the existing tax/incentives have some impact on optimal operation of HEINs, they are determined exactly. Considering the important role of different markets on optimal operation of HEINs, different levels of market participation from getting price curves to participating as a price-maker is modeled. In addition, it is elaborated the HEINs potential makes it possible to use HEIN in the level of smart buildings to smart cities.

9.7 Conclusions

HEINs have proven that gas, heat, and electricity networks management simultaneously has more optimal results than single carrier network management. Therefore, HEINs are growing increasingly. To optimize the operation of a HEIN, it is essential to model all parts of HEINs completely. Therefore, this chapter proposes a high-quality model for each part of HEINs structure including loads, equipment, and markets. According to the fact that an optimized operation of HEINs must minimize cost and pollutants emissions while maximize the flexibility (the services are completely provided for end-users to maximize their satisfaction) and sustainability. As a result, the optimized operation of HEINs is a techno-economic-environmental evaluation. The most important results of HEINs optimized operation are as follows:

- Minimum cost/maximum revenue;
- Supplying end-users services completely to maximize their satisfaction (maximum flexibility);
- Maximizing sustainability to operate continuously without any failure, as a result, the congestion and voltage values outside the standard range is prevented;
- Enhancing reliability and resiliency;
- Operating with maximum efficiency and maximizing the renewable sources exploitation;
- Optimal power/price offering strategies which minimize HEINs cost.

To achieve the optimized operation, it is necessary to identify what potentials are available for HEINs. Therefore, all of internal/external potentials are elaborated.

Also, due to the importance of HEINs participation in markets to increase their revenues, as well as improve their operation and power systems operation, the markets available for HEINs are explained. Thus, the main achievement of this study are as follows:

- Modeling and quantifying the different parts of HEINs to help the optimal results achievements through the HEINs power management;
- Defining the optimal operation in HEINs power management and determining main parameters, such as flexibility, which must be optimized;
- Determining internal potentials of HEINs for optimizing their operation such as peer-to-peer transactions;
- Defining external available tools for optimizing HEINs operation such as ancillary services market;
- Studying the impact of existing tax/incentives on HEINs operation;
- Defining different levels of market participation from getting price curves to participating as a price-maker;
- Structuring the different HEIN levels from smart-building to smart city.

References

1 Ceseña, E.A.M. and Mancarella, P. (2018). Energy systems integration in smart districts: robust optimisation of multi-energy flows in integrated electricity, heat and gas networks. *IEEE Trans. Smart Grid* 10 (1): 1122–1131.

2 Coelho, A., Neyestani, N., Soares, F., and Lopes, J.P. (2020). Wind variability mitigation using multi-energy systems. *Int. J. Electr. Power Energy Syst.* 118: 105755.

3 Gazijahani, F.S., Ajoulabadi, A., Ravadanegh, S.N., and Salehi, J. (2020). Joint energy and reserve scheduling of renewable powered microgrids accommodating price responsive demand by scenario: a risk-based augmented epsilon-constraint approach. *J. Clean. Prod.* 262: 121365.

4 Ding, X., Guo, Q., Qiannan, T., and Jermsittiparsert, K. (2021). Economic and environmental assessment of multi-energy microgrids under a hybrid optimization technique. *Sustain. Cities Soc.* 65: 102630.

5 Lekvan, A.A., Habibifar, R., Moradi, M. et al. (2021). Robust optimization of renewable-based multi-energy micro-grid integrated with flexible energy conversion and storage devices. *Sustain. Cities Soc.* 64: 102532.

6 Chen, X., Dall'Anese, E., Zhao, C., and Li, N. (2019). Aggregate power flexibility in unbalanced distribution systems. *IEEE Trans. Smart Grid* 11 (1): 258–269.

7 Lu, X., Chan, K.W., Xia, S. et al. (2019). A model to mitigate forecast uncertainties in distribution systems using the temporal flexibility of EVAs. *IEEE Trans. Power Syst.* 35 (3): 2212–2221.

8 Liao, H. and Milanović, J.V. (2018). Flexibility exchange strategy to facilitate congestion and voltage profile management in power networks. *IEEE Trans. Smart Grid* 10 (5): 4786–4794.

9 Anwar, M.B., Qazi, H.W., Burke, D.J., and O'Malley, M.J. (2018). Harnessing the flexibility of demand-side resources. *IEEE Trans. Smart Grid* 10 (4): 4151–4163.

10 Kazemi-Razi, S.M., Abyaneh, H.A., Nafisi, H. et al. (2021). Enhancement of flexibility in multi-energy microgrids considering voltage and congestion improvement: Robust thermal comfort against reserve calls. *Sustain. Cities Soc.* 74: 103160.

11 Good, N. and Mancarella, P. (2019). Flexibility in multi-energy communities with electrical and thermal storage: a stochastic, robust approach for multi-service demand response. *IEEE Trans. Smart Grid* 10 (1): 503–513. https://doi.org/10.1109/ TSG.2017.2745559.

12 Good, N., Karangelos, E., Navarro-Espinosa, A., and Mancarella, P. (2015). Optimization under uncertainty of thermal storage-based flexible demand response with quantification of residential users' discomfort. *IEEE Trans. Smart Grid* 6 (5): 2333–2342.

13 Romanchenko, D., Nyholm, E., Odenberger, M., and Johnsson, F. (2021). Impacts of demand response from buildings and centralized thermal energy storage on district heating systems. *Sustain. Cities Soc.* 64: 102510.

14 Kazemi-Razi, S.M., Askarian-Abyaneh, H., Nafisi, H. et al. (2020). Optimization of operation of microgrid by thermal demand response considering enhancement of consumers' thermal comfort. *Iran. Electr. Ind. J. Qual. Product.* 8 (3): 68–77.

15 Ahmad, F., Alam, M.S., and Asaad, M. (2017). Developments in xEVs charging infrastructure and energy management system for smart microgrids including xEVs. *Sustain. Cities Soc.* 35: 552–564.

16 Guo, Q., Liang, X., Xie, D., and Jermsittiparsert, K. (2021). Efficient integration of demand response and plug-in electrical vehicle in microgrid: environmental and economic assessment. *J. Clean. Prod.* 291: 125581.

17 Taşcıkaraoğlu, A., Paterakis, N.G., Erdinç, O., and Catalao, J.P.S. (2018). Combining the flexibility from shared energy storage systems and DLC-based demand response of HVAC units for distribution system operation enhancement. *IEEE Trans. Sustain. Energy* 10 (1): 137–148.

18 Cao, Y., Huang, L., Li, Y. et al. (2020). Optimal scheduling of electric vehicles aggregator under market price uncertainty using robust optimization technique. *Int. J. Electr. Power Energy Syst.* 117: 105628.

19 Elbatawy, S. and Morsi, W.G. (2021). Integration of prosumers with battery storage and electric vehicles via transactive energy. *IEEE Trans. Power Deliv.* 37 (1): 383–394.

20 Ma, H., Zhang, C., Jia, J. et al. (2021). Investigation on human thermal comfort of the ecological community in arid area of Lanzhou, China. *Sustain. Cities Soc.* 72: 103069.

21 Sun, S., Kazemi-Razi, S.M., Kaigutha, L.G. et al. (2022). Day-ahead offering strategy in the market for concentrating solar power considering thermoelectric decoupling by a compressed air energy storage. *Appl. Energy* 305: 117804.

22 Oikonomou, K., Parvania, M., and Khatami, R. (2019). Deliverable energy flexibility scheduling for active distribution networks. *IEEE Trans. Smart Grid* 11 (1): 655–664.

23 Oikonomou, K. and Parvania, M. (2018). Optimal coordination of water distribution energy flexibility with power systems operation. *IEEE Trans. Smart Grid* 10 (1): 1101–1110.

24 Nasiri, N., Yazdankhah, A.S., Mirzaei, M.A. et al. (2020). A bi-level market-clearing for coordinated regional-local multi-carrier systems in presence of energy storage technologies. *Sustain. Cities Soc.* 63: 102439.

25 Fontenot, H., Ayyagari, K.S., Dong, B. et al. (2021). Buildings-to-distribution-network integration for coordinated voltage regulation and building energy management via distributed resource flexibility. *Sustain. Cities Soc.* 69: 102832.

26 Li, L. and Zhang, S. (2021). Techno-economic and environmental assessment of multiple distributed energy systems coordination under centralized and decentralized framework. *Sustain. Cities Soc.* 72: 103076.

27 Müller, F.L., Szabó, J., Sundström, O., and Lygeros, J. (2017). Aggregation and disaggregation of energetic flexibility from distributed energy resources. *IEEE Trans. Smart Grid* 10 (2): 1205–1214.

28 Yazdani-Damavandi, M., Neyestani, N., Shafie-khah, M. et al. (2017). Strategic behavior of multi-energy players in electricity markets as aggregators of demand side resources using a bi-level approach. *IEEE Trans. Power Syst.* 33 (1): 397–411.

29 D. Port (2021). Renewables and demand data. https://dataport.pecanstreet.org/ (accessed 1 September 2021).

30 Good, N.P. (2015). *Techno-Economic Assessment of Flexible Demand*. The University of Manchester (United Kingdom).

31 Kazemi-Razi, S.M. (2020). *Optimal Scheduling of Active Distribution Networks in the Context of Energy and Reserve Market*. Amirkabir Uiversity of Technology (Tehran Polytechnic).

32 Enescu, D. (2017). A review of thermal comfort models and indicators for indoor environments. *Renew. Sust. Energ. Rev.* 79: 1353–1379. https://doi.org/10.1016/J.RSER.2017.05.175.

33 Chaudhuri, T., Soh, Y.C., Li, H. et al. (2017). Machine learning based prediction of thermal comfort in buildings of equatorial Singapore. In: *2017 IEEE International Conference on Smart Grid and Smart Cities (ICSGSC)* (23–26 July 2017), 72–77. Singapore: IEEE.

34 Zhou, X., Xu, L., Zhang, J. et al. (2020). Data-driven thermal comfort model via support vector machine algorithms: insights from ASHRAE RP-884 database. *Energy Build.* 211: 109795.

35 von Grabe, J. (2016). Potential of artificial neural networks to predict thermal sensation votes. *Appl. Energy* 161: 412–424.

36 Jiang, L. and Yao, R. (2016). Modelling personal thermal sensations using C-support vector classification (C-SVC) algorithm. *Build. Environ.* 99: 98–106.

37 Li, Y., Rezgui, Y., Guerriero, A. et al. (2020). Development of an adaptation table to enhance the accuracy of the predicted mean vote model. *Build. Environ.* 168: 106504.

38 Wang, Z., Sun, S., Lin, X. et al. (2019). A remote integrated energy system based on cogeneration of a concentrating solar power plant and buildings with phase change materials. *Energy Convers. Manag.* 187: 472–485.

39 Geidl, M. and Andersson, G. (2007). Optimal power flow of multiple energy carriers. *IEEE Trans. Power Syst.* 22 (1): 145–155.

40 Chicco, G. and Mancarella, P. (2009). Distributed multi-generation: a comprehensive view. *Renew. Sust. Energ. Rev.* 13 (3): 535–551.

41 EPRI (2021). OpenDSS. https://www.epri.com/pages/sa/opendss (accessed 10 September 2021).

42 MathWorks (2021). MATLAB. https://www.mathworks.com/ (accessed 10 September 2021).

43 U. D. of Energy (2012). EnergyPlus. https://energyplus.net/ (accessed 10 September 2021).

44 S. University (2013). esp-r. https://www.strath.ac.uk/research/energysystemsresearchunit/applications/esp-r/ (accessed 10 September 2021).

45 I. E. Solutions (2013). IESVE for Engineers. https://www.iesve.com/ (accessed 10 September 2021).

46 NetLoo (2021). NetLogo. https://ccl.northwestern.edu/netlogo/ (accessed 10 September 2021).

47 EU (2009). Directive 2009/28/EC of the European parliament and of the council of 23 april 2009 on the promotion of the use of energy from renewable sources and amending and subsequently repealing directives 2001/77/EC and 2003/30/EC. *Off. J. Eur. Union*.

48 Tookanlou, M.B., Ardehali, M.M., and Nazari, M.E. (2015). Combined cooling, heating, and power system optimal pricing for electricity and natural gas using particle swarm optimization based on bi-level programming approach: case study of Canadian energy sector. *J. Nat. Gas Sci. Eng.* 23: 417–430. https://doi.org/10.1016/J.JNGSE.2015.02.019.

49 Lehner, M., Tichler, R., Steinmüller, H., and Koppe, M. (2014). *Power-to-Gas: Technology and Business Models*. Springer.

50 He, C., Liu, T., Wu, L., and Shahidehpour, M. (2017). Robust coordination of interdependent electricity and natural gas systems in day-ahead scheduling for facilitating volatile renewable generations via power-to-gas technology. *J. Mod. Power Syst. Clean Energy* 5 (3): 375–388.

51 Kok, K. (2009). Short-term economics of virtual power plants. In: *CIRED 2009-20th International Conference and Exhibition on Electricity Distribution-Part 1* (8–11 June 2009), 1–4. Prague, Czech Republic: IET.

52 Pfenninger, S. and Staffell, L. (2021). Renewables.ninja. https://www.renewables.ninja/ (accessed 1 September 2021).

53 Mohammadi, M., Hosseinian, S.H., and Gharehpetian, G.B. (2012). Optimization of hybrid solar energy sources/wind turbine systems integrated to utility grids as microgrid (MG) under pool/bilateral/hybrid electricity market using PSO. *Sol. Energy* 86 (1): 112–125.

54 Mohammadi, M., Hosseinian, S.H., and Gharehpetian, G.B. (2012). GA-based optimal sizing of microgrid and DG units under pool and hybrid electricity markets. *Int. J. Electr. Power Energy Syst.* 35 (1): 83–92.

55 Mohammadi, M., Hosseinian, S.H., and Gharehpetian, G.B. (2011). Optimal sizing of micro grid & distributed generation units under pool electricity market. *J. Renew. Sustain. Energy* 3 (5): 53103.

56 Flynn, D. (2003). *Thermal power plant simulation and control*, no. 43. Institution of Engineering and Technology IET.

57 P. energetyczna polski do 2040 roku- Projekt, "energy policy of Poland until 2040 (EPP2040) - draft," *warsaw Minist. energy*, 2019.

58 Wu, S., Zhou, C., Doroodchi, E., and Moghtaderi, B. (2019). Thermodynamic analysis of a novel hybrid thermochemical-compressed air energy storage system powered by wind, solar and/or off-peak electricity. *Energy Convers. Manag.* 180: 1268–1280.

59 Tookanlou, M.B., Pourmousavi, S.A., and Marzband, M. (2021). An optimal day-ahead scheduling framework for E-mobility ecosystem operation with drivers preferences. *IEEE Trans. Power Syst.* 36: 5245–5257.

60 Nafisi, H., Agah, S.M.M., Abyaneh, H.A., and Abedi, M. (2015). Two-stage optimization method for energy loss minimization in microgrid based on smart power management scheme of PHEVs. *IEEE Trans. Smart Grid* 7 (3): 1268–1276.

61 Heitsch, H. and Römisch, W. (2003). Scenario reduction algorithms in stochastic programming. *Comput. Optim. Appl.* 24 (2–3): 187–206.

62 Marzband, M., Parhizi, N., Savaghebi, M., and Guerrero, J.M. (2015). Distributed smart decision-making for a multimicrogrid system based on a hierarchical interactive architecture. *IEEE Trans. Energy Convers.* 31 (2): 637–648.

63 Andervazh, M.-R. and Javadi, S. (2017). Emission-economic dispatch of thermal power generation units in the presence of hybrid electric vehicles and correlated wind power plants. *IET Gener. Transm. Distrib.* 11 (9): 2232–2243.

64 Ulbig, A. and Andersson, G. (2015). Analyzing operational flexibility of electric power systems. *Int. J. Electr. Power Syst.* 72: 155–164.

65 Yazdani-Damavandi, M., Neyestani, N., Chicco, G. et al. (2017). Aggregation of distributed energy resources under the concept of multienergy players in local energy systems. *IEEE Trans. Sustain. Energy* 8 (4): 1679–1693.

66 van Leeuwen, G., AlSkaif, T., Gibescu, M., and van Sark, W. (2020). An integrated blockchain-based energy management platform with bilateral trading for microgrid communities. *Appl. Energy* 263: 114613.

67 Shaker, A., Safari, A., and Shahidehpour, M. (2021). Reactive power management for networked microgrid resilience in extreme conditions. *IEEE Trans. Smart Grid* 12: 3940–3953.

68 Zhang, Q., Dehghanpour, K., Wang, Z., and Huang, Q. (2019). A learning-based power management method for networked microgrids under incomplete information. *IEEE Trans. Smart Grid* 11 (2): 1193–1204.

69 Kazemi-Razi, S.M., Mirsalim, M., Askarian-Abyaneh, H. et al. (2018). Maximization of wind energy utilization and flicker propagation mitigation using SC and STATCOM. In: *2018 Smart Grid Conference (SGC)* (28–29 November 2018), 1–6. Sanandaj, Iran: IEEE. https://doi.org/10.1109/SGC.2018.8777744.

70 Silva, J.A.A., López, J.C., Arias, N.B. et al. (2021). An optimal stochastic energy management system for resilient microgrids. *Appl. Energy* 300: 117435.

71 Dehghanpour, K. and Nehrir, H. (2018). A market-based resilient power management technique for distribution systems with multiple microgrids using a multi-agent system approach. *Electr. Power Components Syst.* 46 (16–17): 1744–1755.

72 Good, N., Ceseña, E.A.M., Zhang, L., and Mancarella, P. (2016). Techno-economic and business case assessment of low carbon technologies in distributed multi-energy systems. *Appl. Energy* 167: 158–172.

73 Jiang, B. and Fei, Y. (2014). Smart home in smart microgrid: a cost-effective energy ecosystem with intelligent hierarchical agents. *IEEE Trans. Smart Grid* 6 (1): 3–13.

74 Korkas, C.D., Baldi, S., Michailidis, I., and Kosmatopoulos, E.B. (2015). Intelligent energy and thermal comfort management in grid-connected microgrids with heterogeneous occupancy schedule. *Appl. Energy* 149: 194–203.

75 Hu, B., Wang, H., and Yao, S. (2017). Optimal economic operation of isolated community microgrid incorporating temperature controlling devices. *Prot. Control Mod. power Syst.* 2 (1): 1–11.

76 Pourmousavi, S.A., Nehrir, M.H., and Sharma, R.K. (2015). Multi-timescale power management for islanded microgrids including storage and demand response. *IEEE Trans. Smart Grid* 6 (3): 1185–1195.

77 Wang, Y., Huang, Y., Wang, Y. et al. (2018). Energy management of smart micro-grid with response loads and distributed generation considering demand response. *J. Clean. Prod.* 197: 1069–1083.

78 Pudjianto, D., Ramsay, C., and Strbac, G. (2007). Virtual power plant and system integration of distributed energy resources. *IET Renew. power Gener.* 1 (1): 10–16.

79 Mancarella, P. and Chicco, G. (2013). Real-time demand response from energy shifting in distributed multi-generation. *IEEE Trans. Smart Grid* 4 (4): 1928–1938. https://doi.org/10.1109/TSG.2013.2258413.

80 Jirdehi, M.A., Tabar, V.S., Hemmati, R., and Siano, P. (2017). Multi objective stochastic microgrid scheduling incorporating dynamic voltage restorer. *Int. J. Electr. Power Energy Syst.* 93: 316–327.

81 Shahidehpour, M., Yamin, H., and Li, Z. (2003). *Market Operations in Electric Power Systems: Forecasting, Scheduling, and Risk Management.* Wiley.

82 Kirschen, D.S. and Strbac, G. (2018). *Fundamentals of Power System Economics.* Wiley.

83 Kazemi-Razi, S.M. and Nafisi, H. (2021). Energy markets of multi-carrier energy networks. In: *Planning and Operation of Multi-Carrier Energy Networks*, 87. Springer. https://link.springer.com/chapter/10.1007/978-3-030-60086-0_4.

84 Ofgem (2021). Ofgem reports. https://www.ofgem.gov.uk/electricity (accessed 15 September 2021).

85 Pandžić, H., Morales, J.M., Conejo, A.J., and Kuzle, I. (2013). Offering model for a virtual power plant based on stochastic programming. *Appl. Energy* 105: 282–292.

86 He, G., Chen, Q., Kang, C., and Xia, Q. (2016). Optimal offering strategy for concentrating solar power plants in joint energy, reserve and regulation markets. *IEEE Trans. Sustain. Energy* 7 (3): 1245–1254.

10

Optimal Energy Management of a Demand Response Integrated Combined-Heat-and-Electrical Microgrid

Yan Xu[1], Zhengmao Li[1], Xue Feng[2], and Yumin Chen[3]

[1] *School of Electrical and Electronic Engineering, Nanyang Technological University, Nanyang, Singapore*
[2] *Engineering Cluster, Singapore Institute of Technology, Dover, Singapore*
[3] *China Southern Grid Digital Power Grid Co., Ltd., Digital Power Grid Branch, Engineering and Technology Division, Building C, Yunsheng Science Park, Guangzhou, China*

Nomenclature

A. Acronyms

CHEM   Combined-heat-and-electrical microgrid
CHP   Combined cooling and heat
DHN   District heat network
EES   Electric energy storage
ITC   Indoor temperature control
PBDR   Price-based demand response

B. Sets and Indexes

B/b Set/Index of buildings (heat loads) in DHN
I/i Set/Index of nodes in the power network
J/j Set/Index of PBDR levels
K/k Set/Index of pipelines in heat network
K_{ps}/K_{pr} Sets of all pipelines in supply/return network
K_{hs} Sets of pipelines connected to heat sources
$K_{n,r}^{+}/K_{n,r}^{-}$ The sets of all pipelines start/end at node n in the return network
$K_{n,s}^{+}/K_{n,s}^{-}$ The sets of all pipelines start/end at node n in the supply network
N/n Set/Index of nodes in heat network

Coordinated Operation and Planning of Modern Heat and Electricity Incorporated Networks,
First Edition. Edited by Mohammadreza Daneshvar, Behnam Mohammadi-Ivatloo, and Kazem Zare.
© 2023 The Institute of Electrical and Electronics Engineers, Inc.
Published 2023 by John Wiley & Sons, Inc.

N_{ps}/N_{pr} Sets of connection nodes in the supply, return network

T/t Set/Index of time periods

U/u Set/Index of controllable units in the power network

C. Parameters

ε_j	Demand response level j
A	The cross-sectional area of the pipeline, m^2
ν_P	Price-based demand response constant, kw/¥
C_T	Specific water heat capacity, kWh/°C
C_{air}	Specific air heat capacity, kWh/°C
$C_{CHP}^{st}/C_{CHP}^{sd}$	Unit start-up/shut-down cost of CHP, ¥
C_{EB}^{st}/C_{EB}^{sd}	Unit start-up/shut-down cost of electric boiler, ¥
C_{gas}	The unit price of natural gas, ¥/m^3
C_H	Per unit heat revenue of selling heat to customers, ¥/kWh
$C_{pr,\,t}$	Price of power transactions with the utility grid, ¥/kWh
C_u^{om}	Operation & maintenance cost of controllable unit u, ¥/kWh
L_{HG}	The lower heating value of natural gas, kWh/m^3
L_j^{le}	The demand response rate of level j
l_k	Length of pipe k, m
$P_{ESC}^{MAX}/P_{ESD}^{MAX}$	Maximum charging/discharging power of EES, kW
P_u^{max}/P_u^{min}	Minimum/minimum power output of controllable unit u, kW
$P_{max}/Q_{max,}$	Maximum active/reactive power of a branch, kW/kVar
R_i/X_i	The impedance of branch from node i, Ω/S
R_T	Heat resistance of building shells, °C/kW
R_u^{up}/R_u^{down}	Maximum ramp up/down rate of each controllable unit u, kW/h
$T_{am,\,t}$	Outdoor temperature at time period t, °C
$T_{in,b,t}^{max}/T_{in,b,t}^{min}$	Minimum/maximum temperature of indoor temperature of building b at time t, °C
T_s^{min}/T_s^{max}	Minimum/maximum temperature of water flowing in supply pipes, °C
T_r^{min}/T_r^{max}	Minimum/maximum temperature of water flowing in return pipes, °C
V_s	The voltage of substation
ΔV^{max}	The maximal allowed voltage deviation
λ	Heat transfer coefficient of the pipeline
τ_T	Coefficient of heat conduction
η_{hx}	The efficiency of the primary heat exchanger

η_{ra}	The efficiency of the radiator
η_{CHP}	The power efficiency of the CHP plant
ρ	Water density, kg/m^3
β_T	Unit conversion coefficient
ε	Price elasticity of power demand
$\mu_{\text{EES}}^{\text{MIN}}/\mu_{\text{EES}}^{\text{MAX}}$	The minima/maximal state of charge in BS
$\varphi_{k,t}/\psi_{k,t}$	Node method coefficients, h

D. Variables

$P_{\text{gd},t}/P_{\text{sd},t}$	Power purchased from/sold to the utility grid, kW
$E_{\text{EES},t}^{i}$	Energy stored in EES, kWh
$H_{b,t}^{T}$	Heat conduction power, kW
$H_{b,t}^{\text{load}}$	Design heat load, kW
$H_{\text{EB},t}/H_{\text{CHP},t}$	Heat power produced by electric boiler/CHP plant, kW
$H_{\text{ra},b,t}$	Heat power transferred through the radiator, kW
$H_{s,b,t}^{T}$	Heat energy conduction through building shells of building b, kW
$H_{r,k,t}^{+}/H_{r,k,t}^{-}$	Heat power at the starting, ending point in the return network, kW
$H_{s,k,t}^{+}/H_{s,k,t}^{-}$	Heat power at the starting, ending point in the supply network, kW
$\omega_{s,k,t}/\omega_{r,k,t}$	The mass flow rate of pipeline in the supply/ return network, kg/s
$P_{i,t}/Q_{i,t}$	Active/reactive power flow, kW
$P_{u,t}^{i}$	Power generated (consumed) by controllable unit, kW
$P_{\text{EES},t}^{\text{dis},i}/P_{\text{EES},t}^{\text{ch},i}$	Discharging/charging power of EES, kW
$P_{\text{gd},t}/P_{\text{sd},t}$	Power purchasing from/selling to the utility grid, kW
$P_{\text{load},i,t}/Q_{\text{load},i,t}$	Active/reactive power with PBDR strategy, kW
$P_{\text{load},i,t}^{o}/Q_{\text{load},i,t}^{o}$	Original active/reactive power without PBDR strategy, kW
$W_{k,t}$	The total water mass flowing into the pipeline at the time $\psi_{k,t}$, kg
$V_{k,t}$	The total water mass flowing into the pipeline at the time $\phi_{k,t}$, kg
$T_{s,k,t}^{+}/T_{s,k,t}^{-}$	The temperature at starting/ending point in the supply network, °C
$T_{r,k,t}^{+}/T_{r,k,t}^{-}$	The temperature at the starting, ending point in the return network, °C

$T^{-}_{s,k,t}{'}$ The water mass temperature at the end of supply pipeline without considering the thermal loss

$T^{-}_{r,k,t}{'}$ The water mass temperature at the end of return pipeline without considering the thermal loss

$T^{m}_{n,s}/T^{m}_{n,r}$ The temperature of supply, return network, °C

$T_{in, b, t}$ Indoor temperature, °C

$U^{i}_{CHP,t}$, $U^{i}_{EB,t}$ On-off status of CHP, electric boiler

$U^{ch,i}_{EES,t}$, $U^{dis,i}_{EES,t}$ Binary state variables indicating the charging/discharging stage of EES

$V_{i, t}$ Node voltage

δ_{j} Binary on/off decision of PBDR level

10.1 Introduction

In traditional single energy systems, different types of energy resources including electricity, heat, and gas are individually dispatched without much interaction and coordination. Typically, the energy efficiency of a single energy system can only reach around 45% [1]. There has been a shortage of energy since the mid-1960s, and the situation is worsening. To cope with the energy crisis, strategies are proposed to enhance energy efficiency, one of which is to integrate different energy systems. A microgrid that combines electrical and thermal energy typically consists of distributed generation such as wind turbines, photovoltaic cells, combined heat and power (CHP) plants, and fuel cells, in which electrical and thermal energy are simultaneously supplied and consumed [2–4]. The integration of distinctively featured energy sources comes with operational challenges [5]. Interactions and coordination between different energy sources have also drawn research interests [6].

In the literature for the co-dispatch of the power and thermal energy, a hybrid strategy is proposed in [7] to minimize the excessive production of electrical and thermal energy by CHP plants. Daily net cost is minimized in [8] through the optimal operation of lead-acid batteries subject to a real-time pricing scheme. In addition to component-level focused research, coordinated system-level dispatch strategies of microgrids have also been widely studied. In [9], a system-wide coordinated energy scheduling method is introduced to minimize the total system operation cost. References [10–12] also look into system-level coordination methods for holistic energy utilization enhancement and energy supplying cost reduction. In [13], the power and thermal energy are co-dispatched in a multi-energy system with the objective of reducing the total operation cost; In the studies [14–15], to reduce the power and thermal energy supplying cost from the long run, planning schemes for the distributed generators and energy storage devise are conducted.

It can be seen that intensive research works have been done on the optimal planning and/or operation of the combined power and thermal energy systems or multi-energy microgrid, but most of the existing research only considers specific constraints of the electrical network including power flow, voltage, or phase angle constraints with little emphasis on heat networks. The real-life applications will render impractical if the heat networks are not effectively considered.

With the increasing amount of thermal energy applications, district heat network (DHN) has become popular. Comprehensive mathematical models that capture the dynamics of DHN are proposed in [16–19]. A node method with a CHP network is presented in [20]. An optimal operation method for a microgrid with a CHP network is proposed in [21]. The operation method considers the nodal flow balance and mixing temperature constraints of the DHN with high inertia. The aforementioned work only focuses on heat networks without systematic coordination between heat networks and electrical networks. Considering the inertia which results from the relatively slow mass flow rate of water and delayed consumption of thermal energy, DHN can also be treated as a form of thermal energy storage. All in all, it is essential to investigate the distinct dynamic properties of the DHN and organically take into account its coordination with the electrical to achieve overall enhanced efficiency.

Increasing energy demands pose a great challenge to the balance between the energy source and load [22]. The demand response plays a significant role in regulating supply-demand balances by relying on flexible loads [23]. Day-ahead demand response strategies presented in [24–27] demonstrate their efficacy in increasing the economic performance of electrical networks. In [28], the price-based demand response (PBDR) scheme is utilized in the power system to reduce the cost and mitigate the effect of uncertainties on the system operation. The study in [29] employs the intensive-based demand response scheme in the unit commitment problem of the power system, to reduce the power supplying cost; Reference [30] proposes an optimization model for the economic operation of micro-grid. The model presented in this research mainly aims to minimize the operating cost of a microgrid system and make full use of clean energy under the premise of considering distributed power generation and demand response. Though the demand response scheme for electricity is intensively studied, the thermal demand response has yet to be widely adopted in CHEMs.

From the above, it can be seen that there are several research gaps in the literature:

1) In the current research works, the combined power and thermal network is not comprehensively modeled, which makes them more practical for real-world applications.
2) The multi-energy demand response is not taken into full use to boost the holistic energy dispatch flexibility and further reduce the energy supplying cost.

In this book chapter, an optimal operation method for CHEMs considering demand response is presented. The objective is to minimize the overall daily cost while satisfying technical constraints. The key contributions of this research are summarized as:

1) Differing from most of the existing research works, which either focus on selected key components units or are based on a single-energy network, a comprehensive system-wide dispatch method for CHEMs is proposed where all controllable components are optimally coordinated and DHN models including nodal flow, mixing temperature constraints, storage capacity and transmission delay are systematically integrated into the power network.
2) Few of the existing research on CHEM operations consider demand response. In our proposed method, PBDR and indoor temperature control (ITC) strategies are designed to flexibilize the heat and power loads.
3) Our proposed method also treats DHN as heat storage that can flexibilize system operation and enhance the economic benefits, which wasn't explored in traditional work.
4) The simulation results demonstrate that our method outperforms all the non-coordinated methods or existing methods without demand response in terms of the operating cost.

10.2 CHEM Modeling

10.2.1 CHEM Structure

Figure 10.1a shows the typical diagram of a CHEM with coupled heat and electrical network that is comprised of CHP plants, wind turbines, photovoltaic cells, electrical energy storage (EES), and electric boilers. The electrical network is marked in blue and the DHN in red. The mathematical modeling for all the components in the electrical network can be found in our previous work [9]. The models of DHN and demand response management, which are among the main contributions of our proposed method are detailed in this section.

10.2.2 Modeling for Heat Network

10.2.2.1 District Heating Network Background

A DHN is a system that supplies residential and commercial heat energy at a community scale. The heat generated from sources is transmitted to customers through insulated pipelines. DHN transmission network is usually made up of a primary part and a secondary part, both of which are composed of supply pipelines as well as return pipelines (Figure 10.1b). Generated hot water from the source is firstly delivered through the supply pipelines in the primary network. Upon entering the

(a)

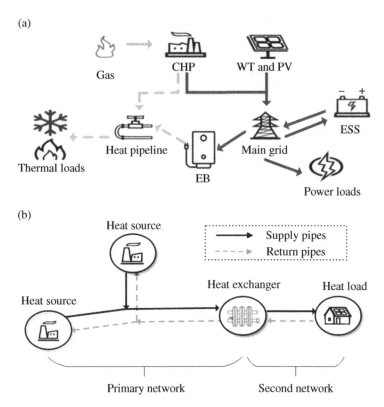

(b)

Figure 10.1 Schematic diagram of the CHEM and thermal network. (a) CHEM. (b) Thermal network.

heat exchanger located in the secondary network, the thermal energy in the generated hot water is delivered to consumers with a notable temperature drop of the hot water. Following that, the exchanged hot water flows back through the return pipelines, which completes the cycle. Note that only the primary network is modeled as secondary pipelines are much shorter. In this way, it can be ignored without losing engineering accuracy [16].

For the DHN, the mass flow rate and temperature of the water in the pipelines are responsible for supplying heat demands. Typically there are three working modes: (i) constant mass flow and variable temperature (CF-VT) strategy; (ii) constant temperature and variable mass flow (CT-VF) strategy; and (iii) variable mass flow and variable temperature (VF-VT) strategy. In the CF-VT strategy, it is assumed that the mass flow rates in the network remain constant while the temperature is adjusted to meet the flexible heat demands; in the CT-VF strategy, the temperature is assumed to be constant while the mass flow rate is regarded as variables to meet the time-varying heat loads; for the VF-VT strategy, both the mass flow rate and temperature can be changed to meet the heat demands.

The constant mass flow and variable temperature (CF-VT) strategy is generally used in the operation of DHNs, especially in northern China [31].

Other constraints for the DHN are formulated as follows [10].

10.2.2.2 Nodal Flow Balance

The water mass rate flowing into one heat node should be equal to that flowing out:

$$\sum_{k \in K_{n,s}^-} \omega_{s,k,t} = \sum_{k \in K_{n,s}^+} \omega_{s,k,t} \tag{10.1}$$

$$\sum_{k \in K_{n,r}^-} \omega_{r,k,t} = \sum_{k \in K_{n,r}^+} \omega_{r,k,t} \tag{10.2}$$

10.2.2.3 Calculation of Heat Energy

The heat power at pipeline starting/ending is defined as the multiplication of the specific heat, temperature, and water flow rate, which is expressed as:

$$H_{r,k,t}^+ = \omega_{r,k,t} \cdot C_T \cdot T_{r,k,t}^+ \cdot \beta_T \tag{10.3}$$

$$H_{r,k,t}^- = \omega_{r,k,t} \cdot C_T \cdot T_{r,k,t}^- \cdot \beta_T \tag{10.4}$$

$$H_{s,k,t}^+ = \omega_{s,k,t} \cdot C_T \cdot T_{s,k,t}^+ \cdot \beta_T \tag{10.5}$$

$$H_{s,k,t}^- = \omega_{s,k,t} \cdot C_T \cdot T_{s,k,t}^- \cdot \beta_T \tag{10.6}$$

10.2.2.4 Mixing Equation for Temperature

According to the first law of thermodynamics, the temperature of each node can be obtained by:

$$\sum_{k \in K_{n,s}^-} T_{s,k,t}^- \cdot \omega_{s,k,t} = T_{n,s}^m \sum_{k \in K_{n,s}^+} \omega_{s,k,t} \tag{10.7}$$

$$\sum_{k \in K_{n,r}^-} T_{r,k,t}^- \cdot \omega_{r,k,t} = T_{n,r}^m \sum_{k \in K_{n,r}^+} \omega_{r,k,t} \tag{10.8}$$

In addition, it is assumed that the mixing temperature at each node is equal to that at the starting point of the pipelines connected to the same node, expressed as:

$$T_{n,s}^m = T_{s,k,t}^+ \tag{10.9}$$

$$T_{n,r}^m = T_{r,k,t}^+ \tag{10.10}$$

10.2.2.5 Heat Dynamics and Loss

In this chapter, a node method [21] is used to model dynamic characteristics and transmission delay in DHN.

If the length and cross-sectional area of pipeline k are l_k and A, the total volume of water mass flowing in the pipeline is denoted as ρAl_k. As in Figure 10.2, cells in the horizontal direction are a sequence of the water mass in the pipeline. $\omega_{s,k,t}\Delta t$ denotes the water mass which flows into the supply pipeline. The blocks towards the right end are the water mass that has flown out of the pipeline by the end of period t. It is assumed that after $\psi_{k,t}$, the water mass starts to flow out. After $\phi_{k,t}$, the water mass of the cell flows out of the pipeline completely. $\psi_{k,t}$ and $\phi_{k,t}$ can be defined as (10.11) and (10.12). Besides $W_{k,t}$ and $V_{k,t}$ denote the total water masses flowing into the pipeline from the time $t - \psi_{k,t}$ to the time t and from the time $t - \varphi_{k,t}$ to the time t respectively. They can be denoted as (10.13) and (10.14).

$$\psi_{k,t} = \min_{n \in N} \left\{ n : s.t. \sum_{\omega = t-n}^{t} (\omega_{s,k,\omega} \cdot \Delta t) \geq \rho Al_k \right\} \tag{10.11}$$

$$\phi_{k,t} = \min_{m \in N} \left\{ m : s.t. \sum_{\chi = t-m}^{t} (\omega_{s,k,\chi} \cdot \Delta t) \geq \rho Al_k + (\omega_{s,k,t} \cdot \Delta t) \right\} \tag{10.12}$$

$$W_{k,t} = \sum_{k = t-\psi_{k,t}}^{t} (\omega_{s,k,t} \cdot \Delta t) \tag{10.13}$$

$$V_{k,t} = \begin{cases} \sum_{k = t-\phi_{k,t}+1}^{t} (\omega_{s,k,t} \cdot \Delta t) & \text{if } \phi_{k,t} \geq \psi_{k,t} \\ W_{k,t}, & \text{otherwise} \end{cases} \tag{10.14}$$

In the node method, two steps are needed to model the dynamic characteristics of the DHN. In the first step, the temperature loss of the fluid is temporarily not considered. Only the temperature at the end of each pipeline at time t is estimated, using the average temperature of the water masses in the shade (Figure 10.2) flowing out of the pipeline during period t.

$$T_{s,k,t}^{-}{}' = \sum_{\chi = t-\phi_{k,t}}^{t-\psi_{k,t}} \xi_{k,t,\chi} T_{s,k,\chi}^{+} \tag{10.15}$$

$$T_{r,k,t}^{-}{}' = \sum_{\chi = t-\phi_{k,t}}^{t-\psi_{k,t}} K_{k,t,\chi} T_{r,k,\chi}^{+} \tag{10.16}$$

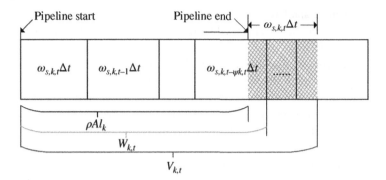

Figure 10.2 The schematic diagram of the pipeline section.

The variable $\xi_{k,\,t,\,\omega}$ in (10.15) and (10.16) is defined as:

$$\xi_{k,t,\chi} = \begin{cases} (\omega_{s,k,t} \cdot \Delta t - V_{k,t} + \rho A l_k)/(\omega_{s,k,t} \cdot \Delta t), & \chi = t - \phi_{k,t} \\ (\omega_{s,k,\chi} \cdot \Delta t)/(\omega_{s,k,t} \cdot \Delta t), & \omega = t - \phi_{k,t} + 1, ..., t - \psi_{k,t} - 1 \\ (W_{k,t} - \rho A l_k)/(\omega_{s,k,t} \cdot \Delta t), & \chi = t - \psi_{k,t} \\ 0, & \text{otherwise} \end{cases}$$

$$(10.17)$$

Note that the coefficients $\phi_{k,\,t}$, $\psi_{k,\,t}$ and $\xi_{k,\,t,\,\omega}$ are constants used in the CF-VT control strategy.

In the second step, considering the existence of the heat loss of the fluid, the temperature drop caused by the heat loss can be used to modify the eventual temperature at the outlet as:

$$T_{s,k,t}^{-} = T_{am,t} + B_{k,t}\left(\sum_{\chi = t - \psi_{k,t}}^{t - \psi_{k,t}} \xi_{k,t,\chi} T_{s,k,t}^{-}{}' - T_{am,t} \right) \tag{10.18}$$

$$T_{r,k,t}^{-} = T_{am,t} + B_{k,t}\left(\sum_{\chi = t - \psi_{k,t}}^{t - \psi_{k,t}} \xi_{k,t,\chi} T_{r,k,t}^{-}{}' - T_{am,t} \right) \tag{10.19}$$

The variable $J_{k,t}$ in Eq. (10.18)–(10.19) can be defined as follow:

$$B_{k,t} = \exp\left[-\frac{\lambda \Delta t}{A\rho C_T} \right]\left(\psi_{k,t} + \frac{1}{2} + \frac{V_{k,t} - W_{k,t}}{\omega_{s,k,t-\psi_{k,t}}\Delta t} \right) \tag{10.20}$$

10.2.3 Indoor Temperature Control

The heat which can be produced by the CHPs and electric boilers could enter the primary heat exchanger and to the supply pipelines as (10.21)–(10.22).

$$H_{s,k,t}^{+} - H_{r,k,t}^{-} = (H_{\mathrm{CHP},t} + H_{\mathrm{EB},t}) \cdot \eta_{\mathrm{hx}} \tag{10.21}$$

$$\omega_{s,k,t} = \omega_{r,k,t} \tag{10.22}$$

According to the ITC strategy, the heat required by the customers is closely related to the difference between the indoor and outdoor temperature [32].

The outdoor temperature could influence the indoor temperature through the thermal conduction at the building envelopes including roofs, walls, doors, and so on. The thermal conduction can be denoted as:

$$H_{b,t}^{T} = H_{s,b,t}^{T}/\tau_{T} = (T_{\mathrm{am},t} - T_{\mathrm{in},b,t})/R_{T} \tag{10.23}$$

Based on the above, a time-varying heat conduction model is shown in (10.24). It demonstrates that the indoor heat energy variations are the combined results of the heat conduction from the outdoor environment and the heat obtained from the DHN. Note that some factors such as customer behaviors and residence thermal comfort may also impact the indoor temperature at the building level. As this work focuses on the system level, the above human factors are neglected [26]. A complete indoor temperature model is shown in [29] and can be applied if needed. Equation (10.23) is then linearized to a state model as (10.25). Furthermore, heat demands $H_{b,t}^{\mathrm{load}}$ are designed to be the minimal required to keep the comfortable indoor temperature in (10.26). The relationship between heat loads and heat power flowing in the pipelines are in (10.27)–(10.28).

$$C_{\mathrm{air}}(dT_{\mathrm{in},b,t}/dt) = H_{\mathrm{ra},b,t} \cdot \eta_{\mathrm{ra}} + H_{b,t}^{T} \tag{10.24}$$

$$C_{\mathrm{air}}(T_{\mathrm{in},b,t} - T_{\mathrm{in},b,t-1})/\tau_{T} = H_{\mathrm{ra},b,t} \cdot \eta_{\mathrm{ra}} + H_{b,t}^{T} \tag{10.25}$$

$$H_{b,t}^{\mathrm{load}} = (T_{\mathrm{in},b,t}^{\mathrm{min}} - T_{\mathrm{am},t}) \Delta t/R_{T} \tag{10.26}$$

$$H_{s,k,t}^{-} - H_{s,k,t}^{+} = H_{b,t}^{\mathrm{load}} \tag{10.27}$$

$$\omega_{s,k,t} = \omega_{r,k,t} \tag{10.28}$$

10.2.4 Price-based Demand Response

PBDR has been widely applied to reshape the load profiles in the electricity market [33]. In PBDR management, the electricity price for the next day is made available to the consumers one day in advance. Consumers can adjust their electrical demand according to price information. The relationship between the electricity price $C_{\mathrm{pr},t}$ and the demands $P_{\mathrm{load},i,t}$ is modeled as:

$$P_{\mathrm{load},i,t} = \upsilon_{P} C_{\mathrm{pr},t}^{\varepsilon} \tag{10.29}$$

Table 10.1 The levels for the PBDR.

PBDR levels	Price rate (%)	Expected response rates (%)
1	70	107.9
2	80	104.8
3	90	102.3
4	100	100.0
5	110	98.0
6	120	96.2
7	130	94.6
8	140	93.1
9	150	91.8
10	160	90.5

According to [34], a power price elasticity of -0.2122 is selected in this chapter. In total, ten price level rates are predefined as shown in Table 10.1.

With the PBDR levels in Table.10.1, the power demands could respond accordingly. The actual demands with the PBDR strategy can be calculated by (10.30)–(10.31):

$$P_{\text{load},i,t} = \sum_{j \in J} \varepsilon_j L_j^{\text{le}} P_{\text{load},i,t}^o \tag{10.30}$$

$$Q_{\text{load},i,t} = \sum_{j \in J} \varepsilon_j L_j^{\text{le}} Q_{\text{load},i,t}^o \tag{10.31}$$

10.3 Coordinated Optimization of CHEM

Based on the comprehensive modeling of the coupled power and heat network, an optimal coordinated dispatch method for the CHEM operation is proposed in this section.

10.3.1 Objective Function

The objective is to minimize the daily operation cost C_{total}. while satisfying all constraints at both the component and the system level. The total cost includes the fuel cost of the CHP plants $F_{f,t}$, start-up/shut-down cost $F_{\text{st},t}/F_{\text{sd},t}$, operation &

maintenance (O&M) cost $F_{om,t}$, cost of energy exchange with the utility grid $F_{e,t}$, and the revenue from selling heat to consumers $F_{s,t}$

$$\text{MIN } F_{\text{total}} = \sum_{t=1}^{T} \left(F_{f,t} + F_{e,t} + F_{st,t} + F_{sd,t} + F_{om,t} - F_{s,t} \right) \Delta t \tag{10.32}$$

$$F_{f,t} = C_{\text{gas}} \cdot \left[(P_{\text{CHP},t} / \eta_{\text{CHP}}) \cdot \Delta t / L_{\text{HG}} \right] \tag{10.33}$$

$$F_{e,t} = C_{\text{pr},t} \cdot \left(P_{\text{gd},t} + P_{\text{sd},t} \right) \Delta t \tag{10.34}$$

$$F_{st,t} = \max \left\{ 0, U_{\text{CHP},t}^{i} - U_{\text{CHP},t-1}^{i} \right\} \cdot C_{\text{CHP}}^{st} \\ + \max \left\{ 0, U_{\text{EB},t}^{i} - U_{\text{EB},t-1}^{i} \right\} \cdot C_{\text{EB}}^{st} \tag{10.35}$$

$$F_{sd,t} = \max \left\{ 0, U_{\text{CHP},t-1}^{i} - U_{\text{CHP},t}^{i} \right\} \cdot C_{\text{CHP}}^{sd} \\ + \max \left\{ 0, U_{\text{EB},t-1}^{i} - U_{\text{EB},t}^{i} \right\} \cdot C_{\text{EB}}^{sd} \tag{10.36}$$

$$F_{s,t} = \sum_{b \in B} \left(C_{H,t} H_{b,t}^{\text{load}} \right) \cdot \Delta t \tag{10.37}$$

$$F_{om,t} = C_{u}^{\text{om}} \cdot \left(\sum_{i \in I} P_{u,t}^{i} \right) \Delta t \tag{10.38}$$

10.3.2 Operational Constraints

The operational constraints are formulated as:

$$P_{u}^{\text{MIN}} \leq P_{u,t}^{i} \leq P_{u}^{\text{MAX}} \tag{10.39}$$

$$-R_{u}^{\text{down}} \Delta t \leq P_{u,t}^{i} - P_{u,t-1}^{i} \leq R_{u}^{\text{up}} \Delta t \tag{10.40}$$

$$0 \leq P_{\text{EES},t}^{\text{dis},i} \leq U_{\text{EES},t}^{\text{dis},i} \cdot P_{\text{ESD}}^{\text{MAX}} \tag{10.41}$$

$$0 \leq P_{\text{EES},t}^{\text{ch},i} \leq U_{\text{EES},t}^{\text{ch},i} \cdot P_{\text{ESC}}^{\text{MAX}} \tag{10.42}$$

$$\mu_{\text{EES}}^{\text{MIN}} \cdot E_{\text{EES}}^{\text{cap}} \leq E_{\text{EES},t}^{i} \leq \mu_{\text{EES}}^{\text{MAX}} \cdot E_{\text{EES}}^{\text{cap}} \tag{10.43}$$

$$U_{\text{EES},t}^{\text{ch},i} + U_{\text{EES},t}^{\text{dis},i} \leq 1 \tag{10.44}$$

$$E_{\text{EES},0}^{i} = E_{\text{EES},T\Delta t}^{i} \tag{10.45}$$

$$P_{i+1,t} = P_{i,t} + P_{G,i,t} - P_{\text{load},i,t} \tag{10.46}$$

$$Q_{i+1,t} = Q_{i,t} + Q_{G,i,t} - Q_{\text{load},i,t} \tag{10.47}$$

$$V_{i+1,t} = V_{i,t} - \frac{r_i P_{i,t} + x_i Q_{i,t}}{V_s} \tag{10.48}$$

$$1 - \Delta V_{\text{max}} \leq V_{i,t} \leq 1 + \Delta V_{\text{max}} \tag{10.49}$$

$$P_{G,i,t} = \sum_{i \in I} P^i_{MT,t} + \sum_{i \in I} P^i_{WT,t} + \sum_{i \in I} P^i_{PV,t} + \sum_{i \in I} \left(P^{dis,i}_{EES,t} - P^{ch,i}_{EES,t} \right)$$
$$- \sum_{i \in I} P^i_{EB,t} \tag{10.50}$$

$$-P_{max} \leq P_{i,t} \leq P_{max} \tag{10.51}$$

$$-Q_{max} \leq Q_{i,t} \leq Q_{max} \tag{10.52}$$

$$T^{min}_s \leq T^+_{s,k,t} \leq T^{max}_s \tag{10.53}$$

$$T^{min}_s \leq T^-_{s,k,t} \leq T^{max}_s \tag{10.54}$$

$$T^{min}_r \leq T^+_{r,k,t} \leq T^{max}_r \tag{10.55}$$

$$T^{min}_r \leq T^-_{r,k,t} \leq T^{max}_r \tag{10.56}$$

$$T_{in,b,t} \geq T^{min}_{in,b,t} \tag{10.57}$$

$$T_{in,b,t} \leq T^{max}_{in,b,t} \tag{10.58}$$

Constrains (10.39) and (10.40) are the capacity limits and ramping limits for distributed generators and ancillary units. The charging/discharging power and energy limits of the ESS are shown as (10.41)–(10.44). Equation (10.45) means that the remaining energy in the ESS at the end of the day should be the same as the beginning. The linearized distribution power flow [34] is presented in (10.46)–(10.52). Note that at the first bus, power purchased from/sold to the utility grid $P_{gd,t}/P_{sd,t}$ is also considered. Constrains (10.53)–(10.56) show that the internal temperature should be kept within a range to protect the heat pipelines. The indoor temperature of each building should be maintained within a suitable range as (10.57)–(10.58) to ensure customers' thermal comfort level. Other constraints of the DHN are given in (10.1)–(10.28).

10.3.3 Solution Method

It is obvious that (10.35) and (10.36) are nonlinear which could result in considerable solution difficulties. To solve this, they can be linearized below:

$$\begin{cases} F_{st,t} \geq 0 \\ F_{st,t} \geq \left(U^i_{CHP,t} - U^i_{CHP,t-1} \right) \cdot C^{st}_{CHP} + \left(U^i_{EB,t} - U^i_{EB,t-1} \right) \cdot C^{st}_{EB} \end{cases} \tag{10.59}$$

$$\begin{cases} F_{sd,t} \geq 0 \\ F_{sd,t} \geq \left(U^i_{CHP,t-1} - U^i_{CHP,t} \right) \cdot C^{sd}_{CHP} + \left(U^i_{EB,t-1} - U^i_{EB,t} \right) \cdot C^{sd}_{EB} \end{cases} \tag{10.60}$$

Our model is a mixed-integer linear programming problem, which can be solved by some commercial solvers such as Cplex and Gurobi. The flowchart is given in Figure 10.3.

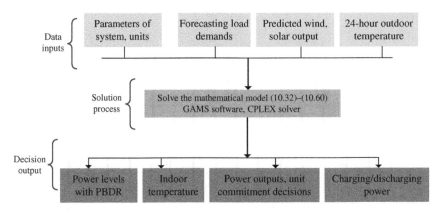

Figure 10.3 The flow chart of the proposed method.

10.4 Case Studies

10.4.1 Simulation Test Setup

Two systems are used to show the economic benefits of applying the proposed method. The first case is on the IEEE 33-bus radial network [26], where CHPs and EBs provide heat in a 13-pipeline DHN. The second is done on a larger scale CHEM with IEEE 69-bus radical system [35] and a 29-pipeline DHN to demonstrate the performance and scalability of our method,

10.4.1.1 33-bus CHEM [26]

The schematic diagram of the first system (IEEE 33-bus network with 13-pipeline DHN) is in Figure 10.4. V_s is set as 1.0 p.u. and ΔV_{max} is set to be $\pm 5\%$ of the nominal level. The unit dispatch time is 1 hour and the total time is 24 hours. It is assumed that every heat load node has 150 residential buildings.

Figure 10.5 shows the day-ahead power transaction prices. Note that data is collected from [9, 11, 22], which is then modified with practical considerations.

Parameters of the first system are shown in Tables 10.2 and 10.3. Table 10.4 shows the parameters of the generators and auxiliary units. The mass flow rate of each pipeline is in Table 10.5.

10.4.1.2 69-bus CHEM [35]

Figure 10.6 shows the second test system. The capacities of the components in the second case are the same as those in the first one. The configuration is summarized in Table 10.6.

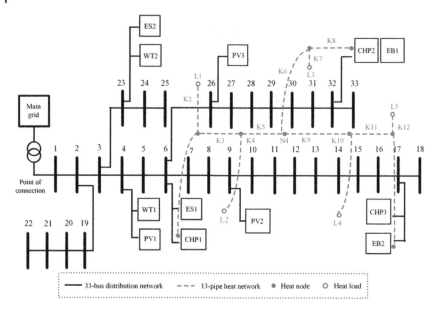

Figure 10.4 Structure of our IEEE-33 bus system. *Source:* Caracterdesign/E+/Getty Images and Caterpillar Energy Solutions GmbH.

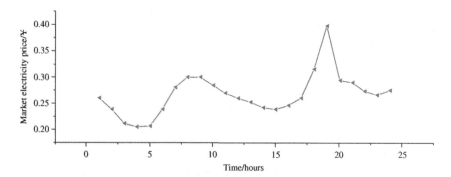

Figure 10.5 Power exchanging price.

Table 10.2 Parameters for the ESS.

Name	Value	Name	Value	Name	Value
μ_{EES}^{MIN}	0.15	μ_{EES}^{MAX}	1.0	C_{EES}^{om}	0.073 ¥/kW
E_{EES}^{cap}	800 kWh	$\eta_{ESC/ESD}$	0.8	α_{ES}	0.001
P_{ESC}^{MAX}	250 kW	P_{ESD}^{MAX}	250 kW	$E_{EES,\,0}$	400 kWh

Table 10.3 Parameters for the whole system.

Name	Value	Name	Value	Name	Value
ρ	1000 kg/m^3	C^{air}	0.53 kWh/°C	η_{hx}	0.9
η_{ra}	0.9	R_T	2.8°C/kW	C_H	0.1
$T_{in,b,t}^{max}$	22	$T_{in,b,t}^{min}$	18	η_{CHP}	0.3
η_h	0.85	η_L	0.15		

Table 10.4 Parameters of all the componentst.

Type	C_{om}	P^{max}	P^{min}	R^{down}	R^{up}	C^{st}	C^{sd}
CHP	0.0990¥/kW	1000 kW	200 kW	5 kW/min	10 kW/min	1.94¥	1.82¥
Photovoltaic cell	0.0133¥/kW	300 kW	0 kW	/	/	/	/
Wind turbine	0.0145¥/kW	350 kW	0 kW	/	/	/	/
Electric boiler	0.0089¥/kW	300 kW	0 kW	2 kW/min	3 kW/min	1.32¥	1.15¥

Table 10.5 Mass flow rate in the DHN (kg/h).

Pipeline	Mass flow	Pipeline	Mass flow	Pipeline	Mass flow
K1	2400	K6	1600	K10	800
K2	800	K7	2400	K11	1600
K3	1600	K8	800	K12	800
K4	800	K9	800	K13	2400
K5	800	/	/	/	/

10.4.2 Discussions on Simulation Results

In this section, three case studies are carried out where Case Study 1 and 2 are used as benchmarks to validate the effectiveness of the proposed method in Case Study 3:

1) *Case Study 1* uses the conventional method without considering the specific DHN models and demand response.
2) *Case Study 2* only considers the DHN constraints.
3) *Case Study 3* uses the proposed method.

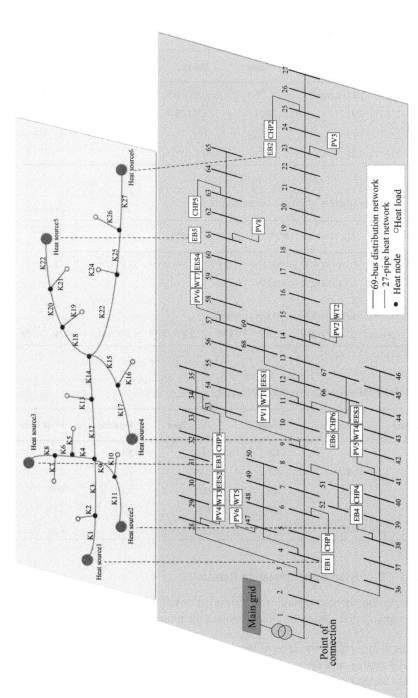

Figure 10.6 The structure of the second CHEM test system. *Source:* note_yn/Adobe Stock, Flavijus Piliponis/Adobe Stock, Rapheephat/Adobe Stock, and aldorado/Adobe Stock.

Table 10.6 Configuration of test systems.

		Test system	
		33-bus CHEM	69-bus CHEM
Electrical power network	Bus	33	69
	Line	32	68
	WT	5	6
	PV	8	8
	EES	3	4
DHN	Node	6	14
	Pipeline	13	29
	Load	5	10
	CHP	3	6
	EB	6	6

All case studies are conducted on both the 33-bus CHEM and the 69-bus CHEM to assess the scalability of the proposed method. The computer used is an Intel(R) Core i5-7200U, 2.50-GHz personal computer with an 8GB memory. The simulation is implemented in the General Algebraic Modelling System platform and then solved by the CPLEX solver.

10.4.2.1 33-bus CHEM
Load Leveling Effect
The obtained power demand profile is in Figure 10.7. It is shown that the original demand peak appears around 10:00–12:00 and 18:00–22:00, which correspond to

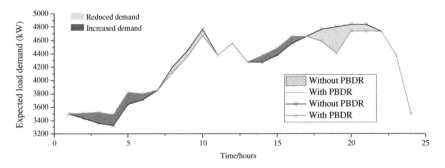

Figure 10.7 Power demands with/without PBDR strategy.

periods with a high probability of human activities. A load valley appears around 1:00–4:00 when human activities are at a minimum. With the PBDR strategy in play, during time periods of 1:00–6:00 and 14:00–16:00, the power demands are slightly increased, which is incentivized by the relatively low electricity prices. In contrast, during the periods of 8:00–10:00 and 18:00–21:00, the power demands are decreased due to the increased electricity prices.

Power Balance Condition

Figure 10.8a–c present the power balance conditions under three cases in the 33-bus CHEM. SOC stands for the state of charge of EES. The proposed method prioritizes the use of the internal generation components including CHP plants, wind turbines, and PV when the unit power generation cost is lower than the electricity price. On the other hand, when the internal generation components are not able to meet the high electricity demand during periods from 19:00–21:00, the proposed method purchases power from the main grid. The proposed method also dispatches the CHP plants to supply surplus power with the purpose of reaping more economic returns in addition to meeting the load demands. The proposed method also dispatched the ESS to store excess energy for use at times when there is a shortage of energy around the period 18:00–22:00. The above results show that the proposed method is effective in dispatching components based on various internal and external constraints.

Heat Balance Condition

Figure 10.9a shows the total heat energy generated by the CHP plants and electric boilers in the three case studies. It can be found that changes in the thermal outputs in Case Study 1 follow changes in the outdoor temperature instantly, which are different from those in Case Study 2 and 3. The reason is that when the specific DHN model is considered, the overall thermal behaviors of the system are subject to the thermal transmission delay, which is a more realistic reflection of reality. In view of this, the DHN can be treated as thermal energy storage. For example, during 1:00–5:00 in the morning, the required heat load is, in theory, high due to the low outdoor temperature, but the electric demand is low because there is minimum human activity. During this period, the heat pipelines, which serve somewhat like thermal energy storage decrease their internal temperature in order to satisfy the heat demands. As a result, the thermal outputs required from the CHPs can be significantly reduced. During 10:00–20:00, the electric demand gradually reaches its peak, and the CHPs generated more in order to meet the high demand. While at the same time, the extra

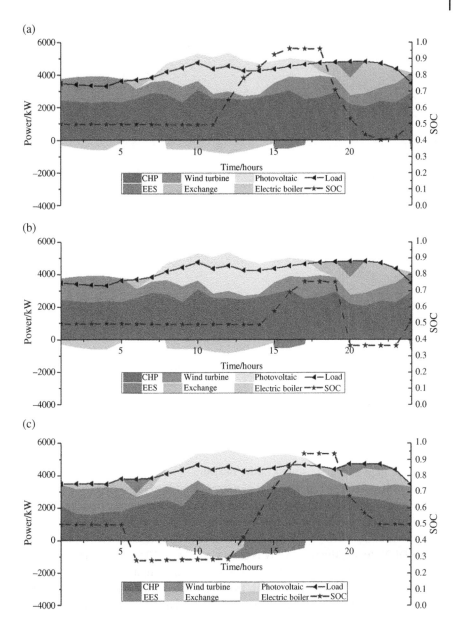

Figure 10.8 Energy balance in the 33-bus CHEM. (a) Case 1. (b) Case 2. (c) Case 3.

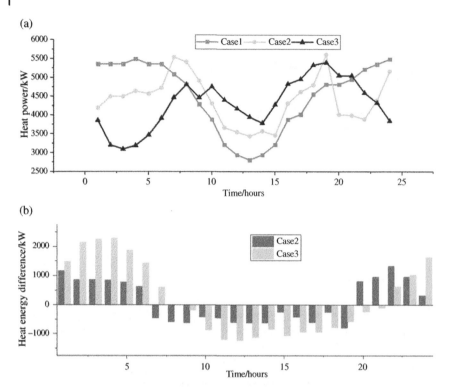

Figure 10.9 Results of Heat generation and consumption. (a) Comparisons of heat and power generation among Case 1–3. (b) Storage state of the DHN.

generated heat gets to be stored in the heat pipelines, which results in a higher internal temperature of the thermal pipelines. That is to say, in Case Study 2 and 3, due to the modeling of the DHN in terms of its thermal inertia, and the DHN is practically treated as a form of thermal energy storage, which can supplement the operations of the CHP plants and electric boilers. The status of the stored energy of the DHN during the whole day is shown in Figure 10.9b. Furthermore, compared with Case 2, the heat output in Case 3 demonstrated more flexible adjustments which follow the variations in the electric demands (Figure 10.7), largely owing to the added ITC strategy. From the above analysis, we can safely conclude that the heat network and electrical network are intrinsically coupled. And the heat network itself has the potential to act as a form of thermal energy storage due to its inertia. This, in practice and if well-coordinated, has a great potential to enhance the overall efficiency.

Table 10.7 The total system operation cost (¥).

Items	Case study 1	Case study 2	Case study 3
Fuel cost	18164.34	17656.67	17042.57
O&M cost	6780.94	6713.23	6290.99
Start-up/Shut-down cost	10.90	12.50	12.50
Power exchange cost	155.52	173.55	161.68
Heat revenue	10861.60	10861.60	10861.60
Net operating cost	14250.10	13694.35	12646.14

Net Operating Cost

The net operating costs of the three case studies are compared in Table 10.7. Compared with Case Study 1, the total operation cost in Case Study 2 is lower, which is due to reduced transmission loss in the DHN. In addition, the transmission delay of the DHN decouples the heat generation and consumption, which contributes to a higher energy utilization efficiency and lower net operating cost. The net operating cost in Case Study 3 is 12646.14¥, which gives a saving rate of 11.2%. The results show that the peak power loads are shifted and heat loads reshaped due to PBDR and ITC. The demand response management proves to significantly contribute to economic savings for CHEM.

10.4.2.2 69-bus CHEM

Figure 10.10 shows the generation and consumption of different components in the 69-bus CHEM for 3 case studies. The net operating cost is presented in Table 10.8. It is observed that in Case Study 3 using the proposed method, the electric peak load has been shifted, which is attributed to the implementation of PBDR. As a result, the amount of power required from the CHPs can be reduced. The proposed method achieves the lowest operating cost, which is 9.3% lower than that in Case Study 1. The simulation results validate that the proposed method enables greater dispatch flexibility and reduces total operating costs in a larger system.

It is worth pointing out that the IEEE 33-bus distribution system and 69-bus system are both standard benchmark test systems. They have been widely used in similar studies. The number of variables in the formulated optimization problem is more than 10 000, whereas the solution process takes only 5.67 seconds for the 33-bus system and 20.34 seconds for the 69-bus system. It is safe to say that the proposed method can be further scaled up to suit larger systems, considering the fast computational speed.

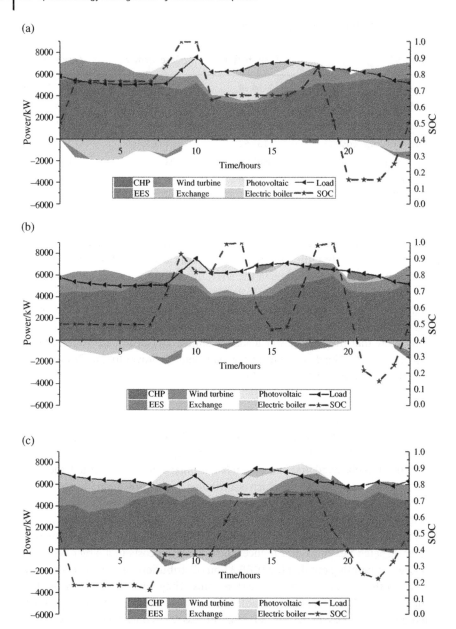

Figure 10.10 Energy balance in the 69-bus CHEM. (a) Case 1. (b) Case 2. (c) Case 3.

Table 10.8 Total system operation cost (¥).

Items	Case 1	Case 2	Case 3
Total cost	25335.82	24497.49	23151.80

10.4.3 Conclusion

In this chapter, an optimal coordinated scheduling method of a CHEM with the demand response is studied, considering practical and comprehensive modeling of the thermal network. The structure and dynamic properties of the heat network are modeled, and demand response is integrated into our proposed method. Three case studies are performed on two systems. Simulation results show that:

1) Operation coordination of the power and heat networks can introduce a greater extent of scheduling flexibility and reduce the overall operating cost.
2) The DHN can be utilized as heat storage owing to its intrinsic transmission delay and high inertia, enabling its participation in the dispatch of CHEM.
3) The PBDR and ITC strategy can effectively flexibilize the CHEM operation and contribute to further operating cost savings.
4) The proposed method can be efficiently solved within a reasonable amount of time which makes it suitable for online applications.

Future work may include modeling the randomness of renewable generations, more detailed thermal comfort levels, and user behaviors at the residential level as well as using more advanced solution algorithms for an even higher computational efficiency.

References

1 Guelpa, E., Bischi, A., Verda, V. et al. (2019). Towards future infrastructures for sustainable multi-energy systems: a review. *Energy* 184: 2–21.
2 Carrasco, J.M., Franquelo, L.G., Bialasiewicz, J.T. et al. (2006). Power-electronic systems for the grid integration of renewable energy sources: a survey. *IEEE Trans. Ind. Electron.* 53 (4): 1002–1016.
3 Quelhas, A., Gil, E., McCalley, J.D. et al. (2007). A multiperiod generalized network flow model of the US integrated energy system: Part I—model description. *IEEE Trans. Power Syst.* 22 (2): 829–836.

4 Quelhas, A. and McCalley, J.D. (2007). A multiperiod generalized network flow model of the US integrated energy system: Part II-simulation results. *IEEE Trans. Power Syst.* 22 (2): 837–844.

5 Chen, C., Duan, S., Cai, T. et al. (2011). Smart energy management system for optimal microgrid economic operation. *IET Renew. Power Gener.* 5 (3): 258–267.

6 Vasebi, A., Fesanghary, M., and Bathaee, S.M.T. (2007). Combined heat and power economic dispatch by harmony search algorithm. *Int. J. Electr. Power Energy Syst.* 29 (10): 713–719.

7 Smith, A.D. and Mago, P.J. (2014). Effects of load-following operational methods on combined heat and power system efficiency. *Appl. Energy* 115: 337–351.

8 Lujano-Rojas, J.M., Dufo-López, R., Bernal-Agustín, J.L. et al. (2017). Optimizing daily operation of battery energy storage systems under real-time pricing schemes. *IEEE Trans. Smart Grid.* 8 (1): 316–330.

9 Li, Z. and Xu, Y. (2018). Optimal coordinated energy dispatch of a multi-energy microgrid in grid-connected and islanded modes. *Appl. Energy* 210: 974–986.

10 Wang, L., Li, Q., Ding, R. et al. (2017). Integrated scheduling of energy supply and demand in microgrids under uncertainty: a robust multi-objective optimization approach. *Energy* 130: 1–14.

11 Kanchev, H., Colas, F., Lazarov, V. et al. (2014). Emission reduction and economical optimization of an urban microgrid operation including dispatched EB-based active generators. *IEEE Trans. Sustain. Energ.* 5 (4): 1397–1405.

12 Yan, B., Luh, P.B., Warner, G. et al. (2017). Operation and design optimization of microgrids with renewables. *IEEE Trans. Autom. Sci. Eng.* 14 (2): 573–585.

13 Zhang, C., Xu, Y., Li, Z. et al. (2018). Robustly coordinated operation of a multi-energy microgrid with flexible electric and thermal loads. *IEEE Trans. Smart Grid.* 10 (3): 2765–2775.

14 Li, Z., Xu, Y., Fang, S. et al. (2019). Optimal placement of heterogeneous distributed generators in a grid-connected multi-energy microgrid under uncertainties. *IET Renew. Power Gener.* 13 (14): 2623–2633.

15 Li, Z., Xu, Y., Feng, X. et al. (2020). Optimal stochastic deployment of heterogeneous energy storage in a residential multienergy microgrid with demand-side management. *IEEE Trans. Industr. Inform.* 17 (2): 991–1004.

16 Li, Z., Wu, L., Xu, Y. et al. (2021). Multi-stage real-time operation of a multi-energy microgrid with electrical and thermal energy storage assets: a data-driven MPC-ADP approach. *IEEE Trans. Smart Grid.* 13 (1): 213–226.

17 Basu, A.K., Bhattacharya, A., Chowdhury, S. et al. (2011). Planned scheduling for economic power-sharing in a CHP-based micro-grid. *IEEE Trans. Power Syst.* 27 (1): 30–38.

18 Sdringola, P., Proietti, S., Astolfi, D. et al. (2018). Combined heat and power plant and district heating and cooling network: a test-case in Italy with integration of renewable energy. *J. Sol. Energy Trans. ASME* 140 (5): 054502.

19 Zheng, J., Zhou, Z., Zhao, J. et al. (2018). Integrated heat and power dispatch truly utilizing thermal inertia of district heating network for wind power integration. *Appl. Energy* 211: 865–874.

20 Li, Z., Wu, L., and Xu, Y. (2021). Risk-averse coordinated operation of a multi-energy microgrid considering voltage/var control and thermal flow: an adaptive stochastic approach. *IEEE Trans. Smart Grid.* 12 (5): 3914–3927.

21 Gu, W., Wang, J., Lu, S. et al. (2017). Optimal operation for integrated energy system considering thermal inertia of district heating network and buildings. *Appl. Energy* 199: 234–246.

22 Balijepalli, V.S.K.M., Pradhan, V., Khaparde, S.A. et al. (2011). Review of demand response under smart grid paradigm/ISGT2011-India. *IEEE*: 236–243.

23 Amini, M.H., Talari, S., Arasteh, H. et al. (2019). Demand response in future power networks: panorama and state-of-the-art. In: *Sustainable Interdependent Networks II: From Smart Power Grids to Intelligent Transportation Networks* (ed. M.H. Amini, K.G. Boroojeni, S.S. Iyengar, et al.), 167–191. Springer.

24 Chen, Z., Wu, L., Fu, Y. et al. (2012). Real-time price-based demand response management for residential appliances via stochastic optimization and robust optimization. *IEEE Trans. Smart Grid.* 3 (4): 1822–1831.

25 Brahman, F., Honarmand, M., and Jadid, S. (2015). Optimal electrical and thermal energy management of a residential energy hub, integrating demand response and energy storage system. *Energy Build.* 90: 65–75.

26 Li, Z. and Xu, Y. (2019). Temporally-coordinated optimal operation of a multi-energy microgrid under diverse uncertainties. *Appl. Energy* 240: 719–729.

27 Bahrami, S., Toulabi, M., Ranjbar, S. et al. (2018). A decentralized energy management framework for energy hubs in dynamic pricing markets. *IEEE Trans. Smart Grid.* 9 (6): 6780–6792.

28 Zhang, C., Xu, Y., Dong, Z.Y., and Wong, K.P. (2018). Robust coordination of distributed generation and price-based demand response in microgrids. *IEEE Trans. Smart Grid.* 9 (5): 4236–4247. https://doi.org/10.1109/TSG.2017.2653198.

29 Mohandes, B., El Moursi, M.S., Hatziargyriou, N.D. et al. (2020). Incentive based demand response program for power system flexibility enhancement. *IEEE Trans. Smart Grid.* 12 (3): 2212–2223.

30 Wang, Y., Huang, Y., Wang, Y. et al. (2018). Energy management of smart micro-grid with response loads and distributed generation considering demand response. *J. Clean. Prod.* 197: 1069–1083.

31 He, P., Sun, G., Wang, F. et al. (2009). *District Heating Engineering*. Beijing, China: Architecture & Building Press.

32 Anvari-Moghaddam, A., Monsef, H., and Rahimi-Kian, A. (2015). Optimal smart home energy management considering energy saving and a comfortable lifestyle. *IEEE Trans. Smart Grid.* 6 (1): 324–332.

33 Mohajeryami, S., Moghaddam, I.N., Doostan, M. et al. (2016). A novel economic model for price-based demand response. *Electr. Power Syst. Res.* 135: 1–9.

34 Levihn, F. (2017). CHP and heat pumps to balance renewable power production: lessons from the district heating network in Stockholm. *Energy* 137: 670–678.

35 Jafari, A., Ganjehlou, H.G., Khalili, T. et al. (2020). A two-loop hybrid method for optimal placement and scheduling of switched capacitors in distribution networks. *IEEE Access* 8: 38892–38906.

11

Optimal Operation of Residential Heating Systems in Electricity Markets Leveraging Joint Power-Heat Flexibility

Hessam Golmohamadi

Department of Computer Science, Aalborg University, Aalborg, Denmark

11.1 Why Joint Heat-Power Flexibility?

In the last decade, renewable electricity generation, e.g. wind and solar, has increased from 4204 TWh in 2010 to 7486 TWh in 2020. At the same time, the CO_2 production and the average temperature of the earth increased. Increasing the environmental concerns, many countries have decided to accelerate capital investments in harvesting renewable energies and decreasing the dependency on fossil fuels. Increasing the penetration of renewable energies, the fluctuation of power grids increases. To hedge against intermittent power, demand-side flexibility is a workable solution. Demand flexibility is defined as the change in energy consumption of consumers in response to an external stimulus, e.g. energy price and financial incentives. The demand flexibility is applicable in all demand sectors, including residential, e.g. intelligent home management systems, commercial, e.g. supermarket refrigerators, agricultural, e.g. farms irrigation pumps, and industrial sectors, e.g. oil refinery plants [1].

In the residential sector, the smart home energy management systems are scheduled not only to provide flexibility for the supply side but also to reduce household energy consumption costs. Energy consumption of household appliances, including heat ventilation and air conditioning, wet appliances, and lighting systems, can be optimized in response to the flexibility requirement of the energy markets [2]. In the commercial sector, the parking lot aggregators use the flexibility potential of the parked electric vehicles to meet the power needs of shopping stores when the electricity price is high and/or the power system reliability is jeopardized [3]. In the agricultural sector, the on-farm solar sites supply the power consumption of the water irrigation system at midday hours when the peak generation of

Coordinated Operation and Planning of Modern Heat and Electricity Incorporated Networks,
First Edition. Edited by Mohammadreza Daneshvar, Behnam Mohammadi-Ivatloo, and Kazem Zare.
© 2023 The Institute of Electrical and Electronics Engineers, Inc.
Published 2023 by John Wiley & Sons, Inc.

the solar panels coincides perfectly with the high evapotranspiration of crops [4]. In this way, not only peak-shaving is provided for the supply-side but also the remote farms are supplied by low electricity prices during critical hours. Regarding the industrial sector, the industrial operations of factories, e.g. cement manufacturing plants [5], can be optimized to integrate power flexibility of the industrial processes into the upstream network. Figure 11.1 explains a schematic diagram about the flexibility potentials of different demand sectors in electricity markets.

By increasing the renewable power penetration, the share of electricity is increasing in heating systems. In this way, electrically operated heat pumps and district heating (DH) are alternative solutions to overcome the fluctuation of renewable energies. The electrical heating systems can facilitate the integration of renewables into buildings and decrease the fossil fuel dependency for heating purposes. To address the flexibility of heating systems in energy markets with high renewable power penetration, three major actions should be taken as follows:

1) Extracting the flexibility potentials of heating systems.
2) Aggregating the flexibility potentials of DH.
3) Integrating the flexibility potentials into energy systems.

To address the first issue, expert knowledge about the potentials of joint power-heat flexibility is required. In this way, two challenges are raised: (i) how to extract the flexibility potentials of heating systems for space heating and domestic hot water (DHW) consumption without exceeding the residents' comfort, and (ii) how to integrate the power-heat flexibility into the energy market. Considering the second issue, demand response aggregators are suggested in recent studies. The aggregators are intermediary entities between the energy markets and the heat consumers, to unlock the flexibility potentials of heating systems. Regarding the third issue, the structure of energy markets is modeled from financial and technical viewpoints. The flexibility potentials of heating systems are classified based on the flexibility requirements of the energy systems. It is discussed how the flexibility of the heating system is unlocked based on the length of advance notices on the day-ahead, adjustment, and near real-time markets.

This chapter aims to discuss the first and third issues. To achieve the aim, first, Section 11.2 illustrates the literature review on residential heating systems. In Section 11.3, the main opportunities and challenges of intelligent heating systems are explained. Section 11.4 describes the flexibility potentials of heating systems. In Section 11.5, different generations of heat controllers are explained from the flexibility point of view. In Section 11.6, the mathematical structures of thermal dynamics are stated for buildings with multi-temperature zones. In this section, differential equations are described to show how indoor air temperatures are affected by energy consumption and weather data. In Section 11.7, an economic heat controller is simulated to reduce the cost of buildings energy consumption

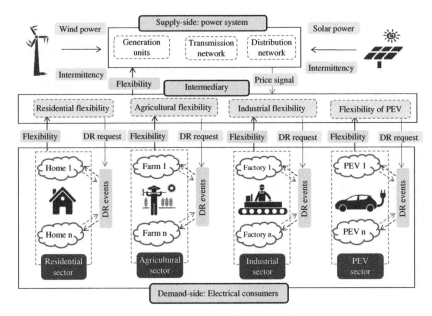

Figure 11.1 Demand-side flexibility in different sectors (PEV: plug-in electric vehicle). *Source:* Golmohamadi [6]/Elsevier/Public Domain CC BY 4.0.

in an electricity market with dynamic prices. In Section 11.8, the heat controller participates in a hierarchical electricity market to provide flexibility for the electricity market on long, mid and short advance notices from 24 hours ahead until near real-time. Section 11.9 addresses a heat controller for the detached houses connected to district heating. This controller aims to optimize the operation of mass valves in the mixing loop not only to reduce the power-heat energy cost but also to provide demand flexibility for the heat network. Finally, Section 11.10 concludes the discussed studies. Also, the important achievements of the suggested approaches are pointed out.

11.2 Literature Review

In the literature, many research and industrial studies are carried out to unlock the flexibility potentials of building heating systems. In [7], a predictive controller is suggested for Norwegian residential buildings to unlock the flexibility of air-source heat pumps. The controller provides demand response on dynamic energy prices. The Model Predictive Control (MPC) is suggested in different buildings with energy storage tanks and photovoltaic panels to reduce energy consumption costs [8].

The study is conducted in three regions and confirms 34%, 54%, and 90% of energy consumption reduction in Helsinki, Strasbourg, and Athens, respectively. In [9], automated demand response services are addressed to estimate the flexibility functions of heat pump systems in real operations. The proposed approach is examined under different ancillary service markets, including Spain, Germany, and Swiss. Bidding strategies are suggested for joint heat-electricity demands of intelligent buildings to decrease energy consumption costs [10]. The simulation results reveal that the joint heat-electricity bidding causes a 2.4% reduction in total energy consumption cost in comparison to independent heat and electricity control. Hybrid heating systems are addressed in the research study [11] including air-source heat pumps and backup heaters as electric resistance and gas boilers. Dynamic simulations are conducted to optimize the seasonal performance of the hybrid heating system in cold seasons. In [12], a novel approach is suggested for low-temperature DH systems to control the mass flow of apartment buildings. The study concludes that the traditional heat networks should change to ring networks to facilitate optimized mass flow control. Demand response programs are investigated for single buildings equipped with heat pumps and thermal storage [13]. The simulation results show that the energy flexibility depends mainly on the capacity of the thermal storage and operation control. In this way, the specific costs of demand response can be reduced between 45% and 75%. Variable speed control of heat pumps is proposed in the research study [14] to unlock the flexibility potentials of buildings. In this study, a real heat pump with a semi-virtual building is addressed to examine the proficiency of the control approaches. In [15], a bi-level MPC approach is addressed for building space heating demands in integrated heat-power energy systems. The results confirm that the bi-level approach not only minimizes the energy consumption cost of the buildings but also maximizes the profit of the integrated energy systems simultaneously. Different control approaches are examined in the research study [16] to minimize the total energy consumption of hybrid ground-coupled heat pump. The control approaches include dynamic programming, non-linear MPC, and linear control. An economic MPC is proposed to coordinate the operation of electric and heating demands in smart buildings [17]. The suggested MPC aims to achieve nearly zero energy consumption in near-real-time and actively participate in demand response programs. In [18], the MPC is discussed in a Swiss office building to optimize the energy consumption and comfort level of occupants. The study states that the rising energy prices and the importance of energy flexibility push the technology to use more sophisticated and advanced heat controllers. Optimal control strategies are discussed for an office building equipped with heat pump, electric resistor, heat storage, i.e. water tanks, phase change material, and thermochemical storage [19]. The study introduces a power flexibility indicator to calculate flexibility potentials of the building heating system. In [20], energy cost analysis is carried out for smart buildings supplied by district heating. Also, advanced control algorithms are suggested. The results show that the novel control approach improves the control accuracy, thermal comfort, and energy cost up to 7.6%, 6.7%, and 24.7%, respectively. The impacts of thermal storage capacity and heat pump power on the demand response potentials of buildings are investigated [21]. Moreover, market incentives and concepts are suggested to unlock the flexibility of the air-to-water heat pump in response to dynamic electricity prices. In [22], a novel dual-source building energy system is discussed to increase reliability and decrease the investment and operation costs of energy systems. The heating system is comprised of water and air source heat pumps and energy storage tanks. The results show that the suggested approach provides peak load shifting for the supply network and up to 10% cost saving for the building.

Based on the literature and to the best of the authors' knowledge, most research studies exploit the flexibility potentials of the heating systems in response to variable prices of single electricity markets. Besides, barely the electricity market uncertainties are addressed to investigate the impact of energy price uncertainty on heat control strategies. Moreover, most studies have concentrated on heat pump controls, while the mixing loop controls of DH are barely discussed. To narrow these gaps, the current study, first of all, optimizes the control strategies of heat pumps in three trading floors of electricity markets with price uncertainty. Stochastic programming and scenario generation approaches are used to cover the electricity price uncertainties. Then, an economic controller is addressed to control the mass flow of mixing valves in residential buildings connected to DH. All in all, the main contributions of the current study can be pointed out as follows:

1) Exploiting potentials of residential heating systems equipped with thermal storage and heat controller.
2) Unlocking flexibility of building heating systems in the day-ahead, intraday, and balancing markets in response to short, mid, and long advance notices.
3) Optimizing the operation of building heating systems supplied by mixing loop of low-temperature district heating.

11.3 Intelligent Heating Systems

Traditional heat controllers keep a balance between the heat demand and supply. In this structure, the network constraints are not addressed. In contrast, the advanced heat controllers incorporate the energy network constraints into the problem. Therefore, the advanced controllers not only supply the heat demand of the buildings but also respond to demand response requests. The demand response programs are scheduled to provide flexibility for the supply network with the purposes of power regulation, enhanced reliability, peak-shaving/valley-filling, frequency control, and voltage balance. For a single-carrier network, the flexibility potentials of the heating system affect the heat network merely, e.g. the DH with fossil fuel heat sources. In contrast, in a multi-carrier network, both the power and heat networks benefit from the flexibility of the intelligent heating systems. To exemplify, the electrically operated heat pumps provide energy flexibility for both the heat and power networks.

These days, the share of heat pumps is increasing in energy systems worldwide. Therefore, the interdependency of heat and power networks is increasing significantly. In this situation, the coordinated operation of the heat and power networks provides cost savings and enhanced reliability for multi-carrier energy systems. To provide a general overview, the main benefits of smart heating systems for power and heat networks can be stated as follows (but not necessarily limited to):

1) Power network
   1.1 Power regulation
   1.2 Voltage balance
   1.3 Frequency control
   1.4 Enhanced power quality/security/reliability
   1.5 Renewable energy integration
   1.6 Power congestion management
   1.7 Power loss reduction

2) Heat network
   2.1 Heat loss reduction (increase efficiency)
   2.2 Reduction in investment cost
   2.3 Facilitating the integration of low-temperature DH
   2.4 Heat waste recovery, e.g. data centers, industrial waste, sewage water.

Although the coordinated operation of multi-carrier networks would benefit the consumers and energy markets, there are still some challenges in establishing intelligent heating systems. Many consumers prefer not to participate in demand response programs to protect their comfort level. In the energy market, more pricing mechanisms and incentive schemes are required to encourage flexible consumers to respond to demand response requests. From the data and IoT viewpoints, big data should be collected by sensors/meters. The collected data should be stored and processed by demand response providers in the power and heat networks. Installing sensors, energy meters, and communication equipment need capital investment. In addition, many consumers are reluctant to reveal their private energy data, e.g. indoor temperature. Consequently, there are many challenges for using the household energy data by third parties, e.g. demand response providers. Although the challenges are not limited to the abovementioned facts, the major challenges can be categorized as follows:

1) Energy markets: energy pricing for responsive consumers.
2) Consumers' behavior: widespread adoption of demand response programs.
3) Data and IoT: data collection, communication, processing.
4) Regulatory: lack of regulation for flexible consumers, tax issues.

In many countries with high wind power penetration, e.g. Germany and Denmark, the Power-to-X (PtX) model is changing the energy system structures. The PtX makes it possible to convert the surplus of renewable energies into alternative energy resources, e.g. hydrogen, methanol, and diesel, that can be utilized, transported, stored, and reconverted to electricity. In this structure, the heat network is a key sector to counterbalance the fluctuations of the renewable energy supply. For example, during high wind power hours, the heat networks store the energy in the thermal storages, e.g. borehole, and water tanks. In contrast, the heat networks are supplied by thermal storage during wind lull hours. The smart heat controller makes it possible to operate the heating system in a multi-carrier energy system and especially in the PtX structure. The increased interdependency of energy systems, i.e. gas, power, and heat, push heating technologies toward smart heat controllers.

11.4 Flexibility Potentials of Heating Systems

The flexibility opportunities of heating systems can be stated as three categories:

1) Thermal dynamics of buildings.
2) Heat carrier of DH.
3) Heat storage of buildings/DH.

The flexibility of thermal dynamics is reflected in the heat capacity and heat resistance of the buildings' air, envelopes, windows, and materials. To extract the flexibility potentials of thermal dynamics, a comfort temperature interval is defined by occupants. The interval is comprised of three items including (i) preferred temperature setpoint, (ii) lower temperature threshold, and (iii) upper-temperature threshold. The flexibility potentials stem from the variations of the indoor air temperature within the comfort bound. Decreasing the gap between the lower and upper thresholds, the flexibility potentials of the thermal dynamic decrease. For the strict temperature setpoint, e.g. constant 21 °C, barely any flexibility is expected. In contrast, for the comfort bound, e.g. between

20 and 23 °C, the heating system can be turned off/down (on/up) during high (low) price hours when the power market encounters a deficit (excess) of electric power.

The second type of flexibility potentials is reflected in the thermal behavior of the heat carrier. The flexibility potentials of heat carriers are addressed in the whole and half heating pipes. For the former, both the forward and return pipes are studied. In the latter, only the forward pipes are considered for flexibility management. In the recent generations of DH systems, low-temperature DH systems are highly motivated. In this structure, the outflow temperature decreases to increase the system efficiency. Besides, it decreases the heat loss and investment cost on the heating systems. This is the main objective of mixing loop controllers to restore the heat energy of the return pipes.

In contrast to the previous heat flexibilities, the third heat flexibility is specifically designed to provide flexibility for the heat consumers and suppliers. The thermal storage can be in different forms, e.g. water tanks, chemical storage, aquifer, borehole, and Phase Change Materials (PCM). The heat storage can provide peak-shaving and/or valley-filling for joint heat-power consumption. Also, they can provide flexibility from near real-time until weekly and seasonal flexibility depending on the type, size, and technology.

As mentioned above, different flexibility potentials can be extracted from heating systems. In order to unlock heat flexibility, expert knowledge, as well as heat controllers, are required. The heat controllers can be installed on the demand side, i.e. apartment blocks, or on the supply side, e.g. heat networks. In the next section, different types of heat controllers are surveyed. Figure 11.2 provides a general overview of flexibility potentials in heating systems.

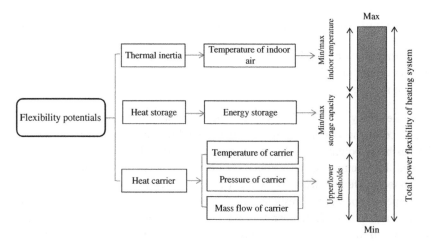

Figure 11.2 Different types of flexibility potentials in heating systems. *Source:* Golmohamadi [6]/Elsevier/Public Domain CC BY 4.0.

11.5 Heat Controllers

Heat controllers are the critical part of heating systems to unlock power-heat flexibility. Advanced heat controllers are an intermediary application tool between the supply and demand sides. On the supply side, the heat controllers have one-/two-way communication with the energy market, heat service providers, and meteorological offices. They receive the data of energy prices, weather forecasts, and flexibility requirements from the supply side and send the energy consumption data to the heat service provider. On the demand side, the heat controller accepts the input from occupants, e.g. temperature setpoints and upper/lower temperature, to meet residents' comfort. Therefore, the controller aims to meet the heating demands of the households while providing flexibility for the heat and power networks. Due to the interdependency of heat consumption and weather conditions, the predictive controllers receive important weather data from meteorological offices.

Generally, heat controllers adjust four variables including:

1) Heat demand on buildings side.
2) Supply temperature at DH substation.
3) Mass flow of the mixing loop/heat network.
4) Differential pressure at DH substations.

From the flexibility point of view, the heat controllers are classified into classic (inflexible) and advanced (flexible) controllers. The classic controllers keep a balance between heat demand and supply without considering the technical requirements of the power/heat networks. Adversely, advanced controllers are developed with the purpose of energy flexibility and efficiency. The advanced controllers optimize the exploitation of the heating system in response to power availability on the supply side and/or technical problems, e.g. unforeseen failures on the heat distribution network. Meanwhile, the predictive controllers are in communication with meteorological offices to optimize the exploitation of the heating systems based on weather data. The economic heat controllers optimize the household heat consumption in response to dynamic electricity prices. In electricity markets with high renewable power penetration, there is a correlation between the electricity price and renewable power availability. Therefore, when the power system encounters a power shortage (excess), the high (low) electricity prices motivate the heat controllers to decrease (increase) energy consumption near the lower (upper) comfort temperature. Consequently, not only heat-to-power flexibility is integrated into the power system but also the household energy consumption cost decreases.

The heat controllers can be centralized or distributed. The centralized controllers are normally located on the DH side adjusting supply temperature and differential pressure. The distributed controllers are installed on the building side to adjust heat demand and mass flow. The heat controller may access offline weather forecasts or use the online weather data. In the former, the operation of the heating system is preoptimized based on the forecasting weather data. In the latter, the model predictive control (MPC) approach is normally applied to re-optimize the operation of the heating system in response to the updated weather data. Moreover, they can address the energy price in the heating strategies. This type of controller is called economic MPC (EMPC). To sum up, Figure 11.3 describes a schematic diagram of a heat controller to supply the space heating and DHW consumption.

11.6 Thermal Dynamics of Buildings

The heat controllers optimize the energy equations of the buildings. In order to present general thermal dynamics for buildings, the mathematical formulations for thermal dynamics with multi-temperature zones are explained. Let consider a residential building with R rooms (R temperature zones) in Figure 11.4. Suppose that room r has common envelopes with rooms $r' \in \{1,2,...,R\}$ and outdoor space $a \in \{1, 2, ..., A\}$. Then, the thermal dynamics of room r is expressed as the three following differential equations [23]:

$$C_i^r \times \frac{dT_i^r}{dt} = \left(\frac{1}{R_{ih}^r} \times \left(T_h^r - T_i^r \right) + \frac{1}{R_{ie}^r} \times \left(T_e^r - T_i^r \right) + \left(A_w^r \times P_{\text{Sun}}^r \right) \right)$$

$$\quad (11.1)$$

$$C_e^r \times \frac{dT_e^r}{dt} = \left(\frac{\left(T_i^r - T_e^r \right)}{R_{ie}^r} + \sum_{\substack{r'=1 \\ r' \neq r}}^{R} \frac{\left(T_i^{r'} - T_e^r \right)}{R_e^{rr'}} + \sum_{a=1}^{A} \frac{\left(T_{\text{am}} - T_e^r \right)}{R_e^{ra}} \right) \quad (11.2)$$

$$C_h^r \times \frac{dT_h^r}{dt} = \frac{\left(T_i^r - T_h^r \right)}{R_{ih}^r} + P_{\text{Floor}}^r \quad (11.3)$$

where T_i^r, T_e^r, and T_h^r are the temperatures of room air, walls (envelopes), and heater (radiator) for room r; C_i^r, C_e^r, and C_h^r are heat capacities of indoor air, envelope, and heater; R_{ie}^r, R_{ih}^r, and R_e^{ra} are heat resistance between room air and envelope, room air and heater, envelope and ambient; $R_e^{rr'}$ is heat resistance between envelopes of room r and r'; A_w^r is the fraction of solar power P_{Sun}^r absorbed by the indoor air; P_{Floor}^r is the energy consumption of the heater in room r; T_{am} is the ambient temperature; t is the time index.

The Eqs. (11.1)–(11.3) state the thermal dynamics of room air, envelope, and heater temperatures, respectively. Equation (11.1) is comprised of three terms. The first and second terms explain the heat exchange between the heater and room

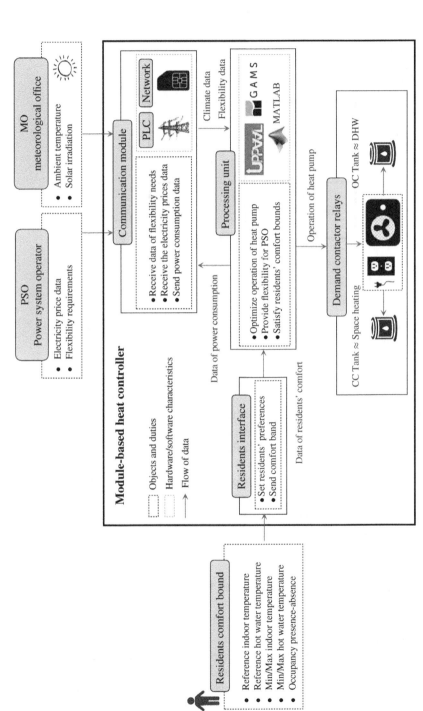

Figure 11.3 Schematic structure of the heat controller (PLC: power line carrier, CC: closed cycle, OC: open cycle). *Source:* Golmohamadi [6]/ELSEVIER.

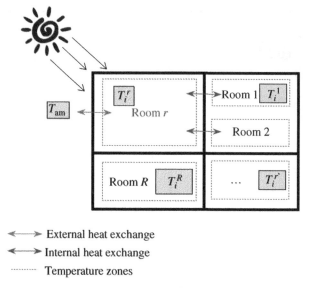

→ External heat exchange
↔ Internal heat exchange
········ Temperature zones

Figure 11.4 General floor plan of a building with R temperature zones. *Source:* Golmohamadi et al. [23]/with permission of Elsevier.

air, and between envelope and room air, respectively. The third term illustrates the energy of solar power. In Eq. (11.2), the first term states the heat flux between the room air and the envelopes. The second and third terms are the heat transfer between the envelopes of room r, which are surrounded by envelopes of rooms r' and ambient, respectively. In Eq. (11.3), the first term is the heat flux between room air and heater, and the second term is the heat extraction from the heating system (radiator). Note that the C, R, and A are constant factors and depend mainly on the physical features of buildings, e.g. layout, insulation quality, and windows dimension. To estimate the constant factors for a specific building, machine learning algorithms are used to train the differential Eqs. (11.1)–(11.3) using historical data.

11.7 Economic Heat Controller in Dynamic Electricity Market

In this section, an EMPC approach is discussed to optimize the heating operation of buildings in response to dynamic energy prices. The residential heating system is comprised of a heat pump, heat buffer, and heat radiators. First of all, the objective function of the EMPC and corresponding constraints are explained. Afterward, the case study and simulation results are stated.

11.7.1 Objective Function of EMPC

The objective function of the controller is to minimize the household energy consumption cost as well as the deviation from the setpoint temperature. The residents' comfort bound is modeled as lower and upper thresholds. Then, the objective function and the constraints are stated as follows [23]:

$$
\min_{\left(T_i^r, P_t^{\text{Elec}}, \varepsilon_k\right)} \sum_{t=1}^{N} \left[\left(w_1 \times \sum_{r=1}^{R} \left[T_i^{r,\text{Ref}} - T_i^r(t) \right]^2 \right) \right.
$$

$$
\left. + \left(w_2 \times \left[\lambda_t^{\text{DA}} \times P^{\text{Elec}}(t) \right] \right) + \left(w_3 \times \sum_{k=1}^{K} \varepsilon_k^2 \right) \right] \tag{11.4}
$$

s.t.

$$(11.1) - (11.3) \tag{11.5}$$

$$P_{\text{Min}}^{\text{Elec}} \leq P^{\text{Elec}}(t) \leq P_{\text{Max}}^{\text{Elec}} \tag{11.6}$$

$$T_{\text{Min}}^r \pm \varepsilon_k \leq T_i^r(t) \leq T_{\text{Max}}^r \pm \varepsilon_k \tag{11.7}$$

$$T_w^{\text{Min}} \pm \varepsilon_k \leq T_w(t) \leq T_w^{\text{Max}} \pm \varepsilon_k \tag{11.8}$$

$$P^{\text{Elec}}(t) - P^{\text{Elec}}(t-1) \leq \gamma_{\text{HP}}^{\text{up}} \tag{11.9}$$

$$P^{\text{Elec}}(t-1) - P^{\text{Elec}}(t) \leq \gamma_{\text{HP}}^{\text{down}} \tag{11.10}$$

where w_1, w_2, w_3 are weighting factors; $T_i^{r,\text{Ref}}$ is the setpoint temperature inserted by residents; λ_t^{DA} is the electricity price of the day-ahead market at hour t; T_w is the water temperature; P^{Elec} is the electricity consumption of the heat pump compressor; ε_k is the slack variable.

The objective function is comprised of three terms. The first term minimizes the deviation from the setpoint temperature. The deviation is not allowed to exceed the lower and upper thresholds set by the residents, i.e. constraint (11.7). The second term minimizes the household energy cost based on hourly energy price. The third term minimizes the slack variable to provide soft constraints for indoor temperature. The slack variable prevents hard constraints and improves the computational burden as well as tractability of the control method.

The objective function of the controller optimizes the indoor air temperature derived from thermal dynamics (11.1)–(11.3). The approach is generalized for a building with R temperature zones.

Regarding the constraints, first of all, inequality constraint (11.6) confines the power demand of the compressor. The minimum and maximum of room and water temperature are stated using constraints (11.7) and (11.8). In these two constraints, the slack variable ε_k provides soft limitations for indoor air and water temperature.

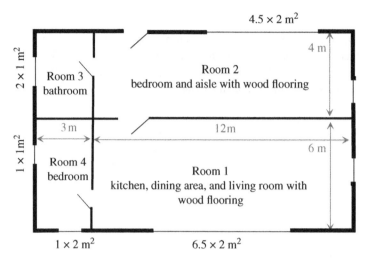

Figure 11.5 Floor plan of the test house with four temperature zones. *Source:* Golmohamadi et al. [23]/with permission of Elsevier.

Note that constraint is set by residents as their comfort bound. The inequality constraints (11.9) and (11.10) meet the ramp-up $\gamma_{\text{HP}}^{\text{up}}$ and ramp-down $\gamma_{\text{HP}}^{\text{down}}$ rates of the compressor.

11.7.2 Case Study of EMPC

In order to examine the applicability of the EMPC, a $150\,\text{m}^2$ detached house is addressed. The house is comprised of one kitchen, two bedrooms, and one bathroom. The EMPC aims to maintain the temperature of the four rooms within the comfort bounds. Also, the controller optimizes the heat consumption to reduce the energy consumption cost. Figure 11.5 sketches the floor plan of the simulated house.

Figure 11.6 shows the optimized operation of the heat pump in four rooms in terms of heat consumption and indoor air temperature for 7 days. Besides, the profile of dynamic electricity prices is provided to study the behavior of the heat controller for peak, shoulder, and valley periods. Based on the graph, different comfort bounds are determined for the rooms. The lower/upper thresholds are scaled-down (-up) by the residents during unoccupied (occupied) hours. As the graphs reveal, the heat controller meets the room temperature within the comfort bounds for the four rooms. Moreover, heat consumption is normally minimized during high electricity prices and/or unoccupied hours. In contrast, when the electricity price is low, or the rooms are occupied, the heat controller increases the heat consumption.

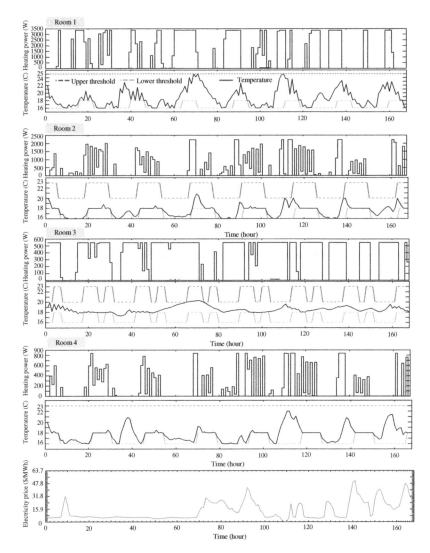

Figure 11.6 Optimized exploitation of heating system in four rooms in response to dynamic electricity prices. *Source:* Golmohamadi et al. [23]/with permission of Elsevier.

11.8 Flexible Heat Controller in Uncertain Electricity Market

In Section 11.7, the controller optimizes the exploitation of the heat pump in a dynamic electricity market with deterministic prices. In electricity markets with high renewable power penetration, there is a close correlation between renewable power availability and energy prices. Therefore, the intermittent nature of renewable power may affect the profile of electricity prices. To overcome the fluctuation of renewable power, three market floors are normally performed. The market floors include day-ahead, intraday, and balancing markets. The three markets are performed 24 hours, 60–10 minutes, and a few seconds before power delivery time. The electricity prices of the market floors are uncertain values and depend on renewable power availability. Approaching the energy delivery time, the uncertainty of electricity prices decreases. Besides, the market floors close to real-time, i.e. balancing markets, have more volatility.

In this section, in contrast to Section 11.7, it is assumed that the heat controller is supplied by the uncertain electricity market with the three market floors. According to market regulations, it is evident that a single-family house is not allowed to participate in the electricity market directly. Therefore, the suggested controller only aims to investigate the impact of uncertain electricity prices on the optimized operation of residential heating systems. In addition to the market economy and MPC approaches, stochastic programming is used in the objective function to address the electricity price uncertainties. Therefore, the controller is called stochastic EMPC (SEMPC).

11.8.1 Objective Function of SEMPC

The objective function of the SEMPC controller is to minimize the household energy cost in three uncertain electricity markets meeting the residents' comfort bounds. The controller uses three-stage stochastic programming to address the energy transactions in the three electricity markets as follows [24]:

$$
\begin{aligned}
\underset{\left(P_{E,t}^{DA}(\omega),P_{E,t}^{IM}(\omega),P_{E,t}^{BM(+)}(\omega),P_{E,t}^{BM(-)}(\omega)\right)}{\text{Minimize}} [\text{Cost}] = \\
\left[\sum_{\omega\in N_\omega}\sum_{t=\tau}^{N_\tau}\left[E_{\omega_1}\left[\lambda_t^{DA}(\omega_1)\times P_{E,t}^{DA}(\omega_1)+E_{\omega_2|\omega_1}\left[\lambda_t^{IM}(\omega_2)\times P_{E,t}^{IM}(\omega_2)\right.\right.\right.\right.\\
\left.\left.\left.\left.+E_{\omega_3|\omega_1,\omega_2}\left[\left(\lambda_t^{BM(+)}(\omega_3)\times P_{E,t}^{BM(+)}(\omega_3)\right)-\left(\lambda_t^{BM(-)}(\omega_3)\times P_{E,t}^{BM(-)}(\omega_3)\right)\right]\right]\right]\right]\right]
\end{aligned}
$$

$$(11.11)$$

where λ_t^{DA}, λ_t^{IM}, λ_t^{BM} are electricity prices of the day-ahead, intraday, and balancing markets; $P_{E,t}^{DA}$, $P_{E,t}^{IM}$, $P_{E,t}^{BM}$ are net power traded in the three market floors; ω_1, ω_2, ω_3 are indices of electricity price scenarios for the three markets accordingly; E is the expectation operator. Note that the superscripts $+$ and $-$ denote the signs of power system imbalance in terms of positive and negative imbalances.

The objective function includes three terms. The first term, i.e. the first stage of stochastic programming, describes the energy procurements from the day-ahead market. The second stage of stochastic programming explains the energy transaction in the intraday market with the second term. Finally, the third stage is stated in the third term. In this term, the two-price scheme is used for balancing market prices. Therefore, the balancing prices for the positive and negative power system imbalances are defined separately.

The objective function of the SEMPC meets the following constraints [24]: (11.1)–(11.3) and (11.6)–(11.10)

$$0 \leq P_{E,t}^{DA}(\omega_1) \leq \alpha \times P_{E,t}^{Rated} \tag{11.12}$$

$$-\beta \times P_{E,t}^{Rated} \leq P_{E,t}^{IM}(\omega_2) \leq \beta \times P_{E,t}^{Rated} \tag{11.13}$$

$$-\gamma \times P_{E,t}^{Rated} \leq P_{E,t}^{BM}(\omega_3) \leq \gamma \times P_{E,t}^{Rated} \tag{11.14}$$

$$\alpha + \beta + \gamma = 1 \tag{11.15}$$

$$P_{E,t}^{IM}(\omega_2) = P_{E,S,t}^{IM}(\omega_2) - P_{E,P,t}^{IM}(\omega_2) \tag{11.16}$$

$$P_{E,t}^{BM}(\omega_3) = P_{E,t}^{BM(+)}(\omega_3) - P_{E,t}^{BM(-)}(\omega_3) \tag{11.17}$$

$$P_t^{Elec}(\omega) = P_{E,t}^{DA}(\omega_1) + P_{E,t}^{IM}(\omega_2) + P_{E,t}^{BM}(\omega_3) \tag{11.18}$$

Note that the market constraints are described in addition to the thermal dynamics (11.1)–(11.3) and demand constraints (11.6)–(11.10). Constraints (11.12), (11.13), and (11.14) confine the energy transactions in the market floors to a fraction of the rated power consumption of the heat pump compressor. The fraction factors α, β, and γ are quantified in descending order $\alpha > \beta > \gamma$. Approaching the power delivery time, the uncertainty of market floors decreases. Therefore, the balancing market has a lower power capacity than the day-ahead market. The day-ahead market is designed for energy transactions. Adversely, the balancing market aims to counterbalance the power fluctuations. For this reason, the capacity of trading power in the balancing market should be lower than in the day-ahead market. This is the main idea behind the descending factors of the controller. Moreover, based on (11.12), the controller is allowed to buy electricity from the day-ahead market. On the contrary, in the intraday and balancing markets, it is allowed to buy power or sell the day-ahead power in response to price signals. The summation of market factors must be equal to 1 in constraint (11.15). Constraints (11.16) and (11.17) describe the net power in the intraday and balancing markets, respectively. The subscripts S and P stand for the sold and purchased power. Finally, the power balance is stated in Eq. (11.18).

Note that the electricity prices convey important signals about renewable power availability. Therefore, the price-responsive controller unlocks the power flexibility of the heating system in the market floors. The households benefit from reduced energy consumption costs and the energy networks take advantage of the demand flexibility.

To elaborate on the three market floors, the three stages of stochastic programming are described as follows.

11.8.2 First Stage

In the first stage of the stochastic programming, the day-ahead market is clearing while the next market floors, i.e. the intraday and balancing markets, are still unclear. Therefore, the variable of day-ahead power is defined as *here-and-now*, and the intraday and balancing power are defined as *wait-and-see* variables. The set of decision variables in the first stage is shown as follows:

$$x^{r=1} = \left\{ \left(P_{E,t}^{DA} \right)_{\text{here-now}}, \left(P_{E,t}^{IM}, P_{E,t}^{BM} \right)_{\text{wait-see}} \right\} \tag{11.19}$$

11.8.3 Second Stage

In the second stage, the day-ahead market has already been cleared. The intraday market is clearing and the balancing market is still unclear. Therefore, the day-ahead power is realized, the intraday power is the *here-and-now* variable and the balancing market is a *wait-and-see* variable. Then, the decision set of the second stage is stated as below:

$$x^{r=2} = \left\{ \left(P_{E,t}^{DA} \right)_{\text{realized}}, \left(P_{E,t}^{IM} \right)_{\text{here-now}}, \left(P_{E,t}^{BM} \right)_{\text{wait-see}} \right\} \tag{11.20}$$

11.8.4 Third Stage

Finally, in the third stage of stochastic programming, the day-ahead and intraday markets have already been realized. In contrast, the balancing market is performing. Therefore, the day-ahead and intraday power are *realized* variables, and the balancing power is *here-and-now* variable. No unclear market floor is left for further processing; therefore, there is no *wait-and-see* variable for the next stages. The decision set of the third stage is stated as follows:

$$x^{r=3} = \left\{ \left(P_{E,t}^{DA}, P_{E,t}^{IM} \right)_{\text{realized}}, \left(P_{E,t}^{BM} \right)_{\text{here-now}} \right\} \tag{11.21}$$

Considering the three stages of stochastic programming (11.19)–(11.21), the compact form of the objective function (11.11) can be reformulated in the extended form as follows:

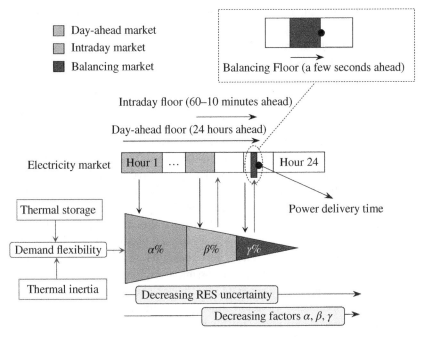

Figure 11.7 Participation of SEMPC in three uncertain electricity markets. *Source:* Golmohamadi et al. [24]/with permission of Elsevier.

$$H_1 = \underset{(P_{E,t}^{DA})}{\text{Minimize}} \left[\sum_{\omega \in N_\omega} \sum_{t=\tau}^{N_\tau} \left[E_{\omega_1} \left[\lambda_t^{DA}(\omega_1) \times P_{E,t}^{DA}(\omega_1) + [H_2] \right] \right] \right] \quad (11.22)$$

$$H_2 = \underset{(P_{E,t}^{IM})}{\text{Minimize}} \left[\sum_{\omega \in N_\omega} \sum_{t=\tau}^{N_\tau} \left[E_{\omega_2|\omega_1} \left[\lambda_t^{IM}(\omega_2) \times P_{E,t}^{IM}(\omega_2) + [H_3] \right] \right] \right] \quad (11.23)$$

$$H_3 = \underset{(P_{E,t}^{BM})}{\text{Minimize}} = \left[\sum_{\omega \in N_\omega} \sum_{t=\tau}^{N_\tau} \left[E_{\omega_3|\omega_1,\omega_2} \left[\left(\lambda_t^{BM(+)}(\omega_3) \times P_{E,t}^{BM(+)}(\omega_3) \right) \right. \right. \right.$$
$$\left. \left. \left. - \left(\lambda_t^{BM(-)}(\omega_3) \times P_{E,t}^{BM(-)}(\omega_3) \right) \right] \right] \right] \quad (11.24)$$

To sum up, Figure 11.7 describes how the SEMPC controller participates in the three market floors.

11.8.5 Scenario Generation

In the suggested approach, the controller participates in three market floors with price uncertainties. To model the price uncertainties, time-series-based ARIMA

(auto-regressive integrated moving average) is discussed as the scenario generation scheme. The ARIMA generates some scenarios for future time slots based
on the historical data of the electricity market. The mathematical formulation
of ARIMA is described as follows [6]:

$$\lambda_t = (\alpha) + \left(\beta_1\lambda_{t-1} + \beta_2\lambda_{t-2} + \cdots + \beta_p\lambda_{t-p}\right) + \left(\phi_1\varepsilon_{t-1} + \phi_2\varepsilon_{t-2} + \cdots + \phi_q\varepsilon_{t-q}\right)$$

(11.25)

Let us suppose λ_t as the forecasting electricity price. Then, the forecasting price
includes three sections. The first term, (α), is the constant factor of the time series.
The second term is the linear summation of p lags of forecasting electricity prices.
The third term indicates the forecasting error with q-lagged error parameters. The
ARIMA approach generates a definite number of price scenarios with associated
probabilities for each time slot in which the total summation of probabilities for
each time slot is equal to 1. The approach is repeated for the three market floors
to generate enough scenarios. Then, the generated scenarios are combined
through a scenario tree to generate all possible trajectories. Suppose that a, b,
and c numbers of scenarios are generated for the day-ahead, intraday, and balancing markets, respectively. Therefore, the total number of $a \times b \times c$ scenarios
is generated by the scenario tree. In this way, some scenarios may be similar.
To improve the computational burden of the problem, the Kantorovich distance
approach is addressed to remove similar scenarios. This approach is also known
as the scenario reduction approach. This is an effective tool when the problem
is intractable due to a significant number of generated scenarios. Further information about the mathematics and formulations of the ARIMA approach and Kantorovich distance can be found in [25].

11.8.6 Case Study of SEMPC

In order to examine the SEMPC controller, the same case study of Figure 11.5 is
addressed. Figure 11.8 explains the optimized exploitation of the controller in the
three market floors. To cover the electricity price uncertainties, the ARIMA
approach is adopted as the scenario generation method. ARIMA generates 10 price
scenarios for each time slot. Regarding the three market floors, 10^3 scenarios are
generated in total. The price data are extracted from the Nord Pool Electricity
Market, which is publicly available [26].

The power consumption of the heat pump is the summation of the demand in
the three market floors. First of all, subfigure (a) explains the power trading in the
day-ahead market. As the graphs reveal, there are two peak duration, i.e. hours
9–11 and 18–19, in the day-ahead market. The heat controller decreases the heat
consumption in peak hours to provide flexibility for the power system on long

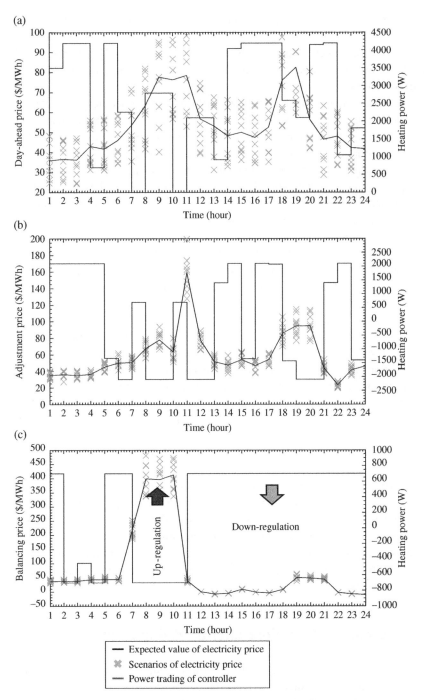

Figure 11.8 Integration of energy flexibility of the single-family house into (a) day-ahead market, (b) intraday market, and (c) balancing market. *Source:* Golmohamadi et al. [24]/with permission of Elsevier.

advance notice. In contrast, the power consumption increases during off-peak hours, e.g. 14–17. In this figure, the green markers show the electricity price scenarios. Also, the expected values of electricity prices are depicted by the black solid line. Therefore, the solid blue line indicates the power consumption of the controller in response to the expected electricity prices.

In the intraday market, there is a considerable jump in the electricity price at hour 11. Besides, a peak price occurs between hours 18–20. In response, the heat controller minimizes the heat consumption of the household during peak hours. In contrast, the energy consumption increases in off-peak hours, e.g. hours 13–17, when the intraday market offers low electricity prices. Regarding the power values, the intraday power takes positive and negative values. The positive values mean that the controller purchases power from the intraday market. On the contrary, the negative values show that the heat controller sells some part of prepurchased energy (from the day-ahead market). It is clear that the heat controller is not a prosumer; therefore, the amount of sold power to the intraday market must be lower than the purchasing power in the day-ahead market. Consequently, the heat controller adjusts the power consumption in the intraday market 60–10 minutes before power delivery time.

In the balancing market, two different patterns are recognized. Firstly, the balancing price encounters a huge jump between hours 8 to 10. Regarding a correlation between balancing price and renewable power availability, the power system faces a severe power shortage. Consequently, the heat controller not only zeros the power consumption but also decides not to consume a part of pre-purchased energy from the day-ahead and intraday markets. The negative power values mean that the controller sells the pre-purchased power in the opposite direction of the power system imbalance. In this way, the controller provides up-regulation to the balancing market. Adversely, the balancing market encounters zero and/or negative electricity prices in hours 12–18. It means that the electricity market encounters renewable power excess. In response, the heat controller maximizes the energy consumption to provide down-regulation for the power system. Finally, the total power consumption of the heat pump is the summation of power consumption in the three market floors.

To sum up, in the peak (off-peak) price hours, the heat controller decreases (increases) the energy consumption to provide up- (down-) regulation for the power system at the opposite direction of the power system imbalance. The household benefits from the low energy cost, and the power system takes advantage of power flexibility.

11.9 Economic Heat Controller of Mixing Loop

In Sections 11.7 and 11.8, the responsive controllers are discussed for individual heat pumps. In many countries, residential buildings are connected to DH systems. In the DH structure, to increase network efficiency and decrease energy loss

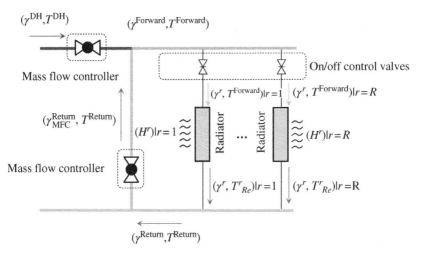

Figure 11.9 The schematic diagram of the control valves for buildings connected to DH. *Source:* Golmohamadi and Larsen [27]/Elsevier/Public Domain CC BY 4.0.

in the heat network, mixing loop controllers are installed. The mixing loop combines the outflow mass (from the buildings) with the inflow mass (from the DH). As a result, the temperature of return mass to the DH decreases significantly.

Figure 11.9 shows the schematic diagram of mass flow control in a mixing loop structure. Based on the graph, on the DH side, the mixing loop controls two mass flow valves, including valves of inflow and outflow water. On the demand side, the controller optimizes the on/off states of the radiators.

11.9.1 Objective Function of Mixing Loop

In this section, an economic controller is suggested for the mixing loop. The economic controller aims to control the inflow and outflow of thermal mass minimizing the household energy consumption cost. The energy consumption is derived from the dynamic energy market. The objective function is stated as follows [27]:

$$
\underset{(T_i^r, H^r, \chi^r)}{\text{Min}} \sum_{t=1}^{T} \left[\left(\mu_1 \times \sum_{r=1}^{R} \left[T_{i,\text{Ref}}^r - T_i^r(t) \right]^2 \right) + \left(\mu_2 \times \left[\lambda^{\text{DH}}(t) \times H^r(t) \right] \times \chi^r(t) \right) \right]
$$

$$(11.26)$$

where $T_{i,\text{Ref}}^r$ is the temperature setpoint determined by residents; T_i^r is the indoor temperature; r is the room index, $r = 1,..., R$; λ^{DH} is the dynamic energy price; H^r is the heat consumption of the radiator r (room r); and χ^r is the binary variable describing the on/off states of the radiators.

The objective function includes two terms. The first term states the residents' comfort, and the second term is related to the household economy. In this way, the first term minimizes the difference between the indoor and reference temperatures. The second term minimizes the cost of heat consumption. The weighting factors μ_1 and μ_2 make a trade-off between the energy cost and comfort level through suppressing or enhancing the role of the corresponding terms in the objective function of the controller.

The objective function meets the following constraints [27]:

$$T_{i,\mathrm{Min}}^r \leq T_i^r(t) \leq T_{i,\mathrm{Max}}^r \tag{11.27}$$

$$\gamma^r(t) = \gamma_{\mathrm{Nominal}}^r \times \chi^r(t) \tag{11.28}$$

$$\chi^r(t) = \begin{vmatrix} 1 & \mathrm{On} \\ 0 & \mathrm{Off} \end{vmatrix} \tag{11.29}$$

$$T_{\mathrm{Re}}^r(t) = \left(\alpha^r \times T_{\mathrm{Fo}}^r(t)\right) + \left(\beta^r \times T_h^r(t)\right) + \left(\kappa^r \times T_i^r(t)\right) + \sigma^r \tag{11.30}$$

$$\left(\gamma^{\mathrm{DH}}(t) \times T^{\mathrm{DH}}(t)\right) + \left(\gamma_{\mathrm{MFC}}^{\mathrm{Return}}(t) \times T^{\mathrm{Return}}(t)\right) = \left(\gamma^{\mathrm{Forward}}(t) \times T^{\mathrm{Forward}}(t)\right) \tag{11.31}$$

$$\sum_{r=1}^R \gamma^r(t) \times T_{\mathrm{Re}}^r(t) = \gamma^{\mathrm{Return}}(t) \times T^{\mathrm{Return}}(t) \tag{11.32}$$

$$H^r(t) = \chi^r(t) \times \left(c_\rho \times \gamma^r \times \left(T_{\mathrm{Fo}}^r(t) - T_{\mathrm{Re}}^r(t)\right)\right) \tag{11.33}$$

where T_{Re}^r, T_{Fo}^r, and T_h^r are the temperature of outflow, inflow, and radiator; α^r, β^r, κ^r, σ^r are constant coefficients; γ^{DH}, $\gamma_{\mathrm{MFC}}^{\mathrm{Return}}$, $\gamma^{\mathrm{Forward}}$ denote the mass flow of forward water on the DH-side, the outflow rate, and inflow rate on the house-side, respectively; and c_ρ is the specific heat of water. The notations of the mathematical model can also be found in Figure 11.9.

Constraint (11.27) meets the minimum and maximum rooms temperature. Equations (11.28) and (11.29) explain the mass flow of radiators. Equation (11.30) describes the outflow temperature as a function of the temperature of inflow, radiator, and rooms. The constant factors $(\alpha^r, \beta^r, \kappa^r, \sigma^r)$ depend on the physical characteristics of the buildings and heating systems. Equations (11.31) and (11.32) describe the heat balance in the nodes of DH-side and house-side, respectively. Finally, Eq.(11.33) explains the heat energy consumption of the radiators.

11.9.2 Case Study of Mixing Loop

In this section, it is assumed that the simulated house of Figure 11.5 is connected to DH by the mixing loop. The mixing loop controller minimizes the household

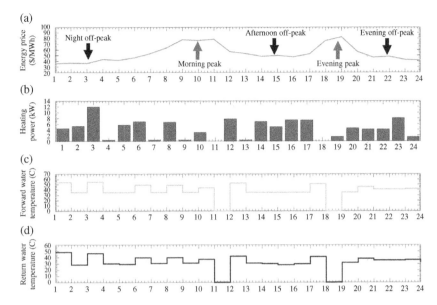

Figure 11.10 Optimized operation of mixing loop. (a) Electricity price, (b) total heat consumption, (c) inflow temperature, and (d) outflow temperature. *Source:* Golmohamadi and Larsen [27]/Elsevier/Public Domain CC BY 4.0.

energy cost restoring the energy of return water. Figure 11.10 explains the optimized exploitation of the mixing loop in terms of energy consumption and inflow/outflow temperatures. Meanwhile, the profile of dynamic electricity prices is depicted. It is supposed that the DH is supplied by electrical power. Regarding the profile of electricity prices, there are two peak durations, i.e. morning and evening peak, and three off-peak periods, i.e. night, afternoon, and evening, during the 24 hours. Based on the graph, the controller reduces the energy consumption during peak hours, e.g. 9–11 and 18–19. Adversely, the energy consumption increases during low price hours including 1–3, 14–17, and 20–23. The outflow temperature is about 30 °C in most time slots; therefore, it increases the energy efficiency (heat loss reduction) in the heat network.

Figure 11.11 shows the mass flows of DH-side and house-side. As the graphs reveal, the mass flow of the heating system increases in the low price hours, e.g. 1–3, 14–17, and 20–23. In contrast, the mass flow decreases during high price hours, e.g. 9–11. In some peak hours, e.g. 18–19, the controller zeros the mass flow to reduce the cost of energy consumption. Assuming a correlation between the energy prices and wind power availability, the simulation results confirm that the mixing loop controller exploits the heating system normally in low price hours

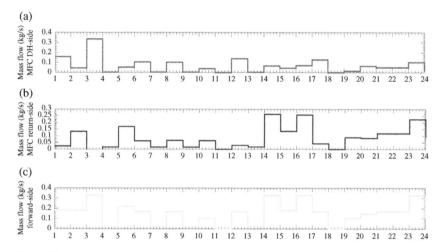

Figure 11.11 Optimized operation of mixing loop. (a) Mass flow of DH-side, (b) outflow rate of thermal mass, and (c) inflow rate of thermal mass. *Source:* Golmohamadi and Larsen [27]/ Elsevier/Public Domain CC BY 4.0.

to reduce the heat consumption cost. Moreover, it integrates flexibility into the power system during renewable power shortages.

11.10 Conclusion

In this chapter, mathematical formulations of thermal dynamics were presented for residential buildings with different temperature zones. Besides, the objective functions and associated constraints were discussed for three types of heat controllers:

1) Economic MPC in dynamic electricity markets.
2) Stochastic economic MPC in uncertain electricity markets.
3) Economic heat control for mixing loop of district heating.

In the dynamic electricity market, the controller minimized the household energy cost. Also, it provided flexibility for the electricity market in response to price variations. In the uncertain electricity markets, the heat controller provided demand flexibility for the day-ahead, intraday, and balancing markets on long, mid, and short advance notices. Considering the correlation between electricity prices and the availability of renewable power, the controller could provide up-/down-regulation for the electricity market in the opposite direction of power system imbalance. The mixing loop controller optimized the operation of inflow and outflow valves in buildings connected to district heating. The optimized heating operation not only minimized the household energy cost but also provided power flexibility for the upstream network. In all heating strategies, the rooms' temperature met the household's comfort.

In real-world studies, the thermal dynamics of buildings are extracted from historical data recorded by measurement sensors. The differential equations of thermal dynamics are trained by machine learning algorithms over the measured data. Generally, such sensor data is not widely available for practical studies. This is one of the main challenges to designing heat controllers. In conclusion, the important achievements of the suggested approaches are pointed out as follows:

1) The heat controllers optimize the joint heat-electricity consumption of the residential buildings considering the flexibility requirements of the supply side. In this way, the households take advantage of reduced energy consumption costs, and the energy networks benefit from enhanced stability and power balance.
2) Flexibility opportunities of building thermal inertia and storage can be coordinated to provide flexibility for the electricity markets on short, mid, and long advance notices of demand response programs. Therefore, the joint heat-power flexibility is firstly scheduled in the energy markets, e.g. the day-ahead market; then it is adjusted in the adjustment markets, e.g. the intraday market; finally, it is regulated in the near real-time markets, e.g. the balancing market.
3) The economic mixing loop control restores the heat of outflow water to increase the efficiency of power and heat networks. In this way, not only the heat loss decreases in the heat networks, but also it facilitates the integration of low-temperature DH into energy systems.

Funding

This study is part of the funded projects FED (Flexible Energy Denmark with grant agreement 892601).

References

1 Golmohamadi, H. and Asadi, A. (2020). Integration of joint power-heat flexibility of oil refinery industries to uncertain energy markets. *Energies* 13 (18). https://doi.org/10.3390/en13184874.
2 Golmohamadi, H., Keypour, R., Bak-Jensen, B., and Radhakrishna Pillai, J. (2019). Optimization of household energy consumption towards day-ahead retail electricity

price in home energy management systems. *Sustain. Cities Soc.* 47: 101468. https://doi.org/10.1016/j.scs.2019.101468.

3 Golmohamadi, H. (2021). Virtual storage plants in parking lots of electric vehicles providing local/global power system supports. *Energy Storage* 43: 103249.

4 Golmohamadi, H. (2020). Operational scheduling of responsive prosumer farms for day-ahead peak shaving by agricultural demand response aggregators. *Int. J. Energy Res.* 45 (1): https://doi.org/10.1002/er.6017.

5 Golmohamadi, H., Keypour, R., Bak-Jensen, B. et al. (2020). Robust self-scheduling of operational processes for industrial demand response aggregators. *IEEE Trans. Ind. Electron.* 67 (2): 1387–1395. https://doi.org/10.1109/TIE.2019.2899562.

6 Golmohamadi, H. (2021). Stochastic energy optimization of residential heat pumps in uncertain electricity markets. *Appl. Energy* 303: 117629. https://doi.org/10.1016/j.apenergy.2021.117629.

7 Clauß, J., Stinner, S., Sartori, I., and Georges, L. (2019). Predictive rule-based control to activate the energy flexibility of Norwegian residential buildings: case of an air-source heat pump and direct electric heating. *Appl. Energy* 237: 500–518. https://doi.org/10.1016/j.apenergy.2018.12.074.

8 Tarragona, J., Pisello, A.L., Fernández, C. et al. (2022). Analysis of thermal energy storage tanks and PV panels combinations in different buildings controlled through model predictive control. *Energy* 239: 122201. https://doi.org/10.1016/j.energy.2021.122201.

9 Mor, G., Grillone, J.C.B., Amblard, F. et al. (2021). Operation and energy flexibility evaluation of direct load controlled buildings equipped with heat pumps. *Energy Build.* 253: 111484. https://doi.org/10.1016/j.enbuild.2021.111484.

10 Dou, X., Shao, Y., Wang, J., and Hu, Q. (2021). Heat-electricity joint bidding strategies for intelligent buildings in intelligent building cluster. *Int. J. Electr. Power Energy Syst.* 129: 106891. https://doi.org/10.1016/j.ijepes.2021.106891.

11 Dongellini, M., Naldi, C., and Morini, G.L. (2021). Influence of sizing strategy and control rules on the energy saving potential of heat pump hybrid systems in a residential building. *Energy Convers. Manag.* 235: 114022. https://doi.org/10.1016/j.enconman.2021.114022.

12 Kuosa, M., Rahiala, S., Tallinen, K. et al. (2019). Mass flow controlled district heating with an extract air heat pump in apartment buildings: a practical concept study. *Appl. Therm. Eng.* 157: 113745. https://doi.org/10.1016/j.applthermaleng.2019.113745.

13 D'Ettorre, F., De Rosa, M., Conti, P. et al. (2019). Mapping the energy flexibility potential of single buildings equipped with optimally-controlled heat pump, gas boilers and thermal storage. *Sustain. Cities Soc.* 50: 101689. https://doi.org/10.1016/j.scs.2019.101689.

14 Péan, T., Costa-Castelló, R., Fuentes, E., and Salom, J. (2019). Experimental testing of variable speed heat pump control strategies for enhancing energy flexibility in buildings. *IEEE Access* 7: 37071–37087. https://doi.org/10.1109/ACCESS.2019. 2903084.

15 Jin, X., Wu, Q., Jia, H., and Hatziargyriou, N.D. (2021). Optimal integration of building heating loads in integrated heating/electricity community energy systems: a Bi-level MPC approach. *IEEE Trans. Sustain. Energy* 12 (3): 1741–1754. https:// doi.org/10.1109/TSTE.2021.3064325.

16 Atam, E., Patteeuw, D., Antonov, S.P., and Helsen, L. (2016). Optimal control approaches for analysis of energy use minimization of hybrid ground-coupled heat pump systems. *IEEE Trans. Control Syst. Technol.* 24 (2): 525–540. https://doi.org/ 10.1109/TCST.2015.2445851.

17 Liberati, F., Giorgio, A.D., Giuseppi, A. et al. (2019). Joint model predictive control of electric and heating resources in a smart building. *IEEE Trans. Ind. Appl.* 55 (6): 7015–7027. https://doi.org/10.1109/TIA.2019.2932954.

18 Sturzenegger, D., Gyalistras, D., Morari, M., and Smith, R.S. (2016). Model predictive climate control of a swiss office building: implementation, results, and cost – benefit analysis. *IEEE Trans. Control Syst. Technol.* 24 (1): 1–12. https://doi. org/10.1109/TCST.2015.2415411.

19 Finck, C., Li, R., Kramer, R., and Zeiler, W. (2018). Quantifying demand flexibility of power-to-heat and thermal energy storage in the control of building heating systems. *Appl. Energy* 209: 409–425. https://doi.org/10.1016/j.apenergy.2017.11.036.

20 Ahn, J., Chung, D.H., and Cho, S. (2018). Energy cost analysis of an intelligent building network adopting heat trading concept in a district heating model. *Energy* 151: 11–25. https://doi.org/10.1016/j.energy.2018.01.040.

21 Bechtel, S., Rafii-Tabrizi, S., Scholzen, F. et al. (2020). Influence of thermal energy storage and heat pump parametrization for demand-side-management in a nearly-zero-energy-building using model predictive control. *Energy Build.* 226: 110364. https://doi.org/10.1016/j.enbuild.2020.110364.

22 Wang, Y., Quan, Z., Jing, H. et al. (2021). Performance and operation strategy optimization of a new dual-source building energy supply system with heat pumps and energy storage. *Energy Convers. Manag.* 239: 114204. https://doi.org/10.1016/ j.enconman.2021.114204.

23 Golmohamadi, H., Guldstrand Larsen, K., Gjøl Jensen, P., and Riaz Hasrat, I. (2021). Optimization of power-to-heat flexibility for residential buildings in response to day-ahead electricity price. *Energy Build.* 232: 110665. https://doi.org/ 10.1016/j.enbuild.2020.110665.

24 Golmohamadi, H., Larsen, K.G., Jensen, P.G., and Hasrat, I.R. (2021). Hierarchical flexibility potentials of residential buildings with responsive heat pumps: a case study of Denmark. *J. Build. Eng.* 41: 102425. https://doi.org/10.1016/j.jobe.2021.102425.

25 Shumway, R.H. and Stoffer, D.S. (2000). Time series regression and ARIMA models BT. In: *Time Series Analysis and Its Applications* (ed. R.H. Shumway and D.S. Stoffer), 89–212. New York: Springer New York.

26 Nordic Electricity Market (2021). Nordpool. https://www.nordpoolgroup.com.

27 Golmohamadi, H. and Larsen, K.G. (2021). Economic heat control of mixing loop for residential buildings supplied by low-temperature district heating. *J. Build. Eng.* 103286. https://doi.org/10.1016/j.jobe.2021.103286.

12

Hybrid Energy Storage Systems for Optimal Operation of the Heat and Electricity Incorporated Networks

Sara Haghifam[1,2], Mojtaba Dadashi[2], Saba Norouzi[2], Hannu Laaksonen[1], Kazem Zare[2], and Miadreza Shafie-khah[1]

[1] School of Technology and Innovations, Flexible Energy Resources, University of Vaasa, Vaasa, Finland
[2] Faculty of Electrical and Computer Engineering, University of Tabriz, Tabriz, Iran

Nomenclature

Acronyms

BESS	Battery energy storage system
CHP	Combined heat and power
DA	Day-ahead
EB	Electric boiler
EHP	Electric heat pump
GFG	Gas fired generator
GST	Gas station
HESS	Hybrid energy storage system
IEHN	Integrated electricity and heat network
NGESS	Natural gas energy storage system
NGFG	Non-gas fired generator
P2X2P	Power-to-X-to-power
PV	Photovoltaic system
RT	Real-time
ST	Solar thermal system
TESS	Thermal energy storage system
WT	Wind turbine

Coordinated Operation and Planning of Modern Heat and Electricity Incorporated Networks,
First Edition. Edited by Mohammadreza Daneshvar, Behnam Mohammadi-Ivatloo, and Kazem Zare.
© 2023 The Institute of Electrical and Electronics Engineers, Inc.
Published 2023 by John Wiley & Sons, Inc.

Sets and Indices

bs $\in$ BS	Set of BESSs
$c \in C$	Set of CHPs
eb $\in$ EB	Set of EBs
$g \in G$	Set of GFGs
gs $\in$ GS	Set of NGESSs
hp $\in$ HP	Set of EHPs
ng $\in$ NG	Set of NGFGs
$p \in P$	Set of PVs
$s \in S$	Set of Scenarios
st $\in$ ST	Set of STs
$t \in T$	Set of Hours
ts $\in$ TS	Set of TESSs
$w \in W$	Set of WTs

Parameters

α, β	Cost coefficients of NGFGs
CP	EHPs' coefficient of performance
D^E, D^H	Electricity/heat demand of IEHN
$E_{\text{BESS}}^{\text{ini}}, E_{\text{BESS}}^{\text{max}}, E_{\text{BESS}}^{\text{min}}$	Initial/maximum/minimum electrical energy stored in BESSs
$E_{\text{NGESS}}^{\text{ini}}, E_{\text{NGESS}}^{\text{max}}, E_{\text{NGESS}}^{\text{min}}$	Initial/maximum/minimum gas energy stored in NGESSs
$E_{\text{TESS}}^{\text{ini}}, E_{\text{TESS}}^{\text{max}}, E_{\text{TESS}}^{\text{min}}$	Initial/maximum/minimum thermal energy stored in TESSs
$G_{\text{NGESS}}^{\text{in, max}}, G_{\text{NGESS}}^{\text{in, min}}$	Maximum/minimum charged gas of NGESSs
$G_{\text{NGESS}}^{\text{out, max}}, G_{\text{NGESS}}^{\text{out, min}}$	Maximum/minimum discharged gas of NGESSs
$H_{\text{EHP}}^{\text{max}}, H_{\text{EHP}}^{\text{min}}$	Maximum/minimum produced heat of EHPs
$H_{\text{HN}}^{\text{max}}$	Maximum heat exchanged with heat network
H_{ST}	Output heat of STs
$H_{\text{TESS}}^{\text{ch, max}}, H_{\text{TESS}}^{\text{ch, min}}$	Maximum/minimum charged heat of TESSs
$H_{\text{TESS}}^{\text{dch, max}}, H_{\text{TESS}}^{\text{dch, min}}$	Maximum/minimum discharged heat of TESSs
HPR	Heat-to-power ratio of CHPs
HV_{gas}	Heat value of natural gas
$P_{\text{BESS}}^{\text{ch, max}}, P_{\text{BESS}}^{\text{ch, min}}$	Maximum/minimum charged power of BESSs

$P_{\text{BESS}}^{\text{dch, max}}$, $P_{\text{BESS}}^{\text{dch, min}}$	Maximum/minimum discharged power of BESSs
$P_{\text{CHP}}^{\text{max}}$, $P_{\text{CHP}}^{\text{min}}$	Maximum/minimum generated power of CHPs
$P_{\text{EB}}^{\text{max}}$, $P_{\text{EB}}^{\text{min}}$	Maximum/minimum generated power of EBs
$P_{\text{EN}}^{\text{max}}$	Maximum power exchanged with electricity network
$P_{\text{GFG}}^{\text{max}}$, $P_{\text{GFG}}^{\text{min}}$	Maximum/minimum generated power of GFGs
$P_{\text{NGFG}}^{\text{max}}$, $P_{\text{NGFG}}^{\text{min}}$	Maximum/minimum generated power of NGFGs
P_{PV}	Output power of PVs
P_{WT}	Output power of WTs
RD_{CHP}, RU_{CHP}	Ramp-down/up ratio of CHPs
RD_{GFG}, RU_{GFG}	Ramp-down/up ratio of GFGs
RD_{NGFG}, RU_{NGFG}	Ramp-down/up ratio of NGFGs
sd_{CHP}, su_{CHP}	Shout-down/start-up price of CHPs
sd_{GFG}, su_{GFG}	Shout-down/start-up price of GFGs
φ	Scenario probability
λ_D^E, λ_D^H	Sold price to electricity/heat demand
λ_{DA}^E, λ_{DA}^H	DA electricity/heat price
λ_{gas}	Natural gas price
λ_{RT}^E, λ_{RT}^H	RT electricity/heat price
$\eta_{\text{BESS-ch}}$, $\eta_{\text{BESS-dch}}$	Charge/discharge efficiency of BESSs
$\eta_{\text{CHP-E}}$, $\eta_{\text{CHP-H}}$	Electricity/heat efficiency of CHPs
$\eta_{\text{EB-H}}$	Heat efficiency of EBs
$\eta_{\text{GFG-E}}$	Electricity efficiency of GFGs
$\eta_{\text{NGESS - in}}$, $\eta_{\text{NGESS-out}}$	Charge/discharge efficiency of NGESSs
$\eta_{\text{TESS-sb}}$, $\eta_{\text{TESS-ch}}$, $\eta_{\text{TESS-dch}}$	Standby/charge/discharge efficiency of TESSs

Decision Variables

E_{BESS}	Electrical energy stored in BESSs
E_{NGESS}	Gas energy stored in NGESSs
E_{TESS}	Thermal energy stored in TESSs
G_{GST}	Gas flow from GST
$G_{\text{GST-CHP}}$	Gas flow from GST to CHPs
$G_{\text{GST-GFG}}$	Gas flow from GST to GFGs
$G_{\text{GST-NGESS}}^{\text{in}}$	Total charged gas of NGESSs from GST
$G_{\text{NGESS}}^{\text{out}}$	Total discharged gas of NGESSs
$G_{\text{NGESS-CHP}}$	Gas flow from NGESSs to CHPs
$G_{\text{NGESS-GFG}}$	Gas flow from NGESSs to GFGs

H_{CHP}	Generated heat of CHPs
H_{DA}	DA exchanged heat with heat network
H_{EB}	Generated heat of EBs
H_{EHP}	Generated heat of EHPs
H_{imb}	Imbalance heat
H_{RT}	RT exchanged heat with heat network
$H_{\text{TESS}}^{\text{ch}}, H_{\text{TESS}}^{\text{dch}}$	Charged/discharged heat of TESSs
OC_{CHP}	Operating cost of CHPs
OC_{GFG}	Operating cost of GFGs
OC_{NGESS}	Operating cost of NGESSs
OC_{NGFG}	Operating cost of NGFGs
$P_{\text{BESS}}^{\text{ch}}, P_{\text{BESS}}^{\text{dch}}$	Charged/discharged power of BESSs
P_{CHP}	Generated power of CHPs
P_{DA}	DA exchanged power with electricity network
P_{EB}	Consumed power of EBs
P_{EHP}	Consumed power of EHPs
P_{GFG}	Generated power of GFGs
P_{imb}	Imbalance power
P_{NGFG}	Generated power of NGFGs
P_{RT}	RT exchanged power with electricity network
SDC_{CHP}	Shout-down cost of CHPs
SUC_{CHP}	start-up cost of CHPs
SDC_{GFG}	Shout-down cost of GFGs
SUC_{GFG}	start-up cost of GFGs
Δ_H, Δ_P	Index for calculation of imbalance heat/power

Binary Variables

$U_{\bullet}$ Binary variable for the performance of $\bullet$

12.1 Introduction

By and large, a considerable amount of energy consumption is provided by fossil fuels. These days, due to the lack of these fuel sources as well as their irreversible environmental impacts, the tendency towards the exploitation of renewable-based resources in energy systems has increased remarkably [1]. Accordingly, the ultimate goal of the European Union is to reach a 100% renewable-based energy system and carbon neutrality by the end of 2050 [2]. While the emergence of renewable energies is able to overcome the mentioned challenges, the stochastic and intermittent nature of these units is increased the need for flexibility in the electricity sector [3]. In recent years, several solutions have been presented to promote flexibility in power systems. One of the most prevalent solutions for flexibility provision is the utilization of real and virtual energy storage systems like compressed air, battery, and pumped hydropower storage units, as well as demand response programs [4]. Apart from using storage resources as enabling technologies, other solutions have been raised currently to provide more flexibility in electricity networks. Out of the novel and pragmatic solutions, coupling of different energy sectors, including electricity and heat, in the form of an incorporated energy system has attracted more attention [5]. Nonetheless, for connecting these two sectors and establishing an integrated electricity and heat network (IEHN), the existence of power-to-X-to-power (P2X2P) conversion technologies is highly required [6]. P2X2P is a set of technologies that provides synergies among different networks and allows for converting an energy carrier into other energy carriers, some of which are combined heat and power (CHP) plants [7], power-to-heat (P2H) units like electric heat pump (EHP) and electric boiler (EB) [8], heat-to-power (H2P) units, power-to-gas (P2G) units [9], gas-to-power (G2P) units like a gas-fired generator (GFG) [10], etc. In this regard, power-to-X (P2X) interface elements provide the opportunity to transfer and consume surplus electricity production in the form of other energies. On the other hand, it is possible to store the excess power generation in other sectors by using the storage capacities. Then, the stored energy can be returned to the electricity system by X-to-power (X2P) interface elements when the network is faced with a shortage. Based on these explanations, it seems that coupling of heat and electricity networks by P2X2P infrastructures as well as exploiting various kinds of electricity, heat, and gas storage units in the form of a hybrid energy storage system (HESS) [11] not only promote flexibility of these systems but also diminish the operating costs of IEHN owing to the reduction in fuel consumption.

In the past few years, several research works have been conducted to assess the impact of sector coupling as well as P2X2P and storage technologies on the optimal operation of the IEHN, some of which are highlighted in the following:

A mixed-integer linear programming model has been used in [12] to evaluate the coordinated operation of heat and power networks in the presence of demand response programs. The objective function of this problem is to minimize the operating cost of the IEHN considering the technical constraints of the whole system. A robust optimization model has been developed in [13] for the economic operation of the IEHN in the presence of an uncertain environment as well as demand response programs. The primary goal of this research work is to examine the impact of CHP units on the total operating cost of the incorporated system. To enhance the use of wind energy, the coupling of power and heat networks, as well as exploiting P2X technologies, including CHP, EB, and EHP, has been provided in [14]. The objective of this study is to minimize the operating cost of dispatchable units and the curtailment cost of wind power. A distributed optimization technique has been suggested in [15] to investigate the optimal operation of the IEHN in a decentralized manner. The objective function of this study is to diminish the total operating cost of the incorporated system taking into account its technical constraints. A two-stage stochastic programming approach has been utilized in [5] for the optimal operation of the IEHN in both day-ahead (DA) and real-time (RT) stages. The raised framework has considered the detailed reserve modeling of P2X2P and storage elements with the aim of expected cost minimization. A distributed optimization approach has been suggested in [16] for the optimal performance of an IEHN, which is operated by separate power and heat system operators. The main goal of this problem is to minimize the operating cost of coupled networks considering privacy-preserving matters of each independent system operator. A distributed optimization model combined with a hybrid stochastic-robust technique has been presented in [17] to cope with various uncertainties in the operation of the IEHN in a decentralized way. The primary objective of this problem is to decrease the operating costs of the IEHN with the minimum information exchange between power and heat system operators. A novel optimization approach has been proposed in [18] to relax non-convexities raised from the coupling of heat and power sectors through CHP units. This study's objective function is the minimization of operating costs and losses of the considered IEHN. A two-stage stochastic programming scheme has been utilized in [19] for the optimal operation of the IEHN with the detailed modeling of reserve provision and heat regulation of the existing P2X infrastructures, i.e., CHP units, EHPs, EBs, etc. This work has attempted to minimize the operating cost of the IEHN in two different stages, namely the DA operational costs as well as RT regulation costs. A two-stage robust optimization model has been executed in [20] for the energy and reserve co-optimization of the IEHN in the DA operation and RT regulation stages, respectively. The considered problem's first stage is associated with the DA energy and reserve costs minimization, while the second stage is related to the minimization of the RT regulation cost caused by the error of wind power forecasting. A two-stage robust model has been provided in [21] to optimize the operation of the IEHN in the presence of integrated demand response programs. The purpose of this problem is to evaluate the effect of integrated demand response programs on the optimal operation of the studied system with the aim of net profit maximization.

Reviewed articles reveal that while the impact of the electricity-heat sector coupling and related conversion technologies on the optimal performance of IEHNs has been assessed in several studies, the simultaneous role of battery (BESS), thermal (TESS), and natural gas (NGESS) energy storage systems in a platform called HESS has not been analyzed in the previous research works. Accordingly, the present chapter tends to model an IEHN in the existence of numerous interface elements in the first step. Next, by modeling BESS, TESS, and NGESS and building a HESS, it seeks to evaluate the importance of these devices on the optimal operation of the incorporated network. In order to implement the mentioned problem and to deal with raised uncertainties in the IEHN operator's decision-making process, a scenario-based two-stage stochastic programming approach is exploited in this chapter.

The rest of this chapter is arranged as follows: the structure of the considered IEHN, the proposed methodology, and problem formulations are explained in more detail in Section 12.2. The implementation of a case study and its discussions are provided in Section 12.3. Ultimately, the chapter is concluded in Section 12.4.

12.2 Methodology

As mentioned earlier, this chapter aims at modeling an IEHN in the presence of various sector coupling and storage technologies and investigating the effect of a HESS, consisting of BESS, TESS, and NGESS, on the optimal performance of the incorporated network. The schematic structure of the considered IEHN is illustrated in Figure 12.1.

As shown in Figure 12.1, the electricity and heat sectors have been coupled with one another via CHP units, EBs, and EHPs. In this regard, CHP plants cogenerate thermal and power energy, while EBs and EHPs consume power to produce thermal energy. The electricity demand of the system can also be provided by dispatchable and non-dispatchable resources, namely non-gas fired generators (NGFGs), gas-fired generators (GFGs), wind turbines (WTs), and photovoltaic systems (PVs). On the contrary, the heat demand of the system can also be provided by solar thermal systems (STs). Besides the available elements, the IEHN operator has the opportunity to compensate for/supply its shortage demands/excess generation from/to the upstream power and heat networks as well. On the other hand, the IEHN contains the combination of BESS, TESS, and NGESS, as a HESS, to

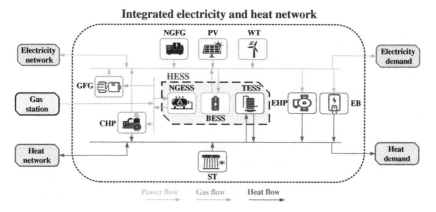

Figure 12.1 Structure of the IEHN.

enhance the flexibility of the whole system and adjust the exchanged energy with the upstream networks through dispatching these resources.

A scenario-based two-stage stochastic programming scheme is executed to achieve the optimal operation of the considered IEHN and handle the existing uncertainties in the decision-making process of the IEHN operator. Electricity and heat demand, electricity and heat price, wind speed, and solar irradiance are the stochastic variables of the problem. It is notable that, for the realization of these uncertain parameters, a high number of scenarios are generated by the Monte Carlo Simulation technique and reduced to an adequate number by utilizing the fast backward/forward scenario reduction algorithm [22]. In two-stage optimization models, two types of decisions are taken, first-stage or here-and-now decisions and second-stage or wait-and-see decisions [23]. In this regard, at the first stage, the IEHN operator determines its optimal DA exchanged power and heat with the upstream networks based on data available at the time of the decisions. At the second stage and after the realization of stochastic factors, the operator tries to decrease its RT imbalance cost and compensate for its forecasted errors by adjusting the available dispatchable as well as flexible units.

The objective function of the considered problem is maximizing the expected profit of the IEHN operator, which is defined as the difference between the operator's income and expenditures. The income includes the revenue from DA exchanging power and heat energy with upstream grids as well as selling power and heat energy to customers. In contrast, expenditures include RT imbalance costs as well as operating costs of available dispatchable and flexible units. In this context, the RT imbalance cost is a penalty cost resulting from the deviation of the IEHN's RT generation/consumption from its DA scheduled energy, see Eq. (12.1).

$$\text{ObjFun}_{\text{IEHN}} = \text{Max} \sum_{t=1}^{T} \left\{ P_{\text{DA}}(t) \cdot \lambda_{\text{DA}}^{E}(t) + H_{\text{DA}}(t) \cdot \lambda_{\text{DA}}^{H}(t) \right. $$

$$ \left. - \sum_{s=1}^{S} \varphi(s) \cdot \begin{bmatrix} P_{\text{imb}}(t,s) \cdot \lambda_{\text{RT}}^{E}(t,s) + H_{\text{imb}}(t,s) \cdot \lambda_{\text{RT}}^{H}(t,s) \\[4pt] + \sum_{c=1}^{C} \text{OC}_{\text{CHP}}(c,t,s) + \sum_{gs=1}^{GS} \text{OC}_{\text{NGESS}}(gs,t,s) \\[4pt] + \sum_{g=1}^{G} \text{OC}_{\text{GFG}}(g,t,s) + \sum_{ng=1}^{NG} \text{OC}_{\text{NGFG}}(ng,t,s) \\[4pt] - D^{E}(t,s) \cdot \lambda_{D}^{E}(t) - D^{H}(t,s) \cdot \lambda_{D}^{H}(t) \end{bmatrix} \right\}, $$

$$(12.1)$$

In Eq. (12.1), the first and second terms are the income from DA exchanging power and heat with the upstream electricity and heat networks, respectively. The third and fourth terms are costs of RT imbalance power and heat, respectively. The fifth, sixth, seventh, and eighth terms are related to the operating costs of CHP, NGESS, GFG, and NGFG, respectively. Ultimately, the ninth and tenth terms are the revenue from selling power and heat energy to the customers, respectively. The mentioned objective function is subject to a set of constraints as follows:

12.2.1 Power and Heat Balance Constraints

$$P_{\text{RT}}(t,s) = \begin{bmatrix} \sum_{c=1}^{C} P_{\text{CHP}}(c,t,s) + \sum_{g=1}^{G} P_{\text{GFG}}(g,t,s) + \sum_{ng=1}^{NG} P_{\text{NGFG}}(ng,t,s) \\[4pt] + \sum_{w=1}^{W} P_{\text{WT}}(w,t,s) + \sum_{p=1}^{P} P_{PV}(p,t,s) + \sum_{bs=1}^{BS} P_{\text{BESS}}^{\text{dch}}(bs,t,s) \\[4pt] - \sum_{bs=1}^{BS} P_{\text{BESS}}^{\text{ch}}(bs,t,s) - \sum_{hp=1}^{HP} P_{\text{EHP}}(hp,t,s) - \sum_{eb=1}^{EB} P_{\text{EB}}(eb,t,s) - D^{E}(t,s) \end{bmatrix}, \forall_{t,s}$$

$$(12.2)$$

$$P_{\text{imb}}(t,s) = [P_{\text{DA}}(t) - P_{\text{RT}}(t,s)] \cdot \Delta_{P}(t,s), \forall_{t,s}$$

$$\Delta_{P}(t,s) = \begin{cases} 1 & , \quad P_{\text{DA}}(t) > P_{\text{RT}}(t,s) \\ -1 & , \quad P_{\text{DA}}(t) < P_{\text{RT}}(t,s) \end{cases} \qquad (12.3)$$

As shown in Eq. (12.3), P_{imb} is calculated as the difference between the IEHN's DA and RT exchanged power with the upstream electricity network. However, since the RT imbalance cost is considered as a penalty cost for the IEHN operator, Δ_{P} is utilized as an auxiliary index to enter P_{imb} as a positive term into the objective function, whether the RT exchanged power is higher or lower than the DA exchanged power.

$$H_{\text{RT}}(t, s) = \begin{bmatrix} \sum_{c=1}^{C} H_{\text{CHP}}(c, t, s) + \sum_{hp=1}^{HP} H_{\text{EHP}}(hp, t, s) + \sum_{eb=1}^{EB} H_{\text{EB}}(eb, t, s) \\ + \sum_{st=1}^{ST} H_{\text{ST}}(st, t, s) + \sum_{ts=1}^{TS} H_{\text{TESS}}^{\text{dch}}(ts, t, s) - \sum_{ts=1}^{TS} H_{\text{TESS}}^{\text{ch}}(ts, t, s) - D^{H}(t, s) \end{bmatrix}, \forall_{t,s}$$

$$(12.4)$$

$$H_{\text{imb}}(t, s) = [H_{\text{DA}}(t) - H_{\text{RT}}(t, s)] \cdot \Delta_{H}(t, s), \forall_{t,s}$$

$$\Delta_{H}(t, s) = \begin{cases} 1 & , & H_{\text{DA}}(t) > H_{\text{RT}}(t, s) \\ -1 & , & H_{\text{DA}}(t) < H_{\text{RT}}(t, s) \end{cases}$$

$$(12.5)$$

Similarly, Δ_{H} is used as an auxiliary index in Eq. (12.5) to enter H_{imb} as a positive term into the objective function, whether the RT exchanged heat is higher or lower than the DA exchanged heat.

12.2.2 Exchanged Power and Heat with Upstream Networks Constraints

$$-P_{\text{EN}}^{\max} \leq P_{\text{RT}}(t, s) \leq P_{\text{EN}}^{\max} \tag{12.6}$$

$$-H_{\text{HN}}^{\max} \leq H_{\text{RT}}(t, s) \leq H_{\text{EN}}^{\max} \tag{12.7}$$

The IEHN's RT exchanged power and heat with the electricity and heat upstream networks are limited by Eqs. (12.6) and (12.7), respectively.

12.2.3 Gas Flow Constraints

$$G_{\text{GST}}(t, s) = \sum_{c=1}^{C} G_{\text{GST-CHP}}(c, t, s) + \sum_{g=1}^{G} G_{\text{GST-GFG}}(g, t, s) + \sum_{gs=1}^{GS} G_{\text{GST-NGESS}}^{\text{in}}(gs, t, s), \forall_{t,s}$$

$$(12.8)$$

$$\sum_{gs=1}^{GS} G_{\text{NGESS}}^{\text{out}}(gs, t, s) = \sum_{c=1}^{C} G_{\text{NGESS-CHP}}(c, t, s) + \sum_{g=1}^{G} G_{\text{NGESS-GFG}}(g, t, s), \forall_{t,s}$$

$$(12.9)$$

Based on Eq. (12.8) and as shown in Figure 12.1, the total amount of gas flow from the gas station (GST) is equal to the required gas of CHPs, GFGs, and NGESSs. On the other hand, NGESSs are able to procure the needed gas of CHPs and GFGs during their discharging mode. In this regard, Eq. (12.9) is utilized to specify the share of NGESSs in supplying the gas of the aforementioned units.

12.2.4 Combined Heat and Power Plants Constraints

$$G_{\text{GST-CHP}}(c, t, s) + G_{\text{NGESS-CHP}}(c, t, s) = \frac{P_{\text{CHP}}(c, t, s)}{\text{HV}_{\text{gas}} \cdot \eta_{\text{CHP-E}}(c)}, \; \forall_{c,t,s} \tag{12.10}$$

$$H_{\text{CHP}}(c, t, s) \leq P_{\text{CHP}}(c, t, s) \cdot \text{HPR}(c) \cdot \eta_{\text{CHP-H}}(c), \; \forall_{c,t,s} \tag{12.11}$$

$$P_{\text{CHP}}^{\min}(c) \cdot U_{\text{CHP}}(c, t, s) \leq P_{\text{CHP}}(c, t, s) \leq P_{\text{CHP}}^{\max}(c) \cdot U_{\text{CHP}}(c, t, s), \; \forall_{c,t,s} \tag{12.12}$$

$$H_{\text{CHP}}(c, t, s) - H_{\text{CHP}}(c, t-1, s) \leq \text{RU}_{\text{CHP}}(c) \cdot U_{\text{CHP}}(c, t, s), \; \forall_{c,t,s} \tag{12.13}$$

$$H_{\text{CHP}}(c, t-1, s) - H_{\text{CHP}}(c, t, s) \leq \text{RD}_{\text{CHP}}(c) \cdot U_{\text{CHP}}(c, t, s), \; \forall_{c,t,s} \tag{12.14}$$

$$0 \leq \text{SUC}_{\text{CHP}}(c, t, s) = \text{su}_{\text{CHP}}(c) \cdot (U_{\text{CHP}}(c, \; t, \; s) - U_{\text{CHP}}(c, \; t-1, \; s)), \; \forall_{c,t,s} \tag{12.15}$$

$$0 \leq \text{SDC}_{\text{CHP}}(c, t, s) = \text{sd}_{\text{CHP}}(c) \cdot (U_{\text{CHP}}(c, \; t-1, \; s) - U_{\text{CHP}}(c, \; t, \; s)), \; \forall_{c,t,s} \tag{12.16}$$

$$\text{OC}_{\text{CHP}}(c, t, s) = G_{\text{GST-CHP}}(c, t, s) \cdot \lambda_{\text{gas}} + \text{SUC}_{\text{CHP}}(c, t, s) + \text{SDC}_{\text{CHP}}(c, t, s), \; \forall_{c,t,s} \tag{12.17}$$

Technical and operational constraints of CHP plants are stated in Eqs. (12.10)–(12.16). Moreover, the operating costs of these units are computed by Eq. (12.17) [24].

12.2.5 Electric Heat Pumps Constraints

$$H_{\text{EHP}}(hp, t, s) = P_{\text{EHP}}(hp, t, s) \cdot \text{CP}(hp), \; \forall_{hp,t,s} \tag{12.18}$$

$$H_{\text{EHP}}^{\min}(hp) \leq H_{\text{EHP}}(hp, t, s) \leq H_{\text{EHP}}^{\max}(hp), \; \forall_{hp,t,s} \tag{12.19}$$

By consuming the electricity, EHPs generate heat energy, as shown in Eq. (12.18). Furthermore, the generated heat of these resources is limited by Eq. (12.19) [20].

12.2.6 Electric Boilers Constraints

$$P_{\text{EB}}^{\min}(eb) \leq P_{\text{EB}}(eb, t, s) \leq P_{\text{EB}}^{\max}(eb), \; \forall_{eb,t,s} \tag{12.20}$$

$$H_{\text{EB}}(eb, t, s) = P_{\text{EB}}(eb, t, s) \cdot \eta_{\text{EB-H}}(eb), \; \forall_{eb,t,s} \tag{12.21}$$

Similar to EHPs, EBs produce thermal energy by consuming electricity as well. Eq. (12.20) limits their power consumption, and Eq. (12.21) shows the amount of their generated heat [20].

12.2.7 Non-Gas Fired Generators Constraints

$$P_{\text{NGFG}}^{\min}(\text{ng}) \leq P_{\text{NGFG}}(\text{ng}, t, s) \leq P_{\text{NGFG}}^{\max}(\text{ng}), \forall_{\text{ng},t,s} \qquad (12.22)$$

$$P_{\text{NGFG}}(\text{ng}, t, s) - P_{\text{NGFG}}(\text{ng}, t-1, s) \leq \text{RU}_{\text{NGFG}}(\text{ng}), \forall_{\text{ng},t,s} \qquad (12.23)$$

$$P_{\text{NGFG}}(\text{ng}, t-1, s) - P_{\text{NGFG}}(\text{ng}, t, s) \leq \text{RD}_{\text{NGFG}}(\text{ng}), \forall_{\text{ng},t,s} \qquad (12.24)$$

$$\text{OC}_{\text{NGFG}}(\text{ng}, t, s) = \alpha(\text{ng}) \cdot P_{\text{NGFG}}(\text{ng}, t, s) + \beta(\text{ng}), \forall_{\text{ng},t,s} \qquad (12.25)$$

Technical and operational constraints of NGFGs are specified by Eqs. (12.22)–(12.24). Also, these sources' operating costs are calculated by Eq. (12.25).

12.2.8 Gas Fired Generators Constraints

$$G_{\text{GST-GFG}}(g, t, s) + G_{\text{NGESS-GFG}}(g, t, s) = \frac{P_{\text{GFG}}(g, t, s)}{HV_{\text{gas}} \cdot \eta_{\text{GFG-E}}(g)}, \forall_{g,t,s} \qquad (12.26)$$

$$P_{\text{GFG}}^{\min}(g) \cdot U_{\text{GFG}}(g, t, s) \leq P_{\text{GFG}}(g, t, s) \leq P_{\text{GFG}}^{\max}(g) \cdot U_{\text{GFG}}(g, t, s), \forall_{g,t,s}$$
$$(12.27)$$

$$P_{\text{GFG}}(g, t, s) - P_{\text{GFG}}(g, t-1, s) \leq \text{RU}_{\text{GFG}}(g) \cdot U_{\text{GFG}}(g, t, s), \forall_{g,t,s} \qquad (12.28)$$

$$P_{\text{GFG}}(g, t-1, s) - P_{\text{GFG}}(g, t, s) \leq \text{RD}_{\text{GFG}}(g) \cdot U_{\text{GFG}}(g, t, s), \forall_{g,t,s} \qquad (12.29)$$

$$0 \leq \text{SUC}_{\text{GFG}}(g, t, s) = \text{su}_{\text{GFG}}(g) \cdot (U_{\text{GFG}}(g, t, s) - U_{\text{GFG}}(g, t-1, s)), \forall_{g,t,s}$$
$$(12.30)$$

$$0 \leq \text{SDC}_{\text{GFG}}(g, t, s) = \text{sd}_{\text{GFG}}(g) \cdot (U_{\text{GFG}}(g, t-1, s) - U_{\text{GFG}}(g, t, s)), \forall_{g,t,s}$$
$$(12.31)$$

$$\text{OC}_{\text{GFG}}(g, t, s) = G_{\text{GST-GFG}}(g, t, s) \cdot \lambda_{\text{gas}} + \text{SUC}_{\text{GFG}}(g, t, s) + \text{SDC}_{\text{GFG}}(g, t, s), \forall_{g,t,s}$$
$$(12.32)$$

Technical and operational constraints of GFGs are illustrated by Eqs. (12.26)–(12.31). In addition, these generators' operating costs are calculated based on Eq. (12.32).

12.2.9 Hybrid Energy Storage System Constraints

As previously stated, the studied IEHN is equipped with a HESS, which consists of three main elements, i.e., BESS, TESS, and finally, NGESS. The operational and technical constraints of these devices are mathematically modeled in the following.

12.2.9.1 Battery Energy Storage System

$$P_{\text{BESS}}^{\text{ch,min}}(\text{bs}).U_{\text{BESS}}(\text{bs}, t, s) \leq P_{\text{BESS}}^{\text{ch}}(\text{bs}, t, s) \leq P_{\text{BESS}}^{\text{ch,max}}(\text{bs}).U_{\text{BESS}}(\text{bs}, t, s), \forall_{\text{bs},t,s}$$

$$(12.33)$$

$$P_{\text{BESS}}^{\text{dch,min}}(\text{bs}).(1 - U_{\text{BESS}}(\text{bs}, \ t, \ s)) \leq P_{\text{BESS}}^{\text{dch}}(\text{bs}, t, s)$$
$$\leq P_{\text{BESS}}^{\text{dch,max}}(\text{bs}).(1 - U_{\text{BESS}}(\text{bs}, \ t, \ s)), \forall_{\text{bs},t,s}$$

$$(12.34)$$

$$E_{\text{BESS}}(\text{bs}, t, s) = E_{\text{BESS}}(\text{bs}, t - 1, s) + P_{\text{BESS}}^{\text{ch}}(\text{bs}, t, s) \cdot \eta_{\text{BESS-ch}}(\text{bs})$$
$$- P_{\text{BESS}}^{\text{dch}}(\text{bs}, t, s)/\eta_{\text{BESS-dch}}(\text{bs}), \forall_{\text{bs},t > 1,s}$$

$$(12.35)$$

$$E_{\text{BESS}}(\text{bs}, t, s) = E_{\text{BESS}}^{\text{ini}}(\text{bs}), \forall_{\text{bs},t = 1,s}$$

$$(12.36)$$

$$E_{\text{BESS}}^{\text{min}}(\text{bs}) \leq E_{\text{BESS}}(\text{bs}, t, s) \leq E_{\text{BESS}}^{\text{max}}(\text{bs}), \forall_{\text{bs},t,s}$$

$$(12.37)$$

The charged and discharged power of BESSs is limited by Eqs. (12.33) and (12.34), respectively. Additionally, the amount of electrical energy stored in these systems and its limitations are demonstrated by Eqs. (12.35)–(12.37) [25].

12.2.9.2 Thermal Energy Storage System

$$H_{\text{TESS}}^{\text{ch,min}}(\text{ts}).U_{\text{TESS}}(\text{ts}, t, s) \leq H_{\text{TESS}}^{\text{ch}}(\text{ts}, t, s) \leq H_{\text{TESS}}^{\text{ch,max}}(\text{ts}).U_{\text{TESS}}(\text{ts}, t, s), \forall_{\text{ts},t,s}$$

$$(12.38)$$

$$H_{\text{TESS}}^{\text{dch,min}}(\text{ts}).(1 - U_{\text{TESS}}(\text{ts}, \ t, \ s)) \leq H_{\text{TESS}}^{\text{dch}}(\text{ts}, t, s)$$
$$\leq H_{\text{TESS}}^{\text{dch,max}}(\text{ts}).(1 - U_{\text{TESS}}(\text{ts}, \ t, \ s)), \forall_{\text{ts},t,s}$$

$$(12.39)$$

$$E_{\text{TESS}}(\text{ts}, t, s) = E_{\text{TESS}}(\text{ts}, t - 1, s) \cdot \eta_{\text{TESS-sb}}(\text{ts}) + H_{\text{TESS}}^{\text{ch}}(\text{ts}, t, s) \cdot \eta_{\text{TESS-ch}}(\text{ts})$$
$$- H_{\text{TESS}}^{\text{dch}}(\text{ts}, t, s)/\eta_{\text{TESS-dch}}(\text{ts}), \forall_{\text{ts},t > 1,s}$$

$$(12.40)$$

$$E_{\text{TESS}}(\text{ts}, t, s) = E_{\text{TESS}}^{\text{ini}}(\text{ts}), \forall_{\text{ts},t = 1,s}$$

$$(12.41)$$

$$E_{\text{TESS}}^{\text{min}}(\text{ts}) \leq E_{\text{TESS}}(\text{ts}, t, s) \leq E_{\text{TESS}}^{\text{max}}(\text{ts}), \forall_{\text{ts},t,s}$$

$$(12.42)$$

The charged and discharged heat of TESSs is limited by Eqs. (12.38) and (12.39), respectively. Also, the amount of thermal energy stored in these systems and its limitations are shown by Eqs. (12.40)–(12.42) [26]. It must be noted that $\eta_{\text{TESS-sb}}(\text{ts})$ in Eq. (12.40) models the efficiency of TESSs for storing thermal energy in their standby period.

12.2.9.3 Natural Gas Energy Storage System

$$G_{\text{NGESS}}^{\text{in,min}}(\text{gs}).U_{\text{NGESS}}(\text{gs}, t, s) \leq G_{\text{GST-NGESS}}^{\text{in}}(\text{gs}, t, s)$$
$$\leq G_{\text{NGESS}}^{\text{in,min}}(\text{gs}).U_{\text{NGESS}}(\text{gs}, t, s), \forall_{\text{gs},t,s}$$

$$(12.43)$$

$$G_{\text{NGESS}}^{\text{out, min}}(gs).(1 - U_{\text{NGESS}}(gs, t, s)) \leq G_{\text{NGESS}}^{\text{out}}(gs, t, s)$$
$$\leq G_{\text{NGESS}}^{\text{out, max}}(gs).(1 - U_{\text{NGESS}}(gs, t, s)), \forall_{gs,t,s} \tag{12.44}$$

$$E_{\text{NGESS}}(gs, t, s) = E_{\text{NGESS}}(gs, t-1, s) + G_{\text{GST-NGESS}}^{\text{in}}(gs, t, s) \cdot \eta_{\text{NGESS-in}}(gs)$$
$$- G_{\text{NGESS}}^{\text{out}}(gs, t, s)/\eta_{\text{NGESS-out}}(gs), \forall_{ts, t > 1, s} \tag{12.45}$$

$$E_{\text{NGESS}}(gs, t, s) = E_{\text{NGESS}}^{\text{ini}}(gs), \forall_{gs, t = 1, s} \tag{12.46}$$

$$E_{\text{NGESS}}^{\text{min}}(gs) \leq E_{\text{NGESS}}(gs, t, s) \leq E_{\text{NGESS}}^{\text{max}}(gs), \forall_{gs,t,s} \tag{12.47}$$

$$\text{OC}_{\text{NGESS}}(gs, t, s) = G_{\text{GST-NGESS}}^{\text{in}}(gs, t, s) \cdot \lambda_{\text{gas}}, \forall_{gs,t,s} \tag{12.48}$$

The amount of charged and discharged gas of NGESSs is restricted by Eqs. (12.43) and (12.44), respectively. Moreover, the amount of gas energy stored in these systems and its limitations are stated by Eqs. (12.45)–(12.47). Finally, their operating costs are calculated by Eq. (12.48) [27].

It is worth noting that, the mathematical model of PVs, WTs as well as STs and their associated expressions are part of data pre-processing, where the values of wind speed and solar irradiation are converted into power and heat values. Then, these values act as input data for the raised optimization problem. In order to calculate the output power and heat of these renewable-based resources, expressions that have been presented in [23] are exploited in this study.

12.3 Numerical Results and Discussions

In this section of the chapter, a typical case study is conducted to investigate the impact of a HESS on the optimal performance of IEHNs. Notably, the optimization problem provided in the previous section has been solved by the CPLEX solver in the GAMS programming software. Before analyzing simulation results, the studied IEHN and its technical specifications are presented in the following.

As stated, to consider the stochastic nature of electricity and heat prices, electricity and heat demands, wind speed, and solar irradiance in the IEHN's optimal operation, a two-stage programming scheme is executed, and at the RT stage, ten reduced scenarios are generated based on the utilized model in [22]. Figure 12.2 shows the electricity and heat demands of the system for these scenarios.

Also, Figure 12.3 depicts the RT electricity and heat prices for the generated scenarios. In this study, the price of natural gas is considered as three tariffs based on Figure 12.4. Moreover, the heat value of natural gas is assumed to be 0.0093 MWh/m³.

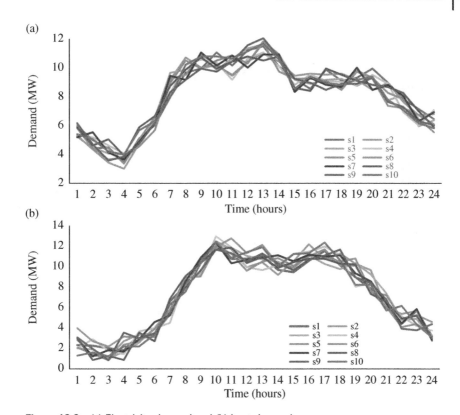

Figure 12.2 (a) Electricity demand and (b) heat demand.

Technical specifications of the IEHN's generation sources, as well as P2X technologies, are summarized in Table 12.1. Characteristics of the available BESS, TESS, and NGESS within the HESS are reported in Table 12.2. In the end, the installed wind power, solar power, and solar heat capacities of the IEHN are 2.0 MW, 3.15 MW, and 3.2 MW, respectively.

As mentioned in Section 12.2, the IEHN's exchanged power and heat energy with the upstream electricity and heat networks is performed in two DA and RT stages. In this context, the DA traded power and heat of the IEHN equipped with the HESS are illustrated in Figure 12.5. Notably, in this figure, positive bars are exported or sold energy and negative bars are imported or purchased energy.

Based on Figure 12.5, the IEHN has imported power from the electricity network in most hours of the day, and merely at hour one, it has exported power to the upstream grid. On the other hand, the IEHN operator has exported heat energy to the upstream network at hours 1–4 and 24. In the rest hours of the day, heat

(a)

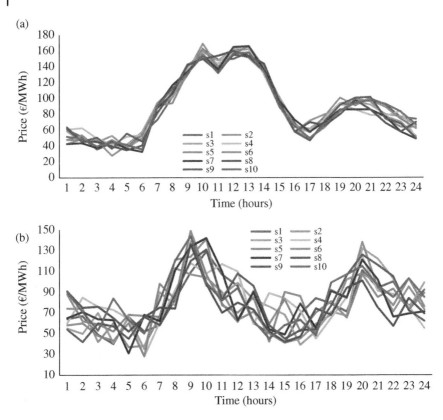

Figure 12.3 (a) Electricity price and (b) heat price.

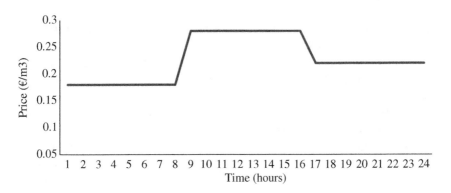

Figure 12.4 Natural gas price.

Table 12.1 Technical specifications of generation units and P2X technologies.

			CHPs		
# Unit	P_{CHP}^{max}	P_{CHP}^{min}	$\eta_{CHP\text{-}E}$	$\eta_{CHP\text{-}H}$	HPR
1	5	0.5	0.45	0.55	1

GFGs					
# Unit	P_{GFG}^{max}	P_{GFG}^{min}	$\eta_{GFG\text{-}E}$	RD_{GFG}, RU_{GFG}	
1	2	0.2	0.85	1, 1	

NGFGs					
# Unit	P_{NGFG}^{max}	P_{NGFG}^{min}	α, β	RD_{NGFG}, RU_{NGFG}	
1	3	0.3	26, 42	1.5, 1.5	

EHPs				
# Unit	H_{EHP}^{max}		H_{EHP}^{min}	CP
1	1.5		0	2.5

EBs			
# Unit	P_{EB}^{max}	P_{EB}^{min}	$\eta_{EB\text{-}H}$
1	2	0	0.8

Table 12.2 Technical specifications of the HESS.

			BESS	
# Unit	$E_{BESS}^{max}, E_{BESS}^{min}$	$P_{BESS}^{ch,max}, P_{BESS}^{ch,min}$	$P_{BESS}^{dch,max}, P_{BESS}^{dch,min}$	$\eta_{BESS\text{-}ch}, \eta_{BESS\text{-}dch}$
1	5, 0.5	1, 0	1, 0	0.95

TESS				
# Unit	$E_{TESS}^{max}, E_{TESS}^{min}$	$H_{TESS}^{ch,max}, H_{TESS}^{ch,min}$	$H_{TESS}^{dch,max}, H_{TESS}^{dch,min}$	$\eta_{TESS\text{-}ch}, \eta_{TESS\text{-}dch}, \eta_{TESS\text{-}sb}$
1	4, 0	2, 0	2, 0	0.98, 0.98, 0.95

NGESS				
# Unit	$E_{NGESS}^{max}, E_{NGESS}^{min}$	$G_{NGESS}^{in,max}, G_{NGESS}^{in,min}$	$G_{NGESS}^{out,max}, G_{NGESS}^{out,min}$	$\eta_{NGESS\text{-}in}, \eta_{NGESS\text{-}out}$
1	1000, 50	250, 0	250, 0	0.98

energy has been imported from the network. For another issue, it is observed that the grid operator has purchased less power from the electricity network during the middle of the day when the electricity price is at its peak value and prefers to utilize its own dispatchable resources. The reason is that at these hours, generation prices

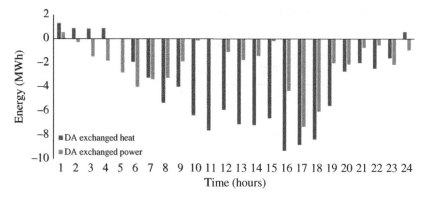

Figure 12.5 IEHN's DA exchanged energy in the presence of the HESS.

of these types of resources are lower than electricity prices. Additionally, the operator expects that available PVs generate more power at the considered time
interval.

To better investigate the HESS's impact on the optimal operation of the IEHN,
the DA traded power and heat of the system in the absence of all storage elements
is demonstrated in Figure 12.6.

By comparing Figures 12.5 and 12.6, it is shown that if the IEHN does not contain the energy storage capacity, it is forced to import more power and heat energy
from the upstream networks at peak hours, which leads to high operating costs.
For another, it can be concluded that in the presence of the HESS, the IEHN
has purchased more energy from the heat and electricity upstream grids during
hours 16–18. That is because at these hours, the energy price in both sectors

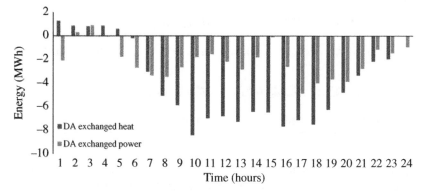

Figure 12.6 IEHN's DA exchanged energy in the absence of the HESS.

has reduced, and the IEHN operator has tended to store more energy in the existing HESS and use it at peak hours.

To assess the performance of the studied IEHN at the RT stage, in the following of this section, the operating points of the system's generation units, P2X elements, and storage systems, as well as its imbalance power and heat energy, are analyzed in more detail for a sample scenario, scenario 4. In this regard, the production and consumption profiles of connected resources to the electricity sector, as well as the amount of charged and discharged power of the BESS are displayed in Figure 12.7.

According to Figure 12.7, the IEHN operator has exploited the BESS's storage capacity to transfer power from off-peak hours to peak hours and hence enhance its expected profit. Moreover, to improve the incorporated network's flexibility and efficiency, the available EB has consumed electricity at off-peak hours to produce thermal energy due to the lower electricity price compared to the heat price. As for the EHP, due to the high CP, this unit has been in service for most hours of the studied day, and it is out of service only at peak hour 12, when both power consumption and electricity prices are at their peak values.

On the other hand, the production profiles of connected resources to the heat sector, as well as the amount of charged and discharged heat of the TESS, are depicted in Figure 12.8.

As clear in Figure 12.8, the IEHN has generated the highest amount of heat energy at hours 9–10 and 20–21, where the price of the heat network is at its peak value. Furthermore, clearly, the TESS has charged at off-peak hours and discharged at peak hours to supply more thermal energy to the IEHN.

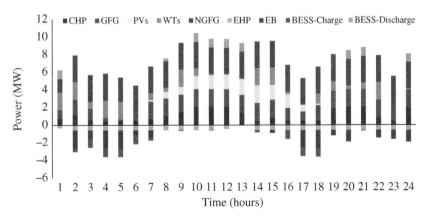

Figure 12.7 Operating points of connected sources to the electricity sector in scenario 4.

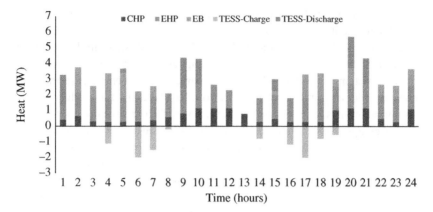

Figure 12.8 Operating points of connected sources to the heat sector in scenario 4.

The amount of gas consumption by the IEHN's gas-based resources, namely GFG and CHP, as well as the amount of gas exchange with the NGESS, are shown in Figure 12.9.

In Figure 12.9, the total amount of discharged gas from the NGESS is equal to the sum of discharged gas to the CHP and GFG. Similar to the BESS and TESS, in the NGESS, gas has stored during off-peak hours and discharged at peak hours as well. Moreover, during the peak of electricity and heat consumption, the IEHN's gas-based resources, i.e., CHP and GFG, have consumed the highest amount of gas. In addition to the gas flow from the GST, the discharged gas of the NGESS has also provided some part of the consumed gas of these units.

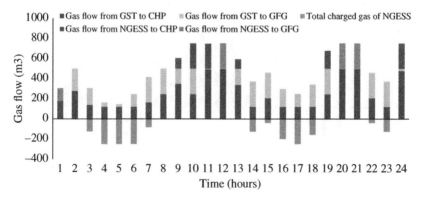

Figure 12.9 Gas flow of resources within the IEHN in scenario 4.

To evaluate the effect of the HESS presence on the IEHN's RT operation, the imbalance power and imbalance heat of the system with and without considering the storage capacity are compared with one another for a sample scenario, scenario 4, in Figure 12.10a and b.

As it is shown in Figure 12.10a and b, in the absence of all storage elements, the imbalance power and heat of the IEHN have increased significantly at the RT stage. The reason is that by adding more flexibility to the IEHN, storage devices allow the system operator to adjust its RT production/consumption more appropriately and minimize its imbalance costs.

To confirm the effective impact of the HESS on the IEHN's low imbalance power and imbalance heat costs, these costs are calculated for all scenarios and compared in two varied cases, with and without considering the HESS. Table 12.3 shows the obtained results.

According to Table 12.3, the system's imbalance costs in both electricity and heat sectors have been reduced remarkably by exploiting the HESS. As stated before, the reason for this reduction is to improve the flexibility of the IEHN at the RT

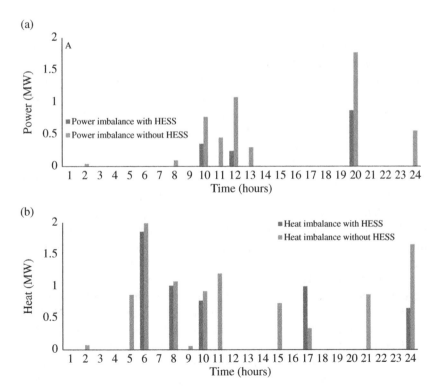

Figure 12.10 (a) Imbalance power in scenario 4 and (b) imbalance heat in scenario 4.

Table 12.3 Comparison of imbalance costs in all scenarios.

Number scenario	Imbalance cost without the HESS (€/day)		Imbalance cost with the HESS (€/day)		Cost variation (%)	
	Heat	Power	Heat	Power	Heat	Power
s1	549.49	313.94	274.39	28.85	−50.06	−90.81
s2	572.22	436.67	292.12	17.78	−48.95	−95.93
s3	544.98	1161.48	339.07	729.10	−37.78	−37.23
s4	707.76	598.55	308.17	167.58	−56.46	−72.00
s5	1156.78	505.14	737.31	289.00	−36.26	−42.79
s6	374.93	758.25	16.11	409.65	−95.70	−45.98
s7	593.11	324.55	376.08	0	−36.59	−100
s8	683.20	515.92	275.68	227.13	−59.65	−55.98
s9	378.91	522.47	46.63	145.33	−87.69	−72.18
s10	562.42	916.88	237.80	598.19	−57.72	−34.76

stage so that the system operator is able to adjust its generation in a way the lowest possible cost is imposed on the system.

Based on Table 12.3, the highest imbalance cost reduction is related to the power sector in scenario 7, where the IEHN has been able to diminish its imbalance power cost to zero by utilizing the HESS. To further analyze this matter, in Figure 12.11a and b, the considered system's imbalance powers with and without the existence of HESS are compared.

As clear in Figure 12.11a and b, by adding the capacity of flexible resources to the system, the IEHN operator has been able to match the DA and RT powers and hence reduce imbalance power and its associated cost to zero.

In the end, the daily profit of the IEHN operator in the presence and absence of the HESS is compared with each other in Table 12.4. Moreover, to evaluate the role of the BESS, TESS, and NGESS, the profit of the IEHN operator without considering one of the mentioned elements is calculated in this table as well.

As expected, the highest amount of profit is associated with the case in which the HESS is equipped with all three storage systems, case 5. Among cases 2, 3, and 4, the best case is related to case 3, in which the HESS contains BESS and NGESS. This matter specifies that the IEHN's electricity sector, as the backbone of the whole incorporated system, has a better situation. The reason is that, on the one hand, the system has the BESS that has procured some part of its required power. On the other hand, by charging gas in the NGESS and discharging it to the CHP and GFG, it has been able to provide some part of its needed power at the peak of electricity consumption and price.

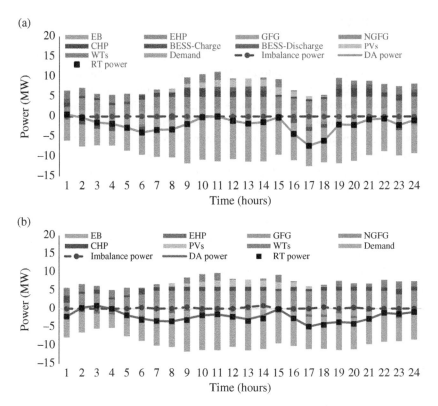

Figure 12.11 (a) Powers with the HESS in scenario 7 and (b) powers without the HESS in scenario 7.

Table 12.4 Expected profit in different cases.

	BESS	TESS	NGESS	Profit (€/day)	Variation (%)
Case 1 (Base)	✗	✗	✗	5432.76	—
Case 2	✗	✓	✓	6662.69	22.64
Case 3	✓	✗	✓	6773.58	24.68
Case 4	✓	✓	✗	6533.93	20.27
Case 5	✓	✓	✓	7226.50	33.02

12.4 Conclusion

This chapter modeled an IEHN in the presence of various generation units, P2X conversion elements, as well as storage devices and evaluated the effect of a HESS, consisting of BESS, TESS, and NGESS, on the optimal operation of the incorporated network. To consider a wide range of uncertainties in the mentioned decision-making problem, a two-stage stochastic programming scheme was suggested in this study. A typical case study was implemented to investigate and compare the performance of the system in the presence and absence of the HESS. Simulation results represented that in the absence of storage devices, the operator is forced to import more power and heat energy from the upstream grids in the DA stage, which leads to high operating costs of the system. On the other hand, by utilizing a HESS with all three elements, the RT imbalance power and RT imbalance heat of the IEHN are decreased remarkably, and hence the benefit of the system operator could be increased up to 33%. For another issue, it was observed that in the studied system, the combination of the BESS and NGESS, as a HESS, is more profitable than other possible cases so that the operator could earn 24% more benefit as opposed to the base case, without the HESS.

Acknowledgment

The work of Sara Haghifam has been supported by the KAUTE Foundation, which supports the best researchers in the field to produce results that reach not just the scientific community but also businesses and decision-makers. This foundation aims to help Finnish society and businesses to reform economically, socially, and ecologically in a sustainable way.

References

1 Zhang, M., Wu, Q., Wen, J. et al. (2021). Optimal operation of integrated electricity and heat system: a review of modeling and solution methods. *Renew. Sust. Energ. Rev.* 135: 110098.

2 Bogdanov, D., Gulagi, A., Fasihi, M., and Breyer, C. (2021). Full energy sector transition towards 100% renewable energy supply: integrating power, heat, transport and industry sectors including desalination. *Appl. Energy* 283: 116273. https://doi.org/10.1016/j.apenergy.2020.116273.

3 Bernath, C., Deac, G., and Sensfuß, F. (2021). Impact of sector coupling on the market value of renewable energies–a model-based scenario analysis. *Appl. Energy* 281: 115985.

4 Kamrani, F., Fattaheian-Dehkordi, S., Gholami, M. et al. (2021). A two-stage flexibility-oriented stochastic energy management strategy for multi-microgrids considering interaction with gas grid. *IEEE Trans. Eng. Manag.* https://doi.org/10.1109/TEM.2021.3093472.

5 Zhang, M., Wu, Q., Wen, J. et al. (2020). Two-stage stochastic optimal operation of integrated electricity and heat system considering reserve of flexible devices and spatial-temporal correlation of wind power. *Appl. Energy* 275: 115357.

6 Skov, I.R., Schneider, N., Schweiger, G. et al. (2021). Power-to-X in denmark: an analysis of strengths, weaknesses, opportunities and threats. *Energies* 14 (4): 913.

7 Ahn, H., Miller, W., Sheaffer, P. et al. (2021). Opportunities for installed combined heat and power (CHP) to increase grid flexibility in the US. *Energy Policy* 157: 112485.

8 Kirkerud, J.G., Bolkesjø, T.F., and Trømborg, E. (2017). Power-to-heat as a flexibility measure for integration of renewable energy. *Energy* 128: 776–784.

9 Xing, X., Lin, J., Song, Y. et al. (2018). Modeling and operation of the power-to-gas system for renewables integration: a review. *CSEE J. Power Energy Syst.* 4 (2): 168–178.

10 Yamchi, H.B., Safari, A., and Guerrero, J.M. (2021). A multi-objective mixed integer linear programming model for integrated electricity-gas network expansion planning considering the impact of photovoltaic generation. *Energy* 222: 119933.

11 Barelli, L., Ciupageanu, D.-A., Ottaviano, A. et al. (2020). Stochastic power management strategy for hybrid energy storage systems to enhance large scale wind energy integration. *J. Energy Storage* 31: 101650.

12 Chen, Y., Xu, Y., Li, Z., and Feng, X. (2019). Optimally coordinated dispatch of combined-heat-and-electrical network with demand response. *IET Gener. Transm. Distrib.* 13 (11): 2216–2225.

13 Majidi, M., Mohammadi-Ivatloo, B., and Anvari-Moghaddam, A. (2019). Optimal robust operation of combined heat and power systems with demand response programs. *Appl. Therm. Eng.* 149: 1359–1369.

14 Yu, Y., Chen, H., Chen, L. et al. (2019). Optimal operation of the combined heat and power system equipped with power-to-heat devices for the improvement of wind energy utilization. *Energy Sci. Eng.* 7 (5): 1605–1620.

15 Chen, Y., Zhang, Y., Wang, J., and Lu, Z. (2020). Optimal operation for integrated electricity–heat system with improved heat pump and storage model to enhance local energy utilization. *Energies* 13 (24): 6729.

16 Liu, T., Pan, W., Zhu, Z., and Liu, M. (2021). Optimal risk operation for a coupled electricity and heat system considering different operation modes. *IEEE Access* 9: 18831–18841.

17 Zhong, J. et al. (2021). Distributed operation for integrated electricity and heat system with hybrid stochastic/robust optimization. *Int. J. Electr. Power Energy Syst.* 128: 106680.

18 Deng, L., Sun, H., Li, B. et al. (2021). Optimal operation of integrated heat and electricity systems: a tightening McCormick approach. *Engineering* 7 (8): 1076–1086.

19 Wu, Q., Tan, J., Jin, X. et al. (2022). Chapter 9 – Day-ahead stochastic optimal operation of the integrated electricity and heating system considering reserve of flexible devices. In: *Optimal Operation of Integrated Multi-Energy Systems Under Uncertainty*, 221–249. Elsevier https://doi.org/10.1016/B978-0-12-824114-1.00011-1.

20 Wu, Q., Tan, J., Zhang, M. et al. (ed.) (2022). Chapter 6 – Adaptive robust energy and reserve co-optimization of an integrated electricity and heating system considering wind uncertainty. In: *Optimal Operation of Integrated Multi-Energy Systems Under Uncertainty*, 145–170. Elsevier https://doi.org/10.1016/B978-0-12-824114-1.00013-5.

21 Tan, H., Yan, W., Ren, Z. et al. (2022). A robust dispatch model for integrated electricity and heat networks considering price-based integrated demand response. *Energy* 239: 121875.

22 Dadashi, M., Zare, K., Seyedi, H., and Shafie-khah, M. (2022). Coordination of wind power producers with an energy storage system for the optimal participation in wholesale electricity markets. *Int. J. Electr. Power Energy Syst.* 136: 107672.

23 Haghifam, S., Zare, K., Abapour, M. et al. (2020). A stackelberg game-based approach for transactive energy management in smart distribution networks. *Energies* 13 (14): 3621.

24 Jadidbonab, M., Mohammadi-Ivatloo, B., Marzband, M., and Siano, P. (2020). Short-term self-scheduling of virtual energy hub plant within thermal energy market. *IEEE Trans. Ind. Electron.* 68 (4): 3124–3136.

25 Haghifam, S., Zare, K., and Dadashi, M. (2019). Bi-level operational planning of microgrids with considering demand response technology and contingency analysis. *IET Gener. Transm. Distrib.* 13 (13): 2721–2730.

26 Habibi, M., Vahidinasab, V., and Sepasian, M.S. (2022). Value of integrated electricity and heat scheduling with considering TSO–DSO cooperation. *Int. J. Electr. Power Energy Syst.* 135: 107526.

27 Mansour-Saatloo, A., Agabalaye-Rahvar, M., Mirzaei, M.A. et al. (2020). Robust scheduling of hydrogen based smart micro energy hub with integrated demand response. *J. Clean. Prod.* 267: 122041.

13

Operational Coordination to Boost Efficiency of Complex Heat and Electricity Microgrids

Ali Jalilian[1], Milad Mohammadyari[2], Sasan Azad[3,4], Mohammad Taghi Ameli[3,4], and Ensieh Ghanbari[5]

[1] Deputy of Operation and Dispatch, Kermanshah Power Electrical Distribution Company (KPEDC), Kermanshah, Iran
[2] Department of Electrical Engineering, University of Tehran, Tehran, Iran
[3] Department of Electrical Engineering, Shahid Beheshti University, Tehran, Iran
[4] Electrical Networks Research Institute, Shahid Beheshti University, Tehran, Iran
[5] Department of Electrical Engineering, Iran University of Science and Technology, Tehran, Iran

Abbreviations

EH	Energy hubs
EMS	Energy management systems
IES	Integrated energy systems
HES	Hybrid energy systems
MES	Multi-energy systems
RES	Renewable energy sources

Sets and Indices

S, s	Set and index for generated scenarios in the stochastic optimization method
T, t	Set and index for time periods
Ψ, ψ	Set and index of components in the system directly connected to the upstream power and urban gas networks

Parameters

R_t	The amount of solar radiation power in the surface area unit
A	PV unit's surface area
α/β	PV/inverter's efficiency

Coordinated Operation and Planning of Modern Heat and Electricity Incorporated Networks,
First Edition. Edited by Mohammadreza Daneshvar, Behnam Mohammadi-Ivatloo, and Kazem Zare.
© 2023 The Institute of Electrical and Electronics Engineers, Inc.
Published 2023 by John Wiley & Sons, Inc.

v	Wind speed
V_{ci}/V_{co}	Cut-in/ cut-out speed of the wind turbine
$P^{FC}_{Capacity}$	The electrical power generation capacity of the fuel cell
$R^{FC}_{Th/El}$	The thermal to electrical energy generation ratio in the fuel cell
LHV^{gas}	Lower heating value of gas
η^{CHP}_{el}	Electrical efficiency of CHP unit
μ_{GT}	Percentage of heat losses in gas turbine
H^{HST}_{Cap}	Capacity of the heat storage tank
$H^{HST}_{in\_max}$	Maximum input heat flow capacity of the heat storage tank
$H^{HST}_{Out\_max}$	Maximum output heat flow capacity of the heat storage tank
$\eta^{HST, in}$	Heat storage tank's input heat efficiency
$\eta^{HST, Out}$	Heat storage tank's output heat efficiency
ζ	Heat storage tank's hourly heat loss ratio
η^{in}/η^{out}	Charge/discharge efficiency of the electrical storage
δ	The value of battery self-discharge
$E^{Bt}_{max}/E^{Bt}_{min}$	The upper and lower limits of the battery's charge
$\hat{P}^{in}/\hat{P}^{out}$	Maximum charging and discharging power of the battery
P^L_t	Expected consuming electrical load
DR^{Elec}_{max}	The costumer's maximum load variation
$Price^{\psi}_{t,s}$	Gas or electricity price
γ_s	Probability of each scenario
$P^{max}_{t,s}$	Maximum power exchange between grid and the upstream network

Variables

P^{PV}_t	Generated power of the PV unit
P^{Wind}	Generated power of the wind turbine
P^{FC}	Generated electrical power of the fuel cell unit
H^{FC}	Generated heat of the fuel cell
P^{CHP}_t	Generated power by gas turbine
G^{CHP}_t	Purchased gas from gas network
$G^{CHP,EX}_t$	Exhaust gas of gas turbine
$H^{HST,in/out}_t$	The input/output heat of the heat storage tank
H^{HST}_t	Available heat of the heat storage tank
P^{in}_t/P^{out}_t	Charge/discharge power of the electrical storage

E_t^{Bt}	Available charge of the energy storage system in each time step
β_t^{ch}	Binary variable showing the charging/discharging state of the battery
P_t^{TOU}	Load's variation due to applying demand response program in the microgrid
C^{Total}	System total operation cost
$P_{t,s}^{\psi}$	Electricity or gas consumption and contains a negative value for generation
$P_{t,s}^{L_{\text{DR}}}$	Microgrid's load consumption after applying the demand response program
$P_{t,s}^{\text{Grid}}$	Power obtained from the upstream electrical grid
$P_{t,s}^{G}$	Sum of electrical power generation
$\beta_{t,s}^{\text{pur}}$	Binary variable for status of purchasing from the upstream grid
Φ	Electrical or thermal power or natural gas flow

13.1 Introduction

The energy crisis, environmental issues such as global warming and air pollution, and also increase in energy demand, have all leaves us with no choice, but to use the available energy resources most efficiently and to move toward renewable energy sources (RESs). In this regard, integrated energy systems (IES) are proposed as a solution to the mentioned issues. Energy system integration is the process of coordinating the operation and planning of energy systems across multiple pathways and/or geographical scales to deliver reliable, cost-effective energy services with minimal impact on the environment [1]. IES provides the opportunity to bring together different energy systems as multi-energy systems (MES) or hybrid energy systems (HES). MES makes the generation, storage, and supply of different energy vectors in different geographical scales possible. This offers great operational flexibility and efficiency since it provides physical links between different energy sources through cogeneration. It also enables the integration of local energy generation, such as renewable resources, that are able to interact with each other and also with upstream networks. Therefore, as smart cities and smart homes are rapidly progressing, the utilization of HESs in this global trend seems inevitable.

Despite all the advantages of the IES, the management of this system introduces serious challenges in terms of planning and operational performance. In other words, the coordinated, economic, reliable, and even environmentally friendly performance of these systems rely on comprehensive studies of optimal planning and operational scheduling. For this reason, a sample HES consisting of the generation, storage, and consumption of electrical and thermal energy is presented in this chapter to illustrate the optimal operational management and coordination of IESs in a microgrid. The performance of different energy components, their coordinated operation with each other, along with their interaction with upstream electrical and gas networks, are modeled through a mixed-integer linear programming optimization framework to minimize the total operation cost of the microgrid while maintaining the system's technical constraints. The inherent uncertainties of renewable energy generation, electrical and thermal load, etc., are taken into account using a stochastic optimization approach. The simulation results of the coordinated hybrid system are further discussed in detail to provide a more intuitive illustration of the system's mechanism.

The effective performance of energy systems depends on the coordinated operation of different energy generation and storage units with load consumption patterns, energy price, etc. Therefore, different approaches have been presented in the literature in order to operate these energy systems optimally. Reference [2] proposes the coordination of microgrid's energy units to reduce total energy cost in an optimization model using genetic algorithm. The optimal operation of multiple energy hubs (EHs) is studied in [3], considering various energy resources, such as renewable energy, combined heat and power (CHP), power-to-gas, and gas-fired generator. In this study, a multi-objective model is proposed to maximize social welfare and minimize CO_2 emission, and the problem is solved using genetic algorithm. In [4–6], the optimal operation framework of a microgrid is presented in a multi-objective model to reduce air pollutants along with energy costs. In addition to the costs of supplying energy, the cost of decreasing the battery's lifespan is considered in [7]. In [8], a solution for determining the optimal size of three types of storage systems in microgrid has been proposed using particle swarm optimization. A robust chance-constrained optimization model is used in [9] for optimal operation management of an EH comprising of electrical, heating, and cooling demands, renewable power generation, and electrical energy storage devices. The considered EH in this study follows a centralized framework, and the EH operator is responsible for the optimal operation of the hub assets based on the day-ahead scheduling.

Also, energy management systems (EMSs) in microgrids are facing problems regarding the management of RESs such as wind or solar energies. This problem is due to the inherently uncertain nature of these energies that causes the difference between the amount of predicted and actual generation of energy [10]. One solution to this problem is using electrical energy storage systems [11]. In most cases, storage systems maintain the balance between generation and consumption by storing energy in off-peak hours and discharging it in peak hours [12]. Various studies have been carried out regarding the use of energy storage systems in microgrids [13]. Multi-objective energy management of the microgrid, considering wind power generation uncertainties, in the presence of electrical energy storage systems has been done in [14] using artificial neural network to predict wind power generation. In [15], the EMS of the microgrid has been analyzed considering battery as the storage system, and the optimization problem has been solved using distributed intelligence and multi-agent systems. Reference [16] solves the optimal management of distributed generation resources and storage systems in the microgrid using matrix real-codded genetic algorithm method. Simultaneous optimization of distributed generation resources and storage system capacities, considering the impact of climate conditions and non-dispatchable energy sources, has been introduced in [17]. The energy system management problem is presented for several microgrids, while the energy storage system has been considered as the network's reserve source.

There are numerous technical and economic incentives for the utilization of distributed generation resources to reduce greenhouse gas emissions, improve the efficiency and reliability of energy systems, energy policies in the competitive markets, and postpone transmission and distribution system reconstruction [18]. In fact, distributed generation resources comprise renewable energy units such as wind turbines, photovoltaic, fuel cells, biomass, and non-renewable units, such as micro-turbines, gas engines, diesel generators, etc. [19]. Since distributed generation resources are installed in proximity of the customers, they need no energy transmission system [20]. Integration and control of distributed generation resources, the storage equipment, and flexible loads in the network can form a low-voltage distribution network called a microgrid that is able to operate in islanding or grid-connected modes [21].

The probabilistic and varying nature in some cases such as load consumption, cost of purchasing energy, and generated power of RESs, issue problems in the operation of these networks. For this reason, scenario-based stochastic optimization methods [22–25] and probability density function [26] have been presented in the literature to cover these uncertainties. The uncertainty modeling through scenario-generation methods requires the knowledge of probability distribution function. In these methods, by dividing the probability distribution function into different segments, each segment is chosen as a scenario with a probability of occurrence proportionate to the probability distribution function [27]. The advent and considerable progress in different technologies, such as cogeneration, electrical heaters, heat pumps, etc., make the interactions and physical links between electrical and thermal energy systems stronger. Also, the advances in electrical and thermal storage systems contribute to the flexibility and reliability improvement of IESs. On the other hand, the implementation of some control and management practices, such as DRPs, has proven to be successful strategies in the operation of energy systems. Thus, some of the conventional energy components and strategies in IESs, along with their modeling consideration, are discussed in the next section.

13.2 Integrated Energy System Resources

13.2.1 Renewable Energy Resources

RESs are important parts of most energy systems since they align with many targets that IESs are based on. Thanks to a great deal of effort put into the enhancement of renewable energy technologies, many types of small-scale electrical/thermal renewable resources, such as photovoltaic panels, wind turbines, solar water heaters, etc., are currently available that can be incorporated into even single-house/small-building nano-grids. Some of the RESs that are commonly used in energy systems are briefly referred to in the following.

13.2.1.1 Photovoltaic Unit

The generated power of the PV unit corresponds with the solar radiation, its surface area, and efficiency. This generated power can be used for the microgrid's internal consumption and charging the battery, or it can be sold to the upstream grid during surplus generation hours. The generated power of the PV unit depends on the amount of solar irradiation and the PV's surface area and can be modeled through (13.1) [23].

$$P_t^{PV} = \alpha \times \beta \times A \times R_t \tag{13.1}$$

R is the amount of solar radiation power in the surface area unit. A is the PV unit's surface area. α and β are the PV and inverter's efficiencies, respectively.

13.2.1.2 Wind Turbine

The generated power of the wind turbine depends on wind speed and its rated power (13.2) [28]. Same as PV unit, the output power of wind turbine can be used in either way stated before.

$$P^{Wind} = \begin{cases} 0 & 0 \le v \le V_{ci} \\ P_{rated}^{Wind} \times \left(\frac{v - V_{ci}}{V_r - V_{ci}} \right)^3 & V_{ci} \le v \le V_r \\ P_{rated} & V_r \le v \le V_{co} \\ 0 & V_{co} \le v \end{cases} \tag{13.2}$$

where, P^{Wind} is the generated power of the wind turbine, and v is the wind speed. V_{ci} and V_{co} are cut-in and cut-out speeds of the wind turbine, respectively.

13.2.2 Combined Heat and Power Technologies

Cogeneration or CHP is the use of a heat engine to generate electricity and heat at the same time. Cogeneration is important not only because it is an efficient use of fuel or heat but also because it establishes an actual physical link between electrical and thermal energy in HESs. Due to the advances in CHP technology, cogeneration of heat and power is no longer limited to large-scale power plants, such as combined cycle power plants, and there are currently different small-scale decentralized CHP technologies such as microturbines or fuel cells that can be utilized in microgrids and residential units. Some of the common CHP technologies are mentioned below.

13.2.2.1 Fuel Cell Unit

Fuel cell is capable of generating heat and electrical energy. The generated electrical energy of this unit can supply the microgrid's load or be stored in the battery. Clearly, the generated electrical power of the fuel cell must not exceed its nominal capacity. This issue has been addressed in the mathematical model by (13.3) [23].

$$P^{\text{FC}} \leq P^{\text{FC}}_{\text{Capacity}} \tag{13.3}$$

$$H^{\text{FC}} \leq P^{\text{FC}} \times R^{\text{FC}}_{\text{Th}-\text{El}} \tag{13.4}$$

where P^{FC} is the generated electrical power, and $P^{\text{FC}}_{\text{Capacity}}$ shows the electrical power generation capacity of the fuel cell. H^{FC} is the generated heat of the fuel cell, and $R^{\text{FC}}_{\text{Th}/\text{El}}$ shows the thermal to electrical energy generation ratio in the fuel cell.

13.2.2.2 Gas Turbine

Gas turbine plays a key role in CHP systems. Gas turbine uses gas fuel to generate electrical energy while at the same time the co-generated heat can be used to supply thermal loads. The generated electrical power of the gas turbine can be modeled through (13.5)–(13.6) [29, 30].

$$P^{\text{CHP}}_t = G^{\text{CHP}}_t \times \eta^{\text{CHP}}_{\text{el}} \times \text{LHV}^{\text{gas}} \tag{13.5}$$

$$P^{\text{CHP}}_t \leq P^{\text{CHP}}_t \leq P^{\text{CHP}}_{\text{max}} \tag{13.6}$$

where P^{CHP}_t and G^{CHP}_t are total generated power by gas turbine (MW) and purchased gas from gas network (Nm^3). LHV^{gas} is the lower heating value of gas (MWh/Nm^3).

The exhaust gas of the gas turbines can be used to generate heat. So, this unused energy can be molded as (13.7).

$$G^{\text{CHP,EX}}_t = P^{\text{CHP}}_t \times \left(\frac{1 - \eta^{\text{CHP}}_{\text{el}} - \mu_{\text{GT}}}{\eta^{\text{CHP}}_{\text{el}}} \right) \tag{13.7}$$

where $G^{\text{CHP,EX}}_t$, $\eta^{\text{CHP}}_{\text{el}}$, and μ_{GT} are exhaust gas of gas turbine (Nm^3), electrical efficiency of CHP unit (%), percentage of heat losses in gas turbine (%), respectively.

13.2.3 Energy Storage Systems

Storage systems add huge flexibility to energy systems. These systems reduce the imbalance between the generated and demanded energy by capturing the energy generated at one period for use at a later time. The thermal and electrical energy storage systems are provided through different components and technologies. The existence of a heat storage system in the microgrid makes the thermal energy supply more reliable. Because, as for electricity storage system, the heat storage system can store thermal energy in off-peak load periods and inject it back to the microgrid during peak-load periods. Therefore, even if the microgrid thermal load surpasses the backup burner's capacity, the heat storage system, to some level, prevents the outage of thermal load.

Some of the thermal and electrical storage systems are stated below.

13.2.3.1 Heat Storage Tank

Heat storage tank (also called a hot water tank, thermal storage tank, hot water thermal storage unit, hot water storage tank, and hot water cylinder) is a water tank that stores heat to be used as a thermal source. Water can be used as a heat storage medium because of its high specific heat capacity. An efficiently insulated tank can retain stored heat for days, reducing fuel costs [30, 31].

$$H_t^{HST} = (1 - \zeta)H_{t-1}^{HST,Out} + H_t^{HST,in} \times \eta^{HST,in} - \frac{H_t^{HST,Out}}{\eta^{HST,Out}} \tag{13.8}$$

$$H_t^{HST} \leq H_{Cap}^{HST} \tag{13.9}$$

$$H_t^{HST,in} \leq H_{in\_max}^{HST} \tag{13.10}$$

$$H_t^{HST,Out} \leq H_{Out\_max}^{HST} \tag{13.11}$$

where $H_t^{HST,in}$, $H_t^{HST,Out}$, H_t^{HST}, and H_{Cap}^{HST} represent the input heat, output heat, available heat, and heat capacity of the heat storage tank, respectively. $H_{in\_max}^{HST}$ and $H_{Out\_max}^{HST}$ are the maximum input and output heat flow capacity of the tank. $\eta^{HST,in}$ and $\eta^{HST,Out}$ are the heat storage tank's input and output heat efficiencies, respectively. ζ shows the heat storage tank's hourly heat loss ratio.

13.2.3.2 Battery

The use of the battery as an electricity storage system, along with DRP, makes the microgrid's electrical energy consumption flexible and operation optimization possible. The use of batteries entails some considerations that are detailed through (13.12)–(13.15) [31].

$$E_t^{Bt} = (1 - \delta) \times E_{t-1}^{Bt} + P_t^{in} \times \eta^{in} - \left(\frac{P_t^{out}}{\eta^{out}} \right) \tag{13.12}$$

$$E_{min}^{Bt} \leq E_t^{Bt} \leq E_{max}^{Bt} \tag{13.13}$$

$$P_t^{in} \leq \beta_t^{ch} \times \hat{P}^{in} \tag{13.14}$$

$$P_t^{out} \leq \left(1 - \beta_t^{ch} \right) \times \hat{P}^{out} \tag{13.15}$$

where P_t^{in} and P_t^{out} show the charge and discharge power of the storage. η^{in} and η^{out} show charge and discharge efficiency. E_t^{Bt} shows the available charge of the energy storage system in each time step. δ is the coefficient used to show the value of battery self-discharge. E_{max}^{Bt} and E_{min}^{Bt} are the upper and lower limits of the battery's charge, respectively. β_t^{ch} is a binary variable showing the charging/discharging state of the battery. $\hat{P}^{in}$ and $\hat{P}^{out}$ show the maximum charging and discharging power of the battery. Therefore, (13.12) calculates the amount of energy stored in the battery. This energy increases by charging the battery and decreases by discharging it. But according to (13.13), in each time step, the store energy must not exceed its allowed range. Equations (13.14)–(13.15) show that only one of the charging or discharging states is permitted in each time step.

13.2.4 Demand Response Program

DR can be defined as the changes in electric usage by end-use customers from their normal consumption patterns in response to changes in the price of electricity over time [32]. DR can be also defined as the incentive payments designed to induce lower electricity use at times of high wholesale market price or when system reliability is jeopardized [33]. DR seeks to adjust the demand for power instead of adjusting the supply. When costumers participate in DR, there are three possible ways in which they can change their use of electricity [34]:

- Reducing their energy consumption through load curtailment strategies.
- Moving energy consumption to a different time period.
- Using onsite standby generated energy, thus limiting their dependence on the main grid.

Different DRPs can be classified into two main categories: incentive-based programs (IBP) and price-based programs (PBS) [33]. PBPs are based on dynamic pricing rates in which electricity tariffs are not flat; the rates fluctuate following the real-time cost of electricity. The ultimate objective of these programs is to flatten the demand curve by offering a high prices during peak periods and lower prices during off-peak periods. These rates include the time of use (TOU) rate, critical peak pricing (CPP), extreme day pricing (EDP), extreme day CPP, and real-time pricing (RTP) [32]. TOU program flattens the load curve by shifting the load from peak load hours to off-peak hours, which are usually the high electricity price hours, and decreases the costs [23, 30].

$$P_t^{L_{DR}} = P_t^L + P_t^{TOU} \tag{13.16}$$

$$\left| P_t^{TOU} \right| \le DR_{max}^{Elec} \times P_t^L \tag{13.17}$$

$$\sum_{t \in T} P_t^{TOU} = 0 \tag{13.18}$$

where P_t^L and P_t^{TOU} are the expected consuming electrical load and the load's variation due to applying DRP in the microgrid, respectively. DR_{max}^{Elec} is the costumer's maximum load variation. Accordingly, (13.16) calculates the consuming power of the microgrid after applying the DRP. In (13.17), the maximum allowable load variation in each time step has been limited to a specified percent of the load. Equation (13.18) expresses that DRP in the overall period, which is 24 hours of a day, has no effect on the total electrical energy consumption and only affects the hourly value.

The existence of the introduced resources and strategies in the microgrid creates several potential capabilities for increasing the performance, and decreasing energy costs that achieving them lies in the optimization and operational scheduling of microgrid's different resources. Thus, the optimal operation of the energy supply system for this microgrid is the subject of this section.

13.3 Optimization Scheme for Energy Management in the Microgrid

Noticing the variety of resources available in microgrid structure, coordinated operation of different units favorably resulting in economic performance entails an optimal operation strategy. Usually, day-ahead schemes in a few minutes to one-hour time period resolution unit scheduling, try to describe optimal operation, especially when energy market price variations call for not only plan for meeting the demand economically but also to store and benefit from price changes.

In this section, an optimization scheme for energy management in the microgrid is presented. This minimization takes place as different system agents' technical constraints, load balance, grid connection considerations, system design constraints, and microgrid performance levels are maintained. Like most power system planning and operation optimization programs, load, generation, and price forecasting are of main challenges imposing uncertainty on the expected results. In this regard, a scenario-based stochastic optimization model is presented.

13.3.1 The Objective

The objective function of the presented model aims to minimize the total operation cost of the hybrid system. Equation (13.19) shows the total cost formulation of this system. According to this equation, the total cost consists of the costs of buying electricity and urban gas from corresponding upstream networks, along with the revenue from selling electrical energy to the upstream grid.

$$\text{Minimize} : C^{\text{Total}} = \sum_{s \in S} \sum_{t \in T} \sum_{\psi \in \Psi} \left(\gamma_s \times P_{t,s}^{\psi} \times \text{Price}_{t,s}^{\psi} \right) \tag{13.19}$$

where S and T represent the generated scenario set in the stochastic optimization method, and the set for time periods, respectively. C^{Total} is the system total operation cost. Ψ shows any components in the system directly connected to the upstream power and urban gas networks. $P_{t,s}^{\psi}$ shows electricity or gas consumption and contains a negative value for generation. $\text{Price}_{t,s}^{\psi}$ shows gas or electricity price and can be different based on contracts for some components such as PVs. γ_s shows the probability of each scenario.

13.3.2 Units Technical Constraints

Technical constraints and special characteristics of different resources have been described in the previous section. Any optimization framework, depending on existing technologies installed in a HES, is subject to these constraints.

13.3.3 Load Balance Constraints

This set of constraints guarantees that the system meets electrical and thermal demands in each time period. Total electrical demand contains electrical loads and charging power for electrical storage, which come from generation units, the power from storage discharge, and the upstream electrical grid. It is noteworthy that the amount of electrical load in each time period is influenced by demand-side management programs (13.16). Thermal load balance also constrains the scheduling program, as stated in (13.19).

$$P_{t,s}^{L_{\text{DR}}} + P_{t,s}^{\text{Ch}} = P_{t,s}^{\text{Grid}} + P_{t,s}^{G} + P_{t,s}^{\text{Dch}} : \forall t \in T, s \in S \tag{13.20}$$

$$H_{t,s}^{L} + H_{t,s}^{\text{ch}} = H_{t,s}^{G} + H_{t,s}^{\text{Dch}} : \forall t \in T, s \in S \tag{13.21}$$

where $P_{t,s}^{L_{\text{DR}}}$ is the microgrid's load consumption after applying the DRP. $P_{t,s}^{\text{Ch}}$ and $P_{t,s}^{\text{Dch}}$ represent the charging and discharging power of electrical storage system, respectively. $P_{t,s}^{\text{Grid}}$ denotes the power obtained from the upstream electrical grid. $P_{t,s}^{G}$ shows the sum of electrical power generation. Similarly, parameters represented with H show the thermal power.

13.3.4 Grid Connection Constraints

Because of different power transaction prices (e.g. when considering renewable support policies) when purchasing from or selling to the upstream grid, it is usually assumed that in each time step, the power exchange between the microgrid and the upstream grid is only unidirectional. This statement means that any generated power in the microgrid is primarily used to supply internal consumption. Furthermore, this power exchange must not exceed the upstream grid's considered connection capacity. These constraints are mathematically modeled in (13.22).

$$P_{t,s}^{max} \times \left(1 - \beta_{t,s}^{pur}\right) \leq P_{t,s}^{Grid} \leq P_{t,s}^{max} \times \beta_{t,s}^{pur} \tag{13.22}$$

where the binary variable $\beta_{t,s}^{pur}$ shows the status of purchasing from the upstream grid. Also, $P_{t,s}^{max}$ is a positive value showing the maximum power exchange.

13.3.4.1 System Design Constraints
These types of constraints generally describe how various components are connected to each other in the design of the specific HES. For example, the input power of the storage system could be provided by the generated power from a number of generation units. Also, the discharge power could feed some other units (see Figure 13.1).

$$\Phi^{in} = \Phi^1 + \Phi^2 + \Phi^3 + \dots + \Phi^n \tag{13.23}$$

$$\Phi^{out} = \Phi'^1 + \Phi'^2 + \dots + \Phi'^m \tag{13.24}$$

where Φ shows the electrical or thermal power or natural gas flow.

13.3.5 Microgrid's Other Performance Constraints

In addition to the constraints mentioned above, microgrid operation is constrained to power flow formulations. Especially when the extent of the microgrid could cause a considerable voltage drop across conductors. Generally, the bigger the extent of the microgrid, the more considerations arise about operational performance, e.g., power quality, reliability, etc.

13.4 Simulations and Analysis of Results

In this section, the simulation results of a HES operation, considering its various aspects, such as generation and storage resources and DRP, have been presented. In this regard, the energy system introduced in [23] and illustrated in Figure 13.2 is considered as a basis for the studies in this section. In Figure 13.2, a sample HES of

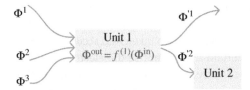

Figure 13.1 System design constraints.

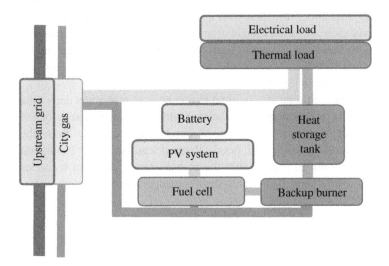

Figure 13.2 Hybrid electrical and thermal energy system.

a microgrid consisting of a photovoltaic/fuel cell/battery system for electrical energy, and a backup burner/fuel cell/heat storage tank system for the thermal energy supply of a residential unit have been considered along with the upstream electrical and urban gas networks.

According to the figure, the electricity storage system can be charged via the upstream network, PV unit, and fuel cell in microgrid light-load periods (the time that the electricity price is lower) to assist in supplying the microgrid's demand during peak-load periods (higher electricity prices). Besides supplying the electrical load and storing PV's generated power, selling energy to the upstream network is provided by the storage system.

The rest of this section deals with numerical simulation of the microgrid's optimal operation problem. In this regard, the input data are first introduced, and a stochastic approach based on scenario generation is provided to solve the optimization problem. Then, the introduced model is solved using the GAMS software, and the results are presented. This section attempts to provide a comprehensive study for assessing the quality of operating the microgrid, in the case of any of the problem's different parameters changes, to identify the impact of each parameter. In this chapter, two cases for uncertain data have been taken into account, namely deterministic and stochastic. The uncertain data comprise the electricity energy price, electrical load value, thermal load value, and solar irradiation value. In this chapter, a scenario-based optimization method has been utilized to model the uncertainties of these parameters. For any of the four introduced uncertain parameters 1000 scenarios were generated using Monte Carlo method assuming normal probability distribution function with a standard deviation of 0.1. Expected hourly profiles of each parameter is taken from [23]. Then the number of scenarios is reduced to 40 scenarios using the fast backward and forward method in SCENRED toolbox of the GAMS software. The details of the scenario-reduction algorithm are discussed in [35, 36].

It is assumed that each time step is equal to one hour of a 24 hours daily time period. In Figure 13.3, the diagrams associated with different scenarios of the uncertain parameters, i.e., hourly price of purchasing electricity from the upstream grid, hourly electrical load, hourly thermal load, and solar irradiation are illustrated. It should be noticed that the probabilities of different scenarios are not the same. Thus, these figures only show the dispersion of different scenarios and not their probability distribution.

13.4.1 Simulation Results

In the following, the model described in the previous section has been simulated in two deterministic and stochastic cases. In each of the deterministic and stochastic cases, both states of with and without DRP have been evaluated. Then, the results of load, energy supply costs, and energy generation costs of different units are analyzed and further discussed.

Simulation results corresponding to costs of purchasing electrical energy, urban gas fuel, revenues from selling generated energy of the PV unit, and finally, the total operation cost of the microgrid during a 24-hour period for different cases are presented in Table 13.1. In this table, Case 1 is without considering DRP, and Case 2 is with considering DRP with the maximum amount of 30%.

As perceived from Table 13.1, DRP can be effective in reducing the energy supply costs of the microgrid. This matter is more clearly understood through Figures 13.4 and 13.5, where the program makes this cost reduction by shifting the electrical load from peak load hours with higher prices to off-peak hours with lower electricity prices. Another important point in Table 13.1 is that by applying DRP, selling energy to the upstream grid, whether in the case of deterministic or stochastic

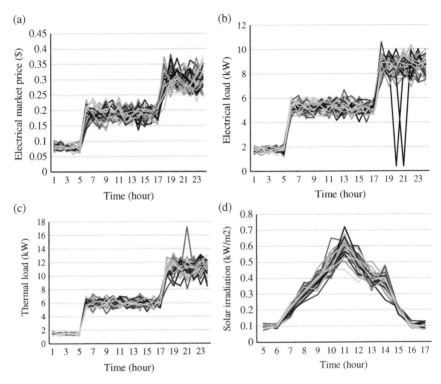

Figure 13.3 Scenarios for hourly (a) price of purchasing electricity from upstream grid, (b) electrical load, (c) thermal load, and (d) solar irradiation.

Table 13.1 Simulation results for microgrid's costs in different cases.

Cost	Deterministic		Stochastic	
	Case 1	Case 2	Case 1	Case 2
Grid`	25.433	22.530	24.186	21.511
Gas	22.085	22.601	22.195	22.494
PV	−4.463	−4.215	−4.203	−4.012
Total	43.055	40.196	42.179	39.993

method, is decreased. The reason for this is that, as observed in Figures 13.4 and 13.5 for hourly electrical load, DRP causes the electrical load in midday hours (5–17) to increase. This period is the one that the PV unit is capable of generating electrical energy.

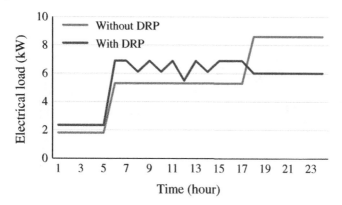

Figure 13.4 Hourly electrical load in deterministic case.

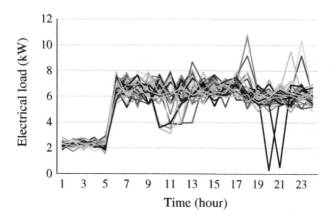

Figure 13.5 Hourly electrical load in stochastic case considering DRP.

Therefore, by applying DRP, the PV unit's generation is more dedicated to microgrid's internal consumptions and less sold to the upstream grid. The diagrams depicted in Figures 13.6 and 13.7 and related to the amount of electrical energy exported to the upstream grid, verify this matter.

In Table 13.2, the details of the PV unit's generated energy consumed in different cases is presented. In stochastic simulation cases, the reported values are the expected values.

13.4.2 Result Analysis for Electrical Storage Unit

Batteries are used usually because of their role in increasing the network's flexibility and providing the capability of shifting the load from peak load hours to off-peak hours, and by doing so, they somehow change the load profile. In

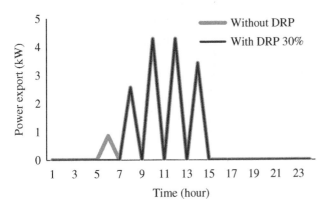

Figure 13.6 The amount of energy sold to upstream grid in deterministic case.

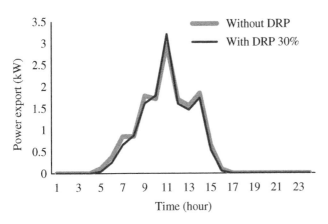

Figure 13.7 The expected value of energy sold to upstream grid in stochastic case.

Table 13.2 Results for consumption of PV unit's generated energy in different cases.

	Deterministic			Stochastic		
	Battery charge (kWh)	Supplying load (kWh)	Selling to grid (kWh)	Battery charge (kWh)	Supplying load (kWh)	Selling to grid (kWh)
Without DRP	11.97	5.985	15.39	3.856	14.796	14.493
With DRP	0.855	17.955	14.535	5.847	13.465	13.834

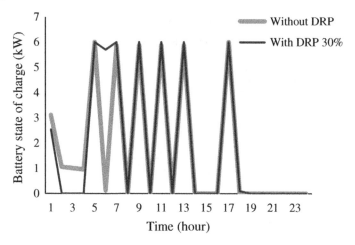

Figure 13.8 Battery charging state in deterministic case.

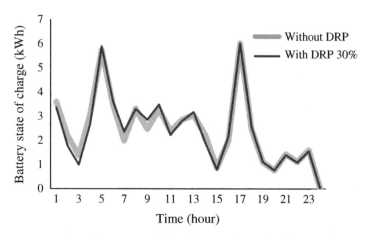

Figure 13.9 Expected value of battery charging state in stochastic case.

Figures 13.8 and 13.9, the amount of battery's stored energy in each hour is shown for deterministic and stochastic methods, respectively. By looking at these figures, it can be conceived that the battery's optimal operation strategy in the microgrid generally consists of charging the battery in off-peak hours and discharging it in peak load hours.

This matter can be perceived more clearly via Figure 13.9, where it shows the weighted sum of the different scenarios. Since in this figure, the two formed climaxes have occurred before load increasing from off-peak to mid-peak and from mid-peak to peak load. The fluctuations in the deterministic method depicted in Figure 13.8 happen because the energy exchange with the upstream grid is unilateral. In this regard, the battery stores some energy in one hour so that the PV can sell its generated energy to the upstream grid in the next hour. The battery discharge in the first hours has occurred due to the assumption of the existing initial charge in the battery.

13.4.3 The Impact of Demand Response Program and Battery on Total Cost Reduction

As observed in the previous results, the electrical energy storage system and DRP, by shifting the electrical load from peak load hours to off-peak hours, and flattening the load profile, reduce the costs of energy supply in the microgrid. In this sub-section, the sensitivity of the model's objective function (total cost of the microgrid's energy supply) to the battery's capacity and the microgrid's maximum contribution in DRP are evaluated. For this reason, the battery's capacity is altered from the minimum possible value up to 10 kW/h for different cases of DRP's maximum contribution, and the microgrid's total cost is calculated for each case. In Figure 13.10, the changes of the total cost versus the change in battery's nominal capacity and also microgrid's maximum contribution to the DRP are plotted. According to this figure, by increasing the microgrid's contribution in DRP and also by increasing the capacity of the battery, the microgrid's total cost of energy supply decreases.

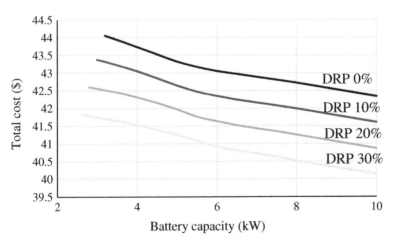

Figure 13.10 The change in microgrid's total cost with respect to change in DRP and battery capacity.

13.4.4 Analysis of the Fuel Cell Results

In Figures 13.11 and 13.12, the output electrical energy of the fuel cell is depicted for deterministic and stochastic simulations, respectively. In the stochastic case, the expected values of the output are plotted. For analyzing these results, first, it should be determined how much it costs for the fuel cell unit to generate each unit of electrical or thermal energy. The input power of the fuel cell is provided by urban gas at the price of 0.12 dollars for each kW/h. Then it is converted to electrical energy with an efficiency of 0.39, and to heat with an efficiency of 0.2184. The thermal energy can be supplied from the alternative path of the backup burner with the same price of 0.12 dollars and the efficiency of 0.95. Therefore, providing the thermal energy generated by the fuel cell from the backup burner unit with a price of 0.0276 dollars is possible. As a result, the break-even price of generating one kW/h of electrical energy from the fuel cell unit will be 0.237 dollars. Comparing this price with the price of the heavy-load period (0.3 dollars) and also the price of the light-load period (0.08), respectively, on and off status of the unit during these two periods are reasonable. But, why do we witness the generation of the fuel cell during the mid-load period (6 to 17), despite the fact that the price of providing power from the upstream grid (0.19 dollars) is less than the price of providing power from fuel cell unit? The reason for this is that according to the problem's assumptions, the power exchange with the upstream grid in each hour is feasible

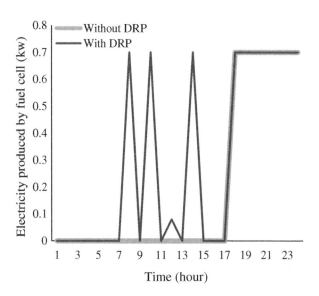

Figure 13.11 Fuel cell's hourly generated electrical energy in deterministic case.

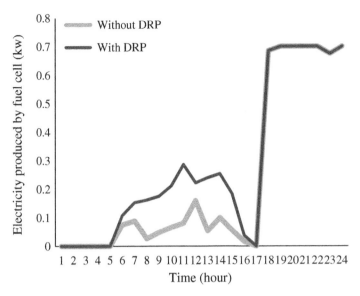

Figure 13.12 Fuel cell's hourly generated electrical energy in stochastic case.

only in one direction. On the other hand, the PV unit's power is being exported to the upstream grid at the same time, so purchasing power is not possible. In fact, the generation cost of the fuel cell during this period should somehow be compared to the opportunity of selling energy to the upstream grid. Because, by electrical energy generation of the fuel cell, and storing it in the battery, the PV unit would be able to sell more energy to the upstream grid.

13.4.5 Analysis of the Backup Burner and the Heat Storage Tank Results

By examining the power generation diagram of the backup burner in Figures 13.13 and 13.14 for deterministic and stochastic cases, respectively, it is noticed that these diagrams, except for the early hours of the day, show approximately the same diagram as the thermal load.

Therefore, the mentioned points cause the output power of the backup burner to show the same thermal load. It is obvious that the small mismatch between these two diagrams is due to the fuel cell's share in supplying thermal energy. Of course, it can be understood from these figures that in the early five hours, the generated thermal energy is equal to 0. It has happened for the assumption of existing initial energy in the heat storage tank.

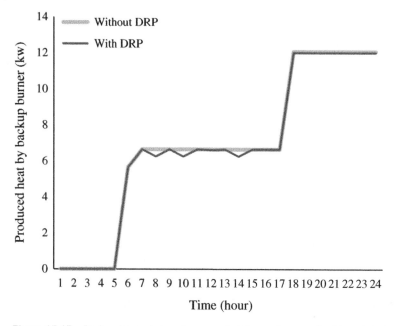

Figure 13.13 Backup burner's hourly generated thermal energy in deterministic case.

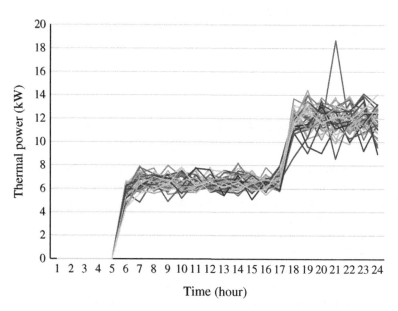

Figure 13.14 Fuel cell's hourly generated electrical energy in stochastic case.

In the following, the backup burner capacity is decreased, and the total cost of the microgrid is calculated in order to study the sensitivity of the problem to backup burner capacity and assessment of the heat storage tank. Figure 13.15 shows the diagram of microgrid total cost versus these changes in the deterministic case. To keep it simple, the results of the stochastic case are not plotted. It can be interpreted that for capacities greater than 12 kW, the backup burner unit always has the required power to supply the load. Hence in this situation, the microgrid's total cost is in the minimum value. But, by further decreasing the capacity of this unit to lower than 12 kW, the total cost will increase due to the indispensable need to use the heat storage tank and fuel cell unit. Also, the reduction of backup burner capacity to lower than 9.6 kW, makes the power resources required to supply the microgrid's load inadequate, and thus, the problem becomes infeasible. In Figure 13.16, the stored energy in the heat storage tank is plotted, where the backup burner has the minimum possible capacity (9.6 kW). It can be noticed in the diagram that the initial energy stored in the tank is discharged during the early hours. It is worth mentioning that since the heat storage tank loses some of its energy every hour, the use of the tank's stored energy is preferred to heat generation by the backup burner. Then, this tank is recharged again to compensate for the backup burner's generation shortage relative to microgrid's consumption,

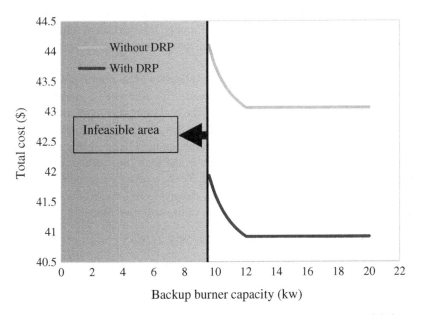

Figure 13.15 Total cost with respect to backup burner's capacity in deterministic case.

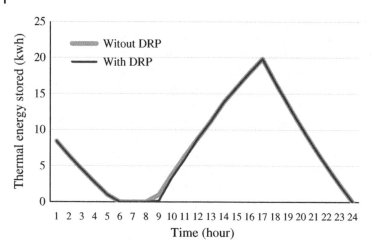

Figure 13.16 Stored energy in heat storage tank in deterministic case.

during heavy-load and late-day hours. As it can be understood from the diagram, the heat storage tank has filled its whole 20 kW/h energy capacity, to supply the required thermal energy of the heavy-load period at 17:00. Therefore, at this operating point, the supply system of the microgrid is at the edge of resource adequacy. Because, any increase in the load, or decrease in thermal energy generation, by any means, will directly lead to customer's lack of supply. It is obvious that the DRP in this situation does not play any role since this program is not capable of shifting thermal load from one hour to the other.

The recent study raises concern about the reliability of microgrid resources in supplying the required thermal load. Therefore, the amount of expected not-supplied thermal load due to the reduction of the backup burner's capacity is evaluated in the following. For this reason, some changes should be applied to the introduced model. Equation (13.20) is reformulated to (13.23). In this equation, if the output energy of the heat storage tank is less than the thermal load, their mismatch is calculated as not-supplied load. Then, a penalty function is added to the objective function as the cost of not supplying the thermal load. Equations (13.25)–(13.26) represent the new form of the objective function.

$$H_{t,s}^{L} + H_{t,s}^{ch} = H_{t,s}^{G} + H_{t,s}^{Dch} + H_{t,s}^{Curtailment} : \forall t \in T, s \in S \tag{13.25}$$

$$\text{Minimize} : C^{Total} = \sum_{s \in S} \sum_{t \in T} \sum_{\psi \in \Psi} \left(\gamma_s \times P_{t,s}^{\psi} \times \text{Price}_{t,s}^{\psi} \right)$$

$$+ \sum_{s \in S} \sum_{t \in T} \sum_{\psi \in \Psi} \left(\gamma_s \times H_{t,s}^{Curtailment} \times \text{Price}_{t,s}^{Curtailment} \right)$$

$$\tag{13.26}$$

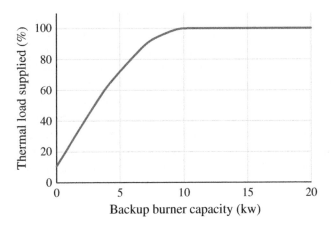

Figure 13.17 Supplied thermal energy with respect to backup burner capacity.

where $H_{t,s}^{\text{Curtailment}}$ is not-supplied thermal load, $\text{Price}_{t,s}^{\text{Curtailment}}$ is the cost of not supplying each kW/h of thermal load. In this study, the cost of not supplying thermal load is assumed to be 1.2 dollar per kW/h. The study of this sub-section is carried out for stochastic case, and the results are plotted in Figure 13.17 for the percentage of supplied thermal load relative to backup burner capacity.

As observed in this figure, because of the fuel cell unit's generation and also the assumption of existing initial thermal energy in the heat storage tank, the supplied load of the microgrid will not be zero even by completely removing the backup burner. But it should be noticed that the thermal energy supply system of the microgrid is highly dependent on the backup burner, and in case it is not being available, the supplied thermal energy is reduced to about 20% of the daily consumption. In this regard, it seems that adding a boiler can be effective in enhancing the reliability of the microgrid's energy supply system. However, making these kinds of investment decisions necessitates cost-benefit studies.

13.5 Conclusion

This chapter attempted to discuss and analyze the optimal operation of microgrids' HES. For this reason, some of the typical components, resources, and strategies used in heat and power integrated systems were briefly described. Then, an energy management scheme for optimal and coordinated operation of the microgrid energy system was introduced. This model receives the input data including microgrid's electrical and thermal loads, energy price, and PV unit's generating power, along with their corresponding uncertainties, from a scenario generation model. The generated scenarios are reduced via a scenario reduction model based on the fast backward-forward method. Finally, this scheduling model handles all the generation, consumption, and energy storage processes in the microgrid. In the simulation results, it was illustrated that increasing microgrid's flexibility, by using a battery or applying DRP, has a notable impact on shifting the load from peak load to off-peak hours, thus, reducing the costs of electrical energy supply. The intended microgrid supplies thermal energy through a backup burner unit and the restored energy of a fuel cell unit. The existence of a heat storage tank in the microgrid improves the reliability of the thermal energy supply system. Because even though the maximum thermal load during the day is assumed to be 12 kW/h, it was demonstrated that by reducing backup burner capacity down to 9.6 kW/h, all the thermal load is still supplied.

References

1 O'Malley, M., Kroposki, B., Hannegan, B. et al. (2016). *Energy Systems Integration. Defining and Describing the Value Proposition*. Golden, CO: National Renewable Energy Lab. (NREL).

2 Wang, Y., Huang, Y., Wang, Y. et al. (2018). Energy management of smart microgrid with response loads and distributed generation considering demand response. *Journal of Cleaner Production* 197: 1069–1083.

3 Eladl, A.A., El-Afifi, M.I., Saeed, M.A., and El-Saadawi, M.M. (2020). Optimal operation of energy hubs integrated with renewable energy sources and storage devices considering CO2 emissions. *International Journal of Electrical Power & Energy Systems* 117: 105719.

4 Brahman, F., Honarmand, M., and Jadid, S. (2015). Optimal electrical and thermal energy management of a residential energy hub, integrating demand response and energy storage system. *Energy and Buildings* 90: 65–75.

5 Majidi, M., Nojavan, S., and Zare, K. (2017). A cost-emission framework for hub energy system under demand response program. *Energy* 134: 157–166.

6 Pashaei-Didani, H., Nojavan, S., Nourollahi, R., and Zare, K. (2019). Optimal economic-emission performance of fuel cell/CHP/storage based microgrid. *International Journal of Hydrogen Energy* 44 (13): 6896–6908.

7 Alramlawi, M., Mohagheghi, E., and Li, P. (2019). Predictive active-reactive optimal power dispatch in PV-battery-diesel microgrid considering reactive power and battery lifetime costs. *Solar Energy* 193: 529–544.

8 Liu, Z., Chen, Y., Zhuo, R., and Jia, H. (2018). Energy storage capacity optimization for autonomy microgrid considering CHP and EV scheduling. *Applied Energy* 210: 1113–1125.

9 Javadi, M.S., Lotfi, M., Nezhad, A.E. et al. (2020). Optimal operation of energy hubs considering uncertainties and different time resolutions. *IEEE Transactions on Industry Applications* 56 (5): 5543–5552.

10 Lidula, N. and Rajapakse, A. (2011). Microgrids research: a review of experimental microgrids and test systems. *Renewable and Sustainable Energy Reviews* 15 (1): 186–202.

11 Pandit, M., Srivastava, L., and Sharma, M. (2015). Environmental economic dispatch in multi-area power system employing improved differential evolution with fuzzy selection. *Applied Soft Computing* 28: 498–510.

12 Garcia-Gonzalez, J., de la Muela, R.M.R., Santos, L.M., and Gonzalez, A.M. (2008). Stochastic joint optimization of wind generation and pumped-storage units in an electricity market. *IEEE Transactions on Power Systems* 23 (2): 460–468.

13 Guo, L., Liu, W., Li, X. et al. (2014). Energy management system for stand-alone wind-powered-desalination microgrid. *IEEE Transactions on Smart Grid* 7 (2): 1079–1087.

14 Motevasel, M. and Seifi, A.R. (2014). Expert energy management of a micro-grid considering wind energy uncertainty. *Energy Conversion and Management* 83: 58–72.

15 Karavas, C.-S., Kyriakarakos, G., Arvanitis, K.G., and Papadakis, G. (2015). A multi-agent decentralized energy management system based on distributed intelligence for the design and control of autonomous polygeneration microgrids. *Energy Conversion and Management* 103: 166–179.

16 Chen, C. and Duan, S. (2014). Optimal allocation of distributed generation and energy storage system in microgrids. *IET Renewable Power Generation* 8 (6): 581–589.

17 Sfikas, E., Katsigiannis, Y., and Georgilakis, P. (2015). Simultaneous capacity optimization of distributed generation and storage in medium voltage microgrids. *International Journal of Electrical Power & Energy Systems* 67: 101–113.

18 Wang, D., Qiu, J., Reedman, L. et al. (2018). Two-stage energy management for networked microgrids with high renewable penetration. *Applied Energy* 226: 39–48.

19 Nikkhah, S. and Rabiee, A. (2019). Multi-objective stochastic model for joint optimal allocation of DG units and network reconfiguration from DG owner's and DisCo's perspectives. *Renewable Energy* 132: 471–485.

20 Aboli, R., Ramezani, M., and Falaghi, H. (2019). Joint optimization of day-ahead and uncertain near real-time operation of microgrids. *International Journal of Electrical Power & Energy Systems* 107: 34–46.

21 Sedighizadeh, M., Esmaili, M., Jamshidi, A., and Ghaderi, M.-H. (2019). Stochastic multi-objective economic-environmental energy and reserve scheduling of microgrids considering battery energy storage system. *International Journal of Electrical Power & Energy Systems* 106: 1–16.

22 Firouzmakan, P., Hooshmand, R.-A., Bornapour, M., and Khodabakhshian, A. (2019). A comprehensive stochastic energy management system of micro-CHP units, renewable energy sources and storage systems in microgrids considering demand response programs. *Renewable and Sustainable Energy Reviews* 108: 355–368.

23 Majidi, M., Nojavan, S., and Zare, K. (2017). Optimal stochastic short-term thermal and electrical operation of fuel cell/photovoltaic/battery/grid hybrid energy system in the presence of demand response program. *Energy Conversion and Management* 144: 132–142.

24 Majidi, M. and Zare, K. (2018). Integration of smart energy hubs in distribution networks under uncertainties and demand response concept. *IEEE Transactions on Power Systems* 34 (1): 566–574.

25 Yuan, W., Wang, X., Su, C. et al. (2021). Stochastic optimization model for the short-term joint operation of photovoltaic power and hydropower plants based on chance-constrained programming. *Energy* 222: 119996.

26 Amir, V., Jadid, S., and Ehsan, M. (2017). Probabilistic optimal power dispatch in multi-carrier networked microgrids under uncertainties. *Energies* 10 (11): 1770.

27 Wang, Z., Shen, C., and Liu, F. (2018). A conditional model of wind power forecast errors and its application in scenario generation. *Applied Energy* 212: 771–785.

28 Daneshvar, M., Mohammadi-Ivatloo, B., Abapour, M., and Asadi, S. (2020). Energy exchange control in multiple microgrids with transactive energy management. *Journal of Modern Power Systems and Clean Energy* 8 (4): 719–726.

29 Liu, S., Zhou, C., Guo, H. et al. (2021). Operational optimization of a building-level integrated energy system considering additional potential benefits of energy storage. *Protection and Control of Modern Power Systems* 6 (1): 1–10.

30 Majidi, M., Mohammadi-Ivatloo, B., and Anvari-Moghaddam, A. (2019). Optimal robust operation of combined heat and power systems with demand response programs. *Applied Thermal Engineering* 149: 1359–1369.

31 Kiptoo, M.K., Lotfy, M.E., Adewuyi, O.B. et al. (2020). Integrated approach for optimal techno-economic planning for high renewable energy-based isolated microgrid considering cost of energy storage and demand response strategies. *Energy Conversion and Management* 215: 112917.

32 Albadi, M.H. and El-Saadany, E.F. (2007). Demand response in electricity markets: an overview. In: *2007 IEEE Power Engineering Society General Meeting*, Tampa, FL, USA (24 June 2007), 1–5. IEEE.

33 Qdr, Q. (2006). Benefits of Demand Response in Electricity Markets and Recommendations for Achieving Them. *US Dept. Energy, Washington, DC, USA, Tech. Rep,* vol. 2006.

34 Siano, P. (2014). Demand response and smart grids—a survey. *Renewable and Sustainable Energy Reviews* 30: 461–478.

35 Dupacová, J., Gröwe-Kuska, N., and Römisch, W. (2000). *Scenario Reduction in Stochastic Programming: An Approach Using Probability Metrics*, Mathematisch-Naturwissenschaftliche Fakultät .... Humboldt-Universität zu Berlin.

36 Heitsch, H. and Römisch, W. (2003). Scenario reduction algorithms in stochastic programming. *Computational Optimization and Applications* 24 (2): 187–206.

14

Techno-Economic Analysis of Hydrogen Technologies, Waste Converter, and Demand Response in Coordinated Operation of Heat and Electricity Systems

Mohammad Hossein Pafeshordeh, Mohammad Rastegar, Mohammad Mohammadi, and Mohammad Reza Dehbozorgi

Department of Power and Control, School of Electrical and Computer Engineering, Shiraz University, Shiraz, Iran

Nomenclature

P_I	New equipment installation cost
P_{FOM}	Fixed operation and maintenance cost
P_V	Equipment's variable cost
P_F	Fuel consumption cost
$E_{Nuclear}$	Electricity production of the nuclear unit
$FAC_{Nuclear}$	Correction factor between production and capacity of the nuclear unit
$C_{Nuclear}$	Capacity of the nuclear power electricity generator
$d_{Nuclear}$	Distribution of the electricity production of the nuclear unit
$e_{Geothermal}$	Electricity production of the geothermal unit
$FAC_{Geothermal}$	Correction factor between production and capacity of the geothermal unit
$C_{Geothermal}$	Capacity of the geothermal power electricity generator
$d_{Geothermal}$	Distribution of the electricity production of the geothermal unit
W_{Hydro}	Hourly distribution of the annual water
S_{Hydro}	Water storage capacity
C_{Hydro}	Capacity of the hydropower plant
μ_{Hydro}	Efficiency of the hydropower plant
$e_{Hydro-average}$	Average hydroelectricity production
e_{Hydro}	Hydropower production

Coordinated Operation and Planning of Modern Heat and Electricity Incorporated Networks,
First Edition. Edited by Mohammadreza Daneshvar, Behnam Mohammadi-Ivatloo, and Kazem Zare.
© 2023 The Institute of Electrical and Electronics Engineers, Inc.
Published 2023 by John Wiley & Sons, Inc.

$C_{\text{ELC-MIN}}$	Minimum electrolyser capacity
f_{Elc}	Hydrogen production of the electrolyser
$f_{\text{H2-Average}}$	Average hydrogen consumption of the sum of all hydrogen demands
s_{Elc}	Hydrogen storage content
α_{Elc}	Fuel efficiency of electrolysers
d_{FRESH}	Hourly demand for freshwater
D_{FRESH}	Annual demand for freshwater
$e_{\text{PUMP-FRESHWATER}}$	Hourly electricity demand of the freshwater pump
$C_{\text{PUMP-FRESHWATER}}$	Minimum necessary capacity of the freshwater pump
$\mu_{\text{FRESHWATERPUMP}}$	Electricity efficiency of the freshwater pump
$w_{\text{FRESH-AVERAGE}}$	Average freshwater production
w_{SALT}	Hourly demand for saltwater
w_{BRINE}	Production of brine
w_{FRESH}	Production of fresh water
$d_{\text{DESALINATION}}$	Electricity demand of desalination plant
$\mu_{\text{DESALINATION}}$	Electricity efficiency of desalination plant
s_{BRINE}	Brine storage content
e_{TURBINE}	Maximum electricity production on the turbine
μ_{TURBINE}	Electric efficiency of the turbine
C_{TURBINE}	Capacity of the turbine
$\varphi_{\text{BRINESTORAGE}}$	Energy equivalent of water in brine storage
$\text{SHARE}_{\text{FRESH}}$	Fresh water share of salt water input
$\text{SHARE}_{\text{BRINE}}$	Brine share of salt water input
C_{CHP}	Capacity of the CHP
μ_{CHP}	Efficiency of the CHP

Abbreviations

BAU	Business as usual scenario
CHP	Combined heat and power
DR	Demand response
FOM	Fixed operation and maintenance
FT	Full technology scenario
MCEN	Multi-carrier energy network
PEM	Polymer electrolyte membrane
RES	Renewable energy source
SOEC	Solid oxide electrolyzer cell

14.1 Introduction

Global warming and air pollution are two of the most important topics in recent years. Lots of plans are devised around the world to control these problems. The 45.4% of generated CO_2 arises from the electrical, thermal, and transport sectors [1]. Hence, fossil fuels will eventually run out and the amount of pollution emitted from these fuels will increase in the future. Therefore, replacing these fossil fuels with clean energies such as hydrogen or supplying energy from RESs has many advantages. MCENs are a solution to these problems. These networks are composed of different energy carriers like electric, heat, and hydrogen and include other energy-generating technologies such as CHP, desalination power plants, hydrogen-based technologies, waste converters, and RESs. This diversity in energy carriers and technologies brings us several benefits such as improving flexibility and the environmental situation and may have a positive effect on the system's total cost. In these types of networks, DR can also play an important role in better operating the system.

Hydrogen-based technologies incorporate a hydrogen cycle that is demonstrated in Figure 14.1. The hydrogen cycle operates by utilizing technologies like electrolyzer, hydrogen storage, and fuel cell. Electrolyzers can be of different types, like polymer electrolyte membrane (PEM), solid oxide electrolyzer cell (SOEC), and alkaline. After converting the electrical energy into hydrogen, hydrogen storage is used to store the produced hydrogen [2]. This storage can be in types of carven,

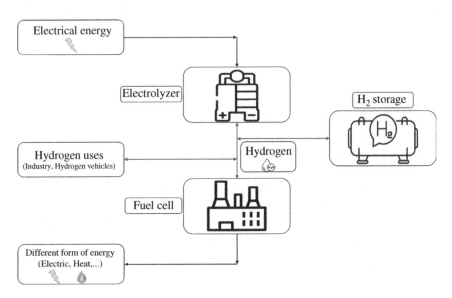

Figure 14.1 Hydrogen technologies cycle.

H_2 steel tanks, and underground storage. The fuel cell will complete this cycle. It converts hydrogen into electrical and thermal forms. The efficiency of fuel cell CHP is approximately 40%. Hydrogen can also supply hydrogen fuel required by the transportation network, industrial use, and supply thermal and electrical demands in the peak time. The importance of this issue can be seen in [3], in which an extensive assessment of the role of hydrogen in the future of Japan has been made. It also reduces curtailments and increases profitability by storing excess electrical energy in the form of hydrogen [4]. Using waste converters and desalination plants can also contribute to providing a part of electrical and thermal energy consumption, reducing the amount of waste in the environment, and supplying clean water, respectively. CHPs, as an energy-efficient technology, can provide both electrical and thermal energies. In addition, they can reduce the emission of greenhouse gases and energy bills because of their high efficiency.

Studies in the field of integrated systems are diverse. In this part, we divide the published papers into two general groups. The first series of articles are papers that have considered only the electrical network. The second series of articles are more related to the concept of MCENs. In addition to the electrical network, they also include other types of networks (gas, thermal, heat). It should be noted that the classification is based on the network type and components analyzed. As an important technology, hydrogen technology is proposed in several articles that add this carrier to networks, each of which uses hydrogen in different studies to achieve different goals. Reference [5] attempts to improve the transport fleet from the fuel consumption and environmental pollution point of view by using electric and hydrogen energy. In [3], extensive assessments of the role of hydrogen in the future of the Japanese network have been made due to the high importance of hydrogen. In [6], the authors try to design and manage a network. The goal is to improve the network in terms of reliability, lower cost, and reduced pollution. As a result, hydrogen is used as an energy carrier and in the transportation fleet. Distributed generation, electric, and hydrogen storage are among the components included in this study. Reference [7] is also one of the studies that targeted the minimum network cost and the highest level of social welfare. In this research, the presence of hydrogen technology, renewable sources, electric and hydrogen vehicles are some of the studied technologies. Reference [8] introduces the concept of interconnected networks for residential users. In this case, consumers will benefit from renewable sources, fuel cells, and energy storage. In this case, they can provide their energy consumption. Also, in [9], a combination of electric power plants with coal fuel and waste converter is considered, which results in improving the efficiency of both power plants. The generated heat from boilers is used to warm the water and reheat power plants' steam. Reference [10] has also optimized the network for profitability by applying demand-side management and classification of the network according to scale and geographical area.

The second series of articles are based on the performance of the electrical network alongside networks such as thermal, gas, etc. In [11], the electrical and thermal networks are analyzed, and the purpose of the study is to optimize the cost of the system and environmental pollution and to show the effect of the storage system. In [12–14], the authors analyze the electrical and thermal networks and the main difference between these three studies is in the analyzed network components. For example, in [12] electric vehicles, demand-side management, hydrogen technology, and waste converters are not mentioned, but in [13], demand-side management is analyzed to minimize annual costs and fuel consumption. In [14], hydrogen technology is added to the problem. In [15], the electrical and gas network is analyzed. Demand-side management, distributed generation, and energy storage are among the components considered in the mentioned paper. In this regard, [16] tries to minimize the cost of the entire system by adding hydrogen technology and considering uncertainty. Reference [17] is also one of the cases that examines the electrical and gas network with the presence of hydrogen technologies. The aim of minimizing carbon dioxide emissions and the presence of renewable sources conduct in this research. References [18–20] study the electrical, thermal, and gas networks altogether. In [18], only distributed generators and storage systems are studied to optimize the network cost. The authors of [19] also consider hydrogen technology. Finally, [20] has a more comprehensive analysis and in addition to the mentioned networks in [18, 19], demand-side management has participated. Such studies have been conducted in Iran. For example, in [21], the authors try to model Iran's network based on renewable sources, and while considering the water shortage crisis, they try to minimize environmental pollution. On the other hand, in [22], the electricity and thermal energy consumption of Taleghan in Iran are supplied by an electrolyzer, and its input electrical energy is fed by PV panels. This study shows such technologies' importance in Iran.

To summarize the research gaps, some articles were limited to only one network or covered a small range of technologies, and also the role of DR was not well represented. On the other hand, the output parameters mentioned in the previous sections, such as the amount of fuel consumed or the amount of carbon dioxide emitted, have not been fully investigated. In addition, the increasing penetration of RESs is among the issues operators face. In this regard, this research examines the issue by having novelties like considering different technologies and DR and modeling of Iran network as an interconnected network. In addition to considering different technologies and penetration RESs, comprehensive studies on output parameters such as emissions, fossil fuel consumption, and total network costs are done. Table 14.1 shows a summary of the literature review. This table helps to better understand the research gaps. In this research, by modeling two scenarios, we analyze an MCEN, which includes electrical, gas, thermal, hydrogen networks. In the first scenario, we examine the future situation of Iran's energy network

Table 14.1 Summary of the literature review.

Reference number	RESs	CHP	Hydrogen technology	Waste converters	DR
[3]	✔	✔	✔	✗	✗
[7]	✗	✗	✗	✔	✗
[8]	✔	✔	✗	✗	✔
[10]	✔	✔	✗	✗	✔
[12]	✔	✔	✔	✗	✔
[16]	✔	✔	✔	✗	✔
[17]	✔	✗	✗	✗	✗
[18]	✔	✔	✔	✗	✗
This paper	✔	✔	✔	✔	✔

considering generation expansion and load growth. In the second scenario, the system includes the technologies and DR mentioned in the introduction section. The aim in both scenarios is the optimal techno-economical operation of the systems. By examining the penetration level of RESs, we will show the effectiveness of these resources. Finally, output parameters such as CO_2 emission, annual system cost, fuel consumption, and energy supply results will be prepared for analysis or relevant decisions. The main contributions of this paper include but are not limited to:

1) Iran's network as an interconnected network is modeled, and different energy-generating technologies' effect on the techno-economical operation of the system is analyzed.
2) The role of DR and its effect on the system's cost and other output parameters is investigated.
3) Output parameters such as emissions, fossil fuel consumption, and total network costs are reported to better show the superiority of integrated energy systems.
4) The effect of the penetration level of renewable sources on output parameters is analyzed.

The rest of this chapter is organized as follows. In Section 14.2, our methodology is discussed. In Section 14.3, the output results of the simulation are analyzed, and the resulting conclusions are accessible. And finally, the chapter is concluded in Section 14.4.

14.2 Methodology

14.2.1 Model

By collecting the required information from the system and its components, we can begin the analysis of the grid by the EnergyPLAN program. EnergyPLAN is a computer model for analyzing and designing energy networks, which was prepared and released in 1999 by Henrik Lund [23]. This model was initially presented as an excel file. Over time, the latest technologies have been added to it, and it has become a computer modeling program. The inputs of this program are demands, sources, capacities, and system costs. Parameters such as the amount of carbon dioxide emitted, the amount of fuel consumed, the annual cost of the system, and dispatches are the outputs of this program. The ability to analyze the system with different energy carriers and components, ease of calculations, open access rights, and comprehensiveness of output parameters are the advantages of this computer program. It should also be noted that research can be done depending on the type of study with economic or technical strategy. Economic simulation is performed to minimize system costs, while the technical simulation is performed to minimize fuel consumption and carbon dioxide emitted. The total cost of the network includes maintenance costs, fuel costs, variable costs, and the cost of installing new devices in the network. In this regard, the following formula will determine the total cost of the network.

$$P_{\text{total}} = P_I + P_{\text{FOM}} + P_V + P_F \tag{14.1}$$

In this formula, P_I is the installation cost, P_{FOM} is the fixed operation and maintenance (FOM) cost, P_V is the variable cost, and P_F is the cost of fuel consumption. All the costs are converted to present value and the interest rate is set to be 6%. Details about costs (FOM, fuel, and CO_2 cost) and the lifetime of technologies are available in Appendix, Tables 14.A, 14.B, and 14.C.

EnergyPLAN minimizes the overall network cost using four steps. First, some small calculations, which are run simultaneously with the typing of input and cost data are performed. The next step consists of some initial calculations, which do not involve electricity balancing. Then, the procedure is divided into either a technical or a market economic simulation. The technical simulation minimizes the import/export of electricity and seeks to identify the least fuel-consuming solution. The market-economic simulation identifies the least-cost solution based on the business-economic costs of each production unit. Finally, the selected outputs are printed.

14.2.2 Case Study

In the simulation and analysis section, we analyze Iran's network as a case study. Iran is a country with a high potential for RESs. Iran also desperately needs

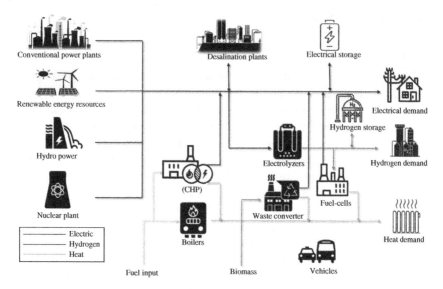

Figure 14.2 Structure of the system.

desalination plants due to consecutive droughts. Here, the studies will be done in a long-term vision until 2050. The study is dynamic, and the results are reported every 10-year. Figure 14.2 indicates the structure of the system. The details of each technology and how they are modeled can be accessed in EnergyPLAN documentation [24]. For the sake of analysis, the model for power plants, CHP, hydrogen technology, desalination plant, and waste converters are presented briefly. The interested reader is encouraged to see the detailed explanation in [24].

- **Nuclear power plant:** Nuclear power station is subject to the condition that it will always be involved in the task of maintaining the grid's stability. Therefore, the power plant does not take part in the active regulation. The electricity production of the nuclear unit (e_{Nuclear}) is simply defined by the capacity, the hourly distribution and the correction factor as stated in Eq. (14.2). In this formula, $\text{FAC}_{\text{Nuclear}}$ is the correction factor between production and capacity, C_{Nuclear} is the capacity of the nuclear power electricity generator, and d_{Nuclear} is the distribution of the electricity production.

$$e_{\text{Nuclear}} = \frac{\text{FAC}_{\text{Nuclear}} * C_{\text{Nuclear}} * d_{\text{Nuclear}}}{\text{MAX}(d_{\text{Nuclear}})} \tag{14.2}$$

- **Geothermal power plant:** Geothermal power station like nuclear power plants does not take part in the active regulation. The electricity production of the geothermal unit ($e_{\text{Geothermal}}$) is simply defined by the capacity, the hourly distribution and the correction factor. In this formula, $\text{FAC}_{\text{Geothermal}}$ is the Correction factor between production and capacity, $C_{\text{Geothermal}}$ is the capacity of the geothermal power electricity generator, and $d_{\text{Geothermal}}$ is the distribution of the electricity production.

$$e_{\text{Geothermal}} = \frac{\text{FAC}_{\text{Geothermal}} * C_{\text{Geothermal}} * d_{\text{Geothermal}}}{\text{MAX}(d_{\text{Geothermal}})} \tag{14.3}$$

- **Hydropower plants:** The hydropower plant is identified by an hourly distribution of the annual water input (W_{Hydro}), a water storage capacity (S_{Hydro}), and the capacity (C_{Hydro}) and efficiency (μ_{Hydro}) of the generator. Based on such inputs, the potential output is calculated simultaneously by the procedure described in the following. First, the average hydroelectricity production ($e_{\text{Hydro-Average}}$) is calculated as the output of the average water supply (Annual water supply divided by 8784 hours/year). Then, the program calculates the hourly modeling of the system, including the fluctuations in the storage content. Furthermore, the hydropower production (e_{Hydro}) is modified in accordance with the generator capacity, the distribution of the water supply, and the storage capacity in using Eq. (14.4).

$$e_{\text{Hydro-Average}} = \frac{\mu_{\text{Hydro}} * W_{\text{Hydro}}}{8784}$$

$$e_{\text{Hydro}} = \text{MAX}\left[e_{\text{Hydro-Average}}, (\text{Hydro storage content} - S_{\text{Hydro}}) * \mu_{\text{Hydro}}\right]$$

$$e_{\text{Hydro}} \leq C_{\text{Hydro}}$$

$$\tag{14.4}$$

- **Hydrogen technology:** The minimum electrolyzer capacity, $C_{\text{ELC-MIN}}$, is calculated in the following way: First, the hydrogen production of the electrolyzer, f_{Elc}, is defined as the average hydrogen consumption of the sum of all hydrogen demands $f_{\text{H2-Average}}$:

$$f_{\text{Elc}} = f_{\text{H}_2\text{-Average}} = \frac{F_{\text{H}_2}}{8784} \tag{14.5}$$

Then, for each hour (x), the hydrogen storage content is calculated, $s_{\text{Elc}}(x)$, as the content of the previous hour plus the average production minus the actual consumption of the hour:

$$s_{\text{Elc}}(x) = s_{\text{Elc}}(x-1) + f_{\text{H}_2\text{-Average}} - f_{\text{H}_2}(x) \tag{14.6}$$

Finally, the minimum electrolyzer capacity, $C_{\text{Elc-MIN}}$, is identified as the maximum production needed to be divided by the fuel efficiency. α_{Elc} is fuel efficiency of electrolysers.

$$C_{\text{Elc-MIN}} = \frac{\text{Hourmax}(f_{\text{Elc}})}{\alpha_{\text{Elc}}} \tag{14.7}$$

- **Desalination plant:** Based on the hourly freshwater demand and storage, the model will calculate the minimum necessary capacity of the freshwater pump and the desalination unit to fulfill the demands as follows:

  First, the hourly demand for freshwater (d_{FRESH}) is calculated from the annual demand (D_{FRESH}) and the hourly distribution. The hourly electricity demand (MW) and minimum necessary capacity (MW) of the freshwater pump are found as the hourly demand for water multiplied by the electricity efficiency of the freshwater pump (kWh/m^3 Fresh Water) divided by 1000 to get the result in MW. In this formula, $\mu_{\text{FRESHWATERPUMP}}$ is the electricity efficiency of the freshwater pump, and $C_{\text{PUMP-FRESHWATER}}$ is the capacity of the freshwater supply pump.

$$e_{\text{PUMP-FRESHWATER}} = \frac{d_{\text{FRESH}} * \mu_{\text{FRESHWATERPUMP}}}{1000}$$
$$C_{\text{PUMP-FRESHWATER-MIN}} = \text{MAX}\left(e_{\text{PUMP-FRESHWATER}}\right) \tag{14.8}$$

Next, the average freshwater production ($w_{\text{FRESH-AVERAGE}}$) is calculated:

$$w_{\text{FRESH-AVERAGE}} = \frac{D_{\text{FRESH}}}{8784} \tag{14.9}$$

The hourly demand for saltwater (w_{SALT}), as well as the production of brine (w_{BRINE}) and the demand for electricity ($d_{\text{DESALINATION}}$), is then found as follows:

$$w_{\text{SALT}} = \frac{w_{\text{FRESH}}}{\text{SHARE}_{\text{FRESH}}}$$
$$w_{\text{BRINE}} = w_{\text{SALT}} * \text{SHARE}_{\text{BRINE}} \tag{14.10}$$
$$d_{\text{DESALINATION}} = w_{\text{FRESH}} * \mu_{\text{DESALINATION}}$$

Afterwards, based on the hourly production of brine (w_{BRINE}), the brine storage content (s_{BRINE}) and the maximum electricity production on the turbine (e_{TURBINE}) is found as follows:

$$e_{\text{TURBINE}} = \min\left[(\mu_{\text{TURBINE}} * s_{\text{BRINE}}), C_{\text{TURBINE}}\right]$$
$$s_{\text{BRINE}} = s_{\text{BRINE}} + w_{\text{BRINE}} * \varphi_{\text{BRINESTORAGE}} - e_{\text{TURBINE}} \tag{14.11}$$

- **Waste converter**: Waste is considered as biomass energy, which cannot be stored but has to be burned regularly. Waste is divided geographically into three district heating groups, and only one hourly distribution can be defined. The following input must be given to the model:
  - The waste resources are divided geographically between the three district heating systems.

– Efficiencies specifying the quantity of the waste input resources converted into the following four energy forms: heat for district heating, electricity, fuel for transport, and fuel for CHP and boilers.
– Hourly distribution of the waste input (heat and electricity output).
• **CHP:** CHPs units are determined by the following inputs; C_{CHP}, μ_{CHP}. The heat demand is supplied by boilers, micro-CHPs, and heat pumps.

Two scenarios are discussed in this paper, as follows:

Business as usual scenario (BAU): In this scenario, the current network of Iran will continue to operate only by increasing the capacity of existing power plants without adding any new type of technologies. This scenario gives us a comprehensive view of the future of Iran's network.

Fully technology scenario (FT): In this scenario, hydrogen technology, waste converter, desalination plant, distributed generation, and the capability of DR are considered in the network over time. This scenario shows the state of the network after the presence of mentioned technologies and DR.

14.2.3 Assumptions

It is assumed that Iran consumes hydrogen in the future for industries or hydrogen cars. Electrical demand has been set based on the predictions and previous years' growth. Also, we assumed a thermal network for Iran by making assumptions. The basis of the work is that by collecting the heat load information of the five different countries listed in Table 14.2 [25], and by measuring factors such as geographical and demographic similarities, a reasonable average thermal load for Iran is considered. Information about technologies capacities and demands are in Tables 14.3 and 14.4. These data are extracted from different sources, such as detailed statistics of the electricity industry, the energy balance sheet of Iran [26], and the Danish energy agency [27].

In this study, due to higher efficiency and larger-scale production, an alkaline electrolyzer is considered. H_2 steel tanks are also considered in this study. The reason for choosing this type of storage is to cover the spatial limitation of other types. However, it is more economically expensive than other types.

Table 14.2 Heat demand of the intended countries.

Country Load (TWh)	Italy	Germany	France	UK	Poland
Thermal	735	1384	758	673	403

Table 14.3 Details of different technologies capacity.

Technologies	Capacity in 2030 (MW)	Capacity in 2040 (MW)	Capacity in 2050 (MW)
CHP	4553.88	7434.68	10 315.48
Power plant	77 522.5	89 062	100 601.6
Geothermal	30	65	100
Nuclear	1076	1127	1178
Hydropower	12 952	13 793.8	14 635.64
Wind	710.72	1479.9	2049.12
Pv	732.292	1700.6	2366.6
Boiler	213 908	248 532	305 139
Electrolyzer	624	1404	2340
Fuel cell	3068	7669	13804

Table 14.4 Details of different demands.

Sector	Energy consumption in 2030 (TWh)	Energy consumption in 2030 (TWh)	Energy consumption in 2030 (TWh)
Electrical	386	481	623
Thermal	910	1110	1376
Hydrogen	4	9	15

Table 14.5 Annual electricity demand for seawater desalination.

	Year 2030	Year 2040	Year 2050
Desalination capacity (Mm^3/y)	500	2000	5000
Electricity demand (TWh_e/y)	2.5	10	25

A desalination plant's capacity is defined as water demand per million square meters. The annual electricity demand for seawater desalination is presented in Table 14.5. The desalination plant is assumed to use the reverse osmosis method.

14.3 Results and Discussion

In this section, we are going to analyze two scenarios introduced in the methodology section. Three factors (fossil fuel consumption, total cost, and CO_2 emission) alongside the generation share of different energy-generation sources are reported for each scenario. At last, the results are discussed, and the two scenarios are compared with each other.

14.3.1 Results

Results of BAU Scenario: As mentioned in the previous section, this scenario shows the current direction of the Iran grid. This scenario is used as the benchmark to make comparisons. As mentioned, the capacity of existing power plants and equipment has been determined based on the forecasts and the trend of their past changes for our analysis period. It is assumed that the turbines in desalination power plants will be able to generate electricity in the second scenario, while in this scenario, they only consume electrical energy. First, we show the changes in carbon dioxide emission value and fossil fuel consumption in three decades. Figure 14.3 shows the amount of carbon dioxide emitted and the amount of fossil fuel consumed.

As we can infer from Figure 14.3, over the decades, the amount of fossil fuel consumption has increased. This is due to using power plants in the network to supply consumers. According to the direct relationship between the amount of fossil fuel consumption and the amount of carbon dioxide produced, it can be seen that the amount of carbon dioxide produced in 2050 will increase to 1282.6 million tons. As extracted from our simulation results, Figure 14.4 shows how electrical demand is supplied. It should be noted that all the thermal demand is supplied by the boilers.

Figure 14.4 indicates that a large share of electricity consumption is provided by fossil power plants. On the other hand, all the required hydrogen energy consumption is provided by an electrolyzer. According to Eq. (14.1), the total cost of the network at the end of the year 2050 will reach \$8691 billion. It should be noted that the cost of installation in the BAU scenario is zero because we assume no new technology will be added to the system, and the system operates by increasing the capacity of existing devices and units.

Results of FT Scenario: In this scenario, the technologies reviewed in the introduction section, such as CHP, fuel cells, and waste converters are added to the network. To show the effectiveness of these technologies, we compare the results of adding these technologies to the BAU scenario. Our comparison is based on carbon dioxide emission, fossil fuel consumption, and the annual cost of the network. These output parameters give us a techno-economical view of the system. First, we will show the results of adding CHP to the network. In this case, the amount of carbon dioxide emission, fossil fuel consumption, and the annual network cost will be reduced by 2.6%, 1.3%, and 0.7% in 2030, 2040, and 2050, respectively. This improvement in the network's output parameters does not apply to the presence of fuel cells, as carbon dioxide emission, fossil fuel consumption, and the annual cost of the network will be increased by 2.64%, 2.8%, and 2.25%, respectively. In Figure 14.5, we see that the majority of electricity is generated by conventional

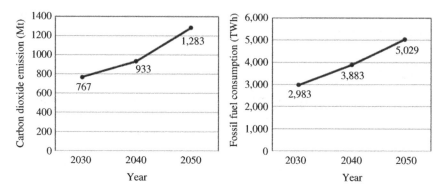

Figure 14.3 Amount of changes in carbon dioxide and fossil fuel consumption.

Electrical producers

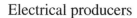

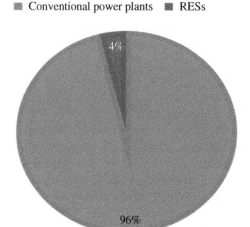

Figure 14.4 The way of supplying electrical and thermal demand.

power plants, where fossil fuels are consumed, so more pollution is produced. Due to the investment cost of this technology, the annual cost of the network is increased. On the other hand, in Figure 14.5, we see that a part of the electrical energy of the network is provided by fuel cell technology.

Considering the existing potential of Iran in the case of RESs, studying the penetration of renewable resources is one of the requirements of such studies. By conducting sensitivity analysis on the penetration of RESs in the network, as it is observed from Figure 14.6, the capacity of 35 GW is the optimal capacity. This sensitivity analysis is based on the minimal cost of the network. It means that by

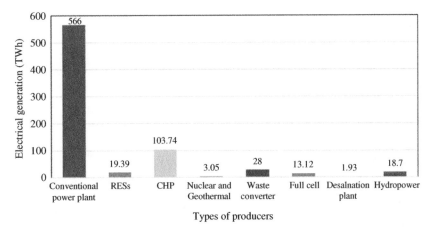

Figure 14.5 Suppliers' share in providing the electrical energy demand in FT scenario with 4.415 GW renewable generation.

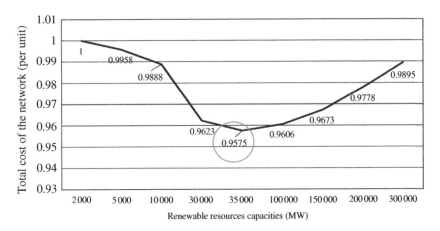

Figure 14.6 Sensitivity analysis results on RESs based on the minimal cost of network.

reaching this capacity by 2050, the total cost of the network will be minimized. Unlike the previous cases where most of the energy production was by fossil fuel power plants, here, RESs play an important role in supplying consumers. After performing the sensitivity analysis, once again, we will examine the impact of the presence of hydrogen technology. In this case, at the end of 2050, emitted carbon dioxide, fossil fuel consumption, and total network costs are reduced by 4%, 4.2%, and 1.48%, respectively. It means that the energy required for devices such as electrolyzers is provided by RESs.

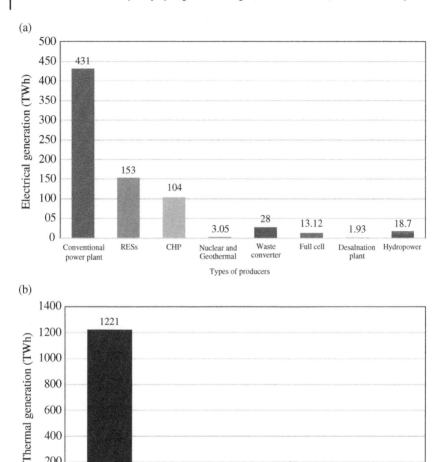

Figure 14.7 (a) Suppliers' share in providing the electrical energy demand in FT scenario with 35 GW renewable generation. (b) Suppliers' share in providing the thermal energy demand in FT scenario.

In the case of the waste converter, the simultaneous supply of thermal and electrical energy also applies, and by consuming biomass energy as the input of these plants, the output parameters will change as follows. As expected, fossil fuel consumption will be reduced by 1.49%, carbon dioxide emission will be reduced by 0.48%, and annual costs will be reduced by 0.75%. In general, by adding all technologies (CHP, fuel cell, and waste converter) to the network, the amount of

Table 14.6   Results of FT scenario with the presence of DR.

Energy reduced	CO₂ (Mt)	Fossil fuel consumption (TWh)	Total cost (Billion $)
Only Electrical	1176	4602	8383
Only thermal	1172	4605	8372
Electrical and Thermal	1153	4526	8120

Table 14.7   Summary of the results of the scenarios.

Name of scenario	CO₂ (Mt)	Fossil fuel consumption (TWh)	Total cost (Billion $)
BAU scenario	1282.6	5029	8691
FT scenario	1153	4526	8120

carbon dioxide emission, fossil fuel consumption, and the total cost of the network after increasing the penetration of RESs are 1194 million tons, 4681 TWh, and $8436 billion, respectively. In other words, compared to the BAU, fossil fuel consumption, carbon dioxide emission, and annual costs will be reduced by 7%, 6.9%, and 2.93%, respectively. Figure 14.7 shows how electrical and thermal energy demand is supplied. This scenario also examines the role and importance of DR. Thus, we assume that 5% of different energy carriers such as electrical and thermal energy decrease in each decade. The reduction of energy consumption has been done only on electric and thermal carriers because the hydrogen energy consumed is too small, and the reduction of this energy will not have a significant effect on the output parameters. Reducing the electrical and thermal energy separately or simultaneously is our approach in this section, the results of which can be seen in Table 14.6.

According to Table 14.6, by decreasing electrical and thermal loads, the amount of carbon dioxide emission is reduced by 3.43%, the amount of fossil fuel consumption is reduced by 3.3%, and the total cost is reduced by 3.75%. Table 14.7 presents a summary of results achieved from the simulation.

14.3.2   Discussion

According to the results, we see that the technologies added to the network can have a positive effect on the operation of the network. For example, waste converters reduced the amount of fossil fuel consumption in the network by using biomass energy as input fuel. In addition to generating electricity, they will also be responsible for providing thermal energy and disposing of waste. Fuel cells and CHPs are

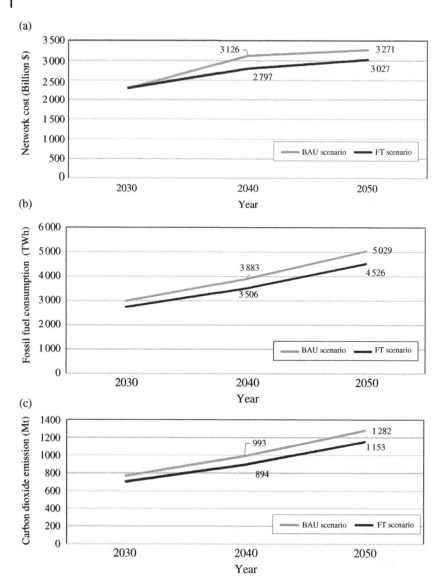

Figure 14.8 (a) Amount of changes in the total cost of the network. (b) Amount of changes in fossil fuel consumption. (c) Amount of changes in carbon dioxide emission of the network.

similar to waste converters that give us the ability to supply a part of the thermal and electrical energy of the network at the same time. An advantage of fuel cells is using pure hydrogen fuel, which can be a good substitute for fossil fuels. Considering the existing renewable resources potential in Iran, provided that the necessary investments are made and the capacity of this type of production increases, the operation of the network will change positively. Hence, RESs can produce a part of the electricity and encourage using technologies such as electrolyzers and fuel cells. As proved, a 5% reduction in electrical and thermal energy consumption reduces 3.75% of the cost and 3.34% of the emitted carbon dioxide. Therefore, having precise and comprehensive management on different network carriers can help to improve the system. The changes in each output parameter such as total cost of the network, carbon dioxide emission, and fossil fuel consumption is shown in Figure 14.8. Reducing costs related to new technologies, increasing fossil fuel costs, and the cost of carbon dioxide emissions reduce the price from 2040 to 2050. According to Figure 14.8, the network's condition is improved by the presence of CHPs, waste converters, RESs, and hydrogen-based technologies in the network. The total cost of the network and the environmental condition meliorate with DR. The total network cost and carbon dioxide emission may be reduced by 6.5% and 10%, respectively, by applying for DR programs and employing technologies.

14.4 Conclusion

With the increasing dependence on fossil fuels, growing concerns over global warming, and the introduction of new technologies, conducting studies in the field of MCENs, has become attractive. These networks are composed of different energy carriers and energy-generating technologies. This feature improves flexibility in supplying consumers, the environmental situation, and techno-economical operation. In this research, we modeled Iran's energy network. We conducted our studies by modeling two scenarios, which were simulated by EnergyPLAN. In the first scenario, we simulated the current path of the country according to the current situation of the network. In the second scenario, we tried to show the positive impact of technologies added to the network on the amount of emitted CO_2, total network cost, and fossil fuel consumption. Also, the importance of DR and the presence of RESs are fully investigated. It is observed that the future of the Iran network will be more dependent on fossil fuels and high carbon dioxide emissions over time. By increasing the capacity of RESs and reaching the desired level, benefits such as reduction in fossil fuel consumption, pollution, and network costs are obtained. In the second scenario, we found that the presence of all technologies in the network caused a reduction in fossil fuel consumption, carbon dioxide emitted, and the total cost of the network. Finally, by integrating DR in the research, the results are improved compared to the first scenario. The output parameters are technically and economically justified. Demonstrating the status of the network in the future with the presence of different technologies, increasing the flexibility of the network in consumer supply, showing the effectiveness of each technology, and finally improving the techno-economic status of the network are the achievements of this study. Nevertheless, other topics in the future could be added to this work to make it more inclusive. One limitation of the scenarios in this paper is the DR plan. Since energy is just reduced and no incentive is paid to the consumer. Dealing with a more detailed and practical type of DR is left out for our future works. Considering energy interactions between different countries such as energy exports and imports and a more detailed look at electrical vehicles can be named as possible improvements.

14.A Appendix

Table 14.A FOM, production cost, and the lifetime of technologies.

	2030		2040		2050		
Technologies	Product (MW/ Million $)	FOM (% of Inv)	Product (MW/ Million $)	FOM (% of Inv)	Product (MW/ Million $)	FOM (% of Inv)	Lifetime (Year)
CHP	2.1	1.63	2.2	1.63	2	1.6	40
Power plant	0.917	1.43	0.917	1.43	0.917	1.43	30
Geothermal	4.47	1.8	4.47	1.8	4.47	1.8	30
Nuclear	5.53	1.68	5.53	1.68	5.53	1.68	30
Hydropower	2.93	0.8	2.93	0.8	2.93	0.8	30
Wind	1.9	1.82	1.5	1.8	1.2	1.7	30
Pv	3.6	1.2	3.2	1.1	2.9	1	25
Fuel cell	2.6	5	2.2	4.8	2	4.5	10
Waste	215	2	185	1.9	150	1.7	20
Boilers	0.8	3.43	0.8	3.43	0.8	3.43	25
Desalination plant	5.8	0.7	5.7	0.7	5.5	0.67	25
Electrolyzer	1	5	0.9	5	0.8	5	25

Table 14.B Fuel cost.

Fuel	2030 ($/GJ)	2040 ($/GJ)	2050 ($/GJ)
Coal	2.2	3.2	5
Fuel oil	6.1	15	20
Petrol	11.9	18.8	23

Table 14.B (Continued)

Fuel	2030 ($/GJ)	2040 ($/GJ)	2050 ($/GJ)
Ngas	6.3	8	16
Waste	0	0	0
Biomass	4.6	5.7	12

Table 14.C Emitted CO_2 price.

	2030 ($/t CO_2)	2040 ($/t CO_2)	2050 ($/t CO_2)
CO_2	11	42	60

References

1 Victor, D., Zhou, D., Mohamed, A.E.H. et al. (2014). Summary for policymakers. In: *Climate Change 2014, Mitigation of Climate Change, United States, Contribution of Working Group III to the Fifth Assessment Report of the Intergovernmental Panel on Climate Change* (ed. A. Grübler and A. Muvundika), 1–33. Intergovernmental Panel on Climate Change (IPCC).

2 Gallandat, N., Romanowicz, K., and Züttel, A. (2017). An analytical model for the electrolyser performance derived from materials parameters. *Power and Energy Engineering* 5: 37–49.

3 Ozawa, A., Kudoh, Y., Murata, A. et al. (2018). Hydrogen in low-carbon energy systems in Japan by 2050: the uncertainties of technology development and implementation. *International Journal of Hydrogen Energy Hydrogen Energy* 43 (39): 18083–18094.

4 Hydrogen and Fuel Cell Technologies Office Office of energy efficiency and renewable energy. https://www.energy.gov/eere/fuelcells/hydrogen-storage (accessed 13 July 2022).

5 Anandarajah, G., McDowall, W., and Ekins, P. (2013). Decarbonising road transport with hydrogen and electricity: long term global technology learning scenarios. *International Journal of Hydrogen Energy Hydrogen Energy* 38 (8): 3419–3432.

6 Farahania, S.S., Bleekerb, C., Wijk, A.V., and Lukszoa, Z. (2020). Hydrogen-based integrated energy and mobility system for a real-life office environment. *Applied Energy* 264: 114695.

7 Tao, Y., Qiu, J., Lai, S., and Zhao, J. (2021). Integrated electricity and hydrogen energy sharing in coupled energy systems. *IEEE Transactions on Smart Grid* 12 (2): 1149–1162.

8 Kotowicz, J. and Uchman, W. (2021). Analysis of the integrated energy system in residential scale: photovoltaics, micro-cogeneration and electrical energy storage. *Energy* 227: 1–25.

9 Chen, H., Zhang, M., Xue, K. et al. (2020). An innovative waste-to-energy system integrated with a coal-fired power plant. *Energy* 194: 116893.

10 Luo, Y., Zhang, X., Yang, D., and Sun, Q. (2020). Emission trading based optimal scheduling strategy of energy hub with energy storage and integrated electric vehicle. *Journal of Modern Power Systems and Clean Energy* 8 (2): 267–275.

11 O'Malley, M.J., Anwar, M.B., Heinen, S. et al. (2020). Multicarrier energy systems: shaping our energy future. *Proceedings of the IEEE* 108 (9): 1437–1456.

12 Chen, Z., Li, Y., Shi, G., and Fu, X. (2020). Optimal planning method for a multi-energy complementary system with new energies considering energy supply reliability. Asia Energy and Electrical Engineering Symposium (AEEES), Chengdu, China (29–31 May 2020).

13 Muhammad Faizan, T., Chen, H., Kashif, M. et al. (2019). Integrated energy system modeling of China for 2020 by incorporating demand response, heat pump and thermal storage. *IEEE Access* 7: 40095–40108.

14 Zhang, W., Han, D. and Sun, W. (2017). Optimal operation of wind-solar-hydrogen storage system based on energy hub. IEEE Conference on Energy Internet and Energy System Integration (EI2), Beijing (26–28 November 2017).

15 Shao, C., Ding, Y., Siano, P., and Song, Y. (2020). Optimal scheduling of the integrated electricity and natural gas systems considering the integrated demand response of energy hubs. *IEEE Systems Journal* 15 (3): 4545–4553.

16 Saatloo, A.M., Rahvar, M.A., Mirzaei, M.A. et al. (2020). Robust scheduling of hydrogen based smart micro energy hub with integrated demand response. *Journal of Cleaner Production* 267: 122041.

17 Shi, T., Huang, R. and Yin, H. (2020). Research on energy management strategy of integrated energy system. IEEE 5th Asia Conference on Power and Electrical Engineering (ACPEE), Chengdu, China (4–7 June 2020).

18 Martinez, C.E.A., Emmanouil, L., Nicholas, G., and Pierluigi, M. (2020). Integrated electricity–heat–gas systems: techno-economic modeling, optimization, and application to multienergy districts. *Proceedings of the IEEE* 108 (9): 1392–1410.

19 Hajimiragha, A., Fowler, M., Geidl, M., and Andersson, G. (2007). Optimal energy flow of integrated energy systems with hydrogen economy considerations. 2007 iREP Symposium – Bulk Power System Dynamics and Control – VII. Revitalizineg Operational Reliability, Charleston, SC (19–24 August 2007).

20 Kholardi, F., Assili, M., Lasemi, M.A., and Hajizadeh, A. (2018). Optimal management of energy hub with considering hydrogen network. International Conference on Smart Energy Systems and Technologies (SEST), Sevilla (10–12 September 2018).

21 Ghorbani, N., Aghahosseini, A., and Breyer, C. (2020). Assessment of a cost-optimal power system fully based on renewable energy for Iran by 2050 – achieving zero greenhouse gas emissions and overcoming the water crisis. *Renewable Energy* 146: 125–148.

22 Shiroudi, A., Taklimi, R.H., Mousavifar, A., and Taghipour, P. (2013). Stand-alone PV-hydrogen energy system in Taleghan Iran using HOMER software: optimization and techno-economic analysis. *Environment, Development and Sustainability* 15: 1389–1402.

23 Lund, H. (2015). Energyplan advanced energy systems analysis computer mode. Aalborg University. https://www.energyplan.eu (accessed 13 July 2022).

24 Lund, H. (2015). Energyplan advanced energy systems analysis computer model. Aalborg University. https://www.energyplan.eu/training/documentation (accessed 13 July 2022).

25 Heat Roadmap Europe (2017). Programme, European Union's Horizon 2020 research and innovation. https://heatroadmap.eu/heating and-cooling-energy-demand-profiles.

26 T. Company (2019). Detailed statistics of the electricity industry. http://www.amar.tavanir.org.ir (accessed 13 July 2022).

27 D.D.E. Agency (2019). Energy statistics. https://www.ens.dk (accessed 13 July 2022).

15

Optimal Operational Planning of Heat and Electricity Systems Considering Integration of Smart Buildings

Mehrdad Setayesh Nazar[1] and Alireza Heidari[2]

[1] Faculty of Electrical Engineering, Shahid Beheshti University, Tehran, Iran
[2] School of Electrical Engineering and Telecommunication, University of New South Wales, Sydney, Australia

Nomenclature

NSBOS	Number of smart building operating scenarios in the day-head horizon
$C_{DA\ Op}^{SB}$	Cost of smart building in the day-ahead horizon
$B_{DA\ Sold}^{SB}$	Benefit of active and reactive power sold to the energy system
$B_{DA\ DRP}^{SB}$	Smart building benefit for implementing demand response program
$Penalty^{Act\ DA}$	Penalty of active power mismatch
$Penalty^{Rea\ DA}$	Penalty of reactive power mismatch
λ_{DA}^{Act}	Price of day-ahead active power
$P_{DA\ Sold}^{SB}$	Smart building day-ahead active power sold to the energy system
λ_{DA}^{Rea}	Price of day-ahead reactive power
$Q_{DA\ Sold}^{SB}$	Smart building day-ahead reactive power sold to the energy system
P_{DEFL}^{SB}	Active power of smart building deferrable electrical load
P_{DISL}^{SB}	Active power of smart building dispatchable electrical load
P_{CL}^{SB}	Active power of smart building critical electrical load
H_{DEFL}^{SB}	Smart building deferrable heating load

Coordinated Operation and Planning of Modern Heat and Electricity Incorporated Networks,
First Edition. Edited by Mohammadreza Daneshvar, Behnam Mohammadi-Ivatloo, and Kazem Zare.
© 2023 The Institute of Electrical and Electronics Engineers, Inc.
Published 2023 by John Wiley & Sons, Inc.

$H_{\text{DISL}}^{\text{SB}}$	Smart building dispatchable heating load
$H_{\text{CL}}^{\text{SB}}$	Smart building critical heating load
$C_{\text{DA Op}}^{\text{ESO}}$	Operating cost of energy system
$C_{\text{DA Pur}}^{\text{ESO}}$	Energy not supplied cost
$C_{\text{DA DRP}}^{\text{ESO}}$	Energy system operator cost for implementing demand response program
$C_{\text{DA MT Op}}^{\text{ESO}}$	Energy system operator distributed generation cost for day-ahead operational planning
$C_{\text{RT MT Op}}^{\text{ESO}}$	Energy system operator distributed generation cost for real-time operational planning
$C_{\text{DA ESS Op}}^{\text{ESO}}$	Energy system operator energy storage cost for day-ahead operational planning
$\eta_{\text{CH ESS}}^{\text{ESO}}$	Energy system operator energy storage charging efficiency
$\eta_{\text{DCH ESS}}^{\text{ESO}}$	Energy system operator energy storage discharging efficiency
LCI	Load commitment index
$C_{\text{RT Op}}^{\text{SB}}$	Cost of smart building in the real-time market
$B_{\text{RT Sold}}^{\text{SB}}$	Smart building's benefit from selling active and reactive power to the energy system
$B_{\text{RT DRP}}^{\text{SB}}$	Smart building benefit for implementing demand response program
$\text{Penalty}^{\text{Act RT}}$	Penalties of smart building real-time market active power mismatch
$\text{Penalty}^{\text{Rea RT}}$	Penalties of smart building real-time market reactive power mismatch
$\zeta_{\text{RT}}^{\text{Act}}$	Price of real-time active power
$P_{\text{RT Sold}}^{\text{SB}}$	Smart building real-time active power that is sold to the energy system
$\zeta_{\text{RT}}^{\text{Rea}}$	Price of real-time reactive power
$Q_{\text{RT Sold}}^{\text{SB}}$	Smart building real-time reactive power sold to the energy system
$C_{\text{RT Op}}^{\text{ESO}}$	Operating cost of energy system in real-time horizon
$C_{\text{RT Pur}}^{\text{ESO}}$	Energy cost purchased from the electricity market in the real-time horizon
$\sum_{\text{NC}} \text{CIC}$	Customer interruption costs in contingent conditions
NC	Number of customers
$C_{\text{RT DRP}}^{\text{ESO}}$	Cost of demand response programs in real-time horizon
W, W'	Weighting factor

15.1 Introduction

Smart buildings technology is widely utilized to increase the efficiency and resiliency of energy system infrastructures. The energy system operator (ESO) can use the smart buildings energy resources to mitigate the impacts of external shocks and system contingencies [1]. The ESO can commit photovoltaic (PV) arrays, plug-in hybrid electric vehicles (PHEVs) parking lots, distributed generation (DG) facilities, boilers, combined heat and power (CHP) units, and wind turbines (WTs) to supply its loads. Further, the ESO can perform a load control process for the smart buildings electrical and heating loads through demand response programs (DRPs) [2]. The DRP process depends on the architecture of the energy system, smart buildings locations, and electrical and heating load profiles.

The literature on the operational scheduling of energy systems can be categorized into the following categories. The first category determines the optimal operational scheduling of DRPs and load commitment strategies for smart building contributions. The second category of papers optimizes the operational scheduling of the energy system considering the combination commitment strategies of distributed energy resources and committable loads.

For the first group, Ref. [3] utilized a linear optimization model to increase the resiliency of office buildings equipped with PV arrays and energy storages facilities. The simulation results revealed that the process increased the resiliency of buildings considering the commitment strategies of energy storage and PV facilities. Reference [4] assessed a tri-level optimization algorithm for an energy management system considering the contingency operating conditions. The energy transaction between the wholesale market and distribution system was optimized first. The optimal values of transacted energy between utility and smart buildings were determined in the second level. At the third level, the energy costs of smart buildings were optimized. Reference [5] proposed a two-stage optimization algorithm for optimal demand response coordination of residential space heating loads. The model considered day-ahead and balancing markets. The first stage model minimized the customers' payments, and the second stage problem maximized the customers' bonuses. Reference [6] introduced a model to optimize the daily operational planning of self-healing buildings energy. The system compromised energy hubs, intermittent electricity generations, DRPs, and PHEVs. The model considered the uncertainties of loads and wind electricity generations. Reference [7] assessed an optimization process to utilize DRPs and enhance the resiliency of the electrical system. The best risk-averse values of system costs were determined by the method. A two-stage chance-constrained optimization process was used to minimize the costs of the system. Reference [8] introduced a two-stage optimization procedure for the optimization process of building microgrids. The building was equipped with intermittent electricity generation, energy storage, and PHEV parking lots. The method increased the resiliency of smart buildings considering consumers comfort level.

Reference [9] presented an optimization algorithm for utilizing the demand response process in smart buildings. The smart building was equipped with roof-mounted PV facilities. The commitment process considered the time-of-use demand response tariff. References [3–9] did not model the optimal scheduling of energy systems considering the smart buildings contribution scenarios.

Based on the above categorization and for the second group, Ref. [10] proposed a real-time load curtailment process. The interactions of multiple agents were modeled. The simulations outputs were successfully converged to the optimal values of demand response set points. Reference [11] introduced a reinforcement algorithm for the day-ahead scheduling load commitment process. The agent-based optimization algorithm determined the optimal sequence of load commitment. Renewable electricity generation facilities were considered in the proposed model. Reference [12] evaluated an optimization process to minimize load curtailments. A robust optimization algorithm determined the optimal commitment of generation facilities, adjustable loads, and energy storage. Reference [13] introduced an optimization algorithm for implementing a demand response process to procure regulation reserve services for contingent conditions. The algorithm utilized a dynamic programming method to optimize the system costs. Reference [14] used a load commitment process to minimize load shedding. The demand response process was performed through committable loads. Reference [15] presented an algorithm to perform the emergency demand response process. The aging of the system facilities and reliability were considered as objective functions. The output results revealed that the proposed demand response process postponed the aging of system facilities and reduced the system's costs. Reference [16] evaluated an integrated optimization model to optimize the operating scheduling of system, distribution generations, intermittent electricity generations, and parking lots. The model minimized emission and operating costs. Reference [17] introduced a regulation reserve-based load control for thermostatically controlled loads. The process was assessed for the heat, ventilating, and air-conditioning units to increase the system's resiliency. Reference [18] presented a stochastic ranking algorithm to procure the regulation reserve services through the demand response process. The procedure was utilized for thermostatically controllable loads. References [10–18] did not explore the impacts of smart buildings bidding scenarios on the optimal scheduling of energy systems.

It can be concluded that the research gaps are:

1) The smart buildings bidding scenarios are not considered in the optimal scheduling of energy systems practices for the day-ahead horizon.
2) The impacts of smart buildings energy generations/consumptions on the optimal real-time scheduling of energy systems are not modeled.

This book chapter is about the optimal operating scheduling of energy systems considering the smart buildings commitment scenarios. The main contributions of this book chapter are:

1) The impacts of smart buildings contribution scenarios on the day-ahead and real-time operational scheduling of energy systems are considered.
2) The heating load commitment index is introduced to assess the volume of served critical heating loads in contingent conditions.

15.2 Problem Modeling and Formulation

As shown in Figure 15.1, the distributed energy resources of the energy system compromise PHEVs parking lots, CHPs, boilers, DGs, WTs, and PV arrays. Further, the energy system imports electrical energy from the electricity market. Smart buildings can participate in the energy system DRPs and change their electrical and heating load profiles. The commitment strategies of smart buildings can change the available energy resources of the energy system. Thus, the energy system should explore the impacts of commitment strategies of smart buildings on its scheduling practices.

The smart buildings are assumed to be equipped with electrical and thermal energy storage facilities, roof-mounted PV panels, small WTs, plug-in electric vehicle parking lots, and smart appliances. Further, the smart building electrical and heating loads are categorized into dispatchable, non-dispatchable, and deferrable loads.

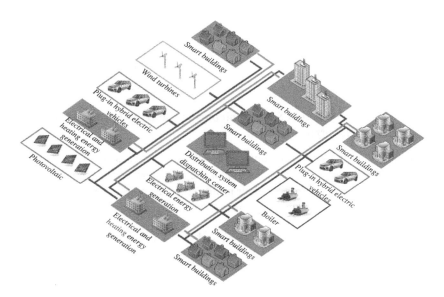

Figure 15.1 Schematic diagram of an electrical and heating energy distribution system.

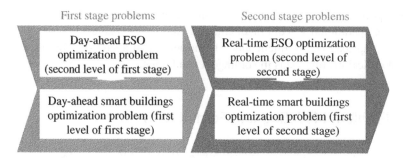

Figure 15.2 The optimization stages and levels.

15.2.1 The Proposed Framework

The optimization machine compromises two stages for optimizing the operational scheduling of energy resources. As shown in Figure 15.2, the first stage problem compromises the day-ahead optimization process of smart buildings and energy system in the first and second levels, respectively. The second stage problem optimizes the real-time scheduling problem of smart buildings and energy system in the first and second levels, respectively.

The uncertainty of the following parameters is modeled in the optimization process using scenario generation/reduction of autoregressive integrated moving average model: smart buildings biddings, active power and reactive power prices, electrical and heating load profiles, WTs electricity generation, PV electricity generation, and parking lots charge and discharges [19, 20]. Further, the Monte Carlo stochastic process is utilized to estimate the intensity and location of the system's contingencies [19].

The formulation of the proposed framework is presented in the following subsections.

15.2.2 Optimal Bidding of Smart Buildings in the Day-Ahead Market (First Level of First Stage Problem)

The smart building owner can submit the values of active power and reactive power bids to the energy system database for the day-ahead horizon. Smart building owners endeavor to maximize their profit. Thus, the objective function of the first level of the problem is proposed as (15.1):

$$
\text{Max } \mathbb{Z}^{SB}_{DA} = \sum_{t=1}^{24} \sum_{NSBOS} \text{Prob.} \left(\begin{array}{l} -C^{SB}_{DA\ Op} + B^{SB}_{DA\ Sold} + B^{SB}_{DA\ DRP} + \\ -\sum \text{Penalty}^{Act\ DA} - \sum \text{Penalty}^{Rea\ DA} \end{array} \right)
$$

(15.1)

The NSBOS parameter is the number of smart building operating scenarios in the day-head horizon. Equation (15.1) is divided into five terms: (i) the cost of smart building ($C_{\text{DA Op}}^{\text{SB}}$), (ii) the benefit of communities sold to the energy system ($B_{\text{DA Sold}}^{\text{SB}}$), (iii) the benefit of demand response program ($B_{\text{DA DRP}}^{\text{SB}}$), (iv) the penalties of day-ahead active power mismatches ($\sum \text{Penalty}^{\text{Act DA}}$), and (v) the penalties of day-ahead reactive power mismatches ($\sum \text{Penalty}^{\text{Rea DA}}$).

The smart building benefit in the day-ahead market can be written as (15.2):

$$B_{\text{DA Sold}}^{\text{SB}} = \left(\sum \lambda_{\text{DA}}^{\text{Act}} \cdot P_{\text{DA Sold}}^{\text{SB}} + \sum \lambda_{\text{DA}}^{\text{Rea}} \cdot Q_{\text{DA Sold}}^{\text{SB}} \right) \tag{15.2}$$

Equation (15.2) compromises the following terms: (i) the profit of active power sold to the energy system ($\sum \lambda_{\text{DA}}^{\text{Act}} \cdot P_{\text{DA Sold}}^{\text{SB}}$), and (ii) the profit of reactive power sold to the energy system ($\sum \lambda_{\text{DA}}^{\text{Rea}} \cdot Q_{\text{DA Sold}}^{\text{SB}}$).

Equation (15.2) has the following constraints: (i) the electrical energy balance equations for the smart building electrical system, (ii) the operating limits of smart building facilities, and (iii) the DRP constraints.

The DRP constraints of electrical loads can be written as (15.3–15.8) based on the fact that the smart building electrical loads consist of dispatchable, non-dispatchable, and deferrable loads [21, 22]:

$$P_{\text{Load}}^{\text{SB}} = P_{\text{DEFL}}^{\text{SB}} + P_{\text{DISL}}^{\text{SB}} + P_{\text{CL}}^{\text{SB}} \tag{15.3}$$

$$\sum_{t=1}^{24} \Delta P_{\text{DEFL}}^{\text{SB}} = 0 \tag{15.4}$$

$$\Delta P_{\text{DLC Max}}^{\text{SB}} = P_{\text{DISL}}^{\text{SB}} \tag{15.5}$$

$$P_{\text{DRP}}^{\text{SB}} = P_{\text{DLC}}^{\text{SB}} + P_{\text{DEFL}}^{\text{SB}} \tag{15.6}$$

$$P_{\text{DLC Min}}^{\text{SB}} \leq P_{\text{DLC}}^{\text{SB}} \leq P_{\text{DLC Max}}^{\text{SB}} \tag{15.7}$$

$$P_{\text{DEFL Min}}^{\text{SB}} \leq P_{\text{DEFL}}^{\text{SB}} \leq P_{\text{DEFL Max}}^{\text{SB}} \tag{15.8}$$

Equation (15.3) denotes that the smart building electrical load compromises deferrable, dispatchable, and critical load. Equation (15.4) presents that the changes in deferrable electrical loads should equal zero in the day-ahead horizon. Equation (15.5) shows that the maximum value of direct load control of electrical loads equals dispatchable electrical load. Equation (15.6) denotes that the active power of the demand response process equals the sum of the changes of active powers of direct load control and deferrable loads. Equation (15.7) presents the limits of electrical loads in the direct load control process. Finally, Eq. (15.8) explains that the deferrable load is constrained by the lower and upper limits.

The DRP constraints of heating loads can be written as (15.9–15.14):

$$H_{\text{Load}}^{\text{SB}} = H_{\text{DEFL}}^{\text{SB}} + H_{\text{DISL}}^{\text{SB}} + H_{\text{CL}}^{\text{SB}} \tag{15.9}$$

$$\sum_{t=1}^{24} \Delta H_{DEFL}^{SB} = 0 \tag{15.10}$$

$$\Delta H_{DLC\ Max}^{SB} = H_{DISL}^{SB} \tag{15.11}$$

$$H_{DRP}^{SB} = H_{DLC}^{SB} + H_{DEFL}^{SB} \tag{15.12}$$

$$H_{DLC\ Min}^{SB} \leq H_{DLC}^{SB} \leq H_{DLC\ Max}^{SB} \tag{15.13}$$

$$H_{DEFL\ Min}^{SB} \leq H_{DEFL}^{SB} \leq H_{DEFL\ Max}^{SB} \tag{15.14}$$

Equation (15.9) presents that the smart building heating load consists of deferrable, dispatchable, and critical loads. Equation (15.10) denotes that the changes in deferrable heating loads equal zero in the day-ahead horizon. Equation (15.11) offers that the maximum value of direct load control of heating loads equals dispatchable heating load. Equation (15.12) shows that the value of heating load in the demand response process equals the sum of the values of direct load control and deferrable loads. Equations (15.13) and (15.14) are the limits of heating loads in the direct load control process and deferrable heating loads, respectively.

15.2.3 Optimal Scheduling of Energy System in the Day-Ahead Market (Second Level of First Stage Problem)

At the second level of the first stage, the energy system dispatches the distributed energy resources considering the submitted bids of smart buildings. Thus, the objective function is written as (15.15):

$$\text{Max } \mathbb{F}_{DA}^{ESO} = \sum_{t=1}^{24} \sum_{NESODAS} \text{prob.} \begin{pmatrix} W_1 \cdot (-C_{DA\ Op}^{ESO} - C_{DA\ Pur}^{ESO} - \text{ENSC} - C_{DA\ DRP}^{ESO} \\ + W_2 \cdot \text{LCI}_{HL} + W_3 \cdot \text{LCI}_{EL} \end{pmatrix} \tag{15.15}$$

The NESODAS is the number of ESO day-ahead scenarios. W is the weighting factor.

Equation (15.15) is divided into seven terms: (i) the operating cost of the energy system ($C_{DA\ Op}^{ESO}$), (ii) the energy cost purchased from the market ($C_{DA\ Pur}^{ESO}$), (iii) the energy not supplied costs (*ENSC*), (iv) the cost of DRPs ($C_{DA\ DRP}^{ESO}$), (v) the *LCI* of heating load, and (vi) the *LCI* of electrical load.

Equation (15.15) has the following constraints: (i) the load flow equations, (ii) the heating energy balance equations, (iii) the operating limits of distributed energy resource facilities, and (iv) the demand response constraints.

Further, the electrical energy storage and PHEV parking lots charge constraints and charge and discharge constraints should be considered in the optimization procedure [21].

The LCI is the load commitment index defined for critical loads. The LCI is utilized to assess the served critical loads in the contingent condition. The $\text{LCI}_{\text{Heating Load}}$ and $\text{LCI}_{\text{Electrical Load}}$ are defined as (15.16) and (15.17), respectively:

$$\text{LCI}_{\text{HL}} = \frac{\sum \text{Served Critical Heating Loads in Contingent Conditions}}{\sum \text{Served Critical Heating Loads in Normal Conditions}}$$

(15.16)

$$\text{LCI}_{\text{EL}} = \frac{\sum \text{Served Critical Electrical Loads in Contingent Conditions}}{\sum \text{Served Critical Electrical Loads in Normal Conditions}}$$

(15.17)

15.2.4 Optimal Bidding Strategies of Smart Buildings in the Real-Time Market (First Level of the Second-Stage Problem)

Same as the day-ahead optimal bidding problem, the smart building submits the values of active and reactive powers bids to the energy system database for the real-time market. Thus, the objective function of this problem is as (15.18):

$$Max\, \mathbb{Z}_{\text{RT}}^{\text{SB}} = \sum_{k=1}^{k+1} \left(\begin{array}{c} - C_{\text{RT Op}}^{\text{SB}} + B_{\text{RT Sold}}^{\text{SB}} + B_{\text{RT DRP}}^{\text{SB}} + \\ - \sum \text{Penalty}^{\text{Act RT}} - \text{Penalty}^{\text{Rea RT}} \end{array} \right)$$

(15.18)

The k parameter is the simulation step for the real-time optimization process. Equation (15.18) is divided into five terms: (i) the cost of smart building in the real-time market ($C_{\text{RT Op}}^{\text{SB}}$), (ii) the smart building's benefit for selling active and reactive power to the energy system ($B_{\text{RT Sold}}^{\text{SB}}$), (iii) the smart building benefit for implementing demand response program ($B_{\text{RT DRP}}^{\text{SB}}$), (iv) the penalties of smart building real-time market active power mismatch ($\sum \text{Penalty}^{\text{Act RT}}$), and (v) the penalties of smart building real-time reactive power mismatch ($\sum \text{Penalty}^{\text{Rea RT}}$). The smart building benefit can be presented as (15.19):

$$B_{\text{RT Sold}}^{\text{SB}} = \left(\sum \xi_{\text{RT}}^{\text{Act}} \cdot P_{\text{RT Sold}}^{\text{SB}} + \sum \xi_{\text{RT}}^{\text{Rea}} \cdot Q_{\text{RT Sold}}^{\text{SB}} \right)$$

(15.19)

Equation (15.19) terms are (i) the profit of active power sold to the energy system ($\sum \xi_{\text{RT}}^{\text{Act}} \cdot P_{\text{RT Sold}}^{\text{SB}}$), and (ii) the profit of reactive power sold to the energy system ($\sum \xi_{\text{RT}}^{\text{Rea}} \cdot Q_{\text{RT Sold}}^{\text{SB}}$).
Equation (15.18) is constrained by (15.2)–(15.14).

15.2.5 Optimal Scheduling of Energy System in the Real-Time Market (Second Level of the Second-Stage Problem)

The ESO optimizes its system's control variables every 15 minutes. The objective function of this problem is written as (15.20):

$$\text{Max } \mathbb{F}_{RT}^{ESO} = \sum_{k=1}^{k+1} \begin{pmatrix} W_1' \cdot (- C_{RT\,Op}^{ESO} - C_{RT\,Pur}^{ESO} - \sum_{NC} CIC - C_{RT\,DRP}^{ESO} \\ + W_2' \cdot LCI_{HL} + W_3' \cdot LCI_{EL} \end{pmatrix} \quad (15.20)$$

The NC parameter is the number of system customers not supplied in contingent conditions. W' is the weighting factor.

Equation (15.20) is divided into six terms: (i) the operating cost of the energy system ($C_{RT\,Op}^{ESO}$), (ii) the energy cost purchased from the market ($C_{RT\,Pur}^{ESO}$), (iii) the customer interruption costs in contingent conditions ($\sum_{NC} CIC$), (iv) the cost of DRPs ($C_{RT\,DRP}^{ESO}$), (v) the *LCI* of heating loads, and (vi) the *LCI* of electrical load.

Equation (15.20) is constrained by Eq. (15.15) constraints.

15.3 Optimization Algorithm

The proposed multi-stage multi-level optimization problem are linear programming problems and can be solved by the CPLEX solver of GAMS. Figure 15.3 depicts the flowchart of the introduced algorithm. The proposed optimization process compromises optimal operational scheduling of smart buildings and energy system.

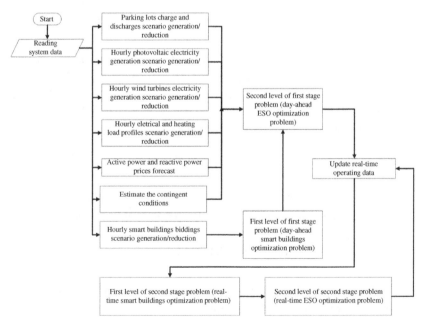

Figure 15.3 The proposed multi-stage multi-level optimization procedure.

The autoregressive integrated moving average models generate the scenarios of smart buildings biddings, active power and reactive power prices, electrical and heating load profiles, WTs electricity generation, PV electricity generation, and parking lots charge and discharges. The Monte Carlo process simulates the intensity and location of external shocks [19, 20]. The detailed formulation of autoregressive integrated moving average model and scenario generation and reduction processes are available in [23] and are not presented for the sack of space.

The optimization process in the day-ahead horizon compromises smart buildings optimal bidding problem and energy system optimal scheduling of distributed energy facilities. At first, the scenario generation and reduction process for intermittent power generations, smart buildings and parking lots contribution scenarios, electrical and heating loads, and electricity prices are carried out. The first level of the first stage problem estimates smart buildings' optimal biddings. The second level of the first stage optimizes the ESO day-ahead scheduling problem. Then, the real-time data is utilized to update the energy system data. In the real-time optimization processes, the optimal bidding of smart buildings and scheduling of ESO resources are determined in the first and second levels, respectively.

15.4 Numerical Results

The simulation process was performed for the 123-bus IEEE test system. Figure 15.4 depicts the configuration of the system [24]. Further, the solar panel data and WTs are available [21].

Figure 15.5 depicts the day-ahead electrical and heating loads. Figure 15.6 shows the prices of the electricity market. The MU stands for the monetary unit.

Table. 15.1 presents the input data of the simulation process.

Table 15.2 presents the parameters of the utility-owned distributed energy resources.

Figures 15.7 and 15.8 present the forecasted values of electricity generation of PV arrays and WTs, respectively. The maximum values of PV arrays and WTs electricity generation were 77.6 and 94.978 kW, respectively.

Figure 15.9 depicts the submitted and accepted values of active and reactive powers of smart buildings. The maximum values of accepted active and reactive power were 1539.62 kW and 416.58 kVAr, respectively.

The aggregated values of submitted active power and reactive power bids were 2428.22 kWh and −1437.32 kVArh, respectively. Further, the aggregated values of accepted active power and reactive power bids were 2735.3 kWh and −1245.19 kVArh, respectively.

Figure 15.10 shows the values of electricity generation of DG facilities. The mean and aggregated value of energy generation of DGs were 118.426 kW and 31264.47 kWh,

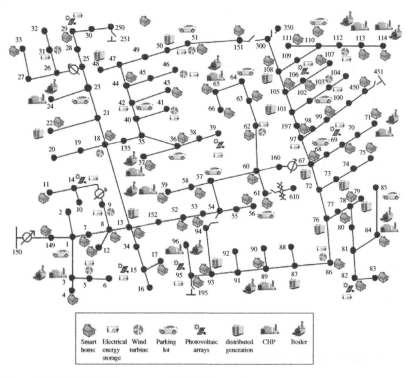

Figure 15.4 The 123-bus energy distribution system.

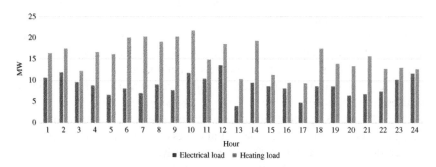

Figure 15.5 The forecasted loads of the energy system.

respectively. The maximum value of generated active power took on a value of 170 kW and was injected into the system by DG 108 at hour 17.

Figures 15.11 and 15.12 present the CHPs' electricity and heating energy generations, respectively. The electricity and heating energy generation of CHPs were

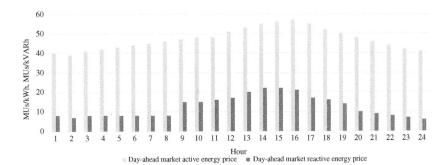

Figure 15.6 The day-ahead prices of the electricity market.

Table 15.1 The input data for simulation.

Scenario generations	Value
Solar irradiation	1000
Wind turbine power generation	1000
PHEV parking lot commitment	1000
Smart buildings bidding	1000
Day-ahead load	100
Day-ahead price	100
Scenario reduction	Value
Solar irradiation	10
Wind turbine power generation	10
PHEV parking lot commitment	10
Smart buildings bidding	10
Day-ahead load	10
Day-ahead price	10
Parameters	
$W, W'\backslash$	1

Table 15.2 The 123-bus system parameters of utility-owned distributed energy resources.

$C_{DA\ MT\ Op}^{ESO}$	0.1 \$/kWh	$C_{DA\ ESS\ Op}^{ESO}$	0.4 \$/kWh
$C_{RT\ MT\ Op}^{ESO}$	0.15 \$/kWh	$\eta_{CH\ ESS}^{ESO}, \eta_{DCH\ ESS}^{ESO}$	0.95
$C_{DA\ ESS\ Op}^{ESO}$	0.4 \$/kWh		

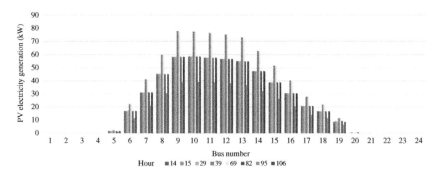

Figure 15.7 The forecasted electricity generation of photovoltaic arrays.

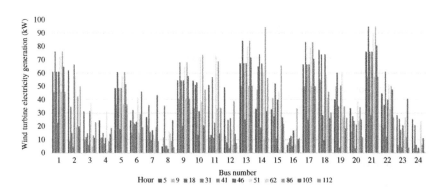

Figure 15.8 The forecasted electricity generation of wind turbines.

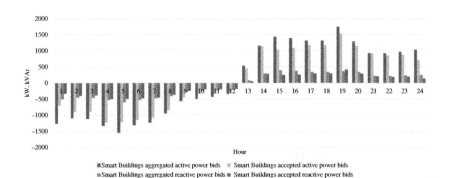

Figure 15.9 The estimated values of smart buildings' submitted and accepted active and reactive power.

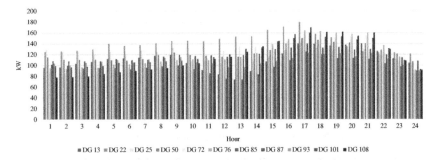

Figure 15.10 The electricity generation of distributed generation facilities.

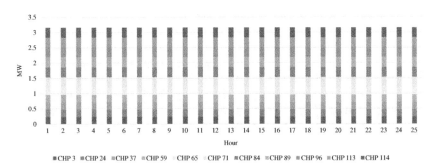

Figure 15.11 The electricity generation of CHPs.

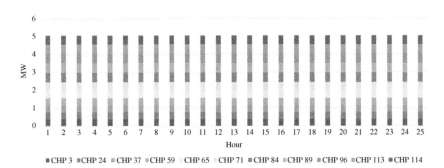

Figure 15.12 The heating energy generation of CHPs.

19.15 and 126.64 MWh, respectively. The average CHPs' electrical and heating energy generation values were 0.2878 and 0.4605 MWh, respectively.

Figure 15.13 shows the electricity transactions of parking lots with the energy system. The aggregated value of electrical energy generation was 18.5036 MWh.

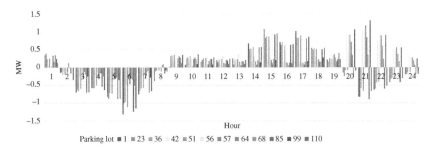

Parking lot ■ 1 ■ 23 ■ 36 ▦ 42 ■ 51 ▦ 56 ■ 57 ■ 64 ■ 68 ■ 85 ■ 99 ■ 110

Figure 15.13 The electricity transactions of parking lots with the energy system for the day-ahead market.

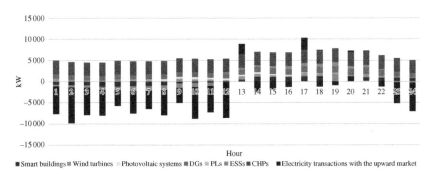

■ Smart buildings ■ Wind turbines ▦ Photovoltaic systems ■ DGs ■ PLs ■ ESSs ■ CHPs   ■ Electricity transactions with the upward market

Figure 15.14 The aggregated electricity generation of distributed energy generation facilities for the day-ahead horizon.

The maximum value of the injected active power took on a value of 1.3265 MW at hour 21 that was injected by PL 99. Further, the minimum value of the active power consumption took on a value of 1.3114 MW at hour 6 that was consumed by PL 1.

Figure 15.14 depicts the aggregated electricity generation of smart buildings, WTs, PV, DG facilities, parking lots, energy storage systems, and CHPs for the day-ahead horizon. The energy system exported electricity for 13, 17, 20, and 21 hours to the electricity market. The aggregated value of imported active power was about 88521.54 MWh. Further, the aggregated values of energy generation of WTs and PV systems were about 9332.17 and 4521.14 kWh, respectively.

Figure 15.15 shows the heating energy generation of CHPs and boilers. The aggregated values of heating energy generation of boilers and CHPs were about 250.46 and 121.57 MWh, respectively.

Figure 15.16 presents the energy system costs for power transactions with the market. The energy system gained profit by exporting electricity for 13, 17, 20,

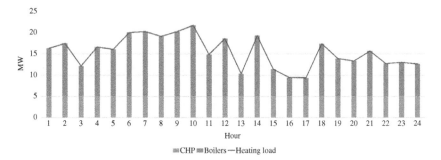

Figure 15.15 The heating energy generation of CHPs and boilers.

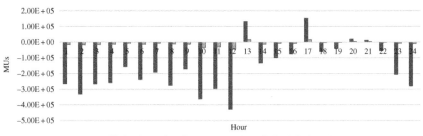

Figure 15.16 The energy system costs for active and reactive power transactions with the electricity market.

and 21 hours to the electricity market. The aggregated values of system costs for active power and reactive power transactions with the wholesale market were about 3927716.38 MUs and 301583.37 MUs, respectively.

Figure 15.17 depicts the expected energy not supplied and operational costs for the most credible contingencies. The average value of the expected operational and energy not supplied costs was 586135.35 MUs. Further, the maximum value of aggregated expected operational and energy not supplied costs took on a value of 4986168 MU that was for contingency 27 and hour 20.

Figure 15.18 presents the estimated values of committed electrical loads for the worst-case contingency. The average value of the served critical electrical loads was about 55.07%, concerning the average value of electrical loads of energy systems. The aggregated value of served critical electrical loads was about 1380.79 kWh.

Further, Figure 15.19 shows the estimated values of the day-ahead load commitment index for the worst-case contingency. The average values of the LCI_{EL} and LCI_{HL} were 0.5517 and 0.5815, respectively.

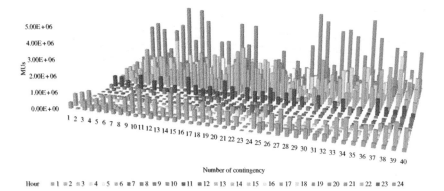

Figure 15.17 The aggregated expected energy not supplied and operational costs for the 40 most credible contingencies for the day-ahead optimization horizon.

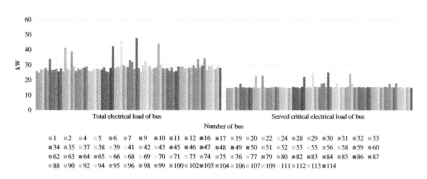

Figure 15.18 The estimated values of served critical electrical loads for the worst-case contingency.

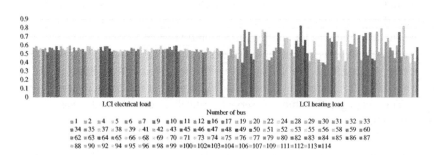

Figure 15.19 The estimated values of LCI$_{EL}$ and LCI$_{HL}$ for the worst-case contingency.

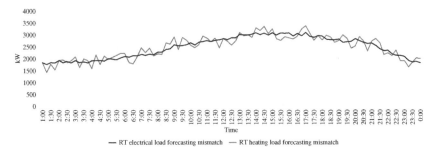

Figure 15.20 The real-time electrical and heating load mismatches for the worst-case contingency.

Figure 15.20 shows the real-time load mismatches for the worst-case contingency. The maximum values of electrical load mismatch and heating load mismatch were about 3101.68 and 3369.72 kW, respectively.

Figure 15.21 presents the real-time market prices. The maximum active and reactive power price values were 60 MUs/kW and 25 MUs/kVAr, respectively.

Figure 15.22 depicts the electricity generation of smart buildings, WTs, PV, DG facilities, parking lots, energy storage systems, and CHPs for compensating the real-time electrical load forecasting mismatch. The average values of electricity generation of smart buildings, DG facilities, and parking lots for compensating the real-time electrical load forecasting mismatch were 498.06, 487.66, and 539.95 kWh, respectively.

Figure 15.23 shows the heating generation of boilers for compensating the real-time heating load forecasting mismatch. The boilers tracked the heating load to compensate for the heating load mismatch. The aggregated and average values of heating power generation of boilers for the real-time horizon were about 2482.59 and 233364.34 kWh, respectively.

Figure 15.21 The real-time market prices.

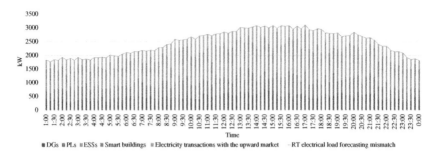

Figure 15.22 The aggregated electricity generation of distributed energy generation facilities for the real-time horizon.

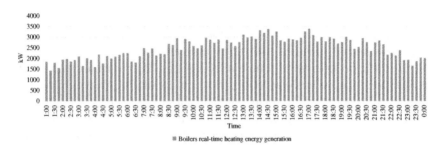

Figure 15.23 The heating generation of boilers for compensating the real-time heating load forecasting mismatch.

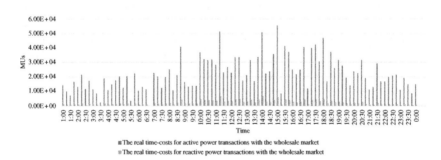

Figure 15.24 The costs of energy transactions with the real-time electricity market.

Figure 15.24 presents the cost of energy transactions with the real-time electricity market. The aggregated values of active power transaction costs and reactive power transaction costs were 22116.32 and 2256 MUs, respectively.

Figure 15.25 presents the electricity generation of DG facilities. The average values of active power injection of DG facilities in the normal and contingent

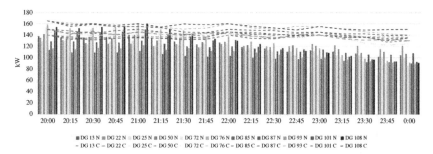

Figure 15.25 The electricity generation of distributed generation facilities in normal and contingent conditions.

conditions were about 120.137 and 143.635 kW, respectively. Further, the aggregated value of active power injection of DG facilities in the normal and contingent conditions were about 26859.79 and 22465.68 kWh, respectively.

Figure 15.26 depicts the smart buildings active power injection into the energy system. The average values of active power injection of smart buildings in the normal and contingent conditions were 886.139 and 1333.43 kW, respectively. Further, the aggregated value of active power injection of smart buildings in the normal and contingent conditions were about 15064.35 and 22668.22 kWh, respectively.

Figure 15.27 shows the parking lots active power injection into the energy system. The average values of active power injection of parking lots in normal and contingent conditions were 1378.19 and 4767.04 kW, respectively. Further, the aggregated value of active power injection of parking lots in the normal and contingent conditions were about 23429.20 and 81039.73 kWh, respectively.

Figure 15.28 presents the active power injection of energy storage facilities into the energy system. The average values of active power injection of energy storage

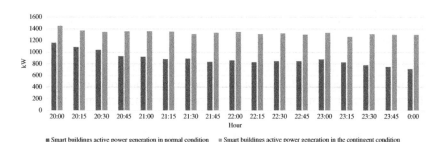

Figure 15.26 The smart buildings active power injection into energy system in normal and contingent conditions.

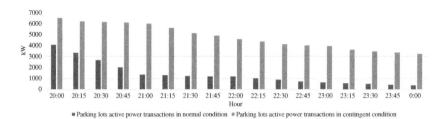

Figure 15.27 The parking lots active power injection into energy system in normal and contingent conditions.

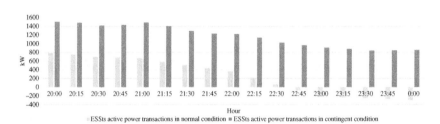

Figure 15.28 The active power injection of energy storage facilities into the energy system in normal and contingent conditions.

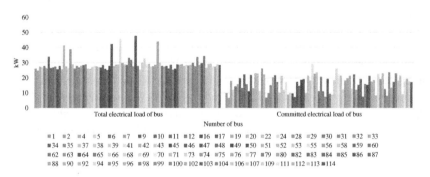

Figure 15.29 The real-time values of committed electrical loads for the worst-case contingency.

systems in normal and contingent conditions were 255.015 and 1169.33 kW, respectively. The aggregated value of active power injection of energy storages in the normal and contingent conditions were about 4335.24 and 19878.54 kWh, respectively.

Figure 15.29 shows the real-time values of committed electrical loads for the worst-case contingency. The average value of served critical electrical load was

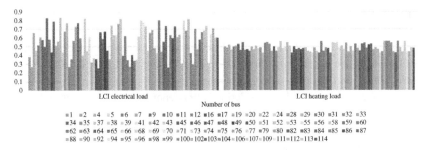

Figure 15.30 The real-time values of the LCI_{EL} and LCI_{HL} for the worst-case contingency.

about 56.48% concerning the average value of the electrical load of the energy system.

Figure 15.30 depicts the real-time values of the LCI_{EL} and LCI_{HL} for the worst-case contingency. The average values of the load commitment index for the electrical and heating loads were 0.5641 and 0.4930, respectively.

15.5 Conclusions

A multi-stage multi-level optimization algorithm for optimal scheduling of energy systems was introduced in the present chapter. The proposed model explored the impacts of smart buildings commitment scenarios in the day-ahead and real-time horizons. The first level of the first-stage problem explored the optimal bidding scenarios of smart buildings. The second level of the first stage problem optimally scheduled the distributed energy resources of the energy system considering the smart building contributions in the day-ahead market. Then, the first level of the second stage problem estimates the optimal bidding of smart buildings in the real-time horizon. Finally, the second level of the second stage problem scheduled the system resources in the real-time market. The main achievements of this book chapter can be presented in the following sentences:

1) The impacts of contribution scenarios of smart buildings on the operational scheduling of energy systems were modeled,
2) The electrical and heating load commitment indices were utilized to explore the value of served critical loads in the contingent conditions.

The proposed model was successfully assessed for the IEEE 123-bus system.

References

1 Yang, Y. and Wang, S. (2021). Resilient energy management with vehicle-to-home and photovoltaic uncertainty. *Int. J. Electr. Power Energy Syst.* 132: 107206.

2 Afrakhte, H. and Bayat, P. (2020). A contingency based energy management strategy for multi-microgrids considering battery energy storage systems and electric vehicles. *J. Energy Storage* 27: 101087.

3 Rosales-Asensio, E., Simón-Martín, M., Borge-Diez, D. et al. (2019). Microgrids with energy storage systems as a means to increase power resilience: an application to office buildings. *Energy* 172: 1005–1015.

4 Mehrjerdi, H. (2019). Multilevel home energy management integrated with renewable energies and storage technologies considering contingency operation. *J. Renew. Sustainable Energy* 11: 025101.

5 Ali, M., Alahäivälä, A., Malik, F. et al. (2015). A market-oriented hierarchical framework for residential demand response. *Int. J. Electr. Power Energy Syst.* 69: 257–263.

6 Mehrjerdi, H. and Hemmati, R. (2020). Coordination of vehicle-to-home and renewable capacity resources for energy management in resilience and self-healing building. *Renew. Energy* 146: 568–579.

7 Guo, Z., Li, G., Zhou, M., and Feng, W. (2019). Resilient configuration approach of integrated community energy system considering integrated demand response under uncertainty. *IEEE Access* 7: 87513–87533.

8 Eseye, A., Lehtonen, M., Tukia, T. et al. (2019). Optimal energy trading for renewable energy integrated building microgrids containing electric vehicles and energy storage batteries. *IEEE Access* 7: 106092–106101.

9 Fotouhi, G.M., Soares, J., Abrishambaf, O. et al. (2017). Demand response implementation in smart households. *Energy Build* 143: 129–148.

10 Srikantha, P. and Kundur, D. (2016). Resilient distributed real-time demand response via population games. *IEEE Trans. Smart Grid.* 8: 2532–2543.

11 Remani, T., Jasmin, E., and Imthias, A.T. (2019). Residential load scheduling with renewable generation in the smart grid: a reinforcement learning approach. *IEEE Sys. J.* 13: 3283–3294.

12 Khodaei, A. (2014). Resiliency oriented microgrid optimal scheduling. *IEEE Trans. Smart Grid.* 5: 1584–1591.

13 Motalleb, M., Thornton, M., Reihani, E., and Ghorbani, R. (2016). Providing frequency regulation reserve services using demand response scheduling. *Energy Convers. Manag.* 124: 439–452.

14 Balasubramaniam, B., Saraf, P., Hadidi, R., and Makram, E. (2016). Energy management system for enhanced resiliency of microgrids during islanded operation. *Electr. Power Syst. Res.* 137: 133–141.

15 Hafiz, F., Chen, B., Chen, C. et al. (2019). Utilising demand response for distribution service restoration to achieve grid resiliency against natural disasters. *IET Gener. Transm. Distrib.* 13: 2942–2950.

16 Hosseinnia, H., Modarresi, J., and Nazarpour, D. (2020). Optimal eco-emission scheduling of distribution network operator and distributed generator owner under employing demand response program. *Energy* 191: 116553.

17 Lu, N. and Zhang, Y. (2012). Design considerations of a centralized load controller using thermostatically controlled appliances for continuous regulation reserves. *IEEE Trans. Smart Grid.* 4: 914–921.

18 Vivekananthan, C. and Mishra, Y. (2014). Stochastic ranking method for thermostatically controllable appliances to provide regulation services. *IEEE Trans. Power Syst.* 30: 1987–1996.

19 Eichhorn, A., Heitsch, H., and Römisch, W. (2010). Stochastic optimization of electricity portfolios: Scenario tree modeling and risk management. In: *Handbook of Power Systems II* (ed. S. Rebennack, P.M. Pardalos, M.V.F. Pereira and N.A. Iliadis). Springer.

20 Heitsch, H. and Römisch, W. (2003). Scenario reduction algorithms in stochastic programming. *Comput. Optim. Appl.* 24 (2–3): 187–206. https://doi.org/10.1023/A:1021805924152.

21 Bostan, A., Setayesh, N.M., Shafie-khah, M.R., and Catalão, J.P.S. (2020). Optimal scheduling of distribution systems considering multiple downward energy hubs and demand response programs. *Energy* 190: 116349.

22 Wang, Z. and Wang, J. (2015). Self-healing resilient distribution systems based on sectionalization into microgrids. *IEEE Trans. Power Sys.* 30: 3139–3149.

23 Rahmani, K. and Setayesh, N.M. (2017). Coordinated bidding of wind and thermal energy in joint energy and reserve markets of Spain by considering the uncertainties. *Energy Environ.* 28: 846–869.

24 Bostan, A., Setayesh, N.M., Shafie-khah, M.R., and Catalão, J.P.S. (2020). An integrated optimization framework for combined heat and power units, distributed generation and plug-in electric vehicles. *Energy* 202: 117789.

16

Coordinated Planning Assessment of Modern Heat and Electricity Incorporated Networks

Hadi Vatankhah Ghadim and Jaber Fallah Ardashir

Department of Electrical Engineering, Tabriz Branch, Islamic Azad University, Tabriz, Iran

16.1 Introduction

Whether we talk about energy hubs, energy internets, community/microgrids, or multi-carrier energy systems [1–6], they are all the sub-clusters of multi-dimensional energy networks (MDEN). Nowadays, MDEN is one of the emerging and prevalent issues in the energy industry. Incorporating different carriers of energy with each other to enhance either the reliability, flexibility, environmental, and cost indicators of a power grid is one of the solutions for the current situation in modern energy networks [7–10], in which integrating intermittent renewable energy sources increases the uncertainty rates in the planning phase of the power grid. Also, there is a burgeoning need for other energy forms such as heating or cooling along with electricity in specific regions. By combining these needs, new form of MDENs is created named as heat and electricity integrated network (HEIN).

Heat and electricity cogeneration was initially introduced by authors of [11] in 1964. The concept was an economical solution to increase the cost-effectiveness of small-scale nuclear power plants so that they get a chance to participate in the energy market that conventional power plants have dominated it. Later on, in 1966, Liebhafsky predicted that although nuclear power plants will dominate the integrated heat and power market, the fuel cell technology might successfully provide power and heat for a small-scale community through a small energy station [12]. That prediction and explanation was the nearest definition to community microgrid and district heating (DH) concepts, and so the HEIN.

Moreover, various technologies were introduced in alignment with the concept of HEIN. However, conventional approaches to heating are known as environmentally inappropriate in different countries such as the United Kingdom, as it

Coordinated Operation and Planning of Modern Heat and Electricity Incorporated Networks,
First Edition. Edited by Mohammadreza Daneshvar, Behnam Mohammadi-Ivatloo, and Kazem Zare.
© 2023 The Institute of Electrical and Electronics Engineers, Inc.
Published 2023 by John Wiley & Sons, Inc.

is responsible for more than 25% of its annual carbon emission. Therefore, the emersion of modernized and optimized heat and power integrated systems is necessary [13]. In conventional thermal power plants, the energy output is usually along with heat losses such as boiler heat loss, pipe heat loss, etc. This issue decreases the efficiency of thermal power plants significantly [14]. However, to overcome this problem, new technologies such as combined heat and power (CHP) and combined cool, heat and power (CCHP) plants have been introduced and implemented in various cases. Also, there are some renewable energy-based power generation technologies such as concentrating solar power (CSP) and integrated collector storage solar (ICSS) plants that use heat energy to supply the heat and power demand in the grid [15, 16]. By introducing these technologies, the planning issues also appeared as new challenges, which necessitated the study on optimal planning of HEINs.

This chapter has an exclusive focus on Heat & Electricity incorporated MDEN and defines the characteristics of optimal planning of power and heat combined energy networks that have both conventional and renewable technologies implemented to generate power. Furthermore, the structures of such scheduling will be discussed. Later on, the benefits and drawbacks of integrating heat and electricity – two different energy carriers – with each other in the planning phase of a grid will be outlined. Eventually, challenges of this decision along with possible research opportunities in this issue will be mentioned. In summary, the goals of this chapter are outlined as below:

- Providing definition for optimal planning of HEINs
- Introducing possible structure and mathematical modeling of assets that can be used in them
- Explaining about advantages that optimal planning can have in HEINs
- Delineating possible challenges in the optimization process of the HEIN planning
- Clarifying future research issues for interested researchers.

16.2 Definition of the Optimal Planning of Heat and Electricity Incorporated Network

HEIN is one of the many variations of MDENs in which both the heat and electrical demands can be supplied. HEIN helps the users to exploit the synergy of two different energy forms in a facility that is usually consisted of various parts such as pure power, pure heat, and cogeneration units [17].

However, to exploit the capabilities of HEINs efficiently, it is necessary to know how to *plan* optimally. Planning is a way to organize a solution for a challenge beforehand for any system. The word planning means to be prepared for the future status of the system as it is forecasted. We assume that in any complex system, there are resources of any kind which are gathered together to pursue one or many common goals. In the case of MDENs, the presence of different energy carriers in the system necessitates appropriate planning to supply the demands in the grid. Nonetheless, any attempt for *optimal planning* in any system means that previous attempts in planning the system were not sufficiently implementable due to the emerging challenges and technologies in the system. The new configuration in a system requires further planning studies to optimize the performance of utilities and meet the demands with lower costs and higher efficiencies under any constraints present in the system.

In HEINs, optimal planning can be done on energy generation, transmission, and consumption facilities. Although having comprehensive information from grid utilities seems necessary for grid planning, having access to detailed information about their specifications is not needed for optimal planning. Different focuses of optimal planning can be found in Figure 16.1. Generation expansion planning, storage capacity reserve planning, and load forecasting are examples for generation, interconnection, and consumption sections of the planning elements in Figure 16.1, respectively.

Currently, researchers are focusing on dynamic planning of the HEINs as the results in dynamic approach is more promising than static or semi-static ones. Dynamic planning usually takes into account the data of previous observations to plan for the long-term future, typically from one to ten years [18]. However, static or semi-static approaches are helpful in the mid-term planning of the

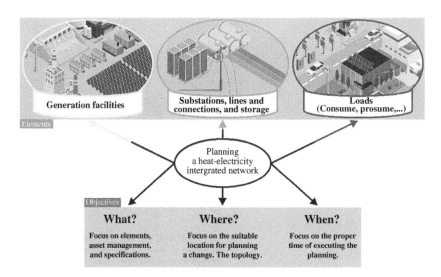

Figure 16.1 Elements and objectives of planning HIENs.

network, as the data from a week to a year is used in this planning approach. It is usually called *operational planning*, and it is an accessible planning solution for the immediate needs of a grid in the mid-term period [19].

16.3 Structure of HEINs for Optimal Planning

HEIN structure is mainly divided into three categories of energy supply, energy conversion, and energy consumption units integrated in a grid to meet the heat and electricity demands simultaneously. Figure 16.2 shows the general scheme of a HEIN.

In this section, different components of a HEIN will be discussed, and mathematical modeling of them will be introduced to provide a basis for further studies in this field.

16.3.1 Natural Gas-Based CHP

The gas consumption equation for natural gas-based CHP is as below [20]:

$$G_{\text{CHP}}(t) = \frac{(s_{\text{CHP}}(t) \cdot P_{\text{out}}(t))}{(\text{LHV} \cdot \eta_{\text{CHP}})} \tag{16.1}$$

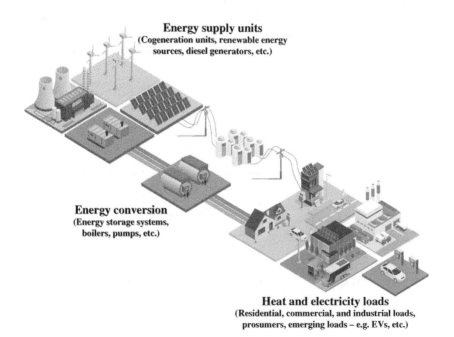

Energy supply units
(Cogeneration units, renewable energy sources, diesel generators, etc.)

Energy conversion
(Energy storage systems, boilers, pumps, etc.)

Heat and electricity loads
(Residential, commercial, and industrial loads, prosumers, emerging loads – e.g. EVs, etc.)

Figure 16.2 Isometric view of a general scheme for a HEIN.

The G_{CHP} is the gas consumption rate, s_{CHP} is the I/O state of CHP in a sampled simulation time, P_{out} is the output power of CHP, LHV is the low heat value of the fuel, which is natural gas here, and η_{CHP} is the efficiency rate of the CHP.

The heat production equation for natural gas-based CHP can be as below:

$$H_{out}(t) = \frac{s_{CHP}(t) \cdot P_{out}(t) \cdot [1 - s_{CHP}(t) - k_1] \cdot \eta_{HR}}{s_{CHP}(t)} \tag{16.2}$$

The H_{out} is the heat energy output of CHP, k_1 is the heat loss factor of CHP, and η_{HR} is the heat recovery efficiency.

The cost function for this kind of CHP's fuel (in here, natural gas) can be expressed as below:

$$Cost_{CHP} = G_{CHP}(t) \cdot G_{price} \tag{16.3}$$

The $Cost_{CHP}$ is the fuel cost of the CHP unit, and the G_{Price} is the price of the fuel (in here, natural gas) that CHP uses to produce heat and power. The price indicator is static due to the fixed-price scheme in gas trading contracts.

16.3.1.1 Operation and Maintenance Costs of CHP

$$Cost_{O \& M_{CHP}}(t) = \eta_{CHP}(t) \cdot C_{O \& M_{CHP}} \cdot P_{out}(t) \tag{16.4}$$

The $C_{O \& M_{CHP}}$ is the constant coefficient of operation and maintenance cost of CHP unit.

16.3.1.2 Environmental Costs of CHP

$$Cost_{E_{CHP}}(t) = \sum_{j=1}^{n} EA_{j_{CHP}} \cdot EE_{j_{CHP}} \cdot P_{out}(t) \tag{16.5}$$

The $EA_{j_{CHP}}$ is the amount of emission in kg per kWh, and $EE_{j_{CHP}}$ is the emission cost in local currency unit (LCU) per kg of any emission gas. This equation should be calculated for any emission gas that is meant to be studied as j is the quantity of these gases such as carbon, sulfur, etc.

16.3.1.3 Startup/Shutdown Costs of CHP

$$\begin{aligned}
I_{CHP}(t) &= Cost_I \cdot (s_{CHP}(t) - s_{CHP}(t) \cdot s_{CHP}(t-1)) \\
O_{CHP}(t) &= Cost_O \cdot (s_{CHP}(t-1) - s_{CHP}(t) \cdot s_{CHP}(t-1))
\end{aligned} \tag{16.6}$$

$Cost_I$ and $Cost_O$ are the startup and shutdown costs in LCU per kWh, relatively [21].

16.3.2 Diesel Generation Set

$$\text{Cost}_{\text{DG}}(t) = k_1 \cdot s_{\text{DG}}(t) + k_2 \cdot s_{\text{DG}}(t) \cdot P_{\text{out}}(t) + k_3 \cdot s_{\text{DG}}(t) \cdot P_{\text{out}}(t)^2 \qquad (16.7)$$

k_1 to k_3 are the coefficients of the cost function, $s_{\text{DG}}(t)$ is the I/O state of DG in a sampled simulation time, and $P_{\text{out}}(t)$ is the output power of DG [22].

16.3.2.1 Operation and Maintenance Costs of DG

$$\text{Cost}_{\text{O\&M}_{\text{DG}}}(t) = \eta_{\text{DG}}(t) \cdot C_{\text{O\&M}_{\text{DG}}} \cdot P_{\text{out}}(t) \qquad (16.8)$$

The $C_{\text{O\&M}_{\text{DG}}}$ is the constant coefficient of operation and maintenance cost of DG unit.

16.3.2.2 Environmental Costs of DG

$$\text{Cost}_{E_{\text{DG}}}(t) = \sum_{j=1}^{n} \text{EA}_{j_{\text{DG}}} \cdot \text{EE}_{j_{\text{DG}}} \cdot P_{\text{out}}(t) \qquad (16.9)$$

The $\text{EA}_{j_{\text{DG}}}$ is the amount of emission in kg per kWh, and $\text{EE}_{j_{\text{DG}}}$ is the emission cost in LCU per kg of any emission gas. This equation should be calculated for any emission gas that is meant to be studied as j is the quantity of these gases such as carbon, sulfur, etc.

16.3.2.3 Startup/Shutdown Costs of DG

$$I_{\text{DG}}(t) = \text{Cost}_I \cdot (s_{\text{DG}}(t) - s_{\text{DG}}(t) \cdot s_{\text{DG}}(t-1))$$
$$O_{\text{DG}}(t) = \text{Cost}_O \cdot (s_{\text{DG}}(t-1) - s_{\text{DG}}(t) \cdot s_{\text{DG}}(t-1)) \qquad (16.10)$$

Cost_I and Cost_O are the startup and shutdown costs in LCU per kWh, relatively [21].

16.3.3 Renewable Energy Utilities (REN)

The fuel of these utilities is natural resources, which is entirely free. However, the uncertainties in the supply of these fuels are an influential factor in determining the cost function of REN utilities. For wind turbines (WT), we have to perceive their speed-power characteristic of them:

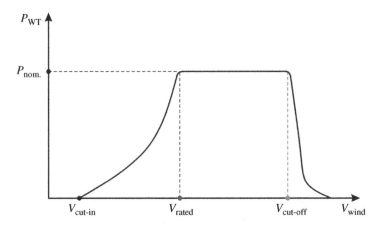

Figure 16.3 Speed-power (V-P) characteristic of a general wind turbine.

As seen from Figure 16.3, the power generation threshold will start from the point in which the wind speed is in its rated value until the cut-off speed. Therefore, the power generation equation will be as below:

$$
P_{out_{WT}} = \begin{cases} 0 & \text{if} \quad V_{Wind} < V_{cut-in} \\ P_{nom.} \times (k_1 + k_2 \cdot V_{Wind} + k_3 \cdot V_{Wind}^2) & \text{if} \quad V_{cut-in} < V_{Wind} < V_{rated} \\ P_{nom.} & \text{if} \quad V_{rated} < V_{Wind} < V_{cut-off} \\ 0 & \text{if} \quad V_{cut-off} < V_{Wind} \end{cases}
$$

$$(16.11)$$

k_1 to k_3 are the coefficients of the generation output function that are defined based on [23]. To find the cost function of the WT, the Building, Operation, and Maintenance (BOM) costs must be included:

$$
\text{Cost}_{WT}(t) = \text{TOC}_{WT} + \frac{\text{Cost}_{O\,\&\,M_{WT}}(t)}{P_{nom.} \times CC_{WT} \times 8760} \times P_{Out_{WT}} \times t \tag{16.12}
$$

The TOC_{WT} is the total overnight cost of the WT in LCU per kW, $\text{Cost}_{O\,\&\,M_{WT}}(t)$ is the operation and maintenance cost of WT in LCU per kWh, and CC_{WT} is the capacity constant of the WT [24].

For Solar PV power plants, the cost function will include the TOC and BOM costs as same as WT:

$$
\text{Cost}_{PV}(t) = \text{TOC} + \text{Cost}_{O\,\&\,M_{PV}}(t) \tag{16.13}
$$

The TOC consists of PV panels, inverters, and installation costs in LCU per kWh.

16.3.4 Fuel Cell

The fuel consumption of Fuel Cell (FC) is calculated as below:

$$G_{FC}(t) = \frac{(s_{FC}(t) \cdot P_{out}(t))}{(LHV \cdot \eta_{FC})} \tag{16.14}$$

The $s_{FC}(t)$ is the I/O state of FC in a sampled simulation time, $P_{out}(t)$ is the power output of the FC, LHV is the low heat value of natural gas that feeds FC, and η_{FC} is the efficiency of the FC. The fuel cost function of the FC is as below:

$$Cost_{FC}(t) = G_{FC}(t) \cdot G_{price} \tag{16.15}$$

The G_{price} is the price of the fuel (in here natural gas) that FC uses to produce heat and power. Price indicator is static due to the fixed-price scheme in gas trading contracts [20].

16.3.4.1 Operation and Maintenance Costs of FC

$$Cost_{O\&M_{FC}}(t) = \eta_{FC}(t) \cdot C_{O\&M_{FC}} \cdot P_{out}(t) \tag{16.16}$$

The $C_{O\&M_{FC}}$ is the constant coefficient of operation and maintenance cost of DG unit.

16.3.4.2 Environmental Costs of FC

$$Cost_{E_{FC}}(t) = \sum_{j=1}^{n} EA_{j_{FC}} \cdot EE_{j_{FC}} \cdot P_{out}(t) \tag{16.17}$$

The $EA_{j_{FC}}$ is the amount of emission in kg per kWh, and $EE_{j_{FC}}$ is the emission cost in LCU per kg of any emission gas. This equation should be calculated for any emission gas that is meant to be studied as j is the quantity of these gases such as carbon, sulfur, etc.

16.3.4.3 Startup/Shutdowns Costs of FC

$$I_{FC}(t) = Cost_I \cdot (s_{FC}(t) - s_{FC}(t) \cdot s_{FC}(t-1))$$
$$O_{FC}(t) = Cost_O \cdot (s_{FC}(t-1) - s_{FC}(t) \cdot s_{FC}(t-1)) \tag{16.18}$$

$Cost_I$ and $Cost_O$ are the startup and shutdown costs in LCU per kWh, relatively [21].

16.3.5 Battery Energy Storage System (BESS)

$$SOC_{BESS}(t) = SOC_{BESS}(t-1) \cdot (1 - \rho_{BESS}) - \Delta t \cdot \left(\frac{P_{DIS}(t)}{\eta_{DIS}}\right) + \Delta t \cdot P_{CH}(t) \cdot \eta_{CH} \tag{16.19}$$

$\text{SOC}_{\text{BESS}}(t)$ and $\text{SOC}_{\text{BESS}}(t-1)$ are the states of the charge for BESS in current and previous period, relatively, ρ_{BESS} is the indicator for self-discharging effect of BESS, $P_{\text{DIS}}(t)$ and $P_{\text{CH}}(t)$ are the discharging and charging power of BESS, η_{DIS}, and η_{CH} are the discharging and charging efficiency values, and Δt is the sampling time for this function. The SOC function is necessary to define the degradation costs of energy storage systems (ESS) such as BESS, as that is the operation and maintenance cost of ESS in a HEIN. The degradation cost function for BESS can be expressed as below:

$$\text{Cost}_{\text{BESS}}(t) = \text{Cost}_{\text{DEG.}} \times P_{\text{DIS/CH}}(t) \times \Delta t \tag{16.20}$$

The $\text{Cost}_{\text{DEG.}}$ is the BESS degradation cost in LCU per kWh [25].

16.3.6 Electrical Heater

This component provides the thermal demand of the HEIN by consuming electrical energy. The heating output of electrical heater can be demonstrated as below:

$$H_{\text{out}}(t) = P_{\text{elec}-\text{in}}(t) \cdot \eta_{\text{heater}} \tag{16.21}$$

The $P_{\text{elec-in}}(t)$ is the consumed electrical power to produce the heat, and the η_{heater} is the efficiency of the electrical heater.

16.3.6.1 Operation and Maintenance Costs

$$\text{Cost}_{\text{O \& M}_{\text{heater}}}(t) = C_{\text{O \& M}_{\text{heater}}} \cdot H_{\text{out}}(t) \tag{16.22}$$

The $C_{\text{O \& M}_{\text{heater}}}$ is the constant coefficient of operation and maintenance cost of electrical heater unit [26].

16.3.7 Natural Gas Boiler

NGBs are other utilities that provide the thermal demands of HEINs. However, due to the usage of natural gas as a fuel, they produce emissions, and it is necessary to consider them in defining their overall cost function. The fuel consumption function for NGB is as below:

$$G_{\text{NGB}}(t) = \frac{H_{\text{out}}(t)}{(\text{LHV} \cdot \eta_{\text{NGB}})} \tag{16.23}$$

The $H_{\text{out}}(t)$ is the heating output of the NGB, and η_{NGB} is the overall efficiency of the NGB. Therefore, the cost function of NGB will be as below:

$$\text{Cost}_{\text{NGB}}(t) = G_{\text{NGB}}(t) \cdot G_{\text{price}} \tag{16.24}$$

The G_{price} is the price of the fuel (in here, natural gas) that NGB use to produce heat. The price indicator is static due to the fixed-price scheme in gas trading contracts [27].

16.3.7.1 Operation and Maintenance Costs of NGB

$$\text{Cost}_{O \& M_{NGB}}(t) = C_{O \& M_{NGB}} \cdot H_{out}(t) \tag{16.25}$$

The $C_{O \& M_{NGB}}$ is the constant coefficient of operation and maintenance cost of NGB unit [26].

16.3.7.2 Environmental Costs of NGB

$$\text{Cost}_{E_{NGB}}(t) = \sum_{j=1}^{n} EA_{j_{NGB}} \cdot EE_{j_{NGB}} \cdot H_{out}(t) \tag{16.26}$$

The $EA_{j_{DG}}$ is the amount of emission in kg per kWh, and $EE_{j_{DG}}$ is the emission cost in LCU per kg of any emission gas. This equation should be calculated for any emission gas that is meant to be studied as j is the quantity of these gases such as carbon, sulfur, etc.

16.3.8 Heat Energy Storage System

HESS is one of the main units to store the excessive thermal output of supply units. Similar to the assumed load leveling role for ESS in energy networks [28], HESS provides thermal energy when the thermal demand reaches a peak. In the case of storage units, the degradation cost function will be defined as the main cost function for it. The state of the charging equation for HESS is as below:

$$\text{SOC}_{HESS}(t) = \text{SOC}_{HESS}(t-1) \cdot (1 - \rho_{HESS}) - \Delta t \cdot \left(\frac{H_{DIS}(t)}{\eta_{DIS}} \right) + \Delta t \cdot H_{CH}(t) \cdot \eta_{CH} \tag{16.27}$$

$\text{SOC}_{HESS}(t)$ and $\text{SOC}_{HESS}(t-1)$ is the state of the charge for HESS in current and previous periods, ρ_{HESS} is the indicator for self-discharging effect of HESS, $H_{DIS}(t)$ and $H_{CH}(t)$ is the discharging and charging heating energy of HESS, η_{DIS}, and η_{CH} are the discharging and charging efficiency values, and Δt is the sampling time for this function. The degradation cost function for HESS can be expressed as below:

$$\text{Cost}_{HESS}(t) = \text{Cost}_{DEG.} \times H_{DIS/CH}(t) \times \Delta t \tag{16.28}$$

The $\text{Cost}_{DEG.}$ is the HESS degradation cost [29].

16.3.9 Power Interchange Between HEIN and Upstream Grid

In some cases, HEINs are microgrids that have interconnection with a large-scale power grid with the aim of helping to balance the demand and supply in HEINs during shortages of energy supply. However, this interaction will have some costs that need to be considered during the optimal planning of HEINs. Therefore, the cost function of power exchange will be as below:

$$\text{Cost}_{P_{\rightleftharpoons}}(t) = \pm P_{\rightleftharpoons}(t) \cdot P_{\text{price}} \tag{16.29}$$

The $P_{\rightleftharpoons}(t)$ is the transferred power between the main grid and HEIN, and P_{price} is the price of exchanged power in LCU per kWh. $P_{\rightleftharpoons}(t)$ can be positive when the power is being injected from the main grid to the HEIN and can be negative when the power is being injected from the HEIN to the main grid [26].

16.3.10 Heat Interchange Between HEIN and Upstream Network

Similar to the power supply from the main grid, HEINs may receive heating energy from the main heating network. However, this interaction will have some costs that need to be considered during the optimal planning of HEINs. Therefore, the cost function of heat exchange will be as below:

$$\text{Cost}_{H_{\rightleftharpoons}}(t) = \pm H_{\rightleftharpoons}(t) \cdot H_{\text{price}} \tag{16.30}$$

The $H_{\rightleftharpoons}(t)$ is the transferred heat between the main grid and HEIN, and H_{price} is the price of exchanged heat. $H_{\rightleftharpoons}(t)$ can be positive when the power is being injected from the main grid to the HEIN and can be negative when the power is being injected from the HEIN to the main grid [26].

Although the mathematical cost functions available in this section can be used in optimal planning studies, differences in utility specification, technical constraints, and even technologies will necessitate the change in these functions. For instance, using biogas CHP units will reduce the G_{Price} as it is much more affordable than natural gas, and therefore, will have different cost functions, which will result in a cheaper energy supply for the grid.

16.4 Advantages and Features of Optimal Planning of HEINs

Optimal planning of HEINs is the process in which the network is optimally planned to reduce the total investment cost [30], increase its reliability of supply [31], reduce the loss of energy in any form [32], or optimize other technical, social, and economic factors, solely or together. Figure 16.4 depicts the summary of features that optimal planning of HEINs provides.

Figure 16.4 Possible features of optimal planning of a HEIN.

16.4.1 Economic Optimization (Revenue & Costs)

Generally, in every real-world system, economic justification is among the most important and influential factors in determining its acceptability rate. Economically feasible systems attract more attention from investors due to their low capital requirements, optimal maintenance costs, and higher revenue rates. Therefore, economic optimization is one of the main optimization goals for energy systems to both rationalize the usage of MDENs and attract more investors to this section, resulting in the burgeoning value of the MDEN market. MDENs have the potential to be operated cost-effectively due to the introduction of emerging technologies into these grids, which can decrease the overall expenditures of these networks. In [33], cost mitigating characteristics of MDENs are discussed. It is concluded that planning, operation, and maintenance costs along with environmental impacts of power systems can be reduced if different energy carriers are integrated within the same power system. Different perspectives are responsible for different approaches in the economic optimization of HEINs. To understand the common views among researchers, and their optimization approaches, other objective functions of optimizations are introduced and explained as below.

16.4.1.1 Generation Costs
Integration of heat and electricity generation units has a significant impact on the overall generation costs of the energy carriers discussed. This issue is due to the fact that the integration of these networks and using technologies to benefit from the waste heat energy of thermal power plants will update the economic dispatch

challenge by creating an opportunity for generation units to find another compensation for their costs, and therefore, the energy generation cost, either in electricity or heat section, will be reduced [34, 35]. The objective function of minimizing heat and electricity generation costs could be as below:

$$
\text{Min} \left[\sum_{i=1}^{\text{power}_i} \text{Cost}_{\text{power}_i} \left(\text{Unit}_{\text{power}_i} \right) + \sum_{j=1}^{\text{heat}_j} \text{Cost}_{\text{heat}_j} \left(\text{Unit}_{\text{heat}_j} \right) \right.
$$

$$
\left. + \sum_{k=1}^{\text{co}-\text{gen}_k} \text{Cost}_{\text{co}-\text{gen}_k} \left(\text{Unit}_{\text{co}-\text{gen}_k} \right) \right]
$$

(16.31)

This minimization objective function can be modified by adding different units, and their generation costs. Readers might consider that in this case, using renewable energy sources for power and/or heat production will reduce the overall cost, much more than what is stated above. However, they should consider that although the overall BOM costs of such units are lower than the conventional utilities [33], there could be limited access to any of them due to geographical constraints. For instance, geothermal power plants are the ideal form of cogeneration units. However, it is not economically or technically possible to establish such power stations anywhere globally as it needs direct access to the heat under the earth plates. Therefore, from a realistic and practical point of view, the objective function above will be efficient enough for initial studies on minimizing the overall generation costs of HEINs. There are numerous studies on modifying and upgrading the aforementioned objective function using heuristic approaches that are available in [36–38].

16.4.1.2 Operational Revenues

Numerous researchers have considered revenue increase as their main objective or one of their optimization objectives. In [39], an algorithm based on mixed-integer linear programming (MILP) approach is introduced to maximize the aggregated capacity, demand response, and energy arbitrage revenue values. The result demonstrated that it is possible to have up to a 60% increase in the total revenue of a HEIN if it is optimally planned and scheduled. Also, in [40], it is stated that there will be a 70–87% growth in revenue for grid operators. This increase is due to the utilization of the planning process with centralized or coordinated optimization approaches. Despite the lower increase in the operating revenue of a HEIN introduced in [41], it is apparent that implementation of optimization techniques in the planning phase of a HEIN based on demand response programs will have beneficial effects on the total revenue of the system.

16.4.1.3 Maintenance Costs

Optimal semi-static planning of HEINs has a reducing impact on the maintenance and control costs of the system. In [42], optimal planning of a HEIN using the

MILP approach caused a 14% reduction in the maintenance costs of the studied system. Also, in [43], it is observed that optimal mid-term operational planning for CHP-based microgrids will reduce the state switching quantity of the grid, and therefore, will lower the maintenance costs as the degradation rate of components fall. Authors in [44] illustrated that using natural gas-based turbines will decrease the costs more than when fuel cells are used. This is due to the fact that natural gas-based turbines cost less in overall BOM costs. In [45], it can be seen that integration of PV utilities with biomass structures to provide the fuel for biogas-based CHP, will result in higher efficiency and lower maintenance costs than the case in which they perform separately. In summary, it is expected that optimal planning of HEINs will have beneficial effects on maintenance costs, as the assets are optimally planned to be used to reduce both the costs of operation and degradation rates, which will have an impact on utilities by prolonging their performance time frame and reducing the need for replacement.

16.4.1.4 Environmental Costs

Environmental costs of HEINs are usually due to the carbon tax policies that grid operators and regulators may propose to help mitigate the impacts of climate change. In [46], the optimal planning of the CCHP-based heat and electricity interconnection of the study resulted in a virtually 20% reduction in emission costs. Also, the authors in [47] concluded that proper planning and integration of ESS in the studied HEIN will decrease the environmental costs by up to 8%. Based on other studies in this field [48, 49], most of the emissions in the HEINs are due to the usage of fossil fuel-based assets such as NGBs, CHPs, DGs, etc. It is noteworthy that higher penetration rates of renewable energy resources along with the optimal implementation of energy storage devices in the HEIN are effective in reducing the overall emissions of the grid, which will eventually reduce the environmental costs in the planning phase [33, 50].

16.4.2 Supply Reliability and Grid Stability Increase

Considering a HEIN as a multi-carrier energy system, it is true that either integration of various energy forms or connection with upstream grids [51] will increase the reliability of the grid and security of energy supply. The presence of storage devices, interconnections between the main power and heat grid, usage of geothermal, electrical, and natural gas boilers for heating, CHP, and CCHP units for cogeneration and renewable power plants such as wind farms and PV plants will increase the overall reliability of the HEIN [52]. One of the main economic indicators to assess the impact of HEIN on the reliability of the grid is the value of load loss (VOLL) [53]. This indicator converts the supply unavailability to financial terms. It is assumed that integration of multiple energy carriers with each other – for instance, heating and electricity – will have a reducing effect on the VOLL, resulting in increased reliability [54]. In [55], a reliability assessment method is proposed to calculate the reliability of the studied HEIN based on the renewable energy supply inadequacy and VOLL. In summary, due to the importance of supply reliability and energy security in grids, HEIN has attracted the attention of researchers as one of the successful approaches of securing the access of society to heating and electricity.

16.4.3 Heat Loss Reduction (Efficiency Increase)

Meeting the heating demand of consumers is one of the leading challenges in HEINs. Most of the integrated networks manage to meet the heating demand inside the boundaries of the network to reduce heat loss [56]. DH is one of the main solutions inside the definition of HEIN to provide heat and power simultaneously with high efficiency. A common example of DH assets is CHP. Optimal planning of DH will result in heat loss reduction along with providing opportunities to use eco-friendly technologies such as biomass CHPs, etc. [57]. Also, in [58], a HEIN is optimally planned to reduce the heat loss rate by using the waste heat of the electricity generation unit, and heating energy from boilers to meet the thermal demand of the system. Authors in [59] have assumed the efficiency of generation unit to a constant value, therefore, solved the optimal planning challenge using the MILP method to minimize the heat waste by implementing waste recovery units. In conclusion, one of the main concerns of the planning process is to increase the efficiency of any HEIN by reducing or reusing the wasted energy, either power or heat, to meet the consumption demand.

16.4.4 Design Complexity Mitigation

Generally, adding new utilities to a grid will increase its complexity. This statement is also true for HEINs. Although adding storage devices, different generation units, extending optimization boundaries to the whole of the system instead of a component, and considering demand-side management issues in HEINs will increase its reliability level, the resultant system will be more complex than before for planning and operation [58]. Authors in [60] have proposed a state-space model based on the Markov method to overcome complexity in planning and system assessment challenge of the studied HEIN. Also, a bi-layer optimization method based on the african buffalo optimization (ABO) approach is introduced in [61] to reduce the complexity of integrating storage systems into a grid. However, the authors of [62] had stated different ideas about reducing complexity in HEINs. They outlined that reduction in quantity and capacity of installed thermal units and using thermal storage systems will have a decreasing impact on both the investment costs and complexity of the system, making it easier for planning and exploitation. After all, what is apparent is that system complexity has been one of the challenges of optimal planning in HEINs due to the existence of conflicting interests between reducing complexity and increasing reliability. However, exploiting from numerous new technologies can help researchers to find a balance in this problem and propose optimal solutions.

16.5 Challenges and Future Research Opportunities in Optimal Planning for HEINs

As stated before, HEINs are networks of integrated heating and electricity to meet the thermal and electrical demand of consumers. Integration of heating and electricity into a grid is due to many objectives such as economic benefits, reliability expectations, and regulatory constraints. However, there are challenges in this integration process even after numerous available scientific research reports about optimizing the planning and operation of these networks. Table 16.1 provides different clusters of challenges in HEINs that need to be addressed [44, 63–65].

Table 16.1 Challenges of HEINs.

Challenges			
Economic	**Environmental**	**Technical**	**Regulatory**
Energy purchasing cost increase	Greenhouse gas emissions of natural gas facilities	Optimal energy resource allocation	Energy-related climate policies
Investment costs of the utilities	Uncertainties in climatological data for renewable generation	Flexibility and reliability issues in a short-term operational period	Uncertainties of demand-side management
Lack of economic resources for development	The temperature increase of the environment	The adaptability of utilities with smart grid communication devices and protocols	International development considerations
Limited access to natural gas and other necessary fossil fuels		Efficiency and adequacy issues of power and heat utilities	Energy security constraints

16.5.1 Economic Challenges

One of the main economic challenges in considering the optimal planning for HEINs is the energy purchasing cost. Objectives of an optimization in energy resource planning can vary between environmental, technical, social, political, or economic aims. It is possible that optimizing one or more indicators in any of the aforementioned objectives can result in besetting of other indicators. For example, reduction in emission and power loss rates, can create an increasing trend for the energy price. In that case, new studies should focus on keeping everything in balance, which means keeping the price of the energy in mind while optimizing.

Investment costs are also another economic challenge in optimal planning of HEINs. Integrating networks of different energy carriers will need sufficient hardware and a communication platform to be controllable. These modifications will require adequate capital to be implemented. However, to attract enough economic resources to initiate an integration of heat and electricity networks, there must be an economic justification using different indicators such as return on investment (ROI) so that the investment party (either being the public or private sector) would find that participation in the project will be a rational decision. Conceptualizing this rationality should be one of the focus fields of the optimization process, and sufficient indicators must be introduced to show that the investment cost is in its optimal point. Also, the investor should be satisfied with ROI rate and/or any other incentivizing factor to partake in the project. To provide an initial insight on this issue, studying [66–68], or any other similar studies are suggested.

Providing required fuel resources for HEINs – specifically natural gas – can be a great challenge in the path of optimal planning. Historical data shows that it is not impossible for a utility to experience a fuel shortage. Therefore, planning for integration of networks that are in need of fuel sources should take these kinds of risks seriously. For example, type of the contracts that power plants have with natural gas providers must be considered in planning for the use of natural gas in both the heating and electricity sectors. A spot market contract is one of the uncertain types of contracts that can be present between power plants and natural gas providers.

This type of contract supports the financial transaction of buying/selling natural gas with the immediate price of the current moment, known as spot market price. However, due to the uncertainties in weather conditions, technical and political settlements (either in domestic or international scale), there is a chance of supply shortage that will create a fuel security issue, and it will eventually become as an energy security challenge in the HEIN. Thus, optimal planning of integration in HEINs must assess the certainty and firmness of the fuel supply from economic perspective [69].

16.5.2 Environmental Challenges

Environmental challenges in optimal planning of HEINs are among the most uncertain ones. Issues such as extreme weather conditions and temperature fluctuations can present many obstacles in the path of planning. A recent example of environmental challenges is the 2021 Texas Power Crisis. In February 2021, the Uri winter storm crawled down to the southern states of the United States, caused a nine-day length power outage in Houston, TX, along with threatening the water distribution system. The main reason for load-shedding in the Houston power grid, was the shortage of gas supply for the natural gas-based power units such as thermal power plants and CHPs. This shortage of supply was because of two reasons:

- Frozen natural gas distribution equipment such as pipelines, pressure stations, etc. have prevented the distribution of sufficient amount of natural gas to feed the power plants.
- The natural gas consumption for heating purposes rose rapidly in residential and commercial section due to the cold weather.

These reasons have caused a power crisis in Houston for nine days that was almost preventable by proper planning on winterizing the grid equipment, and providing improvement on the present energy network [70].

Although GHG emissions of natural gas-based power plants are not a serious concern in comparison to coal or oil-fired power plants, it is necessary to consider carbon capture & sequestration (CCS) technologies to be implemented in natural gas-fired power plants. This consideration will create opportunity for the innovative technologies in this field to reduce the GHG emission of natural gas-based utilities and promote the goal of decarbonizing energy networks [33]. Also, there will be a chance to consider other fuels such as biofuels to be used in customized CHPs to provide power and heat, while reducing the emission rates [71].

16.5.3 Technical Challenges

Technical issues are among the most favored objective functions in planning optimization process for engineering researchers. Resource allocation, reliability, flexibility, demand-side uncertainties, interoperability between assets and communication network in smart grids, and other indicators in power system are among the technical research focus of the planning optimization process. For example, one of the main challenges in integration of heat and power networks, is maintaining flexibility in both of the integrated networks. Focusing on cost-effectiveness of the HEIN optimal planning might bring out vulnerabilities in operational phase in which the HEIN will not be capable of providing a balance between the demand and supply of thermal or electrical energy. Therefore, security constraints of the HEIN such as economic dispatch and unit commitment limitations of distributed energy resources must be considered in planning optimization [72].

Also, with the emergence of smart grids and shaping cyber-physical systems, connection between outdated technologies and new monitoring systems must be assessed thoroughly. These kinds of studies will ensure optimal planning for cyber-physical HEINs along with providing access to data that were never accessible in older HEINs. Therefore, consideration of compatibility between emerging technologies such as smart grids and present HEIN assets as a constraint is necessary to have an optimal planning scheme. Technologies such as internet of things (IoT) and digital twins (DT) can provide options such as forecast data or incident impact evaluation on the HEIN more accurately, which will assist researchers on optimal planning [73].

16.5.4 Regulatory Challenges

Considering regulatory constraints such as climate, security, international, and managerial policies in optimal planning process is a necessity as they can have a significant impact on operation HEINs in the future. For instance, having limited regulation on demand-side management will reduce the flexibility of the grid. Not considering behavioral, geographical, climatological or even technical studies and circumstances in anticipating the necessary demand response measures during the integration for these networks will increase the economic burden of the integration process, and therefore, can have besetting effects on energy prices, investor attraction, and so on [33].

Also, international regulatory actions such as Paris Agreement, can have impacts on gradual integration of heat and electricity networks in developing countries as they are not fully committed to the agreement, and can decide to use environmentally inappropriate fuels such as Mazut to meet the demands in either heat or electricity section. This decision will prevent the optimal planning attempts in optimizing the environmental indicators in HEIN. Therefore, a thorough analysis of local and international regulations and constraints for planning is necessary [74].

Most of these challenges are resolved in academic studies. However, researchers should switch their perspectives toward real-world constraints and improve their studies until there is a realistic study concerning the optimal planning of HEINs. Also, with the emersion of new technologies such as upgraded thermal power plants, storage systems, and energy harvesters, planning objectives should adapt themselves to these findings. Therefore, the necessity of performing optimal dynamic planning gets more. Each of the challenges outlined in Table 16.1 has the potential to become a future research stream for researchers who are willing to contribute to the implementation of renewable-based HEINs.

16.6 Summary

This chapter aimed to provide an accurate definition of optimal planning of HEINs, and demonstrate possible structures and components of these networks. Later on, we have outlined the features of optimally planned HEINs as implementation objectives. Also, the challenges of optimizing the planning process of these networks are briefly reviewed. After all, it is noteworthy that HEINs are becoming necessary energy distribution structures in the current world with emerging economic, environmental, and social constraints. Technical improvements in the utilities of these networks are another positive point in the planning of these systems, which will provide more flexibility for researchers to advance in this field.

To sum up, there are a lot of opportunities available in the development and enhancement of these systems in a way that they become more compatible with regional and international constraints, though many studies have been done in the field of optimal planning of HEINs. Therefore, this chapter provided a basic knowledge of optimal planning issues for interested researchers to pursue more advanced methods in this field.

References

1 Li, C., Wang, Q., Zhu, H. et al. (2021). Coordinative optimal operation mode of regional energy internet. *Journal of Physics: Conference Series* 1871 (1): IOP Publishing: 012021.

2 Daneshvar, M., Asadi, S., and Mohammadi-Ivatloo, B. (2021). Data management in modernizing the future multi-carrier energy networks. In: *Grid Modernization—Future Energy Network Infrastructure* (ed. M. Daneshvar, S. Asadi and B. Mohammadi-Ivatloo), 117–174. Cham: Springer. https://doi.org/10.1007/978-3-030-64099-6_4.

3 Wang, J., Zhong, H., Ma, Z. et al. (2017). Review and prospect of integrated demand response in the multi-energy system. *Applied Energy* 202: 772–782.

4 Liang, J., Luo, Y., Yang, D., and Guo, X. (2017). Overview on implementation planning of energy internet based on energy hub. *2017 5th International Conference on Machinery, Materials and Computing Technology (ICMMCT 2017),* Beijing, China (25–26 March 2017). Atlantis Press, pp. 738–744.

5 Hussain, A., Arif, S.M., Aslam, M., and Shah, S.D.A. (2017). Optimal siting and sizing of tri-generation equipment for developing an autonomous community microgrid considering uncertainties. *Sustainable Cities and Society* 32: 318–330.

6 Xu, X., Jia, H., Wang, D. et al. (2015). Hierarchical energy management system for multi-source multi-product microgrids. *Renewable Energy* 78: 621–630.

7 Qaeini, S., Nazar, M.S., Varasteh, F. et al. (2020). Combined heat and power units and network expansion planning considering distributed energy resources and demand response programs. *Energy Conversion and Management* 211: 112776.

8 Ghaffarpour, R., Mozafari, B., Ranjbar, A.M., and Torabi, T. (2018). Resilience oriented water and energy hub scheduling considering maintenance constraint. *Energy* 158: 1092–1104.

9 Maroufmashat, A., Fowler, M., Khavas, S.S. et al. (2016). Mixed integer linear programing based approach for optimal planning and operation of a smart urban energy network to support the hydrogen economy. *International Journal of Hydrogen Energy* 41 (19): 7700–7716.

10 Mancarella, P. (2014). MES (multi-energy systems): an overview of concepts and evaluation models. *Energy* 65: 1–17.

11 Gaussens, J., Moulle, N., Dutheil, F., and Aldebert, J. (1964). Economic aspects of electricity and industrial heat generating reactors, CEA Saclay.

12 Liebhafsky, H. (1966). Fuel cells and fuel batteries–an engineering view. *IEEE Spectrum* 3 (12): 48–56.

13 Zhang, X., Strbac, G., Shah, N. et al. (2018). Whole-system assessment of the benefits of integrated electricity and heat system. *IEEE Transactions on Smart Grid* 10 (1): 1132–1145.

14 Zhang, T. (2020). Methods of improving the efficiency of thermal power plants. *Journal of Physics: Conference Series* 1449 (1): IOP Publishing: 012001.

15 Souliotis, M., Singh, R., Papaefthimiou, S. et al. (2016). Integrated collector storage solar water heaters: survey and recent developments. *Energy Systems* 7 (1): 49–72.

16 Nazari-Heris, M., Mohammadi-Ivatloo, B., Zare, K., and Siano, P. (2020). Optimal generation scheduling of large-scale multi-zone combined heat and power systems. *Energy* 210: 118497.

17 Yao, S., Gu, W., Lu, S. et al. (2020). Dynamic optimal energy flow in the heat and electricity integrated energy system. *IEEE Transactions on Sustainable Energy* 12 (1): 179–190.

18 Samadi Gazijahani, F. and Salehi, J. (2017). Stochastic multi-objective framework for optimal dynamic planning of interconnected microgrids. *IET Renewable Power Generation* 11 (14): 1749–1759.

19 Seifi, H. and Sepasian, M.S. (2011). *Electric Power System Planning: Issues, Algorithms and Solutions*. Springer Science & Business Media.

20 Li, Y., Zou, Y., Tan, Y. et al. (2017). Optimal stochastic operation of integrated low-carbon electric power, natural gas, and heat delivery system. *IEEE Transactions on Sustainable Energy* 9 (1): 273–283.

21 Parisio, A., Rikos, E., Tzamalis, G., and Glielmo, L. (2014). Use of model predictive control for experimental microgrid optimization. *Applied Energy* 115: 37–46.

22 Liu, H., Ji, Y., Zhuang, H., and Wu, H. (2015). Multi-objective dynamic economic dispatch of microgrid systems including vehicle-to-grid. *Energies* 8 (5): 4476–4495.

23 Gen, B. (2005). *Reliability and Cost/Worth Evaluation of Generating Systems Utilizing Wind and Solar Energy*. University of Saskatchewan.

24 Alavi, S.A., Ahmadian, A., and Aliakbar-Golkar, M. (2015). Optimal probabilistic energy management in a typical micro-grid based-on robust optimization and point estimate method. *Energy Conversion and Management* 95: 314–325.

25 Gu, W., Wang, Z., Wu, Z. et al. (2016). An online optimal dispatch schedule for CCHP microgrids based on model predictive control. *IEEE transactions on smart grid* 8 (5): 2332–2342.

26 Al-Saadi, M.K. (2021). Economic operation planning of combined heat and power smart distribution system. *Journal of Engineering Science and Technology* 16 (1): 025–043.

27 Dong, J., Nie, S., Huang, H. et al. (2019). Research on economic operation strategy of CHP microgrid considering renewable energy sources and integrated energy demand response. *Sustainability* 11 (18): 4825.

28 Komarnicki, P., Lombardi, P., and Styczynski, Z. (2017). Electric energy storage system. In: *Electric Energy Storage Systems*, 37–95. Springer. https://link.springer.com/book/10.1007/978-3-662-53275-1.

29 Zhong, Y., Xie, D., Zhai, S., and Sun, Y. (2018). Day-ahead hierarchical steady state optimal operation for integrated energy system based on energy hub. *Energies* 11 (10): 2765.

30 Fan, H., Yuan, Q., Xia, S. et al. (2020). Optimally coordinated expansion planning of coupled electricity, heat and natural gas infrastructure for multi-energy system. *IEEE Access* 8: 91139–91149.

31 Chen, Z., Shi, G., Li, Y., and Fu, X. (2020). Optimal planning method for a multi-energy complementary system with new energies considering energy supply reliability. *2020 Asia Energy and Electrical Engineering Symposium (AEEES)*, Chengdu, China (29–31 May 2020). IEEE, pp. 952–956.

32 Naderipour, A., Abdul-Malek, Z., Nowdeh, S.A. et al. (2020). Optimal allocation for combined heat and power system with respect to maximum allowable capacity for reduced losses and improved voltage profile and reliability of microgrids considering loading condition. *Energy* 196: 117124.

33 Fallah Ardashir, J. and Vatankhah Ghadim, H. (2021). Introduction and literature review of cost-saving characteristics of multi-carrier energy networks. In: *Planning and Operation of Multi-Carrier Energy Networks* (ed. M. Nazari-Heris, S. Asadi and B. Mohammadi-Ivatloo), 1–37. Cham: Springer International Publishing.

34 Ara, A.L., Shahi, N.M., and Nasir, M. (2019). CHP economic dispatch considering prohibited zones to sustainable energy using self-regulating particle swarm optimization algorithm. *Iranian Journal of Science and Technology, Transactions of Electrical Engineering* 44: 1–18.

35 Nguyen, T.T., Vo, D.N., and Dinh, B.H. (2016). Cuckoo search algorithm for combined heat and power economic dispatch. *International Journal of Electrical Power & Energy Systems* 81: 204–214.

36 Chen, X., Li, K., Xu, B., and Yang, Z. (2020). Biogeography-based learning particle swarm optimization for combined heat and power economic dispatch problem. *Knowledge-Based Systems* 208: 106463.

37 Nasir, M., Sadollah, A., Aydilek, İ.B. et al. (2021). A combination of FA and SRPSO algorithm for combined heat and power economic dispatch. *Applied Soft Computing* 102: 107088.

38 Yang, Q., Liu, P., Zhang, J., and Dong, N. (2021). Combined heat and power economic dispatch using an adaptive cuckoo search with differential evolution mutation. *Applied Energy* 307: 118057.

39 Mohamed, A.A., Sabillon, C., Golriz, A., and Venkatesh, B. (2021). Value-stack aggregator optimal planning considering disparate DERs technologies. *IET Generation, Transmission & Distribution* 15.

40 Wang, H., Xu, H., Wang, X. et al. (2020). Research on energy management of integrated energy systems considering multi-agent. *2020 IEEE Sustainable Power and Energy Conference (iSPEC)*, Chengdu, China (23–25 November 2020). IEEE, pp. 1397–1403.

41 Ju, L., Tan, Q., Lu, Y. et al. (2019). A CVaR-robust-based multi-objective optimization model and three-stage solution algorithm for a virtual power plant considering uncertainties and carbon emission allowances. *International Journal of Electrical Power & Energy Systems* 107: 628–643.

42 Cheng, H., Wu, J., Luo, Z. et al. (2019). Optimal planning of multi-energy system considering thermal storage capacity of heating network and heat load. *IEEE Access* 7: 13364–13372.

43 Ghadimi, P., Kara, S., and Kornfeld, B. (2015). Real-time operation management of CHP system in manufacturing industry. *Modern Applied Science* 9 (2): 158.

44 Zidan, A., Gabbar, H.A., and Eldessouky, A. (2015). Optimal planning of combined heat and power systems within microgrids. *Energy* 93: 235–244.

45 Li, C., Yang, H., Shahidehpour, M. et al. (2019). Optimal planning of islanded integrated energy system with solar-biogas energy supply. *IEEE Transactions on Sustainable Energy* 11 (4): 2437–2448.

46 Home-Ortiz, J.M., Melgar-Dominguez, O.D., Pourakbari-Kasmaei, M., and Mantovani, J.R.S. (2019). A stochastic mixed-integer convex programming model for long-term distribution system expansion planning considering greenhouse gas emission mitigation. *International Journal of Electrical Power & Energy Systems* 108: 86–95.

47 Xiang, Y., Cai, H., Gu, C., and Shen, X. (2020). Cost-benefit analysis of integrated energy system planning considering demand response. *Energy* 192: 116632.

48 Ma, T., Wu, J., Hao, L. et al. (2018). The optimal structure planning and energy management strategies of smart multi energy systems. *Energy* 160: 122–141.

49 Cheng, Y., Zhang, N., Kirschen, D.S. et al. (2020). Planning multiple energy systems for low-carbon districts with high penetration of renewable energy: an empirical study in China. *Applied Energy* 261: 114390.

50 Liu, J., Xu, Z., Wu, J. et al. (2021). Optimal planning of distributed hydrogen-based multi-energy systems. *Applied Energy* 281: 116107.

51 Mendes, G., Ioakimidis, C., and Ferrão, P. (2011). On the planning and analysis of integrated community energy systems: a review and survey of available tools. *Renewable and Sustainable Energy Reviews* 15 (9): 4836–4854.

52 Fallah Ardashir, J. and Vatankhah Ghadim, H. (2021). Chapter 14 - Large-scale energy storages in joint energy and ancillary multimarkets. In: *Energy Storage in Energy Markets* (ed. B. Mohammadi-Ivatloo, A. Mohammadpour Shotorbani and A. Anvari-Moghaddam), 265–285. Academic Press.

53 Shivakumar, A., Welsch, M., Taliotis, C. et al. (2017). Valuing blackouts and lost leisure: estimating electricity interruption costs for households across the European Union. *Energy Research & Social Science* 34: 39–48.

54 Heendeniya, C.B., Sumper, A., and Eicker, U. (2020). The multi-energy system co-planning of nearly zero-energy districts–Status-quo and future research potential. *Applied Energy* 267: 114953.

55 Baneshi, E. and Dehkordi, A.B. (2019). Microgrid optimal planning in two functional modes grid connected and the intentional islanding. *2019 5th Conference on Knowledge Based Engineering and Innovation (KBEI)*, Tehran, Iran (28 February–1 March 2019). IEEE, pp. 857–863.

56 Mirzaei, M.A., Zare Oskouei, M., Mohammadi-Ivatloo, B. et al. (2020). Integrated energy hub system based on power-to-gas and compressed air energy storage technologies in the presence of multiple shiftable loads. *IET Generation, Transmission & Distribution* 14 (13): 2510–2519.

57 Volkova, A., Latõšov, E., Lepiksaar, K., and Siirde, A. (2020). Planning of district heating regions in Estonia. *International Journal of Sustainable Energy Planning and Management* 27: 5–15.

58 Guo, L., Liu, W., Cai, J. et al. (2013). A two-stage optimal planning and design method for combined cooling, heat and power microgrid system. *Energy Conversion and Management* 74: 433–445.

59 Li, Y., Zou, B., Zhu, F., and Fu, J. (2018). An optimal planning method for CCHP systems based on operation simulation. *2018 IEEE International Conference on Smart Energy Grid Engineering (SEGE)*, Oshawa, ON, Canada (12–15 August 2018). IEEE, pp. 40–45.

60 Aval, S.M.M., Ahadi, A., and Hayati, H. (2015). Adequacy assessment of power systems incorporating building cooling, heating and power plants. *Energy and Buildings* 105: 236–246.

61 Singh, P., Meena, N.K., Slowik, A., and Bishnoi, S.K. (2020). Modified african buffalo optimization for strategic integration of battery energy storage in distribution networks. *IEEE Access* 8: 14289–14301.

62 Guelpa, E. and Verda, V. (2019). Thermal energy storage in district heating and cooling systems: a review. *Applied Energy* 252: 113474.

63 Pan, G., Gu, W., Lu, Y. et al. (2020). Optimal planning for electricity-hydrogen integrated energy system considering power to hydrogen and heat and seasonal storage. *IEEE Transactions on Sustainable Energy* 11 (4): 2662–2676.

64 Ghanbari, A., Karimi, H., and Jadid, S. (2020). Optimal planning and operation of multi-carrier networked microgrids considering multi-energy hubs in distribution networks. *Energy* 204: 117936.

65 Mohammadi, M., Noorollahi, Y., Mohammadi-Ivatloo, B., and Yousefi, H. (2017). Energy hub: from a model to a concept–a review. *Renewable and Sustainable Energy Reviews* 80: 1512–1527.

66 Van Der Veen, R.A. and Kasmire, J. (2015). Combined heat and power in Dutch greenhouses: a case study of technology diffusion. *Energy Policy* 87: 8–16.

67 Ancona, M., Bianchi, M., Branchini, L. et al. (2019). Combined heat and power generation systems design for residential houses. *Energy Procedia* 158: 2768–2773.

68 Guo, X., Ndiaye, I., Yan, M. et al. (2020). Feasibility analysis of converter-interfaced combined heat and power system. *2020 IEEE Power & Energy Society General Meeting (PESGM)*, Montreal, QC, Canada (2–6 August 2020). IEEE, pp. 1–5.

69 Freeman, G.M., Apt, J., and Moura, J. (2020). What causes natural gas fuel shortages at US power plants? *Energy Policy* 147: 111805.

70 Webber, M. E. (2021). Interruptible power *Mechanical Engineering* 143 (5): 45–50. https://doi.org/10.1115/1.2021-Sept3.

71 Ji, C., Cheng, K., Nayak, D., and Pan, G. (2018). Environmental and economic assessment of crop residue competitive utilization for biochar, briquette fuel and combined heat and power generation. *Journal of Cleaner Production* 192: 916–923.

72 Wang, J., You, S., Zong, Y. et al. (2019). Flexibility of combined heat and power plants: a review of technologies and operation strategies. *Applied Energy* 252: 113445.

73 Inderwildi, O., Zhang, C., Wang, X., and Kraft, M. (2020). The impact of intelligent cyber-physical systems on the decarbonization of energy. *Energy & Environmental Science* 13 (3): 744–771.

74 Teske, S. (2019). *Achieving the Paris Climate Agreement Goals*. Springer.

17

Coordinated Planning of Thermal and Electrical Networks

Reza Gharibi and Behrooz Vahidi

Department of Electrical Engineering, Amirkabir University of Technology (Tehran Polytechnic), Tehran, Iran

17.1 Introduction

17.1.1 Motivation and Problem Description

Today, energy is one of the main challenges and needs of human beings. These challenges appear in various dimensions, including production, transmission, and consumer demand. Due to population growth, rise in global demand for energy, shortage of fossil fuels, and concerns environmental issues, energy security has become a significant issue for all governments and has forced them to plan for this critical problem [1]. The connection of energy with economic and environmental dimensions multiplies the importance of this human need. In addition, governments have looked at the security aspect of it as energy security, which includes the equitable provision of energy to all committed consumers and the availability of energy to these consumers. Also, the reliability and energy efficiency provided in this section are considered [2].

Hence, access to sustainable energy with high reliability is one of the most critical challenges of all governments and industries, large and small. The fossil fuels burned with very low efficiency in these power plants and converted into another form of energy. The energy produced from the fuel is transferred from these power plants to farther distances from the power plant and is delivered to the distribution network through dedicated lines and transmission networks of each energy (electric, thermal). Finally, it is distributed among the final consumers by this network, which has certain complexities. The use of this system faces many problems, including losses in power lines and transmission networks and the emission of greenhouse gases from fossil fuels. Also, due to the shortage of fossil fuels, their

Coordinated Operation and Planning of Modern Heat and Electricity Incorporated Networks,
First Edition. Edited by Mohammadreza Daneshvar, Behnam Mohammadi-Ivatloo, and Kazem Zare.
© 2023 The Institute of Electrical and Electronics Engineers, Inc.
Published 2023 by John Wiley & Sons, Inc.

use in less efficient plants does not seem logical [3]. Such large systems, along with transmission lines and networks, require high investment in infrastructures. These systems face problems in protection and control that significantly increase the ancillary costs. So, these systems are not suitable for future energy systems. Other energy systems, gas and district heating (DH), also suffer from these problems. In addition, separate planning of these systems poses another challenge [4].

The advent and development of new technologies such as heat pumps, combined heat and power units (CHP), electrical appliances, etc., cause the inevitable interaction of these devices. However, to achieve sustainable energy, it is necessary to pay attention to the role of multi-energy systems (MES) that different energy carriers at different levels interact with each other in an optimal state [5]. Paying attention to MES creates a variety of advantages such as technical, environmental, reliability, energy security, and the most important one, economic. Also, for the success of these integrated devices, the collective management structure of different system components is needed optimally and simultaneously. The concept of energy hub (EH) Introduced to achieve a comprehensive concept of interactive and sustainable systems [6]. The introduction of EH created a new context for researchers to study and plan for different fuel and energy carriers.

17.1.2 Literature Review

So far, much research has been carried out in energy optimization [7, 8]. Some of them have studied the performance of CHP units and their practical limitations [9]. In [10], a single operational CHP model is used alongside energy storage. In [11, 12] use special evolutionary algorithms to optimize independent energy carriers. In [13–15], the EH is considered as the interface between different upstream energy networks, electricity and gas, with their consumers. Reference [16] has presented a new model for optimization of the EH considering its environmental impacts. Reference [17] presents an economically optimal model for reducing the costs of cogeneration units and gas furnaces (GFs) when supplying the consumers. Reference [18] has proposed a linear programming model to control a household system. Some studies have investigated demand response programs on a small scale. In [19], a mathematical optimization model for supplying residential homes equipped with smart systems in the presence of automated response networks has been introduced. Reference [20] presents an optimization method for meeting the load demand of a hub supplying residential consumers. Taking into account the heat pump, the coordination of sources, and carbon emissions, [21] and [22] have suggested an EH model for a residential building. In [23], programming for small EH in a network with diesel generators (DGs) is proposed. Reference [24] has investigated the maintenance and costs of an EH, Considering water as an output in addition to the rest of the electrical and thermal energy. In [25], the game theory is deployed to optimize multiple EHs.

The use of the EH concept has its advantages when it comes to planning and coordinating between independent energy networks such as electricity and gas. The EH is used here as an interface between upstream networks and thermal and electrical energy consumers. The energy carriers in the upstream networks are considered the available inputs of the hub. Using these inputs and internal components of the hub, planning between different carriers to meet the heat and electrical demands of the consumers are performed. The main part of this chapter is planning to meet the needs of energy consumers in large dimensions, not as a building but in a larger dimension of a neighborhood and part of a city. The subsequent considerations are also taken into account.

1) In reality, not only are co-generating units (CHP) restricted to a minimum and maximum, but also they must operate in a specific area of operation in which the amount of heat and electricity generated are interdependent. Therefore, the units used in this chapter work in their own area of operation, and their specifications are deemed close to reality.
2) Units in the EH have marginal costs. Therefore, we try to apply these costs to the internal components of the hub in order to get closer to more realistic conditions.
3) In addition to the independent networks of electricity and gas carriers, wind turbines are also considered another EH input in the planning.
4) Part of the electricity generated by cogeneration units and DGs can be sold to the national grid. In addition to planning to satisfy the load demand of EH consumers, in a part of the discussion, optimization of these conditions is also considered.

This chapter deals with the optimization of different energy carriers through the concept of the EH in order to supply the amount of thermal and electrical load demand of the consumers. The problem is investigated in large dimensions with considerable amounts of loads.

17.1.3 Chapter Organization

In the continuation of this section, after introducing the concept of EHs and the components used in it, also examining the limitations of different components and carriers used in the EH, the mathematical formulation of these components and constraints has been provided. Then, using these mathematical relations, the desired system is optimized, and the results are observed at the end. Also, because the main goal in this section is to optimize the EH, the internal components of the EH are briefly explained. For further study, refer to the references acknowledged.

17.2 The Concept of Energy Hub

Different consumers, including industrial, residential, and commercial, worldwide need different types of energy, which they meet from different energy-related infrastructures. For example, industrial consumers get their energy from coal, biomass, oil-related products, and grid-connected energy carriers, including natural gas, electricity, and district heating. Most different energy infrastructures are programmed and managed independently. Combining different infrastructures and simultaneous and integrated management and planning of these infrastructures can have many advantages. For example, for electricity, it can transfer energy to the consumer at different distances with relatively low losses. Combining infrastructure means exchanging power between various energy carriers [6]. Converting energy into other forms is done by converter devices. The important issue here is where the devices used in the combined energy carriers are used and to what extent each carrier is related to these devices. There are various criteria for planning and responding to this issue, including economic, environmental, availability, and other parameters. In this section, the economic criterion of the problem is examined.

One of the main parts in the future project of energy networks is the concept of energy hub [26–28]. A hub is a unit of energy in which input carriers can be converted and stored, ultimately providing the carrier needed by the consumer. EH is, actually, an interface for various energy infrastructures and consumer loads. EH consumes electricity and gas carriers in the input section and provides energy services to the consumer in the output section, including electricity, heating, cooling, compressed air, etc. Inside the power hub, consisting of components such as CHP technology, transformers to change the voltage and current level for the consumer, heat exchangers (HEs), storage devices, and other equipment. Examples of such EHs are industrial plants, large buildings (hospitals, airports), energy systems in urban and rural areas, also energy systems that are island and off-grid in small dimensions (aircraft, Train, ship). Figure 17.1 is an example of an EH.

The EH and its components may make additional connections to the input and output. For example, to provide electricity demand can supply it directly from the grid, part or all of it from natural gas through the equipment inside the hub. These connections in providing can create two main advantages in using the EH. First, the reliability of the load increases because it is no longer dependent on a network to supply the load. For example, when repairing the infrastructure of an individual network, it can meet the consumer's needs by converting another carrier inside the hub. Second, it provides an extra degree of freedom for optimizing the EH. Carriers entering the hub can be determined by cost, emissions, availability at a particular time, and other optimization metrics. Finally, optimization can determine the amount of input for a given time.

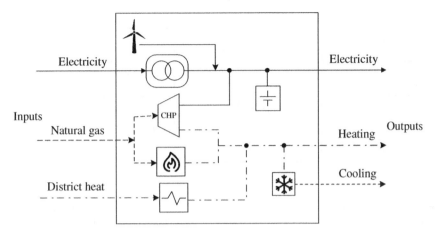

Figure 17.1 Example of an energy hub including transformer, CHP, furnace, absorption cooler, and battery.

In general, the concept of EH is defined as the interface between electricity producers, consumers, and transportation infrastructure, which in terms of the system includes the first two. The third depends on the type of EH and may not be used.

- Inputs and outputs
- Energy converters
- Storage

17.3 Power Flow Modeling of Energy Hub

In this section, a power flow sample is included with the various energy carriers extracted. The challenge in presenting this type of display is that it must be predominantly general, and different energy carriers can be applied to its inputs and outputs. The EH programmer must also be able to use it for real systems [4]. The main goal is that in addition to the model's generality, it should be in a way that explicitly considers the connections between the offending energy carriers in the model. In the following, power flow modeling within an EH is extracted and investigated by providing an example. EH models are considered according to Figure 17.2.

Different energy carriers $\alpha, \beta, \dots \varepsilon$ enter the hub through the m inputs port, and inside the hub is converted to the power requested by the consumer. Current transfer from m inputs port to n outputs port as follows can be expressed as (17.1) [29]:

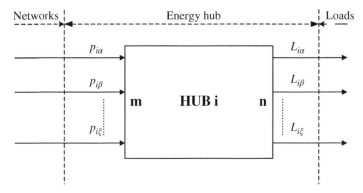

Figure 17.2 Model nomenclature for the energy carrier exchanging energy hub, P indicates the energy flow input to the hub, L Indicates the output energy flow required by the consumer.

$$\begin{bmatrix} L_a \\ \vdots \\ L_\varepsilon \end{bmatrix} = \begin{bmatrix} c_{aa} & \cdots & c_{\varepsilon a} \\ \vdots & \ddots & \vdots \\ c_{a\varepsilon} & \cdots & c_{\varepsilon\varepsilon} \end{bmatrix} \begin{bmatrix} P_a \\ \vdots \\ P_\varepsilon \end{bmatrix} \tag{17.1}$$

where C is called the forward coupling matrix, this matrix describes how to convert power from m inputs to n outputs in hub i. The inputs and components of the coupling matrix are called coupling coefficients. The efficiency of the converters usually depends on the power converted by them. The main reason for extracting Eq. (17.1) is that power flows from input to output. However, it is possible to use reverse power flow with reversible technologies. For example, boilers deliver heat power from the input to the output, but the transformer can transfer power in two directions.

As mentioned, the C matrix describes the power conversion from the input port to the output and shows which power consumption each input carrier plays. However, to determine the inputs required for a given output, we need an inverse relation, which is defined as (17.2):

$$\begin{bmatrix} P_a \\ \vdots \\ P_\varepsilon \end{bmatrix} = \begin{bmatrix} d_{aa} & \cdots & d_{\varepsilon a} \\ \vdots & \ddots & \vdots \\ d_{a\varepsilon} & \cdots & d_{\varepsilon\varepsilon} \end{bmatrix} \begin{bmatrix} L_a \\ \vdots \\ L_\varepsilon \end{bmatrix} \tag{17.2}$$

where matrix D is called backward coupling matrix whose members can be determined based on the forward matrix of Eq. (20.3):

$$d_{\beta a} = \begin{cases} c_{a\beta}^{-1}, & \text{if } c_{a\beta} \neq 0 \\ 0, & \text{else} \end{cases} \tag{17.3}$$

In other words, the elements of the inverse matrix are obtained by inverting the non-zero members of the forward coupling matrix.

According to the backward coupling matrix relations, for finds, all the power required an EH with n output ports, which is supplied by a hub input port, can be expressed as (17.4):

$$P_m = \sum_{\substack{n=1 \\ n \neq m}}^{N} D_{nm} * L_n \tag{17.4}$$

Example: In this section, the forward matrix for the EH shown in Figure 17.3 is extracted. This system includes a HE, a micro-turbine with CHP technology, and a GF.

According to the figure, the power flow from input port 1 to output port 2 is considered. First, to find the elements of the forward connection matrix, we pay attention to the nodes created between the different converters, which are modeled by V to model this part of the dispersion coefficient. The efficiency of the converters (η_{ij}) also determines the elements of the matrix.

- V contains a value between zero and one.

With these explanations, the power relations of each output are determined as follows

Power demand:

$$L_{2e} = P_{1e} + \eta_{12} * P_{1g} \tag{17.5}$$

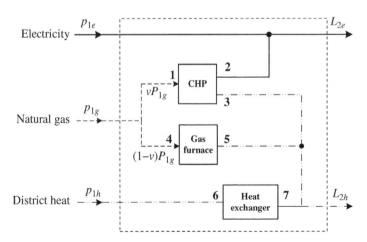

Figure 17.3 Example of a hybrid energy hub including micro turbine, heat exchanger, and gas furnace.

Thermal power demand:

$$L_{2h} = P_{1h} * \eta_{67} + v * \eta_{13} * P_{1g} + (1 - v) * \eta_{45} * P_{1g} \tag{17.6}$$

Finally, if the above relations are expressed as a matrix, the forward connection matrix can be presented as follows.

$$\begin{bmatrix} L_{2e} \\ L_{2h} \end{bmatrix} = \begin{bmatrix} 1 & v * \eta_{12} & 0 \\ 0 & v * \eta_{12} + (1 - v) * \eta_{45} & \eta_{67} \end{bmatrix} \begin{bmatrix} P_{1e} \\ P_{1g} \\ P_{1h} \end{bmatrix} \tag{17.7}$$

17.4 Electricity Market Modeling

Energy consumers and producers participate in the electricity market to buy/sell this energy. This market consists of two parts:

17.4.1 Real-Time Market

Real-time (RT) market transactions are based on the prices of the pool. Due to the high cost of electricity in peak hours and its changes, it is difficult for EH managers to make decisions and plan for the hub based on this market. For these reasons, hub managers refer to the RT market during off-peak hours when prices are low or when it confronts load uncertainty. The cost of electricity supplied by the RT market is formulated as (17.8):

$$C_t^{RT} = \lambda_t^{RT} * P_t^{RT} \tag{17.8}$$

where C_t^{RT}, λ_t^{RT}, and P_t^{RT} are the cost of power purchased from the RT market at time t, the price of electricity in the RT market at time t, and the power purchased from this market at time t, respectively.

17.4.2 Day-Ahead Market (DA)

Hub managers can enter into bilateral agreements with energy producers outside the RT market and agree to have the specified amount of energy purchased and consumed by the EH at a specific time generated by the generator. These contracts can be medium or long-term. The amount of energy purchased in this market by the consumer is determined by the predictions made from the load and economic optimizations. The cost of electricity supplied by the market on the day ahead formulated as (17.9):

$$C_t^{DA} = \lambda_t^{DA} * P_t^{DA} \tag{17.9}$$

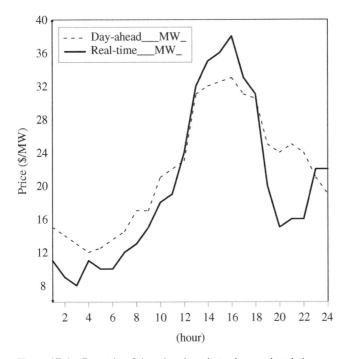

Figure 17.4 Example of day-ahead market prices and real-time.

where C_t^{DA}, λ_t^{DA} and, P_t^{DA} are the cost of power purchased from the day-ahead market at time t, the price of electricity in the DA market at time t, and electricity purchased from this market at time t, respectively.

Figure 17.4 shows an example of RT prices and the day-ahead [30].

17.5 Introduction and Modeling of Components of the Energy Hub

17.5.1 CHP

In combined heat and power technology, the heat generated by power generation in the generator actuators provides the energy needed for heating. CHP is classified according to the types of propulsion generators, including steam turbines, combustion engines, gas turbines, and other generators. The primary sources used in this technology are fossil fuels, biomass, geothermal, and so on.

The term cogeneration, from a thermodynamic point of view, means the production of at least two forms of energy using a primary energy source. There are two

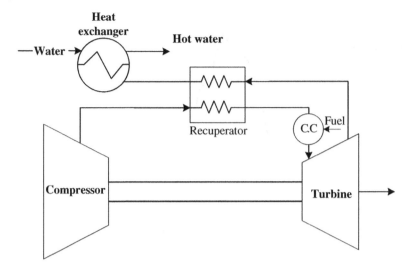

Figure 17.5 Schematic and structure of combined heat and power.

general types of energy, thermal and mechanical. Mechanical energy is most often used to drive an electric generator. For this reason, this technology is known as co-generating power and heat. Speaking of CHP technology, we refer to another type of this technology, the combined cooling heating, and power (CCHP). If an absorption chiller is used to produce refrigeration and use the recyclable heat of the power plant to supply the energy required by the cooling generator, it will be possible to generate electricity, heat, and cold at the same time. Figure 17.5 shows the schematic and structure of CHP technology operation [31].

17.5.1.1 Parametric Modeling and CHP Constraints

At first, for CHP modeling, the modeling of the power generated by the CHP is discussed. The electrical power generated by each CHP unit is calculated as (17.10):

$$P_{eit}^{CHP} = \eta_{ei}^{CHP} * P_{gasit}^{CHP} \tag{17.10}$$

where P_{eit}^{CHP} ' η_{ei}^{CHP} and P_{gasit}^{CHP} are respectively the generated electric power, the coefficient of electrical efficiency, and the amount of natural gas introduced into unit i of the CHP units at time t.

If the EH planner prefers to sell some of its electrical power generations in the RT market and wants to incorporate this action into its desired optimization, then (17.11) the generating power will be separated:

$$P_{eit}^{CHP} = P_{edeit}^{CHP} + P_{egit}^{CHP} \tag{17.11}$$

where $P_{\text{edeit}}^{\text{CHP}}$ and $P_{\text{egit}}^{\text{CHP}}$ are the CHP-generated power to meet the demand of the hub consumer and the power injected into the network at time t, respectively. Otherwise, it will consider its upper limit in planning for this parameter.

The heat power generated by each CHP unit is calculated as (17.12):

$$Q_{\text{it}}^{\text{CHP}} = \eta_{\text{hi}}^{\text{CHP}} * P_{\text{gasit}}^{\text{CHP}} \tag{17.12}$$

where $Q_{\text{it}}^{\text{CHP}}$, $\eta_{\text{hi}}^{\text{CHP}}$, and $P_{\text{gasit}}^{\text{CHP}}$ are the generated thermal power, the efficiency coefficient of unit i, the amount of gas consumed by unit i is from CHP units at time t, respectively.

Finally, the total electrical and thermal production capacity of all CHP units injected into the EH consumers is calculated as (17.13), (17.14):

$$P_{\text{edet}}^{\text{CHP}} = \sum_{i}^{N} P_{\text{edeit}}^{\text{CHP}} \tag{17.13}$$

$$Q_{t}^{\text{CHP}} = \sum_{i}^{N} Q_{\text{it}}^{\text{CHP}} \tag{17.14}$$

- N is the number of CHP units in the EH.

In the optimal economic discussion, there is a need to model the unit cost of simultaneous production. The operating cost of production units includes the cost of fuel consumed and the cost of maintenance. Equation (17.15) shows the cost of fuel consumed unit i of CHP units:

$$C_{\text{gasit}}^{\text{CHP}} = \lambda_{\text{gast}} * P_{\text{gasit}}^{\text{CHP}} \tag{17.15}$$

where $C_{\text{gasit}}^{\text{CHP}}$ and λ_{gast} are the fuel consumption unit i of simultaneous production units at time t, and the price of gas input to unit i CHP at time t, respectively.

After the cost of fuel of the simultaneous production unit, other costs that are considered to be maintenance costs, maintenance costs can be divided into the following three parts:

- Start-up costs
- Shutdown cost
- No-load operation cost

The third part, or the operating cost without load, is a fixed cost, which is considered per hour of operation of the synchronous production unit. Where arises from the cost of periodic repairs, taking into account the operating hours in that period. For modeling, maintenance costs are considered as (17.16)–(17.18):

$$\text{TSUC}_{i}^{\text{CHP}}(t) = \text{SUC}_{i}^{\text{CHP}} * v(t) * (1 - v(t-1)) \tag{17.16}$$

$$\text{TSDC}_{i}^{\text{CHP}} = \text{SDC}_{i}^{\text{CHP}} * v(t-1) * (1 - v(t)) \tag{17.17}$$

$$\text{TNLP}_i^{\text{CHP}}(t) = \text{NLC}_i * v(t) \tag{17.18}$$

In which $\text{SUC}_i^{\text{CHP}}$ ' $\text{TSUC}_i^{\text{CHP}}$ ' $\text{SDC}_i^{\text{CHP}}$ ' $\text{TSDC}_i^{\text{CHP}}$ ' NLC_i, and $\text{TNLC}_i^{\text{CHP}}$ respectively. The cost of turning on at time t is the cost of turning off, the cost of turning off at time t, the cost of maintenance without load and the cost of maintenance without load at time t for the unit i of CHP units.

- v_t is a binary variable that specifies the ON / OFF state of each CHP unit.

Finally, considering Eqs. (17.15–17.18) to calculate the cost, all units of simultaneous production at time t are calculated as (17.19):

$$\text{COCHP}_t = \sum_i^N \left(\lambda_{\text{gast}} * P_{\text{gasit}}^{\text{CHP}} + \text{TSUC}_{it}^{\text{CHP}} + \text{TSDC}_{it}^{\text{CHP}} + \text{TNLC}_{it}^{\text{CHP}} - P_{\text{egit}}^{\text{CHP}} * \lambda_{\text{RTt}} \right)$$

$$\tag{17.19}$$

As mentioned above, if the planner does not decide to sell some of its production to the RT market, it will consider the value of $P_{\text{egit}}^{\text{CHP}}$ in the planning to be a fixed value of zero.

17.5.1.2 CHP Operation Area

It should be noted that the thermal and electrical power outputs of simultaneous production units are not independent and depend on each other, and one of the outputs affects the other. That is, by changing the output of one of the outputs, the other also changes. In general, each cogeneration unit operates in one operating area, and the production of heat and electrical power depends on this operating range. Figure 17.6 shows an example of the operating range of a cogeneration unit.

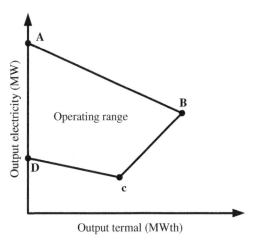

Figure 17.6 An example of a cogeneration unit (CHP) operating range.

To model the range of CHP operations, the breakpoints are considered as the range, then limited by the equation between the two points. For example, modeling the line equation between points A and B is as follows (17.20) [32]:

$$P_{ei}^{CHP} + \left(\frac{P_A - P_B}{Q_B}\right) * Q_i^{CHP} \leq P_A * v_t \tag{17.20}$$

17.5.2 Furnace Gas

Gas stoves are part of the EH that is responsible for generating thermal energy by burning gas. As observed in (17.12), simultaneous generation units also produce thermal energy and generate electrical energy. Also, in Figure 17.6, we saw that the production of heat and electrical energy in CHP are not independent of each other, and the production of each depends on the amount of production of the other, so in the EH, it is necessary to have a unit that can produce thermal energy independently. Compared to simultaneous production units, it has a higher heat efficiency coefficient.

Modeling of GF heat output is done as (17.21) and (17.22):

$$Q_t^{furnace} = \eta_h^{furnace} * P_{gast}^{furnace} \tag{17.21}$$

$$Q_t^{furnace} \leq Q_{max}^{furnace} \tag{17.22}$$

In which $Q_t^{furnace}$ ' η_h^{frnace} ' $P_{gast}^{furnace}$, and $Q_{max}^{furnace}$, respectively, the heat output power of the GF at time t, the coefficient of gas-to-heat efficiency of the GF, the gas input capacity to the GF at time t, and the maximum heat capacity production of GFs.

GF, like simultaneous production units in modeling operating costs, consists of two parts, fuel cost and maintenance cost. GF fuel cost is calculated as (17.23).

$$c_t^{furnace} = \lambda_{gast} * P_{gast}^{furnace} \tag{17.23}$$

where $c_t^{furnace}$ is the cost of fuel-consuming the furnace at time t.

Modeling of maintenance costs can be done as (17.24–17.26):

$$TSUCF_t^{furnace} = SUCF^{furnace} * h(t) * (1 - h(t-1)) \tag{17.24}$$

$$TSDCF_t^{furnace} = SDCF^{furnace} * h(t-1) * (1 - h(t)) \tag{17.25}$$

$$TNLPF_t^{furnace} = NLCF^{furnace} * h(t) \tag{17.26}$$

where $SUCF^{furnace}$ ' $TSUCF_t^{furnace}$ ' $SDCF^{furnace}$ ' $TSDCF_t^{furnace}$ ' $NLCF^{furnace}$, and $TNLCF_t^{furnace}$, respectively, are considered start-up costs. To turn on at time t is the cost of switching off, the estimated cost to turn off at time t is the cost of maintenance without AR and the cost of maintenance without load at time t is the GF.

- h_t is a binary variable that specifies the ON/OFF mode of the GF.

Finally, considering Eqs. (17.23–17.26) to calculate all the costs of the GF at time t is done as (17.27):

$$TC_t^{\text{furnace}} = \left(\lambda_{\text{gast}} * P_{\text{gast}}^{\text{furnace}} + TSUCF_t^{\text{furnace}} + TSDCF_t^{\text{furnace}} + TNLCF_t^{\text{furnace}} \right)$$

(17.27)

where TC_t^{furnace} is all furnace costs at time t.

17.5.3 Diesel Generator

A DG consists of a diesel engine and an electric generator that is responsible for generating electrical energy by burning diesel fuel or natural gas. As stated in the case of gas stoves, simultaneous generation units also produce thermal energy in and the production of electrical energy, and the production of each of these energies is interdependent, which is responsible for supplying electricity independently. This task in the EH is the responsibility of the DG.

Figure 17.7 shows the schematic and diagram of the DG.

The electric power generated by the DG is expressed as (17.28) and (17.29):

$$P_{tj}^{DG} = \eta^{DG} * P_{\text{gasjt}}^{DG}$$

(17.28)

$$P_{tj}^{DG} \leq P_{\max}^{DG}$$

(17.29)

where P_{tj}^{DG}, η^{DG}, P_{gast}^{DG}, and $P_{\max}^{DG}$ respectively are DG power generation, DG efficiency coefficient and fuel input power, maximum DG capacity of unit j of DG units.

In the study and operation of DGs, as in simultaneous production units, the EH planner also prefers to sell some of his electrical power generations in the RT

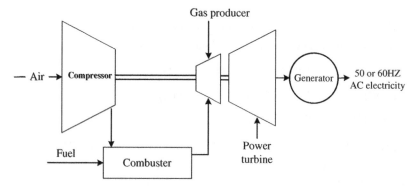

Figure 17.7 Diesel generator diagram.

market and wants to include this measure in his desired optimization as (17.30) individually production capacity:

$$P_{tj}^{DG} = P_{detj}^{DG} + P_{gtj}^{DG} \tag{17.30}$$

where P_{detj}^{DG} and P_{gtj}^{DG} are the electric power injected by the DG to the consumer of the EH and the network, respectively.

Finally, the total electrical generating capacity of all DG units injected into EH consumers is calculated as (17.31):

$$P_t^{DG} = \sum_j^J P_{detj}^{DG} \tag{17.31}$$

where P_t^{DG} is equivalent to the total electrical power of the DG, injection into EH consumers.

- J is the number of DG units available in the EH.

DGs in modeling operating costs consist of two parts, fuel cost and maintenance cost. DG fuel cost is calculated as (17.32) [33].

$$c_{tj}^{DG} = \lambda_{gast} * P_{gastj}^{DG} \tag{17.32}$$

where c_{tj}^{DG} is the fuel cost of unit i of DG units at time t.

Modeling the maintenance costs of each DG unit can be done as (17.33–17.35):

$$\text{TSUCG}_{tj}^{DG} = \text{SUCG}^{DG} * k_j(t) * \left(1 - k_j(t-1)\right) \tag{17.33}$$

$$\text{TSDCG}_{tj}^{DG} = \text{SDCG}^{DG} * k_j(t-1) * \left(1 - k_j(t)\right) \tag{17.34}$$

$$\text{TNLPG}_{tj}^{DG} = \text{NLCG}^{DG} * k_j(t) \tag{17.35}$$

In which SUCG^{DG}"TSUCG_{tj}^{DG}"SDCG^{DG}"TSDCG_{tj}^{DG}"NLCG^{DG} , and TNLCG_{tj}^{DG} are respectively charged start-up costs, to turn on at time t, the cost of switching off, the cost to turn off at time t, the cost of maintenance without load and the cost of maintaining the load body of unit j of DG units at time t.

- K_j is a binary variable that specifies the ON / OFF state of the DG unit j.

Finally, considering Eqs. (17.32–17.35) to calculate the cost, all DG units at time t are calculated as (17.36):

$$\text{CT}_t^{DG} = \sum_J^L \left(\lambda_{gast} * P_{gastj}^{DG} + \text{TSUCG}_{tj}^{DG} + \text{TSDCG}_{tj}^{DG} + \text{TNLCG}_{tj}^{DG} - P_{gtj}^{DG} * \lambda_t^{RT}\right) \tag{17.36}$$

where CT_t^{DG} is the total cost of all DG units at time t.

17.5.4 Wind Turbine

Wind turbines convert wind kinetic energy into electrical energy. These turbines are made in two axes, vertical and horizontal axis. The horizontal axis type is a common type of wind turbine. These turbines are also available in different sizes and capacities. In order to operate and increase the production capacity, wind turbines are used collectively or in the so-called wind farm. The production capacity of wind turbines depends on wind speed.

The relationship between wind turbine output power and wind speed is formulated as (17.37) [34, 35]:

$$
P_{tl}^{\text{wind}} = \begin{cases} 0 & \text{if } v_t \leq v_{\text{in}} \text{ or } v_t \geq v_{\text{out}} \\ P_S * \left(\frac{v_t - v_{\text{in}}}{v_r - v_{\text{in}}} \right)^3 & \text{if } v_{\text{in}} \leq v_t \leq v_r \\ P_S & \text{if } v_r \leq v_t \leq v_{\text{out}} \end{cases}
\tag{17.37}
$$

where $v_t v_{\text{in}} v_r v_{\text{out}} P_S$, and P_{tl}^{wind} show wind speed at time t, cut-in speed, rated speed, cut-out speed, rated power, and electrical power generated by wind turbines at time t, respectively.

As mentioned, to increase capacity, wind turbines are used collectively or wind farms. the production capacity of the wind farm is calculated as (17.38):

$$
P_t^{\text{wind}} = \sum_{l=1}^{L} P_{tl}^{\text{wind}}
\tag{17.38}
$$

where L is the number of wind turbines in the wind farm.

17.5.5 Load

The load requested by the consumer is provided by the sum of electrical and thermal energy generated by the various components of the EH and the electricity markets. Modeling and limiting the electric charge concerning the relations (17.8), (17.9), (17.13), (17.31), and (17.38) are formulated as (17.39):

$$
P_{\text{edet}}^{\text{CHP}} + P_t^{\text{DG}} + P_t^{\text{wind}} + P_t^{\text{DA}} + P_t^{\text{RT}} = P_t^D
\tag{17.39}
$$

where P_t^D is the electrical load of the hub consumers.

Modeling and relation of thermal load according to the equations, (17.14) and (17.21) are formulated as (17.40):

$$
Q_t^{\text{CHP}} + Q_t^{\text{furnace}} = Q_t^D
\tag{17.40}
$$

where Q_t^D is the heat load consumed.

17.5.6 The Objective Function

Managers and operators of the EH, in addition to providing demand, try to minimize operating costs and fuel carriers consumed in the hub.

By considering the limitations of different components of the EH and considering the price of different energy carriers entering the hub and also taking the maintenance costs of the components inside the hub, the objective function is defined, which is the objective function in order to minimize the cost of the minimum material. The objective function of the EH is formulated as (17.41) according to the components and constraints introduced above [33, 36, 37].

$$f_b = \text{Min} \sum_{t=1}^{T} \left[C_t^{RT} + C_t^{DA} + CT_t^{DG} + TC_t^{furnace} + CCHP_t \right] \tag{17.41}$$

where f_b is equal to the minimum cost of the EH to supply energy to the consumer in a given period, and T is equal to the period of study. For example, if the period is one day, T is equal to 24 hours.

17.6 Energy Hub Analyzing and Discussions

After introducing the various components of the EH and modeling the structure and limitations of these components, in this section, we have tried to apply this modeling and numerical simulation on an EH. As mentioned above, EH planners and managers about a DG and a cogeneration unit can decide to sell some of their output to the grid within hours of the period under study, so two scenarios are considered.

- No electricity sold to the grid
- Selling a certain amount of DG and CHP electricity to the grid

17.6.1 Scenario Without Selling Electricity to the Grid

The EH intended for the simulation is shown in Figure 17.8. The hub uses electricity and gas as input and electricity and heat as output. This EH uses the two markets of the day ahead and RT to supply electricity directly. DGs and CHP also indirectly contribute to the supply of electricity by converting fuel and gas. In addition, wind turbines collectively and wind farms provide some of the electrical energy required by the EH. The heat output of the CHP and the GF are responsible for supplying thermal energy to the EH, both of which meet the heat needs of the consumer by converting the gas into heat.

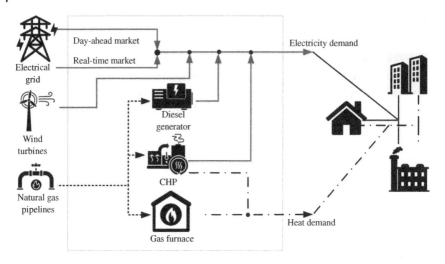

Figure 17.8 The energy hub under study.

The specifications of the DG unit, GF, and gas price are available in Table 17.1. The thermal efficiencies of the GF are selected from [38]. The specifications of CHP units are given in Table 17.2.

The functional area of all three CHP units, which are surrounded by four points A, B, C, D, is shown in Figure 17.9.

In the studied EH, ten identical wind turbines have been used as wind farms. The specifications of this wind turbine are given in Table 17.3. The specifications of wind turbines have been selected from [39]. The wind speed characteristics used for the wind farm in one day are given in Figure 17.10 [40].

The electric load and heat load demand of the EH in Figure 17.11 show that the electric load is selected from [30] and is equal to 2.5% of the electric charge of New

Table 17.1 Specifications of the gas furnace and diesel generator.

Parameter	Value	Parameter	Value
$\lambda_{gas}(\$/MW)$	23	$NLCF^{furnace}$	70
$\eta_h^{furnace}$	0.8	$SUCG^{DG}$	40
η^{DG}	0.4	$SDCG^{DG}$	40
$SUCF^{furnace}$	40	$NLCG^{DG}$	150
$SDCF^{furnace}$	40	P_{max}^{DG} (MW)	25

Table 17.2 Specifications of CHP units.

Parameter	Value	Parameter	Value
η_{e1}^{CHP}	0.4	SUC_2^{CHP}	55
η_{h1}^{CHP}	0.45	SDC_1^{CHP}	55
η_{e1}^{CHP}	0.4	SDC_2^{CHP}	55
η_{h1}^{CHP}	0.45	NLC_1^{CHP}	125
SUC_1^{CHP}	55	NLC_2^{CHP}	135

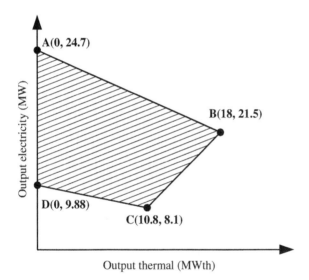

Figure 17.9 CHP operating area used in the hub.

Table 17.3 Technical specifications of wind turbines used in the energy hub.

Turbine model	Rated output (MW)	Cut in speed (m/s)	Rated speed (m/s)	Cutout speed (m/s)	Hub height (m)	Rotor diameter (m)
Bonus 2.3/82.4	2.3	3	15	25	60	82.4

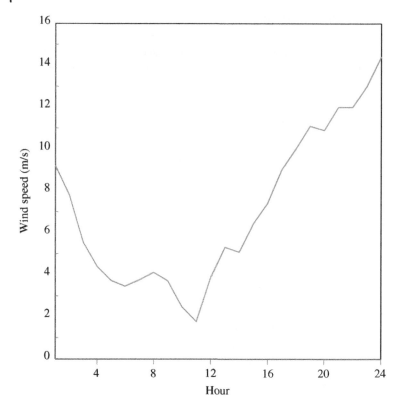

Figure 17.10 Wind speed used in the energy hub.

York City. The heat load demand pattern of the EH has been selected from the Homer software datasheet [41].

The forecast of electricity prices for the day ahead and in RT is given in Figure 17.12. As shown in this figure, the RT electricity market price during peak hours is higher than the day-ahead market price. Price data for these two electricity markets are selected from [30].

The simulation of the desired EH has been performed by considering the objective function (17.41) and the data as mentioned above in GAMS software and using Solver Baron. The simulation results are presented in the following figures. The electricity purchased from the DA and RT electricity markets is shown in Figure 17.13. As can be seen, at times when the RT electricity market prices are higher than the DA, the EH prefers to plan the electricity it needs to purchase from the DA electricity market and at other times to supply its energy from the RT market [42].

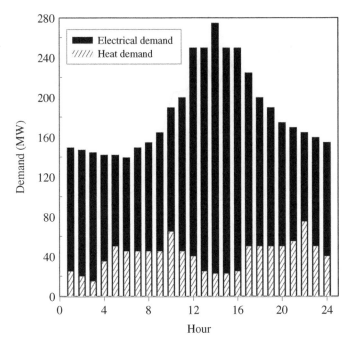

Figure 17.11 Electric and thermal load demand.

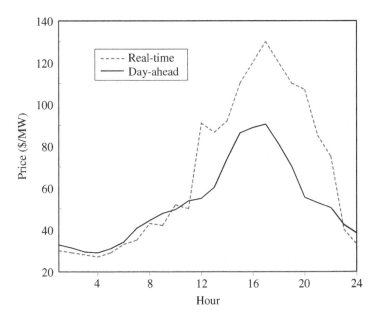

Figure 17.12 Real-time and day-ahead electricity prices.

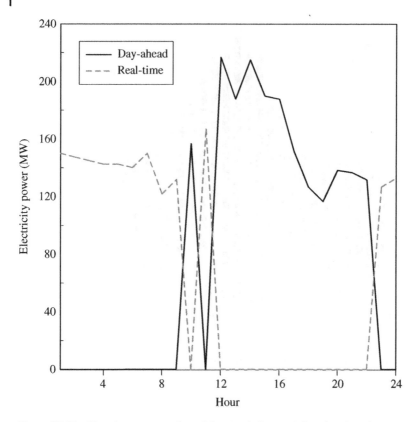

Figure 17.13 Electric power purchased from real-time and day-ahead markets.

The production rate of DGs during the EH review period is shown in Figure 17.14. As can be seen in this figure, the EH of the DG is used during peak load times and meets part of its consumer demand from the production of DGs during the hours of rising prices in direct electricity markets.

The amount of electrical energy production of CHP cogeneration units is shown in Figure 17.15. The amount of thermal energy production of cogeneration units is also shown in Figure 17.16. As shown in Figure 17.15, CHP units start operating in times of high load and rising electricity market prices. As stated in the CHP constraints section, the production of thermal and electrical power in these units is interdependent, and in return for producing a certain amount of electrical energy, it produces a certain amount of thermal power. This limitation can be seen in Figure 17.11. In the middle hours of the day, heat demand is less than the other hours of the day, but the electric demand is the opposite, and in the middle hours of the day, more demand is required by the consumer. The generation power of

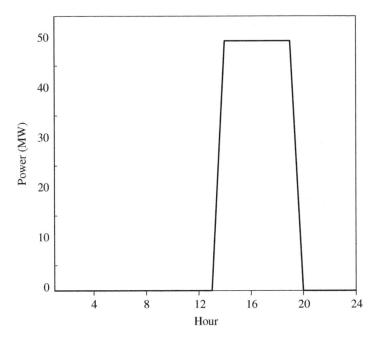

Figure 17.14 Electrical power generated by the diesel generator.

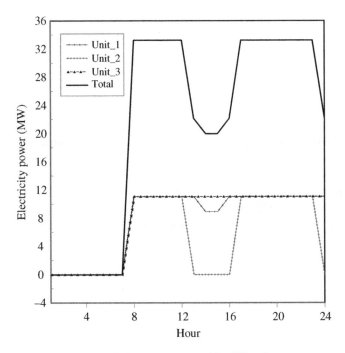

Figure 17.15 Electrical power generated by CHP units.

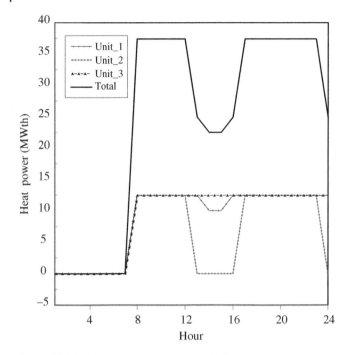

Figure 17.16 Heat power generated by CHP units.

CHP units is limited by the thermal load during these hours. Figure 17.16 also shows that units 1,2 are active in the middle of the day, but unit 2 is off due to the NLC cost of unit 2, which is higher than other units and prefers to turn it off during optimization.

The production of thermal energy by a GF is shown in Figure 17.17 As, shown in the figure, and the GF starts operating during the hours of increasing the energy requested by the EH consumer. As shown in Figure 17.17, in the middle of the day, due to the reduction of heat load, the consuming heat load is supplied by the simultaneous production unit, and the GF is switched off during hours. Also, when the synchronous production units are off, the thermal energy requested by the consumer of the whole hub is supplied by the GF.

The generation of electrical power by wind turbines is considered a wind farm as shown in Figure 17.18. In this wind farm, ten similar turbines with the specifications of Table 17.3 are considered.

The above figures show how the various components of the power hub work in the first scenario, without the sale of electricity to the grid. Also, the final cost of the EH for a period of 24 hours in this scenario is 259316 ($).

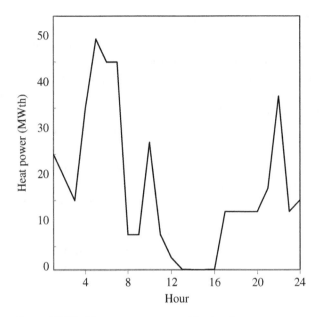

Figure 17.17 Heat power generated by gas furnace.

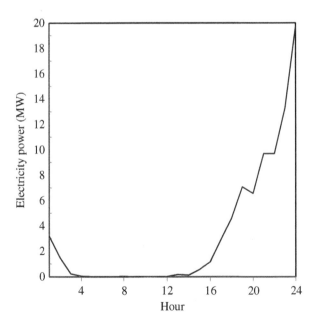

Figure 17.18 Electrical power generated by wind turbines.

17.6.2 Scenario with the Sale of Electricity to the Grid

As mentioned above, the management and planning of the EH are going to decide to sell some of their electricity production in the electricity market. In the desired EH, CHP and Diesel Generated components can sell part of their production in the electricity market. According to Eqs. (17.11) and (17.30), the management of the EH is the maximum amount that it intends to sell from its production. In this section, for each unit of DG ($P_{gtj}^{DG} \leq 5\,\text{MW}$) and for each unitSimultaneous production of ($P_{egit}^{CHP} \leq 5\,\text{MW}$) considered.

Figure 17.19 shows the sales of CHP units to the network. According to the figure, it can be seen that the sales of CHP units are at times of rising electricity market prices, and the limitation of dependence on the thermal power produced by the CHP also manifests itself.

Figure 17.20 also shows the output of CHP units for the EH load. According to Figures 17.19 and 17.20, it can be seen that during the hours when the CHP unit sells to the market is zero, the EH planner prefers to allocate its production to the EH consumer for economic reasons.

Figure 17.21 shows the total sales of electric power generated by DG units. As shown in the figure, both DG units sell the maximum amount of energy set by the EH planner when the electricity market price increases.

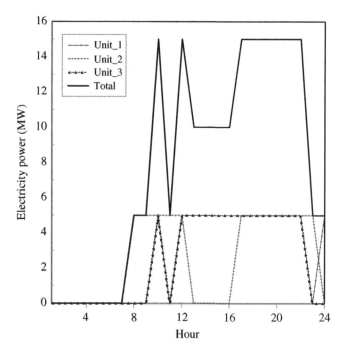

Figure 17.19 Electrical power sales of CHP units.

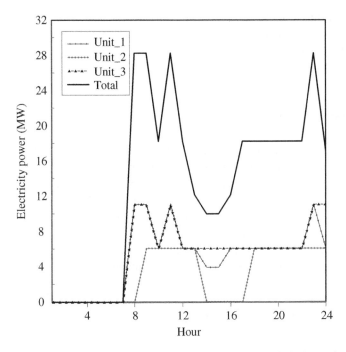

Figure 17.20 Electrical power of CHP units for energy hub demand.

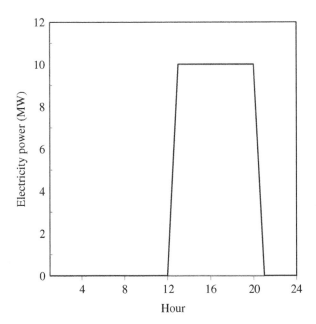

Figure 17.21 Total electrical power sales of diesel generator units.

In the second scenario, whit selling some of the electrical power generated by the EH during peak load and increasing the price of the electricity market, the cost of the EH is reduced to 252121 ($).

17.7 Conclusion

The objective of this chapter was to optimize the combined energy systems. To that end, an EH was introduced, which allowed simultaneous planning of different energy infrastructures and carriers. Simultaneous utilization of different energy carriers improves the reliability of supplying consumers' demand. Moreover, planning of the combined energy systems leads to Economic optimization of consumers' demand. Introducing the concept of the EH was followed by being optimized in an example. According to the results, the EH concept is advantageous in planning combined energy systems. Furthermore, using components such as CHP, GF, and DGs can financially help to meet the consumers, demands, particularly in peak hours in which energy price is high. In the discussion of modeling the internal components of the EH and implementing the model of these components, the marginal costs of these devices were also applied, which made more economical decisions. CHP units were also modeled as they operate in their own functional area, which made the performance of these units more practical and the results closer to reality. The proposed EH was examined in two modes, participating in the electricity market and without it. The results showed that in the hours in which electricity prices increase, the EH, by converting gas into electrical energy by its internal components and selling to the current electricity market, creates more economic conditions than without participating in the electricity market.

References

1 Sovacool, B.K., Valentine, S.V., Bambawale, M.J. et al. (2012). Exploring propositions about perceptions of energy security: an international survey. *Environmental Science & Policy* 16: 44–64.

2 Sovacool, B.K. and Brown, M.A. (2010). Competing dimensions of energy security: an international perspective. *Annual Review of Environment and Resources* 35: 77–108.

3 Winzer, C. (2012). Conceptualizing energy security. *Energy Policy* 46: 36–48.

4 Geidl, M. (2007). *Integrated Modeling and Optimization of Multi-Carrier Energy Systems*. ETH Zurich.

5 Mancarella, P. (2014). MES (multi-energy systems): an overview of concepts and evaluation models. *Energy* 65: 1–17.

6 Geidl, M., Koeppel, G., Favre-Perrod, P. et al. (2007). The energy hub–a powerful concept for future energy systems. *Third Annual Carnegie Mellon Conference on the Electricity Industry*, Carnegie Mellon, State of Pennsylvania (March 2007), vol. 13, p. 14.

7 Ren, H. and Gao, W. (2010). A MILP model for integrated plan and evaluation of distributed energy systems. *Applied Energy* 87 (3): 1001–1014.

8 Ortega-Vazquez, M.A. and Kirschen, D.S. (2010). Assessing the impact of wind power generation on operating costs. *IEEE Transactions on Smart Grid* 1 (3): 295–301.

9 Kim, J.S. and Edgar, T.F. (2014). Optimal scheduling of combined heat and power plants using mixed-integer nonlinear programming. *Energy* 77: 675–690.

10 Wang, H., Yin, W., Abdollahi, E. et al. (2015). Modelling and optimization of CHP based district heating system with renewable energy production and energy storage. *Applied Energy* 159: 401–421.

11 Shabanpour-Haghighi, A., Seifi, A.R., and Niknam, T. (2014). A modified teaching–learning base optimization for multi-objective optimal power flow problem. *Energy Conversion and Management* 77: 597–607.

12 Shabanpour-Haghighi, A. and Seifi, A.R. (2015). Multi-objective operation management of a multi-carrier energy system. *Energy* 88: 430–442.

13 Geidl, M. and Andersson, G. (2007). Optimal power flow of multiple energy carriers. *IEEE Transactions on Power Systems* 22 (1): 145–155.

14 Zhang, X., Shahidehpour, M., Alabdulwahab, A., and Abusorrah, A. (2015). Optimal expansion planning of energy hub with multiple energy infrastructures. *IEEE Transactions on Smart Grid* 6 (5): 2302–2311.

15 Favre-Perrod, P., Kienzle, F., and Andersson, G. (2010). Modeling and design of future multi-energy generation and transmission systems. *European Transactions on Electrical Power* 20 (8): 994–1008.

16 Hemmati, S., Ghaderi, S., and Ghazizadeh, M. (2018). Sustainable energy hub design under uncertainty using Benders decomposition method. *Energy* 143: 1029–1047.

17 Moeini-Aghtaie, M., Abbaspour, A., Fotuhi-Firuzabad, M., and Hajipour, E. (2013). A decomposed solution to multiple-energy carriers optimal power flow. *IEEE Transactions on Power Systems* 29 (2): 707–716.

18 Ashouri, A., Fux, S.S., Benz, M.J., and Guzzella, L. (2013). Optimal design and operation of building services using mixed-integer linear programming techniques. *Energy* 59: 365–376.

19 Bozchalui, M.C., Hashmi, S.A., Hassen, H. et al. (2012). Optimal operation of residential energy hubs in smart grids. *IEEE Transactions on Smart Grid* 3 (4): 1755–1766.

20 Rastegar, M., Fotuhi-Firuzabad, M., Zareipour, H., and Moeini-Aghtaieh, M. (2016). A probabilistic energy management scheme for renewable-based residential energy hubs. *IEEE Transactions on Smart Grid* 8 (5): 2217–2227.

21 Brahman, F., Honarmand, M., and Jadid, S. (2015). Optimal electrical and thermal energy management of a residential energy hub, integrating demand response and energy storage system. *Energy and Buildings* 90: 65–75.

22 Setlhaolo, D., Sichilalu, S., and Zhang, J. (2017). Residential load management in an energy hub with heat pump water heater. *Applied Energy* 208: 551–560.

23 Roldán-Blay, C., Escrivá-Escrivá, G., Roldán-Porta, C., and Álvarez-Bel, C. (2017). An optimisation algorithm for distributed energy resources management in micro-scale energy hubs. *Energy* 132: 126–135.

24 Ghaffarpour, R., Mozafari, B., Ranjbar, A.M., and Torabi, T. (2018). Resilience oriented water and energy hub scheduling considering maintenance constraint. *Energy* 158: 1092–1104.

25 Fan, S., Li, Z., Wang, J. et al. (2018). Cooperative economic scheduling for multiple energy hubs: a bargaining game theoretic perspective. *IEEE Access* 6: 27777–27789.

26 Geidl, M., Koeppel, G., Favre-Perod, P. et al. (2006). Energy hubs for the future. *IEEE Power and Energy Magazine* 5 (1): 24–30.

27 Favre-Perod, P. (2005). A vision of future energy networks. *2005 IEEE Power Engineering Society Inaugural Conference and Exposition in Africa*, Durban, South Africa (11–15 July 2005). IEEE, pp. 13–17.

28 Müller, E.A., Stefanopoulou, A.G., and Guzzella, L. (2006). Optimal power management of hybrid fuel cell systems: A feedback strategy to minimize the system's warm-up time. *Research Frontiers in Energy Science and Technology: Contributions to the Latsis Symposium 2006*, ETH, Zürich, Switzerland (11–13 October 2006). vdf.

29 Geidl, M. and Andersson, G. (2005). A modeling and optimization approach for multiple energy carrier power flow. *2005 IEEE Russia Power Tech*, St. Petersburg, Russia (27–30 June 2005). IEEE, pp. 1–7.

30 nyiso.com (2021). Pricing & load data of New York city. [Online]. https://www.nyiso.com/markets (accessed 14 July 2021).

31 Khetrapal, P. (2020). Distributed generation: a critical review of technologies, grid integration issues, growth drivers and potential benefits. *International Journal of Renewable Energy Development* 9 (2): 189–205.

32 Wu, C., Jiang, P., Gu, W., and Sun, Y. (2016). Day-ahead optimal dispatch with CHP and wind turbines based on room temperature control. *2016 IEEE International Conference on Power System Technology (POWERCON)*, Beijing, China (10–12 July 2016). IEEE, pp. 1–6.

33 Najafi, A., Falaghi, H., Contreras, J., and Ramezani, M. (2016). Medium-term energy hub management subject to electricity price and wind uncertainty. *Applied Energy* 168: 418–433.

34 Daneshvar, M., Mohammadi-Ivatloo, B., Abapour, M., and Asadi, S. (2020). Energy exchange control in multiple microgrids with transactive energy management. *Journal of Modern Power Systems and Clean Energy* 8 (4): 719–726.

35 Rabiee, A. and Soroudi, A. (2013). Stochastic multiperiod OPF model of power systems with HVDC-connected intermittent wind power generation. *IEEE Transactions on Power Delivery* 29 (1): 336–344.

36 Najafi, A. et al. (2019). Uncertainty-based models for optimal management of energy hubs considering demand response. *Energies* 12 (8): 1413.

37 Dolatabadi, A., Jadidbonab, M., and Mohammadi-ivatloo, B. (2018). Short-term scheduling strategy for wind-based energy hub: a hybrid stochastic/IGDT approach. *IEEE Transactions on Sustainable Energy* 10 (1): 438–448.

38 Shahmohammadi, A., Moradi-Dalvand, M., Ghasemi, H., and Ghazizadeh, M. (2014). Optimal design of multicarrier energy systems considering reliability constraints. *IEEE Transactions on Power Delivery* 30 (2): 878–886.

39 Akpinar, E.K. and Akpinar, S. (2005). An assessment on seasonal analysis of wind energy characteristics and wind turbine characteristics. *Energy Conversion and Management* 46 (11–12): 1848–1867.

40 nrel.gov (2021). Wind speed data. [Online]. https://www.nrel.gov/wind/data-tools. html (accessed 18 July 2021).

41 homerenergy.com (2021). Heat load data from Homer datasheet. [Online]. http://www.homerenergy.com (accessed 25 July 2021).

42 Hemmati, M., Abapour, M., Mohammadi-Ivatloo, B., and Anvari-Moghaddam, A. (2020). Optimal operation of integrated electrical and natural gas networks with a focus on distributed energy hub systems. *Sustainability* 12 (20): 8320.

18

Hybrid Energy Storage Systems for Optimal Planning of the Heat and Electricity Incorporated Networks

Hamid HassanzadehFard[1], Seyed Mehdi Hakimi[2], and Arezoo Hasankhani[3]

[1] Department of Electrical Engineering, Miyaneh Branch, Islamic Azad University, Miyaneh, Iran
[2] Department of Electrical Engineering and Renewable Energy Research Center, Damavand Branch, Islamic Azad University, Damavand, Iran
[3] Department of Computer and Electrical Engineering and Computer Science, Florida Atlantic University, Boca Raton, FL, USA

List of Symbols

$P_{\text{FC}}^{E,t}$	The electricity generated by FC at the time t
μ_{FC}^{E}	Conversion efficiencies from fuel to electrical power through the FC
$P_{\text{FC}}^{\text{gas},t}$	Fuel consumption of the FC
$P_{\text{FC}}^{T,t}$	The heat produced by the FC at the time t
μ_{FC}^{T}	Coefficient of the heat conversion over the FC
P_{Boiler}^{t}	The heat generated by boiler at the time t
μ_{Boiler}	Conversion coefficient from gas to thermal through the boiler
$P_{\text{Boiler}}^{\text{gas},t}$	Consumption fuel by the boiler at the time t
P_{PV}^{t}	The power produced by PV at the time t
γ_r	PV arrays efficiency
γ_{pt}	Power tracking equipment efficiency
T^c	Temperature of the PVs
T^r	Reference temperature of the PVs
β	Efficiency temperature coefficient
A_m	Area (m^2) of modules
G^t	The global irradiance
N	Number of modules
V_r	Nominal wind speed
V	Wind speed

Coordinated Operation and Planning of Modern Heat and Electricity Incorporated Networks,
First Edition. Edited by Mohammadreza Daneshvar, Behnam Mohammadi-Ivatloo, and Kazem Zare.
© 2023 The Institute of Electrical and Electronics Engineers, Inc.
Published 2023 by John Wiley & Sons, Inc.

V_{co}	Cut out wind speed
V_{ci}	Cut in wind speed
P_{WT}^{max}	Maximum power of the wind turbine
$E_{el\_Tank}^{t}$	The transferred power from the electrolyzer to the HES
η_{el}	Efficiency of the electrolyzer
$P_{Surplus}^{t}$	Excess generated electricity of renewable energy resources
P_{WT}^{t}	Electrical power generated by wind turbines
$P_{Load}^{E,t}$	Total electrical loads at the time t
η_{BS}^{disch}	Battery discharging efficiency
η_{BS}^{ch}	Battery charging efficiency
$\eta_{inverter}$	Inverter efficiency
$E_{Thermal}^{t}$	Heat loss rate of TES
$P_{Load}^{T,t}$	Total heating and demand at the time t
$P_{Load}^{C,t}$	Total cooling demand at the time t
ε_{el}^{t}	Related coefficients for electrolyzer to H2 storage tank at the time t
ε_{FC}^{t}	Related coefficients for H2 storage tank to FC at the time t
CC^{i}	Capital cost of each applied DG
RC^{i}	Replacement cost of each applied DG
OM^{i}	Operation and maintenance cost of each applied DG
ω_{Fuel}	Fuel cost
ω_{E}	Penalty for environmental emissions
ω_{ENS}	Penalty for interrupted loads
ϕ_{Boiler}	Emission factor for boiler
ϕ_{CHP}	Emission factor for FC
μ_{CHP}	Energy factor of FC
μ_{Boiler}	Energy factor of boiler
N^{i}	Optimum number of each DG
L_{i}	Lifetime of each DG
i	Interest rate
L_{R}	Lifetime of the project
E_{BS}^{t}	Stored energy in battery storages at the time t
$P_{Interrupted}^{t}$	Interrupted loads at the time t
$E_{Thermal}^{t}$	Stored energy in TES at the time t
η_{chi}	Efficiency of the chiller
$P_{FC}^{Th,\,max}$	Maximum generated thermal power of FC
$P_{FC}^{E,\,max}$	Maximum generated electrical power of FC
P_{PV}^{max}	Maximum generated power of PV
P_{Boiler}^{max}	Maximum generated power of boilers
$E_{H_2}^{max}$	Maximum capacity of the H2 storage tank
E_{BS}^{max}	Maximum capacity of battery storage
$E_{Thermal}^{max}$	Maximum capacity of thermal energy storage
$E_{H_2}^{min}$	Minimum rated capacity of the H2 energy storage

$E_{\text{Thermal}}^{\min}$ Minimum rated capacity of thermal energy storage
$E_{\text{BS}}^{\min}$ Minimum rated capacity of battery storage

Abbreviations

ESS   Energy storage system
RES   Renewable energy source
CCHP   Combined cooling, heat and power
DG   Distributed generation
PV   Photovoltaic
FC   Fuel cell
LPSP   Loss of power supply probability
TES   Thermal energy storage
EES   Electrical energy storage
HES   Hydrogen energy storage
GA   Genetic algorithm
PSO   Particle swarm optimization
QPSO   Quantum behaved particle swarm optimization
MG   Microgrid
DR   Demand response
DOD   Depth of discharge
NPC   Net present cost

18.1 Introduction

Providing the energy demands and overcoming the environmental problems are two conflicting troubles that are becoming highly serious in the rapidly developing modern world [1]. Among the different methodologies and technologies being suggested to resolve such a dilemma, CCHP systems can provide both electrical and thermal power for microgrid (MG) systems. In addition, the power system has demonstrated a sharp rise in the utilization of RESs, which have less greenhouse gas emissions and they recognized as clean energy sources [2, 3]. The generated power of RESs is intermittent, stochastic, and variable, and their integration in the power system may negatively affect the system performance, including poor load tracking, load mismatch, frequency deviation, voltage instability, and reliability issues. The utilization of ESSs is one of the main solutions to increase the penetration of RESs in the islanded hybrid systems. The ESS applications consist of peak shaving, as well as renewable energy management and integration [4]. The ESSs can store excess energy during surplus production hours and discharge it appropriately later when needed.

The optimization of the multi-carrier network has been widely evaluated in the literature. In [5], stochastic strategy has been presented for determining the optimum operation of different components, including CHP and RESs, while considering the uncertainties. A deep reinforcement learning-based methodology has been proposed to address energy management strategy for a multi-carrier network [6]. In a similar study [7], a residential home energy management framework has been designed to reduce the load deviation in an MG system. Energy transfer between different residential buildings has been addressed in the presence of RESs and thermal resources, where different loads have been modeled [8, 9]. A combination of hydrogen and electricity network has been applied to meet an optimized energy management system for a multi-carrier network [10]. An optimum energy management schedule has been planned for a multi-carrier network considering its participation in the thermal market, electrical market, and cooling market [11]. An optimal plan for the multi-carrier network has been determined, where the optimal size of RESs has been calculated [12]. In [13], an objective function has been defined for minimizing the total cost and emission in a multi-carrier network.

Furthermore, a reliability index has been considered along with finding an optimal operation of the multi-carrier network, where the dynamic model of the thermal loads has been formulated [14]. The optimum planning of the energy hub has been formulated considering the high penetration of stochastic RESs [15]. The role of the electricity market has been evaluated in the optimal planning of an MG, where the RESs have been highly installed in the network [16]. Another similar study has analyzed the uncertainty of different renewable resources and studied its impact on the optimal planning of the MG [17]. The importance of modeling a multi-carrier network has been justified in comparison with a single-carrier network, where the network has equipped with RESs, diesel generators, FCs, boiler, and energy storage [18]. Two-stage optimal planning has been proposed for a multi-carrier network, where the renewable resources and ESSs have been optimally sized [19]. A dynamic model of the multi-carrier MG has been evaluated considering the intermittent resources in the network, including RESs [20]. An optimum energy management plan has been investigated for an MG in the high penetration of the RESs and electric vehicles [21, 22].

The application of different ESSs in multi-carrier networks has been investigated to achieve optimal and stable operation, especially in high penetration of RESs. A HES has been applied to catch the optimal energy plan and maximum profit, using a mixed heuristic-based optimization [23]. An optimal energy planning has been done by applying mixed-integer nonlinear programming for an energy hub including CHP, EES, and heating/cooling devices [24]. A multi-carrier energy network has been optimally designed considering ESS and demand response (DR) programs [25]. An economic optimization has been addressed for ESSs in the energy hub considering the DR programs [26].

Meta-heuristic algorithms are applied in many complex optimization problems because they are capable of achieving optimal results [27]. The implementation of PSO for solving the optimization problem is studied in [28]. The results showed that the PSO is very effective compared to the other methods. In addition, in [29], the benefits of QPSO are revealed. This algorithm shows high efficiency and strong searchability. Furthermore, QPSO has been widely developed to improve the effectiveness of PSO [30]. Besides the QPSO and PSO, the GA offers significant advantages over many typical searches of optimization methodologies to solve large-scale linear and nonlinear problems [31]. Consequently, the present chapter utilizes different meta-heuristic optimization methods, including QPSO, PSO, and GA algorithms for optimum planning of the heat and electricity incorporated networks with hybrid energy storage systems. Therefore, the primary innovations of the present chapter are as follows:

- Determining the optimal planning and design of a CCHP system to satisfy the required electrical, heating, and cooling demands while the total costs of the system during its' project lifetime is minimized.
- Proposing an efficient objective function including an environmental and economic approach for optimum planning of the heat and electricity incorporated networks.
- Developing a planning and management methodology to supply the total electrical, heating, and cooling demands while the reliability of the system is satisfied.
- Implementing different meta-heuristic algorithms for determining the optimum capacity and operation of different DGs and hybrid ESSs.

The structure of the present chapter is outlined as follows. Section 18.2 explains the description of the proposed model for heat and electricity incorporated networks. Then, the problem formulation is introduced in Section 18.3. The proposed strategy to satisfy total demands is described in Section 18.4. The results and discussion are introduced in Section 18.5. Finally, the conclusion is discussed in Section 18.6.

18.2 Description of the Proposed Model for Heat and Electricity Incorporated Networks

The proposed hybrid system consists of different types of DGs and ESSs, which should be defined first to explain the optimization problem.

18.2.1 Fuel Cell

In the proposed system, Fuel cell units are implemented to produce electricity and thermal energy. The electricity produced by FCs can be introduced as [32]:

$$P_{\text{FC}}^{E,t} = \mu_{\text{FC}}^E \times P_{\text{FC}}^{\text{gas},t} \tag{18.1}$$

in which, μ_{FC}^E and $P_{\text{FC}}^{\text{gas},t}$ are the conversion efficiencies from fuel to electrical power through the FCs and the fuel consumption of the mentioned DG, respectively.

The produced heat of the FC is calculated using (18.2):

$$P_{\text{FC}}^{T,t} = \mu_{\text{FC}}^T \times P_{\text{FC}}^{\text{gas},t} \tag{18.2}$$

where μ_{FC}^T is the coefficient of the heat conversion over the FC.

18.2.2 Boiler

A boiler is a kind of heat generator that can be applied in the CCHP system to provide the thermal demands [33]. In the proposed system, the boiler is applied as a backup system to satisfy the thermal demands. The relation between the fuel consumption and the generated heat of the boiler can be defined as:

$$P_{\text{Boiler}}^t = \mu_{\text{Boiler}} \times P_{\text{Boiler}}^{\text{gas},t} \tag{18.3}$$

where P_{Boiler}^t, μ_{Boiler} and $P_{\text{Boiler}}^{\text{gas},t}$ are the heat generated by boiler, the conversion coefficient from gas to thermal through the boiler, and consumption fuel by the boiler, respectively.

18.2.3 PV

Environmental and economic concerns over fossil fuels encourage the utilization of PVs in hybrid systems [34]. PVs can generate electricity by converting solar irradiation into direct current electricity, applying semiconductors that display the PV effect. The power generated from a PV highly depends on the weather conditions, especially the solar irradiance [35]. The electricity produced by PV is calculated using the following formula:

$$P_{\text{PV}} = \gamma_r \times \gamma_{\text{pt}} \times [1 - \beta(T^c - T^r)] \times A_m \times N \times G^t \tag{18.4}$$

where γ_r and γ_{pt} are the PV arrays and power tracking equipment efficiency, respectively. T^c and T^r denote the temperature and reference temperature of the PVs. β is the efficiency temperature coefficient. A_m and N are the area (m^2), and the number of modules, and G^t(W/m^2) is the global irradiance.

18.2.4 Wind Turbine

The wind turbine is implemented to convert the kinetic energy of the wind into electricity. Wind turbine technology has developed over the years and become the most favorable and reliable RESs today. Wind energy is a clean fuel source and doesn't contaminate the air like power plants that depend on the combustion of various kinds of fossil fuels, including natural gas or coal. Several models are proposed in different studies to estimate the electricity produced by wind turbines. In the present paper, the following equation [33] is considered in the simulation procedure to describe the produced power of the wind turbines:

$$
\begin{cases}
0 & V < V_{ci}, V > V_{co} \\
P_{WT}^{max} \times ((V - V_{ci})/(V_r - V_{ci}))^3 & V_{ci} \leq V < V_r \\
P_{WT}^r & V_r \leq V \leq V_{co}
\end{cases}
\tag{18.5}
$$

in which V, V_{ci}, V_{co}, and V_r are the wind speed, cut-in wind speed, cut-out wind speed, and nominal wind speed, respectively. P_{WT}^{max} is the maximum electricity produced by wind turbines.

18.2.5 Electrolyzer

In the proposed CCHP system, the excess produced power of RESs is utilized in the electrolyzer to produce H_2 for FCs. The electrolyzer implements DC electrical power to break apart water into H_2 and O_2 [36]. The transferred power from the electrolyzer to hydrogen storage is introduced as follows:

$$
E_{el\_Tank}^t = \eta_{el} \times P_{Surplus}^t
\tag{18.6}
$$

in which $E_{el\_Tank}^t$ is the transferred power from the electrolyzer to the HES, and η_{el} is the efficiency of the electrolyzer. $P_{Surplus}^t$ is the excess generated electricity of renewable energy resources and is achieved as:

$$
P_{Surplus}^t = \left[P_{PV}^t + P_{WT}^t \right] - \frac{P_{Load}^{E,t}}{\eta_{inverter}}
\tag{18.7}
$$

where P_{PV}^t, P_{WT}^t, $P_{Load}^{E,t}$ and $\eta_{inverter}$ are the electrical power generated by PVs and wind turbines, total electrical loads, and inverter efficiency, respectively.

18.2.6 Electrical Energy Storage

In the proposed system, the surplus generated electricity of RESs is stored in the electrical storage system, which is then reused when the electrical demand is higher than the generated electricity in the system. The EES capacity depends on different factors such as maximum DOD [37], rated battery capacity, etc. During the charging process, the available energy in electrical energy storage can be obtained as follows:

$$
E_{BS}^{t+1} = E_{BS}^t + \left\{ P_{PV}^t + P_{WT}^t - P_{Load}^{E,t}/\eta_{inverter} \right\} \times \eta_{BS}^{ch}
\tag{18.8}
$$

If the total demand is more than the total produced electricity of PVs and wind turbines, the EESs are in discharging mode. In this condition, the available energy in electrical energy storage is obtained using the following equation:

$$E_{BS}^{t+1} = E_{BS}^t - \left\{ P_{Load}^{E,t}/\eta_{inverter} - \left(P_{PV}^t + P_{WT}^t \right) \right\}/\eta_{BS}^{disch} \tag{18.9}$$

in which η_{BS}^{ch} and η_{BS}^{disch} are the battery bank charging and discharging efficiency [38].

18.2.7 Thermal Energy Storage

In this work, the surplus generated heat of FCs is stored in the TES. When the generated heat of FCs cannot satisfy total cooling and thermal demand, the available heat in TES is applied [39]. The stored heat in the TES at each hour can be obtained as follows:

$$E_{Thermal}^{t+1} = \left(1 - \eta_{Thermal} \right) \times E_{Thermal}^t + P_{FC}^{T,t} - \left(P_{Load}^{T,t} + P_{Load}^{C,t} \right) \tag{18.10}$$

in which $E_{Thermal}^t$ is the heat loss rate of TES, $P_{FC}^{T,t}$ is the generated heat of FCs, $P_{Load}^{C,t}$ and $P_{Load}^{T,t}$ are the cooling and heating demands in the studied system.

18.2.8 Hydrogen Energy Storage

Given that H_2 is known as the main promising solution for storing energy, the HES is utilized for the proposed system to deal with the intermittences of RESs [40]. In the proposed system, the excess electricity produced by PVs and wind turbines is applied to charge the EES. After reaching the maximum allowable capacity of the electrical storage, the excess power is used in the electrolyzer to generate hydrogen for FCs. The generated H_2 of the electrolyzer is stored in the HES to use in FC units [41]. The stored H_2 in the HES at each time is evaluated by:

$$E_{H_2}^{t+1} = E_{H_2}^t + \varepsilon_{el}^t \times E_{el\_Tank}^t - \varepsilon_{FC}^t \times E_{Tank\_FC}^t \quad \forall t \quad \varepsilon_{el}^t, \varepsilon_{FC}^t$$
$$\in \{0, 1\}, \quad \varepsilon_{el}^t + \varepsilon_{FC}^t \le 1 \tag{18.11}$$

ε_{el}^t and ε_{FC}^t introduce the related coefficients for electrolyzer to H_2 storage tank and from H_2 storage tank to FC at the time t.

18.2.9 Loads

The CCHP system is responsible for satisfying the total demands, including electrical, heating, and cooling loads. In the proposed system, the thermal demands consist of space heating/cooling and water heating loads.

18.3 Problem Formulation

The total demands of the system, including electrical, heating, and cooling loads, should be provided by the optimal design of the system, while the total costs of the system during its' project lifetime is minimized. The developed objective function and the applied constraints are introduced as follows.

18.3.1 Objective Function

The developed objective function consists of the total cost of the system, including costs of each DG, cost of fuel, penalty for environmental pollution, and penalty for interrupting demands. It should be emphasized that the total costs of each type of DG consist of investment, replacement, and operating and maintenance costs. In the present chapter, the net present cost (NPC) methodology [42–44] is applied to assess the total cost of the proposed system in the lifetime of the project. However, the primary aim of this chapter is to minimize the TNPC of the system, and can be determined as:

$$OF = \min \left\{ \sum N^i \times CC^i + \sum N^i \times RC^i \times \Psi^i + \frac{1}{CRF^i_{L_R}} \times \sum N^i \times OM^i + \right.$$

$$\frac{\omega_{Fuel}}{CRF^i_{L_R}} \times \left[\sum_{t=1}^{T\,max} \left(\frac{P^t_{FC}}{\mu_{FC}} + \frac{P^t_{Boiler}}{\mu_{Boiler}} \right) \right] + \frac{\omega_E}{CRF^i_{L_R}} \times \left[\sum_{t=1}^{T\,max} \left(P^t_{FC} \times \phi_{FC} + P^t_{Boiler} \times \phi_{Boiler} \right) \right] +$$

$$\left. \frac{\omega_{ENS}}{CRF^i_{L_R}} \times \left[\sum_{t=1}^{T\,max} ENS^t \right] \right\} \tag{18.12}$$

where CC^i, RC^i, and OM^i are the capital, replacement, and operation and maintenance cost of each applied DG. ω_{Fuel}, ω_E, and ω_{ENS} are fuel cost $\left(\frac{\$}{m^3}\right)$, the penalty for environmental emissions $\left(\frac{\$}{kg}\right)$, and penalty for interrupted loads $\left(\frac{\$}{kWh}\right)$, respectively. ϕ_{CHP} and ϕ_{Boiler} are the emission factor $\left(\frac{kg}{kWh}\right)$ for FCs and boilers, respectively. μ_{CHP} and μ_{Boiler} denote the energy factor $\left(\frac{kWh}{m^3}\right)$ of FCs and boilers, respectively. N^i is the optimum number of each DG.

CRF is the capital recovery factor and can be obtained using eq. (18.13):

$$CRF^{ir}_{L_R} = \frac{ir \times (1 + ir)^{L_R}}{(1 + ir)^{L_R} - 1} \tag{18.13}$$

where ir and L_R are interest rate and the project lifetime.

The NPC of all the replacement costs occurring over the lifetime of the project is also considered in the objective function. Therefore, the amount of Ψ_i can be achieved as follows:

$$\Psi_i = \sum_{n=1}^{q_i} \frac{1}{(1+\text{ir})^{L_i^* n}} \tag{18.14}$$

where

$$q_i = \left[\frac{L_R}{L_i}\right] - 1 \quad \text{if } L_R \text{ is dividable to } L_i \tag{18.15}$$

$$q_i = \left[\frac{L_R}{L_i}\right] \quad \text{if } L_R \text{ is not dividable to } L_i \tag{18.16}$$

q_i is the number of each type of DG replacements during the project lifetime. L_i is the lifetime of each DG.

18.3.2 Constraints of Objective Function

The applied components have some operational constraints, which should be considered in the simulation procedure [45, 46]. Electrical and thermal balance constraints along with the energy storage and reliability constraints are explained as follows.

18.3.2.1 Electrical Power Balance Constraint

The total produced power of DGs should satisfy the total electrical demand:

$$P_{\text{PV}}^t + P_{\text{WT}}^t + P_{\text{FC}}^{\text{Electrical},t} \pm E_{\text{BS}}^t + P_{\text{Interrupted}}^t = P_{\text{Load}}^{\text{Electrical},t} \tag{18.17}$$

$$E_{\text{BS}}^t \quad \begin{array}{l} +: \text{Discharging mode} \\ -: \text{Charging mode} \end{array}$$

where P_{PV}^t, P_{WT}^t are the generated electricity of PVs and wind units; $P_{\text{FC}}^{E,t}$ is the generated electricity by FCs; E_{BS}^t is the energy stored in EES; $P_{\text{Load}}^{\text{Electrical},t}$ and $P_{\text{Interrupted}}^t$ are the electrical demand and the interrupted demands at each hour.

18.3.2.2 Thermal and Cooling Power Balance Constraint

The total produced heat of FCs, boilers, and stored heat in TES should satisfy total thermal and cooling demands. In the CCHP system, the absorption chiller is also applied to supply the cooling demand. In this condition, the generated heat is transformed into cooling considering the efficiency of the chiller:

$$P_{\text{FC}}^{\text{Th},t} + P_{\text{Boiler}}^t \pm E_{\text{Thermal}}^t = P_{\text{Load}}^{\text{Thermal},t} + \frac{P_{\text{Load}}^{\text{Cooling},t}}{\eta_{\text{chi}}} \tag{18.18}$$

$E_{\text{Thermal}}^t \quad \begin{array}{l} + : \text{Discharging mode} \\ - : \text{Charging mode} \end{array}$

where P_{Boiler}^t and $P_{\text{FC}}^{\text{Th},t}$ are the heat generated by boilers and FCs; E_{Thermal}^t is the energy stored in TES; $P_{\text{Load}}^{\text{Thermal},t}$ and $P_{\text{Load}}^{\text{Cooling},t}$ are the thermal and cooling demand of the system at each time; and η_{chi} is the efficiency of the chiller.

18.3.2.3 Operational Constraints of Each Type of DG Unit

The produced power of each type of DG should not exceed its rated power as follows:

$$0 \le P_{\text{FC}}^{\text{Th},t} \le P_{\text{FC}}^{\text{Th, max}} \tag{18.19}$$

$$0 \le P_{\text{FC}}^{E,t} \le P_{\text{FC}}^{E, \text{max}} \tag{18.20}$$

$$0 \le P_{\text{PV}}^t \le P_{\text{PV}}^{\text{max}} \tag{18.21}$$

$$0 \le P_{\text{WT}}^t \le P_{\text{WT}}^{\text{max}} \tag{18.22}$$

$$0 \le P_{\text{Boiler}}^t \le P_{\text{Boiler}}^{\text{max}} \tag{18.23}$$

where $P_{\text{FC}}^{\text{Th, max}}$ and $P_{\text{FC}}^{E, \text{max}}$ are the maximum generated thermal energy and electricity of FC unit; $P_{\text{PV}}^{\text{max}}$, $P_{\text{WT}}^{\text{max}}$, and $P_{\text{Boiler}}^{\text{max}}$ are the maximum generated power of PVs, wind units, and boilers, respectively.

18.3.2.4 Energy Storage Constraint

The stored energy in battery storages, thermal storage tank, and sored H_2 in hydrogen storage tank should remain between their acceptable limits as follows [47]:

$$E_{H_2}^{\text{min}} \le E_{H_2}^t \le E_{H_2}^{\text{max}} \tag{18.24}$$

$$E_{\text{BS}}^{\text{min}} \le E_{\text{BS}}^t \le E_{\text{BS}}^{\text{max}} \tag{18.25}$$

$$E_{\text{Thermal}}^{\text{min}} \le E_{\text{Thermal}}^t \le E_{\text{Thermal}}^{\text{max}} \tag{18.26}$$

where $E_{H_2}^{\text{max}}$, $E_{\text{BS}}^{\text{max}}$, and $E_{\text{Thermal}}^{\text{max}}$ are the maximum capacity of the H_2 storage tank, battery storage, and thermal storage tank, respectively. $E_{H_2}^{\text{min}}$, $E_{\text{BS}}^{\text{min}}$, and $E_{\text{Thermal}}^{\text{min}}$ are the allowed minimum rated capacity of the H_2 storage tank, battery storage, and thermal storage tank, respectively.

18.3.2.5 Reliability Constraint

The LPSP is considered as the reliability index in the optimization procedure [48, 49]. An LPSP of zero introduces that the required electrical demands of the system are satisfied, and one represents that all demands are not satisfied [50, 51]. The reliability constraint is considered in the simulation procedure as follows:

$$\text{LPSP} \leq \text{LPSP}^{\text{max}} \tag{18.27}$$

18.4 CCHP Strategy to Satisfy Total Demands

In the studied heat and electricity incorporated networks, the applied components in the system are responsible for providing total demands. It should be mentioned that three optimization algorithms, including QPSO, PSO, and GA, are used to obtain the optimum planning and operation strategy of the proposed system. The proposed approach to provide the total demands is based on the optimum planning and operation of the system. From the point of view of electrical/thermal energy balance, the obtained optimal size realizes the economic operation of the system under the condition of meeting the total demands of the system. However, the proposed strategy to provide total demands of the CCHP system contains the following two sections:

18.4.1 Electrical Power Demand

In the studied CCHP system, the produced electricity of PVs, wind turbines, FCs, and available energy in EESs is used for supplying the electrical demands. The proposed strategy to satisfy the electrical demand is introduced as follows:

At first, the generated electricity by PVs and wind turbines is implemented to provide the required electrical demand. If the output power of PVs and wind turbines is more than the electrical demands, the excess power is implemented to store in the EESs, at first. After reaching the maximum capacity of the EESs, the surplus electricity is considered to implement in the electrolyzer to produce H_2 for FCs. When the total electrical demand is more than the electrical power generated by the PVs and wind turbines, the stored energy in EES is applied at first. Then, the FCs generate electrical power to provide the required electrical demands. Finally, if the total electrical power of DGs cannot satisfy the electrical loads, the remaining loads are interrupted with respect to the reliability index.

18.4.2 Thermal Power Demand

The thermal demands, including heating and cooling loads, are applied in the proposed system. In the studied system, FCs, boilers, and stored energy in TES are considered to provide the thermal demand, according to the following steps.

At the first stage, the heat produced by FCs is used to provide the total thermal demands. When the heat produced by FCs is more than the required thermal demands, the surplus energy is utilized to store in TES. If the generated heat of FCs cannot meet total thermal demands, the stored energy in TES is implemented to supply the remaining demand. Finally, if the total heat generated by FCs with the stored thermal energy in thermal storage cannot satisfy demand, the boilers provide the remaining thermal demand.

18.5 Results and Discussion

In the present chapter, the developed optimization method for optimum planning of the heat and electricity incorporated networks considering different types of ESSs is numerically studied. The schematic of the studied system is presented in Figure 18.1.

Having the complex optimization model developed in MATLAB software, the outcomes demonstrating the effectiveness of the implemented models and approaches are presented and compared. It should be mentioned that three optimization algorithms, including QPSO, PSO, and GA, are applied to solve the proposed optimization problem. The proposed CCHP system consists of four types of demands: electrical, water heating, and space cooling/heating demand. Therefore, various electrical and thermal demands should be provided in the CCHP system. The yearly load profile of electrical demand is shown in Figure 18.2.

In addition, the yearly load profile of space cooling demand is also shown in Figure 18.3.

The proposed heating demand of the system consists of water heating and space heating demands and can be seen in Figures 18.4 and 18.5.

From Figure 18.4, it can be observed that the water heating demand is required in the system at all times during the year. According to Figure 18.5, the space heating load during the colder seasons is more than the warmer seasons. Furthermore, the space heating load is not required on some days during the warm seasons. Similarly, the space cooling demand is equal to zero during colder seasons.

In the proposed CCHP system, the annual solar irradiation and wind velocity are also considered in the optimization procedure, as can be depicted in Figures 18.6 and 18.7.

It should be mentioned that the average value of solar irradiation in summer is higher than in other seasons, which is equal to 604 (W/m^2). This is because the highest solar irradiation happens in the mentioned season. In the same way, the average amount of wind velocity in the fall is higher than other seasons, which is equal to 8.42 (m/s). However, the maximum amount of solar irradiation and wind velocity is equal to 1000 (W/m^2) and 28 (m/s), respectively.

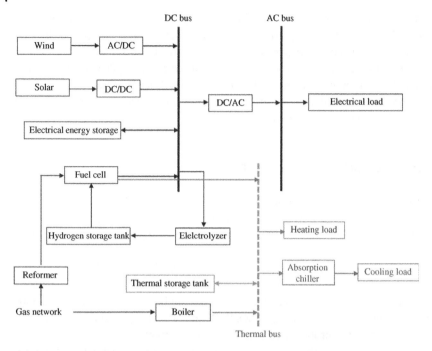

Figure 18.1 The schematic of the studied heat and electricity incorporated networks.

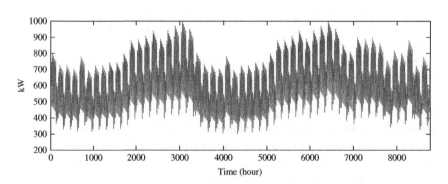

Figure 18.2 The yearly electrical demand.

As mentioned before, the cost of each type of DG consists of capital, operation and maintenance, and replacement costs. The costs for each applied component are summarized in Table 18.1 [38, 52–56].

Meta-heuristic algorithms, including QPSO/PSO/GA, are implemented for the optimum design of the CCHP system considering various types of ESSs.

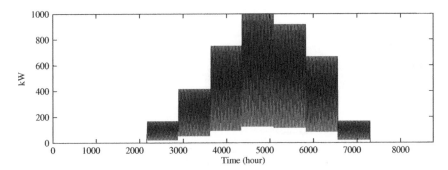

Figure 18.3 The yearly space cooling demand.

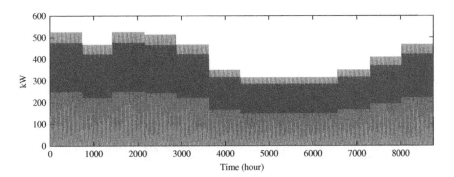

Figure 18.4 The yearly water heating demand.

The simulation results are depicted in Table 18.2. Meanwhile, Figure 18.8 compares the convergence curves of the applied algorithms.

According to Figure 18.8, QPSO is the fastest, and this algorithm has the optimum results among other methodologies. In addition, the QPSO reaches the optimal results in fewer iterations than the PSO and GA algorithms. Consequently, only more detailed results of QPSO are assessed in the remaining sections.

18.5.1 Providing Electrical Demands

As mentioned before, at the first step, the total generated electricity of PVs and wind turbines is used to provide the electrical demand. The produced electricity of PVs and wind turbines are highly affected by weather conditions and can be calculated based on eqs. (18.1) and (18.2). Figures 18.9 and 18.10 depict the produced electricity of these renewable energy resources.

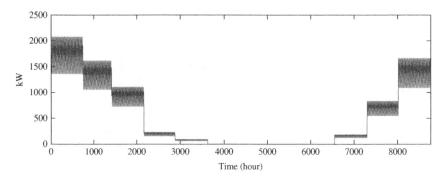

Figure 18.5 The yearly space heating demand.

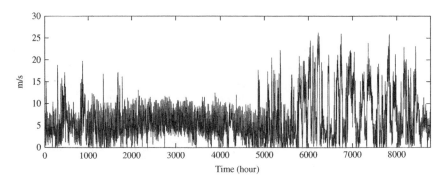

Figure 18.6 The yearly wind velocity data.

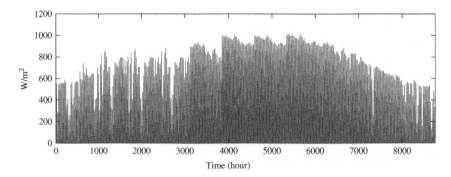

Figure 18.7 The yearly solar irradiation.

If the total generated electricity of RESs is more than the required electrical demands, the surplus energy is implemented to store in the EESs. The excess generated electricity of the mentioned resources at each time is shown in Figure 18.11.

Table 18.1 The costs for each applied component.

	WT (kW)	PV ($/kW)	EES ($/kWh)	TES ($/kW)	HES ($/kg)	FC ($/kW)	Boiler ($/kW)	Electrolyzer ($/kW)
Capital cost	1100	1200	280	20	1.5	600	40	150
O&M cost	10	3	16	0.01	0.6	0.01	0.03	8
Replacement cost	1100	1200	280	20	0.5	600	40	150

The available energy in EES is shown in Figure 18.12.

If the EES is fully charged, the surplus energy is implemented in the electrolyzer to generate H_2 for FCs. Figure 18.13 shows the hourly amount of electrical power used in the electrolyzer to generate H_2 for FCs.

After transferring the surplus power to the electrolyzer, if excess power is available, this power is considered to be dumped. Figure 18.14 shows the hourly amount of dump load during a year.

From Figure 18.14, it is clear that the amount of dump load at times 5000, 6000, and 7000 is the highest. The reason behind it is that the amount of excess energy of RESs at the mentioned times is higher than other times during a year.

If the total produced electricity of the PVs and wind turbines, with the available energy in EES, cannot supply the electrical demands, the FCs start to produce power to supply the remaining electrical demands. The hourly generated electricity of FCs is depicted in Figure 18.15.

Figure 18.16 depicts the available H_2 in the HES at each hour. It can be observed that a high amount of H_2 can be stored during a year, which highlights the significance of HES.

For more visualization, a sample day is chosen to represent the generation of different resources in more detail. The power generation of PVs, wind turbines, FCs, available energy in EES, used energy in the electrolyzer, and interrupted demands are shown in Figure 18.17.

According to Figure 18.17, PVs start to produce electrical power from 07:00 AM to 07:00 PM, and wind turbines can produce power at times from 01:00 AM to 02:00 PM and 11:00 PM to 00:00 AM. In addition, at times between 09:00 AM and 02:00 PM, the total power produced by RESs is more than the total required electrical demand. Hence, this excess power is implemented to store in the EESs at first. Then, the surplus electricity is considered to implement in the electrolyzer to produce hydrogen for FCs (from 12:00 PM to 01:00 PM). When the total electrical demand is more than the electricity produced by the PVs and wind turbines, the available energy in EES is implemented at first and, then the FCs start to generate electricity to supply the required demands. At the final stage, if the CCHP's

Table 18.2 Simulation results of QPSO/PSO/GA.

	WT (kW)	PV (kW)	EES (kWh)	TES (kW)	HES (kg)	FC (kW)	Boiler (kW)	Electrolyzer (kW)
QPSO	730	780	656	3000	60	790	2300	400
PSO	738	770	650	3060	67	807	2250	420
GA	761	800	688	3018	58	801	2265	390

produced power cannot supply the electrical demands (from 07:00 PM to 10:00 PM), the remaining loads are interrupted with respect to the reliability index.

The amount of interrupted demand, LPSP, and dump energy during a year are shown in Table 18.3.

According to Table 18.3, the excess energy generated by PVs and wind turbines is equal to 339.8 MWh during a year. Therefore, this excess energy should be dumped in the absence of ESSs. According to this Table, utilizing the ESSs, including electrical energy storage and hydrogen storage, reduces the amount of dump energy from 339.8 MWh to 10.45 MWh during a year. It can be concluded from the results that there is about a 97% reduction in the amount of dump energy when the ESSs are optimally implemented in the system. In the studied CCHP system, if the total power produced by the applied DGs cannot supply the total electrical demands, the remaining loads are interrupted (46.7 MWh) with respect to the reliability index. According to Table 18.3, the amount of LPSP is equal to 0.857, which is within an acceptable limit.

18.5.2 Providing the Thermal Demands

In the proposed CCHP system, the produced heat of FCs, boilers, and stored energy in TES is implemented to provide the total thermal demands. In the proposed thermal energy management strategy, when the total produced electricity of PVs, wind turbines, and available energy in electrical energy storage cannot meet the required loads, FCs start to produce electricity. At these hours, the FC's produced heat is also utilized for supplying the thermal demands. However, if the heat generated by FCs is more than total thermal demands, the excess thermal energy is implemented to store in TES. In addition, if the FC's produced heat cannot meet the required total thermal demand, the available energy in TES is implemented to supply the remaining demand. Figure 18.18 shows the variations of energy in TES during a year.

According to Figure 18.18, at times between 0–2100 and 7200–8760, the thermal energy storage is empty. The reason behind this is that excess produced heat of FCs is equal to zero at the mentioned times. In the CCHP system, if the FC's produced heat with the stored energy in thermal energy storage cannot meet total demands,

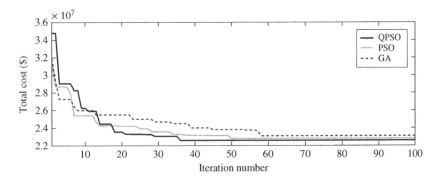

Figure 18.8 The convergence curves of QPSO, PSO, and GA algorithms.

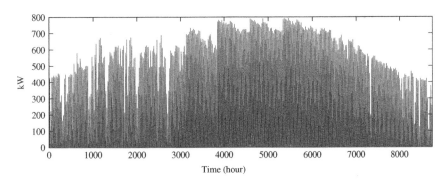

Figure 18.9 The generated electricity of PVs.

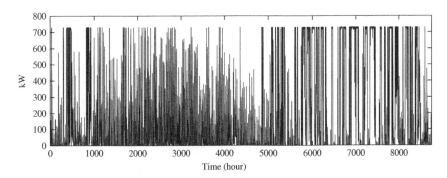

Figure 18.10 The generated electricity of wind turbines.

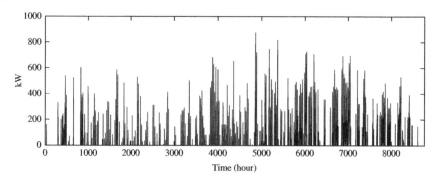

Figure 18.11 The excess generated electricity of RESs.

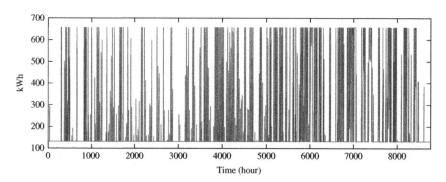

Figure 18.12 The available energy in EES.

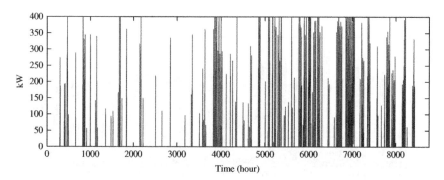

Figure 18.13 The hourly amount of electrical power used in the electrolyzer to generate H_2 for FCs.

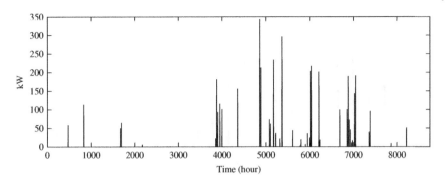

Figure 18.14 The dump load at each hour during a year.

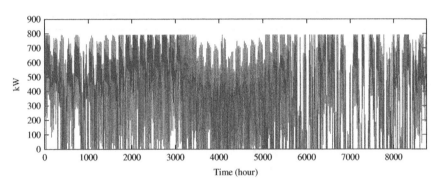

Figure 18.15 The generated electricity of FCs at each hour.

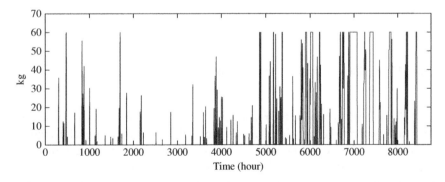

Figure 18.16 The available H_2 in hydrogen storage.

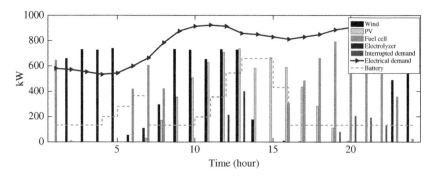

Figure 18.17 The optimal scheduling of electrical demand during a typical day.

Table 18.3 The amount of interrupted demands, LPSP, and dump energy during a year.

Excess energy (MWh)	Interrupted demands (MWh)	Dump energy (MWh)	LPSP (%)
339.8	46.7	10.45	0.857

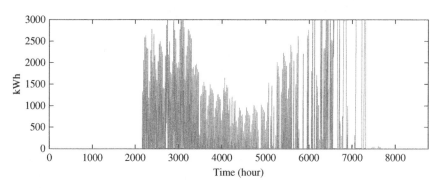

Figure 18.18 The variations of energy in TES during a year.

the boiler is used to provide the remaining thermal demand. Figure 18.19 depicts the yearly generated energy by boilers to supply the remaining thermal demands.

It is obvious that the generated heat of boilers is equal to zero at some times. At the mentioned times, the generated heat of FCs with available energy in TES can meet thermal demand. In addition, the boiler's produced heat at times between 500 and 700 is more than other times. The reason behind this is that the required thermal demand is the highest at the mentioned times. For more visualization, the optimal scheduling of thermal power during an atypical day is shown in Figure 18.20.

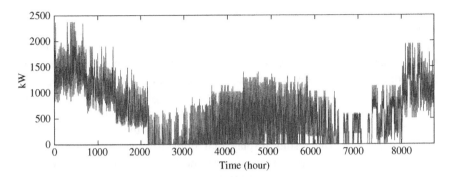

Figure 18.19 The yearly generated energy by boilers.

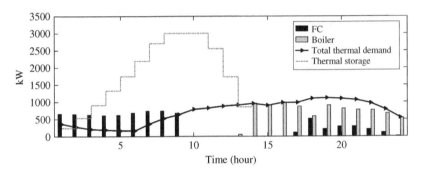

Figure 18.20 The optimal scheduling of thermal power during a typical day.

According to Figure 18.20, at times between 01:00 AM and 09:00 AM, the FC's produced heat is more than the total thermal demand. Thus, the surplus heat is stored in the TES during these times. Meanwhile, when the FC's produced heat cannot meet thermal demand, the available energy in TES is utilized at the first step, then the remaining required thermal demand is supplied by a boiler. In the mentioned day, the boiler starts to generate energy at times between 01:00 PM and 00:00 AM because the heat generated by FCs and stored energy in TES cannot supply the total thermal demand.

18.6 Conclusion

In the present chapter, a novel optimization method is developed for optimum planning of the heat and electricity incorporated networks considering the most cost-efficient RESs and various ESSs, including electrical, thermal, and hydrogen storage. Different types of demands, including electrical, space heating/cooling, and water heating loads, are considered to be provided in the CCHP system. The LPSP is considered as a reliability index to enhance the efficiency of the proposed design. In the studied system, EES is utilized to store the excess generated power of renewable energy resources. In addition, after reaching the maximum capacity of this type of EES, the remaining excess energy is used in the electrolyzer to produce hydrogen for FCs. The HES is also implemented to store the generated H_2 of the electrolyzer. The TES is applied in the system to store the excess heat produced by FCs. The results display that implementing the energy storage devices, including electrical energy storage and hydrogen storage, reduces the amount of dump energy from 339.8 MWh to 10.45 MWh during a year. Consequently, there is about a 97% reduction in the amount of dump energy when the ESSs are optimally applied in the system. To validate the proposed methodology, different meta-heuristic algorithms, including QPSO, PSO, and GA, are is successfully implemented, and their performance is compared. The comparative results show that the QPSO is the fastest, and this algorithm has the optimum results among other methodologies.

Briefly, the main achievements of this work can be listed as follows:

- Implementation of hybrid ESSs is one of the main solutions to increase the penetration of RESs in heat and electricity incorporated network.
- As the FC's produced heat is utilized, higher fuel savings can be achieved.
- Utilization of the TES plays a crucial role in storing the excess heat produced by FCs in heat and electricity incorporated network.
- The developed methodology can be applied as a robust algorithm by the MG system operator. The proposed approach is efficient in solving the optimization of hybrid energy storage systems in the heat and electricity incorporated networks.
- QPSO is a robust meta-heuristic algorithm that can be implemented for solving the optimization problem.

References

1 Kwan, T.H., Shen, Y., and Yao, Q. (2019). An energy management strategy for supplying combined heat and power by the fuel cell thermoelectric hybrid system. *Applied Energy* 251: 113318.

2 Gomes, J.G., Xu, H.J., Yang, Q., and Zhao, C.Y. (2021). An optimization study on a typical renewable microgrid energy system with energy storage. *Energy* 234: 121210.

3 Daneshvar, M., Pesaran, M., and Mohammadi-ivatloo, B. (2018). Transactive energy integration in future smart rural network electrification. *Journal of Cleaner Production* 190: 645–654.

4 Akram, U., Khalid, M., and Shafiq, S. (2018). Optimal sizing of a wind/solar/battery hybrid grid-connected microgrid system. *IET Renewable Power Generation* 12 (1): 72–80.

5 Kazemdehdashti, A., Mohammadi, M., Seifi, A.R., and Rastegar, M. (2020). Stochastic energy management in multi-carrier residential energy systems. *Energy* 202: 117790.

6 Ye, Y., Qiu, D., Ward, J. and Abram, M. (2020). "Model-free real-time autonomous energy management for a residential multi-carrier energy system: A deep reinforcement learning approach. *IJCAI*, Japan, pp. 339–346.

7 Rastegar, M., Fotuhi-Firuzabad, M., and Zareipour, H. (2015). Centralized home energy management in multi-carrier energy frameworks. *2015 IEEE 15th International Conference on Environment and Electrical Engineering (EEEIC)*, Italy, pp. 1562–1566.

8 Hasankhani, A., Hakimi, S., and Vafaeizadeh, A. (2021). Developing energy management system considering renewable energy systems in residential community. *In Design, Analysis, and Applications of Renewable Energy Systems*, pp. 275–299.

9 Hakimi, S. and Hasankhani, A. (2020). Intelligent energy management in off-grid smart buildings with energy interaction. *Journal of Cleaner Production* 244: 118906.

10 Arnone, D., Bertoncini, M., Paternò, G. et al. (2016). Smart multi-carrier energy system: Optimised energy management and investment analysis. *In 2016 IEEE International Energy Conference (ENERGYCON)*, Belgium, pp. 1–6.

11 Khorasany, M., Najafi-Ghalelou, A., Razzaghi, R., and Mohammadi-Ivatloo, B. (2021). Transactive energy framework for optimal energy management of multi-carrier energy hubs under local electrical, thermal, and cooling market constraints. *International Journal of Electrical Power & Energy Systems* 129: 106803.

12 HassanzadehFard, H., Hasankhani, A., and Hakimi, S.M. (2021). Optimal planning and design of multi-carrier energy networks. In: *Planning and Operation of Multi-Carrier Energy Networks*, 209–234. Cham: Springer.

13 Shabanpour-Haghighi, A. and Seifi, A. (2015). Multi-objective operation management of a multi-carrier energy system. *Energy* 88: 430–442.

14 Shariatkhah, M., Haghifam, M., Parsa-Moghaddam, M., and Siano, P. (2015). Modeling the reliability of multi-carrier energy systems considering dynamic behavior of thermal loads. *Energy and Buildings* 103: 375–383.

15 Shahrabi, E., Hakimi, S., Hasankhani, A. et al. (2021). Developing optimal energy management of energy hub in the presence of stochastic renewable energy resources. *Sustainable Energy, Grids and Networks* 26: 100428.

16 Hakimi, S., Bagheritabar, H., Hasankhani, A. et al. (2019). Planning of smart microgrids with high renewable penetration considering electricity market conditions. *In 2019 IEEE International Conference on Environment and Electrical Engineering and 2019 IEEE Industrial and Commercial Power Systems Europe (EEEIC/I&CPS Europe)*, Italy, pp. 1–6.

17 Hakimi, S., Hasankhani, A., Shafie-khah, M., and Catalão, J. (2021). Stochastic planning of a multi-microgrid considering integration of renewable energy resources and real-time electricity market. *Applied Energy* 298: 117215.

18 Ghanbari, A., Karimi, H., and Jadid, S. (2020). Optimal planning and operation of multi-carrier networked microgrids considering multi-energy hubs in distribution networks. *Energy* 204: 117936.

19 Vahid, A., Jadid, S., and Ehsan, M. (2018). Optimal planning of a multi-carrier microgrid (MCMG) considering demand-side management. *International Journal of Renewable Energy Research (IJRER)* 8 (1): 238–249.

20 Amir, V. and Azimian, M. (2020). Dynamic multi-carrier microgrid deployment under uncertainty. *Applied Energy* 260: 114293.

21 Emrani-Rahaghi, P., Hashemi-Dezaki, H., and Hasankhani, A. (2021). Optimal stochastic operation of residential energy hubs based on plug-in hybrid electric vehicle uncertainties using two-point estimation method. *Sustainable Cities and Society* 72: 103059.

22 Proietto, R., Arnone, D., Bertoncini, M. et al. (2014). Mixed heuristic-non linear optimization of energy management for hydrogen storage-based multi carrier hubs. *In 2014 IEEE International Energy Conference (ENERGYCON)*, Croatia, pp. 1019–1026.

23 Javadi, M., Anvari-Moghaddam, A., and Guerrero, J. (2017). Optimal scheduling of a multi-carrier energy hub supplemented by battery energy storage systems. *In 2017 IEEE International Conference on Environment and Electrical Engineering and 2017 IEEE Industrial and Commercial Power Systems Europe (EEEIC/I&CPS Europe)*, Italy, pp. 1–6.

24 Liu, T., Zhang, D., Wang, S., and Wu, T. (2019). Standardized modelling and economic optimization of multi-carrier energy systems considering energy storage and demand response. *Energy Conversion and Management* 182: 126–142.

25 Keihan Asl, D., Hamedi, A., and Reza Seifi, A. (2020). Planning, operation and flexibility contribution of multi-carrier energy storage systems in integrated energy systems. *IET Renewable Power Generation* 14 (3): 408–416.

26 Najafi, A., Falaghi, H., Contreras, J., and Ramezani, M. (2017). A stochastic bilevel model for the energy hub manager problem. *IEEE Transactions on Smart Grid* 8 (5): 2394–2404.

27 Hakimi, S.M., Tafreshi, S.M.M., and Rajati, M.R. (2007). Unit sizing of a stand-alone hybrid power system using model-free optimization. *International Conference on Granular Computing*, USA.

28 Musa, H. and Ibrahim, S.B. (2015). A review of Particle Swarm Optimization (PSO) algorithms for optimal distributed generation placement. *International Journal of Energy and Power Engineering* 4 (4): 232–239.

29 Zhang, Q., Wang, Z., Tao, F. et al. (2014). Design of optimal attack-angle for RLV reentry based on quantum particle swarm optimization. *Advances in Mechanical Engineering* 6: 352983.

30 Xia, Y., Feng, Z.K., Niu, W.J. et al. (2019). Simplex quantum-behaved particle swarm optimization algorithm with application to ecological operation of cascade hydropower reservoirs. *Appl. Soft Comput.* 84: 105715.

31 Ameri, A.A., Nichita, C., Riouch, T., and El-Bachtiri, R. (2015). "Genetic algorithm for optimal sizing and location of multiple distributed generations in electrical network. *IEEE Modern Electric Power Systems (MEPS)*, Wroclaw, Poland.

32 Tooryan, F., HassanzadehFard, H., Collins, E.R. et al. (2020). Smart integration of renewable energy resources, electrical, and thermal energy storage in microgrid applications. *Energy* 212: 118716.

33 Hasankhani, A., Hakimi, S.M., Bodaghi, M. et al. (2021). Day-Ahead optimal management of plug-in hybrid electric vehicles in smart homes considering uncertainties. *In 2021 IEEE Madrid PowerTech*, Spain. IEEE, pp. 1–6.

34 HassanzadehFard, H. and Jalilian, A. (2018). Optimal sizing and location of renewable energy based dg units in distribution systems considering load growth. *International Journal of Electrical Power & Energy Systems* 101: 356–370.

35 HassanzadehFard, H. and Jalilian, A. (2018). Optimal sizing and siting of renewable energy resources in distribution systems considering time varying electrical/heating/cooling loads using PSO algorithm. *International Journal of Green Energy* 15 (2): 113–128.

36 Konstantinopoulos, S.A., Anastasiadis, A.G., Vokas, G.A. et al. (2018). Optimal management of hydrogen storage in stochastic smart microgrid operation. *International Journal of Hydrogen Energy* 43 (1): 490–499.

37 HassanzadehFard, H., Moghaddas-Tafreshi, S.M., and Hakimi, S.M. (2021) Effect of energy storage systems on optimal sizing of islanded micro-grid considering interruptible loads. *Proceedings of the 2011 3rd International Youth Conference on Energetics (IYCE)*, Portugal, pp. 1–7.

38 Tooryan, F., HassanzadehFard, H., Collins, E.R. et al. (2020). Optimization and energy management of distributed energy resources for a hybrid residential microgrid. *Journal of Energy Storage* 30: 101556.

39 Zhang, D., Evangelisti, S., Lettieri, P., and Papageorgiou, L.G. (2015). Optimal design of CHP-based microgrids: multiobjective optimisation and life cycle assessment. *Energy* 85: 181–193.

40 Daneshvar, M., Mohammadi-Ivatloo, B., Zare, K., and Asadi, S. (2021). Transactive energy management for optimal scheduling of interconnected microgrids with hydrogen energy storage. *International Journal of Hydrogen Energy* 46 (30): 16267–16278.

41 HassanzadehFard, H., Tooryan, F., Collins, E.R. et al. (2020). Design and optimum energy management of a hybrid renewable energy system based on efficient various hydrogen production. *International Journal of Hydrogen Energy* 45 (55): 30113–30128.

42 Berk, S.D. and DeMarzo, J.P. (2014). *Corporate Finance*, 3e. Canada: Pearson.

43 HassanzadehFard, H. and Jalilian, A. (2021). Optimization of dg units in distribution systems for voltage sag minimization considering various load types. *Iranian Journal of Science and Technology, Transactions of Electrical Engineering* 45 (2): 685–699.

44 HassanzadehFard, H., Tooryan, F., Dargahi, V., and Jin, S. (2021). A cost-efficient sizing of grid-tied hybrid renewable energy system with different types of demands. *Sustainable Cities and Society* 73: 103080.

45 Daneshvar, M., Mohammadi-Ivatloo, B., Zare, K., and Asadi, S. (2020). Two-stage stochastic programming model for optimal scheduling of the wind-thermal-hydropower-pumped storage system considering the flexibility assessment. *Energy* 193: 116657.

46 Hakimi, S.M., Bagheritabar, H., Hasankhani, A. et al. (2019). Planning of smart microgrids with high renewable penetration considering electricity market conditions. *In 2019 IEEE International Conference on Environment and Electrical Engineering and 2019 IEEE Industrial and Commercial Power Systems Europe (EEEIC/I&CPS Europe)*, Italy. IEEE, pp. 1–5.

47 Hasankhani, A. and Hakimi, S.M. (2021). Stochastic energy management of smart microgrid with intermittent renewable energy resources in electricity market. *Energy* 219: 119668.

48 Lorestani, A., Gharehpetian, G.B., and Nazari, M.H. (2019). Optimal sizing and techno-economic analysis of energy- and costefficient standalone multi-carrier microgrid. *Energy* 178: 751–764.

49 HassanzadehFard, H., Bahreyni, S.A., Dashti, R., and Shayanfar, H.A. (2015). Evaluation of reliability parameters in micro-grid. *Iranian Journal of Electrical and Electronic Engineering* 11 (2): 127–136.

50 Ghazvini, A.M. and Olamaei, J. (2019). Optimal sizing of autonomous hybrid PV system with considerations for V2G parking lot as controllable load based on a heuristic optimization algorithm. *Solar Energy* 184: 30–39.

51 Hakimi, S.M., Hasankhani, A., Shafie-khah, M. et al. (2022). Optimal sizing of renewable energy systems in a microgrid considering electricity market interaction and reliability analysis. *Electric Power Systems Research* 203: 107678.

52 Ghenai, C., Bettayeb, M., Brdjanin, B., and Hamid, A.K. (2019). Hybrid solar PV/PEM fuel cell/diesel generator power system for cruise ship: a case study in stockholm. *Sweden. Case Studies in Thermal Engineering* 14: 100497.

53 Mohammadi, M., Ghasempour, R., Astaraei, F.R. et al. (2018). Optimal planning of renewable energy resource for a residential house considering economic and reliability criteria. *Int J Electr Power Energy Syst* 96: 261e73.

54 Singh, S., Chauhan, P., and Singh, N. (2020). Capacity optimization of grid connected solar/fuel cell energy system using hybrid ABCPSO algorithm. *Int J Hydrogen Energy* 45 (16): 10070–10088.

55 Ramli, M.A., Bouchekara, H., and Alghamdi, A.S. (2018). Optimal sizing of pv/wind/diesel hybrid microgrid system using multi-objective self-adaptive differential evolution algorithm. *Renewable Energy* 121: 400–411.

56 Moradi, M.H., Hajinazari, M., Jamasb, S., and Paripour, M. (2013). An energy management system (EMS) strategy for combined heat and power (CHP) systems based on a hybrid optimization method employing fuzzy programming. *Energy* 49 (1): 86–101.

Index

Coordinated Operation and Planning of Modern Heat and Electricity Incorporated Networks,
First Edition. Edited by Mohammadreza Daneshvar, Behnam Mohammadi-Ivatloo, and Kazem Zare.
© 2023 The Institute of Electrical and Electronics Engineers, Inc.
Published 2023 by John Wiley & Sons, Inc.

IEEE Press Series
on Power and Energy Systems

Series Editor: Ganesh Kumar Venayagamoorthy, Clemson University, Clemson, South Carolina, USA

The mission of the IEEE Press Series on Power and Energy Systems is to publish leading-edge books that cover a broad spectrum of current and forward-looking technologies in the fast-moving area of power and energy systems including smart grid, renewable energy systems, electric vehicles and related areas. Our target audience includes power and energy systems professionals from academia, industry and government who are interested in enhancing their knowledge and perspectives in their areas of interest.

Marius Rosu, Ping Zhou, Dingsheng Lin, Dan M. Ionel, Mircea Popescu, Frede Blaabjerg, Vandana Rallabandi, David Staton

Printed and bound by CPI Group (UK) Ltd, Croydon, CR0 4YY

16/04/2025

14658596-0004